Current Trends in Hardware Verification and Automated Theorem Proving

G. Birtwistle   P.A. Subrahmanyam
Editors

Current Trends in Hardware Verification and Automated Theorem Proving

With 95 Figures

Springer-Verlag
New York Berlin Heidelberg
London Paris Tokyo

Graham Birtwistle
Department of Computer Science
University of Calgary
Calgary, Alberta
Canada T2N 1N4

P.A. Subrahmanyam
AT&T Bell Labs
Holmdel, NJ 07733
USA

Library of Congress Cataloging-in-Publication Data
Current trends in hardware verification and automated theorem proving
    /P.A. Subrahmanyam, Graham Birtwistle, editors.
        p.   cm.
    Includes index.
    1. Integrated circuits—Very large scale integration—Design and
construction.   2. Automatic theorem proving.   I. Subrahmanyam, P.
A.   II. Birtwistle, G.M. (Graham M.)
TK7874.C85 1989
621.395—dc19   89-6049

Printed on acid-free paper.

Camera-ready copy prepared by the authors.
Printed and bound by Edwards Brothers, Inc., Ann Arbor, Michigan.
Printed in the United States of America.

9 8 7 6 5 4 3 2 1

ISBN 0-387-96988-8 Springer-Verlag New York Berlin Heidelberg
ISBN 3-540-96988-8 Springer-Verlag Berlin Heidelberg New York

Current Trends in
Hardware Verification and Automated Theorem Proving

This book provides perspectives on the state of the art in hardware verification and automated theorem proving, and is the outgrowth of a workshop on hardware verification held in Banff, Canada, from June 12 through June 18, 1988.

The first part of the book contains examples of different approaches to hardware verification and, to a lesser extent, complementary approaches to formal specification. The examples include verifications of chips that have actually been fabricated (Chapter 1); designs that are being prototyped (Chapters 2 and 3); simplified abstractions of existing designs (Chapter 4); and simple components (Chapter 5).

The second part of the book deals with the current technology and research trends in the area of automated theorem proving systems. The first three chapters in this part describe, respectively, a "commercial" theorem proving system that attempts to consolidate the state of the art in theorem proving (Chapter 6); an interactive proof editor (Chapter 7); and a logical framework for constructing proof systems tailored to specifiable theories (Chapter 8). The next three chapters discuss experiments conducted in the context of existing verification systems, as well as some enhancements of these systems. Specifically, an enhancement of a proof system for Higher Order Logic (HOL) to deal with recursively defined types is described in Chapter 9. Pragmatic issues in the mechanization of programming logics in HOL are discussed in Chapter 10. Finally, a tutorial overview of the use of rewriting techniques to do equational reasoning, and a discussion of some of the issues involves in structuring large proofs are contained in Chapter 11.

A synopsis of the chapters in the book is given below, following which we summarize the deliberations at the workshop relating to architectures and testability.

Hardware Verification

In Chapter 1, Avra Cohn discusses the "Correctness properties of the Viper Block Model". Viper is a microprocessor designed by W. J. Cullyer, C. Pygott, and J. Kershaw at the Royal Signals and Radar Establishment in Malvern, England, and is now commercially available. It is one of the few microprocessors specifically designed with formal verifiability in mind from the very beginning, and is intended for use in safety-critical applications such as civil aviation and nuclear power plant control. Avra Cohn describes the partially completed correctness proof, in the Higher Order Logic system HOL, of the Viper archietectural 'block model' with respect to Viper's top level functional specifications. This chapter also contains a discussion of the scope and limits of the word 'verification', and cautions against careless uses of the term. The chapter provides an important case study in verification, and illustrates some of the problems the arise in managing large HOL proofs.

In Chapter 2, Paliath Narendran and Jonathan Stillman describe an approach to hardware verification in the context of their success in formally verifying the description of an image processing chip currently under development at Research Triangle Institute. They demonstrate that their approach, which uses an equational approach to theorem proving developed by Kapur and Narendran, can be a viable alternative to simulation. In particular they are able to take advantage of the "recursive nature" of many circuits, such as n-bit adders, and their techniques allow the verification of sequential circuits. They describe the discovery of several design errors in the circuit description, detected during the verification attempt (the actual verification could only take place once these errors were removed) and discuss directions that future work will take. This is an example of the verification of a complex sequential circuit which was *not* designed with verification specifically in mind.

In Chapter 3, Richard Fujimoto and Ganesh Gopalakrishnan discuss the "Specification Driven Design of Custom Hardware in HOP". HOP embodies a simple process model for lock-step synchronous processes. The chapter presents examples of specifications using HOP, and discusses the design of the Utah Simulation engine being developed to speed up distributed discrete event simulation using the time warp technique. Issues in the specification driven design of this system are discussed and illustrated using HOP.

In Chapter 4, R. C. Sekar and M. K. Srivas discuss a method for formally proving the correctness of a microprocessor using an equational technique. The behavioral and structural specifications of the processor are expressed using and equational language, and the asynchronous interaction between the memory and CPU is expressed in the specification by using a nondeterministic function *random*, as opposed to encoding into the specification the anticipated asynchronous interaction between the memory and the CPU. The method is applied to verify an abstract, simplified version of the processor architecture used in Wirth's Modula-II Lilith machine.

In Chapter 5, Joseph A. Goguen states, justifies, and illustrates some new techniques for using an equational logic based programming language like OBJ as a theorem prover. The technique is illustrated in proofs involving simple hardware components, induction and parameterization.

Automated Theorem Proving Systems

In Chapter 6, Bill Pase and Mark Saaltink describe a new tool for formally verifying software. The tool, called m-EVES, consists of a new language called m-Verdi; a sound logic for proving properties of m-Verdi programs; and a theorem prover, called m-NEVER, which integrates many state-of-the-art techniques drawn from the theorem proving literature. Examples are used to illustrate the ideas embodied in the system.

In Chapter 7, Brian Ritchie and Paul Taylor describe aspects of the Edinburgh Interactive Proof Editor (IPE). IPE is a program that enables logically sound proofs to be developed at a computer workstation chiefly by means of pointing and clicking with a mouse. The logic implemented is Intuitionistic First Order Logic extended with some higher order pattern matching to facilitate proofs by induction. There are also facilities for organizing collections of axioms and lemmas into theories. Proofs are based upon Natural Deduction and are generated in a top down (goal directed) style with goals being decomposed into subgoals. Internally the program records the proof as an attributed tree, enabling the proof to be rapidly updated when the goal is modified, an alternative instantiation for a variable is decided upon, or a different proof step is selected, etc. This facility gives exploring a proof some of the quality of using a spreadsheet. The IPE has proved itself to be a very attractive environment for developing simple proofs but has shortcomings when more challenging proofs are attempted. This chapter discusses some of the benefits of developing proofs interactively, and describes proposals for an enhanced implementation of an interactive proof environment (the IPE2).

In Chapter 8, Furio Honsell and Ian Mason describe how one it is possible to use a typed lambda calculus to encode systems for reasoning about machines. The motivation for this work is the growing interest in using computers as an aid for correctly manipulating logical systems, especially those systems for reasoning about hardware and software. Implementing a proof environment for a particular logical system is however, both complex and time consuming. This fact, together with a proliferation of logics, suggest that a uniform and reliable alternative is desirable. One such alternative is the Edinburgh Logical Framework (LF), currently under development at Edinburgh University. The LF is a logic-independent tool which, given a specification for a logical system, synthesizes a proof editor and checker for that system. Its specification language is based on a typed lambda calculus. This language can capture the uniformities and idiosyncrasies of a large class of logics without sacrificing generality for tractability. Peculiarities (such as side conditions on rule application, variable occurrence or formula formation) are expressed at the level of the specification. This chapter summarizes the main meatures of the Edinburgh Logical Framework.

In Chapter 9, Tom Melham discusses the automatic formulation of recursive type definitions in a proof system for Higher Order Logic (HOL). The expressive power of higher order logic makes it possible to define a wide variety of types within the logic, and to prove theorems that state the properties of these types concisely and abstractly. This chapter contains a tutorial introduction to the logical basis for such type definitions. Examples are given of the formal definitions in logic of several simple types. A method is then described for systematically defining any instance of a certain class of commonly-used recursive types. The automation of the method in HOL is also discussed.

In Chapter 10, Mike Gordon discusses the mechanization of programming logics in Higher Order Logic. Formal reasoning about computer programs can be based directly on the semantics of the programming language, or done in a special pur-pose logic like Hoare logic. The advantage of the first approach is that it guarantees that the formal reasoning applies to the language being used. (It is well known, for example, that Hoare's assignment axiom fails to hold for most programming languages.) The advantage of the second approach is that the proofs can be more direct and natural. In this chapter, Gordon describes an attempt to get the advan-tages of both approaches. The rules of Hoare logic are mechnically derived from the semantics of a simple imperative programming language, using the HOL system.

In Chapter 11, Dave Musser explores two aspects of automated theorem prov-ing: the use of term rewriting techniques for equational and inductive reasoning; and the construction of proof trees representing key steps of a proof, as an aid to interactive proof search and as a basis for the synthesis of computations. Musser discusses interactive tools which help the user to visualize and manage the large number of theorems and subgoals that typically arise in formal hardware or soft-ware verification. These topics are illustrated with example proofs carried out with two existing automated theorem prover (RRL and Affirm-85).

Architectures, Verification, and Testability

At the workshop, the discussions on verification, theorem proving, and architectures were deliberately interleaved. The intent was to enable researchers in hardware verification to get a reasonable perspective on architectures, sub-systems, and design styles worth proving, and to enable the architects to comment on some of the formally based verification tools they might be using to design systems.

In Chapter 12, Al Davis gives a provocative answer to the question: *What do computer architects design anyway?* The intent of this chapter is convey to the hardware "verifier" the nature of the circuits that are being designed today, and some of the factors that influence the design process. The goal is to highlight the practical difficulties that must be addressed by automated verifiers in order for such design tools to be routinely accepted by designers. A small catalog of real design fragments is presented, intended to spur verification efforts to tackle a superset of this class of circuits.

In addition to the topics mentioned above, issues relating to processor architectures and testability were also discussed at the workshop. Al Davis reviewed many of the commercially available parallel processors, highlighting the idiosyncrancies of their architectures and relative cost-performance trade-offs. Vinod Agarwal pointed out that hardware testing has always been treated as one of those "shop" problems that is best treated in an ad hoc manner. However, the increasing complexity of application specific integrated circuits and printed circuit boards with surface mount devices is creating a situation wherein manufacturing testing cost (including equipment, maintenance, set up, labour, test programming etc.) is rapidly becoming the most expensive component of total product cost. To reverse this trend, the concepts of design for testability (DFT) and built-in self-test (BIST) are being explored for and adopted in commercial products. Agarwal described these concepts and how they create additional demands on design, design verification, and other related issues. He argued that, in the near future, testing requirements must become a part of the system's functional specification.

This collection of papers should prove useful for practitioners, researchers and graduate students interested in hardware verification and theorem proving. We also hope that the examples and problems documented here will stimulate further research in formally based design techniques.

Graham Birtwistle   *P. A. Subrahmanyam*
University of Calgary   *AT&T Bell Laboratories*

Contents

Correctness Properties of the Viper Block Model: The Second Level

Avra Cohn

University of Cambridge Computer Laboratory
New Museums Site, Pembroke Street
Cambridge, CB2 3QG, England.

Abstract:

Viper [7,8,9,11,23] is a microprocessor designed by W. J. Cullyer, C. Pygott and J. Kershaw at the Royal Signals and Radar Establishment in Malvern, England (RSRE), and is now commercially available. Viper was intended for use in safety-critical applications such as civil aviation and nuclear power plant control. It is currently being evaluated (as part of the "VENOM" project [10]) for use as an input-output controller in the deployment of weapons from tactical aircraft. To satisfy requirements of safety-criticality, Viper has a particularly simple design about which it is relatively easy to reason using current techniques and models. The designers at RSRE, who deserve much credit for the promotion of formal methods, intended from the start that Viper be formally verified. The verification project has been carried out at the University of Cambridge. The University was and is not involved with any of the applications of Viper, civil or otherwise, and the whole verification project is and has been fully in the public domain. This report describes the partially completed correctness proof, in the HOL system, of the Viper 'block model' with respect to Viper's top level functional specification. The (fully completed) correctness proof of the Viper 'major state' model has already been reported in [5]. This paper describes the analysis of the block model in some detail (in Sections 6 to 9), so is necessarily rather long. Section 2 is a discussion of the scope and limits of the word 'verification', and cautions against careless use of the term. The paper includes a very brief introduction to HOL (Section 4), but does not attempt a description or rationalization of Viper's design. The possible uses of the paper are as follows:

- It includes enough detail to support an attempt to repeat the proof in HOL, or possibly in other theorem-proving systems;
- It serves as a guide for future analyses of Viper;
- It complements the existing Viper documentation;
- It covers some general issues in hardware verification;
- It illustrates some problems in managing large HOL proofs.

Contents

Table of Figures

1   Introduction

This report describes the partially completed correctness proof of the Viper 'block model'. Viper [7,8,9,11,23] is a microprocessor designed by W. J. Cullyer, C. Pygott and J. Kershaw at the Royal Signals and Radar Establishment in Malvern, England, (henceforth 'RSRE') for use in safety-critical applications such as civil aviation and nuclear power plant control. It is currently finding uses in areas such as the deployment of weapons from tactical aircraft. To support safety-critical applications, Viper has a particulary simple design about which it is relatively easy to reason using current techniques and models.

The designers, who deserve much credit for the promotion of formal methods, intended from the start that Viper be formally verified. Their idea was to model Viper in a sequence of decreasingly abstract levels, each of which concentrated on some aspect of the design, such as the flow of control, the processing of instructions, and so on. That is, each model would be a specification of the next (less abstract) model, and an implementation of the previous model (if any). The verification effort would then be simplified by being structured according to the sequence of abstraction levels. These models (or levels) of description were characterized by the design team. The first two levels, and part of the third, were written by them in a logical language amenable to reasoning and proof. The top level model was a simple, direct state transformation function – a conditional expression specifying the effect on Viper's registers of processing each class of machine instructions (see [7] and [5]). The lowest level model in the sequence was the circuit structure itself, expressed in the hardware description language ELLA; and there were several levels in between.

To give due credit all around, *all* of the design work for Viper was carried out by the RSRE team. So also was the plan for verifying Viper and for structuring the verification effort, as well as the design, into a sequence of abstraction levels. The RSRE team produced the first two levels of specifications of Viper (and part of the third) in a logical language suitable for verification purposes. Further, W. J. Cullyer carried out a (substantially correct) informal paper-and-pencil correctness proof up to the second abstraction level [8]. Viper was not designed in any part or aspect by the Hardware Verification Group at the University of Cambridge, nor does the University have any involvement with the applications of Viper. The task of formally and mechanically verifying the Viper design (up to register-transfer level) was sub-contracted by RSRE to members of the Hardware Verification Group at the University of Cambridge, for a period of about two years (see Acknowledgements), during which time the manufacture of the chip was already underway or indeed completed at various U.K. sites. All of the verification work is and has been fully in the public domain.

Verification was intended to be done in HOL (Higher Order Logic) [2,14,15], a theorem-proving system derived from R. Milner's LCF system (Logic for Computable Functions) [12,22] and based on higher order logic as set out by A. Church [3]. HOL was implemented by M. Gordon at Cambridge University and is currently in use by the Hardware Verification Group at Cambridge University and at various sites throughout the world. (HOL was developed from an earlier system by M. Gordon called LCF_LSM.) 'Verification' was understood by the designers (as by the LCF and HOL communities) to mean complete, formal, logical proof in an explicit and well-understood logic. That is, it means proof in the usual mathematical sense of a sequence of inference steps, and not just simulation or some other non-formal process. Proofs of this sort are constructed interactively in HOL with machine assistance and user-guidance, and not (usually) fully automatically.

A case study in the methodology for the Viper proof was carried out by the author and M. Gordon in 1986; this treated a simple hardware device (in fact, a component of Viper) at several abstraction levels down to and including gate level [4]. The first level of the Viper correctness proof was carried out by the author in 1986-7 [5]. (As mentioned, an earlier informal proof had been done by W. J. Cullyer.) The formal proof, fully completed, confirmed that the second level of description, the 'major state machine', with certain corrections made, faithfully implemented the top level specification of Viper (again, with certain corrections made). The major state machine was designed to implement each top level state transformation (i.e. the processing of each instruction type) by a sequence of lesser steps, each of which determined the *next* step or else indicated the end of a sequence. The steps were called 'major states'. The major state model concerned itself only with the flow of control in Viper, and not with arithmetic or logical computations. That proof, assisted by the HOL system, consisted of about a million primitive inference steps[1] and took about six person-months to complete. As mentioned, it revealed errors (which were subsequently corrected) in both the major state model and in the top level specification, as presented. These errors, however, did not manifest themselves in the actual Viper chips, so that although the major state model was intended to be a useful link in the chain of abstraction levels, it was of no direct concern to the fabricators of Viper chips.

The block model, described here, *is* of concern to the manufacturers because it directly relates to the circuit design. The block model can be previewed in Figure 1 (Section 6.3); it is a partly pictorial and partly textual (and functional) model consisting of 'blocks' (such as Viper's instruction decoder, its ALU and its memory), with information passing between blocks, and to/from the outside world, at fixed clock cycles. The functional specifications concern the internal combinational logic of the various blocks, but not their time-behaviour, nor the connection between separate blocks; the pictorial specification fills in the rest of that information. Much as at the major state level, the concept of single instruction types being processed by sequences of steps (major states) is built into the block model; specifically, one of the blocks is a counter representing the major state. In addition, several *minor* states implement each major state in the block model; another block is a counter for *minor* states. Thus there is a yet-finer time-scale at the block level than at the major state level.

The first task in the verification effort is to derive a functional expression of the block model in a formal logic which is suitable for reasoning and proof. This is necessary because it is difficult to reason formally about a schematic diagram indicating the transfer of information to and from its sub-units simultaneously[2].

The second task is to analyze the behaviour of the block model *using* its func-

---

[1]On the issue of counting inference steps, see Footnote 10.

[2]It is possible to imagine doing this by reasoning about sequences of annotated pictures, but the real problem is not so much the obvious awkwardness of such a method as the lack of a formal semantics of pictures. The 'proof' would not be formal without a clear semantics; that is, we would have to invent and justify a logical calculus whose terms were pictures.

tional representation. What one ultimately wants to know is (for each instruction type) how *many* minor and major steps must be cycled through before the instruction is fully processed, and what the accumulated effects are on the 'state' of the block model after that number of cycles. These involve extracting from the formal representation (i) the concept of the state of the block machine; (ii) the conditions under which one state leads to another; and (iii) and the assumptions which must be made about initial states and 'normal'[3] behaviour in order to resolve the state transitions. These concepts are implicitly determined once the functional representation is constructed. It must also be shown that the state transition conditions cover all logical possibilities, to ensure that no possible instruction types have been omitted.

The third task is to deduce the results at the *higher* level for each instruction type, under the same conditions that drive the block machine through its major and minor steps for that particular instruction type. Then, case by case, the results are compared at the two levels. This requires (i) relating the block level state to the higher level state; and (ii) relating the conditions which drive the block machine to the more abstract conditional choices at the higher level. Whether the third task is achieved, as intended, via the major state machine, or is achieved directly by comparing the results to the top level results, is a technical question discussed in Section 10; in this case the latter seems less complicated.

The first task has been completed and is treated in Section 6: from the pictorial and textual block information provided by RSRE we have derived a fully formal expression of the block model from which we can logically infer the block model's behaviour on the various classes of Viper instuctions. This is done using techniques now standard in hardware verification. This task is perhaps the most interesting part of the analysis, and is an important achievement in itself.

The second task has also been completed (see Sections 7 and 8): the functional expression has been used to describe the state transition conditions of the block machine, and then to logically derive the cumulative behaviour of Viper for each instruction type. There are about 120 sequences of major state transitions to consider, each requiring certain assumptions to be made concerning normal behaviour and initial conditions. Every major state transition of the sequences comprises a sequence of minor state transitions. The large number of major state sequences includes all possible types of computation – additions, shifts, comparisons, and so on – with elaborations such as indexed or looked-up operands, and so on. We have also proved that among this multitude of cases no case has been omitted.

The development of the proof up to this point is not fixed in advance; rather, the major and minor state transition conditions are determined by the various block definitions. These are repeatedly unfolded (under the transition conditions for each instruction type) to produce results not necessarily foreseen. This means that the proof (up to this point) makes rather unsophisticated use of HOL; this is the use indicated by the nature of the problem[4].

---

[3]'Normal' behaviour means behaviour which is within the scope of the high level specification. For example, we have to assume in the Viper block proof that no resetting signals come in from the outside world during the course of the block-level processing of instructions; and that the block machine's timeout facility is never invoked. This is because the high level specification itself does not treat these block level contigencies.

[4]Like LCF, HOL provides a repertoire of tools (and facilities for designing one's own tool-set) for

The third task has *not* been completed, although some preliminary analysis has been done which indicates that the block results are at least plausible. This is discussed in Section 9. It has proved to be impractical to pursue the third task at present in the absence of (i) better support in HOL for advanced reasoning about intricate bit-string manipulations; and (ii) closer interaction with the design team in order to explain and relate the differing computations at the high and low levels.

In short, what we have achieved is to have proved some useful *properties* of the block model, without having managed to prove the block model to be an adequate implementation of the top level specification. By describing the cumulative effects on the state of the Viper block model of each instruction class, we have essentially *symbolically executed* the block model (considered as a finite-state machine). This in itself is valuable; the results of the symbolic execution could be used to build a simulator able to jump (in a provably correct way) from result to result without having to derive the minor and major transitions of the block machine again. Of course, in *proving* that each derivation is correct and that all possible cases have been treated, we have done rather more than a symbolic execution; and a symbolic execution in itself was not the aim of the work. Nonetheless, the proof can be considered as a kind of 'quality control' for the symbolic evaluation, assuring that the results are dependable. Furthermore, the functional specification of the block model could be used to prove properties of the Viper block model not manifest in the high level specification; for example, we could infer from our representation the consequences of a reset signal or of a 'bad' initial state.

The proof is very hierarchically structured, as one would expect. First the minor state transitions are analyzed (Section 7); then the major state transitions (Section 8.1), themselves composed of minor state sequences; and finally, the *cumulative* results for major state sequences, which process single instruction types (Section 8.3). The many lemmas required throughout the development form a layered dependence structure. These lemmas are described as they arise. Although it is difficult to find a slice through the proof structure which conveys its interlocking complexity and total bulk, it is clearly impossible (and would be hopelessly boring and repetitious, anyway) to describe the proof as a whole; so as far as possible, a particular instruction type is taken as typical: this is addition with overflow detection.

In the process of performing the proof thus far, a great deal has been learned about managing and properly structuring massive proof efforts. The techniques required did not go beyond anything already standard in HOL circles; and aside from space problems, the proof did not tax the HOL system at all, which is encouraging news about HOL. To date, the block level proof comprises about seven million inference steps and required about one person-year to generate. So far, no real surprises have emerged about the behaviour of the block machine – but then, surprises would seem likelier to turn up in the third (uncompleted) phase of the project than in the first two. (The preliminary correctness results are described in Sections 8.3 and 9.)

The reader will notice (probably with increasing distress) that this paper is a very straightforward, chronological and (deliberately) detailed account of the proof effort, intended so that anyone wishing to reproduce the proof, or part of it, could use the paper as a reference; and so that dedicated HOL users can examine details. No attempt has been made to explain the design of Viper or the parts and operation

proving conjectures by applying strategies in a subgoaling fashion.

of the block machine beyond what is really necessary for describing the formal proof in HOL. Anyone interested in Viper apart from this is referred to the RSRE Viper literature.

Although the paper is meant to stand on its own, it will obviously make more sense in the context of the RSRE and HOL literature in general and [5] in particular. A *very* brief introduction to Viper is given in Section 3, and to HOL in Section 4. Though actual HOL tactics and procedures are only suggested in this report, it should be understood that for every theorem and lemma mentioned, a full formal proof in HOL was performed – by the application of a procedure in HOL's meta-language ML. From a research point of view, none of the HOL tactics or methods used are particularly original or interesting.

For an overview of the proof effort, Sections 3, 4, 5, 6.1, 6.2 and the Conclusions section should give some idea of what was proved, and a bit about how.

Appendices containing the HOL versions of the original RSRE definitions of the top level and the block level of Viper are provided for reference. The author claims full responsibility for any typographical or other errors in the translation of the original text (which was in an informally annotated, now-obsolete early prototype of the HOL logic, with some misprints) into the HOL logic. Such errors can be difficult to find. The appendices are supplemented by Figure 1 (Section 6.3), a figure compiled by the author from ten separate RSRE figures. To construct Figure 1, the ten separate figures were connected together according to the coincidence of the names of lines amongst the original figures. They were then topologically rearranged, and completed (for type-correctness) by the addition of an extra block (FSELECT_COMB) as well as six extra internal line names: ram, count, preg, xreg, yreg and areg. The author also claims full responsibility for any errors in the translation into Figure 1 of the original ten figures. All subsequent reasoning about the block model is based on Figure 1.

Finally, Section 2 contains a discussion of the problems involved in asserting that a chip has been 'verified'. There are great dangers in making such claims if their scope and limitations are misunderstood, as they often are. There is little room for misunderstanding in real-life safety-critical applications. The author strongly urges that Section 2 be read and pondered even if the technical sections are skimmed or skipped.

Several other people have worked on the formal verification of processor designs. W. Hunt [17] used the Boyer-Moore theorem prover to prove correct the 'FM8501', a microcoded microprocessor that he developed as part of his Ph.D. research. The FM8501 is a machine invented for the purpose of the proof (and not implemented), which therefore has a cleaner specification than Viper; it is roughly as complex as Viper. J. Joyce [18] has verified Tamarack, another machine invented for proof purposes. Tamarack, though very much simpler than Viper, *has* actually been implemented and fabricated. The proof was also done in HOL. This work is noteworthy because the proof goes all the way down to the transistor level, and is currently being extended up to software levels. An early version of Tamarack was verified by H. Barrow using his VERIFY system [1], and again by M. Gordon using LCF_LSM [13]; neither of these two latter systems were based on a clearly delineated logical calculus.

2   The Scope and Limitations of the Proof

When we hear that a chip such as Viper has been 'verified', it is essential to under-stand exactly what is meant. Several important points sharpen and limit the senses in which a chip (and Viper in particular) can be called verified.

Ideally, one would like to prove that a chip correctly implemented its intended behaviour in all circumstances; we could then claim that the chip's behaviour was predictable and correct. In reality, neither an actual *device* nor an *intention* are objects to which logical reasoning can be applied. The intended behaviour rests in the mind of the architects and is not itself accessible. It can be reported in a formal language, but not with checkable accuracy. At the same time, a material device can only be observed and measured, not verified. It can be described in an abstracted way, and the simplified description verified, but again, there is no way to assure the accuracy of the description. Indeed, the description is bound to be inaccurate in some respects, since it cannot be hoped to mirror an entire physical situation even at an instant, much less as it evolves through time. In short, verification involves two or more *models* of a device, where the models bear an uncheckable and possibly imperfect relation both to the intended design and to the actual device. This point is not merely a philosophical quibble; errors were found both in the top level specification of Viper *and* in its major state model, none of which was either intended by the designers *or* evident in the manufactured Viper chips (these errors are discussed in [5]). The errors were fairly minor and quickly repaired, but their presence throws into relief the rather limited sense in which an actual product can be said to have been verified against the architect's design.

That the actual chips appeared not to suffer from the errors found in the models also illustrates the rather academic nature of the research described in [5] and in the present paper. The chips were already in the process of being built by the time the sub-contracted verification work began on the major state proof at Cambridge; and they *had* been built and were being advertised by the time the work described in this report was undertaken. While it is possible in theory that an error in an abstract specification had been reflected in the circuit design given by RSRE to the manufacturers – the abstract specifications were no doubt in the architects' minds while they designed the circuit – it seems more likely that because of the weak links between the abstract specifications, the circuit design process and manufacturers, that problems in the specification would *not* propagate down to chip problems.

At more concrete levels of description, the problem may be further complicated by there not even being provided a description in a formal language. For exam-ple, Viper's top level specification and its major state model were both supplied in a logical language; but at the block level, the subject of this report, the de-scription given was partly formally (see the Appendix) and partly pictorially (see Figure 1, Section 6.3). Combining these two parts required some human ingenuity and guesswork (see latter part of Section 6.3 and the **Aside** within Section 6.3.1). Before verification can be meaningfully applied in such cases, a fully formal descrip-tion must be produced. Once again, accuracy cannot be checked; the new formal description may be a flawed translation of the pictorial specification, or a flawed combination of picture and text, but there is no sense in which this can be tested. One might thus very well end up proving properties of a formal description bearing an imperfect relation to the intended design. In fact, this *was* a problem in the block level representation of Viper; the author's first attempt at a formal representation of the Viper block diagram involved interchanged line names whose presence was only discovered (rather later in the proof) by an unsystematic inspection. This additional problem of the accuracy of a representation could appear at the gate level, the transistor level or any other level at which a linguistic description has to be constructed creatively from a pictorial one. It further limits the sense in which a chip can be called verified.

Another point which must be made explicit, given that verification relates a more to a less abstract model, is the level of abstraction and the degree of completeness of the models in question. We say that a device has been verified 'at the major state level' or 'at the register transfer level'; it is not enough to say simply 'verified'. For example, Viper's major state machine has been fully verified with respect to its top level specification, where the major state machine captures the flow of control (implicit at the top level) through the fetch-decode-execute cycle, but does *not* concern itself with any arithmetic or logic computations. The block machine concerns itself with Viper's arithmetic and logical operations, and with the transfer of information between registers and memory; and not with gate connections, transistors, electrical effects, timing problems, and similar areas in which unsuspected errors seem most likely to be found[5]. In addition, the models may be incompletely specified. For example, Viper's highest level model is complete only as regards the processing of instructions, and does not cover resetting or timing-out the machine, or other possible behaviours described at the block level. This restricts any analysis to the high level behaviours alone, again ignoring the more subtle issues. In view of all of this, Viper should not be called verified without reference to the nature of the models used to represent it. Further, as mentioned, the verification of the block level with respect to the high level has for practical reasons not been completed.

Various of these limitations on the use of the word 'verified' are obscured in claims such as the following (both taken from promotional material):

> "VIPER is the first commercially available micropressor with ... a formal specification and a proof that the chip conforms to it." [24,25]

> "One unique feature of Viper is that the instruction set is specified mathematically using the language LSF-LSM [*sic*], and the gate-level logic design has been proven to conform to this specification." [16]

The first is perhaps merely sloppy, but the second is simply not true; no proofs have been done at or anywhere near the gate level. Such assertions, taken as assurances of the impossibility of design failure in safety-critical applications, could have catastrophic results. To summarize,

- A physical *chip* is not an object to which proof meaningfully applies.

- The top level formal specification of Viper (and hence any verification effort) is itself incomplete, covering only the fetch-decode-execute cycle of Viper .

- Viper has been analyzed at best at register-transfer level, i.e. still very abstractly, and not yet at levels at which problems seem likeliest to occur.

---

[5]In those areas, enormous amounts of research remain to be done on finding useful, tractable models, even before we begin to verify them.

- *At* register transfer level, the proof has been only partially completed .

Finally, the correctness of an abstract representation of a chip must be placed in context when we talk about the reliability of physical systems in safety-critical applications. The author claims no expertise in the field of reliability, but it would be irresponsible not to point out the obvious: that this very abstract and limited sense of correctness (the equivalence of a register transfer level model to a functional specification of the fetch-decode-execute cycle) is only one of *many* issues which have to be considered collectively. Aside from possible problems at more concrete levels of description, which have already been mentioned, safety will also depend on factors as yet outside of the world of formal description: these range all the way from issues of social administration and communication, as well as staff training and group behaviour, at one end, to the performance of mechanical and chemical parts, and so on, at the other. One has only to list the mass catastrophes of the last ten years or so to perceive the predominant role played by these extra-logical factors. It is the author's guess (albeit, again, not an expert opinion) that the sort of *a posteriori* abstract design correctness discussed in this paper, though of undoubted importance, forms a relatively small contribution to the overall reliability of real machinery. (This seems so at least at the present stage of research into representation and proof, and with the present weak links between designer, verifier and manufacturer.) That is, using a hardware design verified only down to register-transfer level (and there only partially verified and only in 'normal' situations) as part of the control system in extraordinarily hazardous applications (in which large populations or land masses may be destroyed) does not seem significantly safer than using any other design. The use of the word 'verified' must *under no circumstances* be allowed to confer a false sense of security.

These remarks should be taken as evidence, in our short-sighted times, of the need for further basic research (i.e. the *funding* of further basic research), and not as pessimism. After all, the HOL system, currently one of very few theorem-provers capable of handling realistic hardware proofs, is directly based on research in pure mathematics and philosophy by Frege, Russell and most directly, Church, many decades ago; and on R. Milner's theorem prover for denotational semantics, a very different application area. The remarks pertain to the current early (but thoroughly optimistic) state of research into the representation and verification of hardware.

3   The Design of Viper

Viper [7,8,9,11,23] is a microprocessor designed at the Royal Signals and Radar Establishment and now commercially available. It is intended for use in safety-critical applications, and has several design features supporting such applications. Viper is hard-wired rather than microcoded, to minimize the number of gates. As mentioned in the introduction, no attempt is made in this report to describe the design of the machine or its unique features; for that, the reader is referred to the Viper literature. In this section we introduce only those aspects of the architecture required for a discussion of the correctness proof of the block level model. Indeed, of all the features at that level, only those concerned with (uninterrupted) instruction processing are described. This is because, as mentioned, the top level specification of Viper provided by the designers itself only covers instruction processing.

3.1 Viper Instructions

Viper has a 32-bit memory. Addresses are 20-bit words, but the memory is addressed by 21-bit words whose the most significant bit distinguishes main from peripheral memory. (Peripheral memory is for input-output operations.) The registers visible to the user are: a 20-bit program counter, a 32-bit accumulator, two 32-bit index registers, a boolean flag (for holding the results of comparisons, etc) and a stopping flag (which normally indicates an error condition).

Instructions are 32-bit words, of which the top twelve bits are the instruction code and the bottom twenty the address. The twelve bits of the instruction encode the following fields:

- **Bit 4**: A 1-bit indication of whether the instruction is a comparison;

- **Bits 0 to 3**: A 4-bit function selector indicating the ALU operation to be computed, according to whether a comparison has been indicated;

- **Bits 10 and 11**: A 2-bit register source selector for the computation, indicating either the program counter, the accumulator or one of the two index registers as the source of one of the operands;

- **Bits 8 and 9**: A two-bit memory source selector for the computation, indicating either literal addressing, content addressing, or addressing offset by the value in one of the two index registers as the method of accessing the operand in memory;

- **Bits 5 to 7**: A 3-bit destination selector to choose a destination for the computation from amongst the accumulator, the two index registers, and the program counter conditionally or unconditionally on the boolean flag.

(Some of these fields can double for other purposes.) The operations are as one would expect: comparisons test numerical less-than and equality between operands; non-comparisons (involving one or two operands) include addition, subtraction, shifts, logical operations, procedure calls, and so on. As mentioned, the example used in this report is addition with overflow detection. In that case, the comparison field holds the value 0 (indicating a non-comparison) and the appropriate function selector value happens to be 5.

3.2 Design Features of Viper

Certain internal registers are used (in the course of executing instructions) which are not accessible from the outside: these include a 32-bit temporary register; a 20-bit register for storing the address field of instructions; and a 12-bit register for storing the instruction code.

Viper accepts certain inputs from the outside world which control its behaviour but are not modelled at the top level. (These are all shown in Figure 1, Section 6.3, prefaced by 'e_'.) They include a signal for resetting the machine (e_resetbar), one for single-stepping it (e_stepbar), one for forcing an error (e_errorbar), and one for extending read/write cycles (e_reply). There are also outputs *to* the world, for viewing certain state values directly – for example, the indications of whether the machine is stopped (e_stopped), is fetching an instruction (e_fetchbar) or is performing a computation (e_perform). The boolean flag and the major state can also be read externally, on e_bflag and e_majorstate respectively.

Internally, there is a fixed limit (recorded by a counter, count) to the number of cycles in which the memory can respond, after which an exception occurs (to prevent deadlock due to memory failure)[6]. As mentioned earlier, for establishing the correspondence between the block level and top level models we must assume that certain of these signals are well-behaved; e.g. we assume that the reset signal is false throughout the execution of an instruction, and that at the initial stage in processing an instruction, the timeout-counter is not already at its maximum value.

The Viper chip contains no memory aside from its registers. The block level model, however, includes a simple memory model, since that is the minimum configuration in which it makes sense to talk about executing an instruction (for purposes of verification). The memory model provided is simplified in that it responds in a fixed and minimal number of cycles. The Viper design supports other memory protocols, but these are not modelled by RSRE. This means that the timeout facility (with its counter) is not exercised in the correctness proof.

4 The HOL System

The verification described in this report was carried out in the HOL system ('HOL' standing for 'higher order logic'). In this section we attempt to give just enough information about HOL to make the rest of the report readable; readers curious to know more about the system are referred to [2,14,15].

4.1 An Outline of the HOL System

The HOL system is a version of LCF ('logic for computable functions'). LCF was designed by R. Milner in association with C. Wadsworth, M. Gordon, M. Newey and L. Morris [12,22]. HOL, like LCF, is designed to facilitate the interactive generation of formal proofs. In both systems, a *logic* in which problems can be expressed is interfaced to a *programming language* in which proof procedures and strategies can be encoded. The combination enables deductions in the logic (in the sense of chains of primitive inference steps) to be produced by invocation of programming constructs at a higher level of abstractness. This makes it possible for very long, detailed, complex proofs in the logic to be produced by use of procedures meaningful to the user of the system – yet without compromising the formality and completeness of the underlying proof. Examples of procedures meaningful to a user might include: unfolding definitions on fixed parameters, normalizing form, case analysis, and rewriting left-to-right using axioms and previously proved theorems.

HOL differs from LCF in the particular logic used. The logic part of HOL is conventional higher-order logic as set out by Church [3]. The version used in the HOL system is oriented towards proofs about hardware only insofar as it provides built-in types, constants and axioms for representing *bit strings*[7]. New types, constants and axioms can be introduced by the user, and organised in logical *theories*, as in

---

[6]This is the device verified to gate level as a case study [4].

[7]They are built in only for convenience; the machinery for defining them 'from scratch' exists.

LCF. Theorems once proved can be saved in and retrieved from theories. Theories themselves are organized into hierarchies in which the types, constants, axioms and theorems of an ancestor theory are accessible from within a descendent theory.

The programming language of HOL is ML (for 'meta-language'), which orginated as the meta-language in the LCF system (though is now well-known in its own right). The *type discipline* of ML ensures that the only way to create *theorems* in the object logic is by performing proofs; theorems have the ML type *thm*, objects of which can only be constructed by the application of inference rules to other theorems or axioms. (Theorems are written with a turnstile, ⊢, in front of them, and with assumptions to the left of the tunstile.) LCF-style proof is explained more fully in [22].

4.2 The Logic

The HOL system uses the ASCII characters ~, \/ and /\, ==>, !, ?, and \ to represent the logical symbols ¬, ∨, ∧, ⊃, ∀, ∃ and λ respectively.

For the purposes of this paper a *term* of higher-order logic can be one of the following kinds.

- A **variable**;

- A **constant** such as T or F (which represent the truth-values *true* and *false* respectively);

- A **function application** of the form $t_1 t_2$ where the term t_1 is called the *operator* and the term t_2 the *operand*;

- An **abstraction** of the form \x.t where the variable x is called the *bound variable* and the term t the *body*;

- A **negation** of the form ~t where t is a term;

- A **conjunction** of the form t_1/\t_2 where t_1 and t_2 are terms;

- A **disjunction** of the form t_1\/t_2 where t_1 and t_2 are terms;

- An **implication** of the form t_1==>t_2 where t_1 and t_2 are terms;

- A **universal quantification** of the form !x.t where the variable x is the bound variable and the term t is the body;

- An **existential quantification** of the form ?x.t where the variable x is the bound variable and the term t is the body;

- A **conditional** of the form t=>t_1|t_2 where t, t_1 and t_2 are terms; this has *if-part* t, *then-part* t_1 and *else-part* t_2;

- A **local declaration** of the form let x=t_1 in t_2, where x is a variable and t_1 and t_2 are terms; this is provably equivalent to (\x.t_2)t_1 (see Section 4.4); and

- A **list** of the form [t_1:ty;t_2:ty;...;t_n:ty] where ty is a *type* (see below). Here t_n is called the zero[th] element, t_1 the $(n-1)^{th}$ or head[8] and [t_2;...;t_n] the tail.

---

[8]The 'reverse' numbering is used to correspond with conventional significance-order of bit-strings

All terms in HOL have a *type*. The expression t:ty means t has type ty; for example, the expressions T:bool and F:bool indicate that the truth-values T and F have type bool for boolean, and 3:num indicates that 3 is a number.

If ty is a type then (ty)list (also written ty list) is the type of lists whose components have type ty. If ty_1 and ty_2 are types then ty_1->ty_2 is the type of functions whose arguments have type ty_1 and whose results have type ty_2. The cartesian product operator is represented by #, so that ty_1#ty_2 is the type of pairs whose first components have type ty_1 and second, ty_2.

In this paper, the logical constants AND, OR and NOT used by RSRE are used interchangeably with the corresponding HOL constants.

As indicated earlier, the HOL system provides a number of predefined types and constants for reasoning about hardware. The types include $word_n$, the type of n-bit words and $mem_{n1,n2}$ for memories of n_2-bit words addressed by n_1-bit words. The expression #$b_{n-1}\cdots b_0$ (where b_i is either 0 or 1) denotes an n-bit word in which b_0 is the least significant bit.

The predefined constants used in this paper are shown below.

- V:bool list->num converts a list of truth-values to a number;

- VAL_n:$word_n$->num converts an n-bit word to a number;

- $BITS_n$:$word_n$->bool list converts an n-bit word to a list of booleans;

- $WORD_n$:num->$word_n$ converts a number to an n-bit word;

- $FETCH_{n1}$:$mem_{n1,n2}$->($word_{n1}$->$word_{n2}$) looks up a word at an address in memory;

- $STORE_{n1}$:$word_{n1}$->($word_{n2}$->($mem_{n1,n2}$->$mem_{n1,n2}$)) stores a word at an address in memory;

- EL:num->(bool list->bool) selects the specified element of a boolean list, where the last (rightmost, in the notation) element of the list is the $zero^{th}$;

- CONS:ty->(ty list->ty list), for any type ty, constructs a new list $[t;t_1;t_2;\ldots;t_n]$ from a list $[t_1;t_2;\ldots;t_n]$ and an element t:ty, so that t is the nth element of the new list;

- HD:ty list->ty, for any type ty, maps a list $[t_1;t_2;\ldots;t_n]$ to t_1, i.e. to its $(n-1)^{th}$ element;

- TL:ty list->ty list, for any type ty, maps a list $[t_1;t_2;\ldots;t_n]$ to $[t_2;\ldots;t_n]$;

- SEG:(num#num)->(bool list->bool list) returns the specified segment of a *boolean* list between and including the elements whose numbers are given;

- NOT_n:$word_n$->$word_n$ inverts the bits of a word;

- AND_n:$word_n$->$word_n$->$word_n$ conjoins two words bit-wise;

- OR_n:$word_n$->$word_n$->$word_n$ disjoins two words bit-wise; and

- ARB is a constant whose type is instantiable to any type; it denotes an arbitrary unspecified value at each type – this is useful in certain formal expressions.

To make terms more readable, HOL uses certain conventions. One is that a term $t_1 t_2 \cdots t_n$ abbreviates ($\cdots(t_1 t_2) \cdots t_n$); that is, function application associates to the left. The product operator * associates to the right and binds more tightly than the operator ->. For example, the type of SEG could be written simply as num#num->bool list->bool list; and the function REGSELECT used in the Viper block model to select the register operand for a computation (Sections 4.4 and 6.3) has the type

    word32#word32#word32#word20#word2->word32

which abbreviates

    (word32#(word32#(word32#(word20#word2))))->word32

4.3   The Framework for Expressing the Block Model in HOL

The registers of Viper are represented in HOL using the HOL logic's special types for bit-strings. We choose variable names for the various registers so that:

- areg:word32 is the accumulator (a-register);

- xreg:word32 is the first index register (x-register);

- yreg:word32 is the second index register (y-register);

- preg:word20 is the program counter (p-register);

- bflag:bool is the boolean flag (bflag);

- ram:mem21_32 is the memory;

- treg:word32 is the temporary register (t-register);

- addr:word20 is the address register; and

- inst:word12 is the instruction register.

The constants (predicates) pertaining to hardware are used to manipulate these; for example, if FETCH_ABBR(ram,preg) is an abbreviation we introduce that denotes the 12-bit instruction code part of the word in the memory ram pointed to by the program counter preg, then

    ¯EL 4(FETCH_ABBR(ram,preg))

means that the new instruction is not a comparison instruction (see Section 3). If in addition

    WORD4(V(SEG(0,3)(FETCH_ABBR(ram,preg)))) = #0101

then (given that the instruction is a non-comparison), the value 5 is defined to indicate the ALU operation: addition with overflow detection. (This is the example computation used throughout this report.)

As mentioned, FETCH_ABBR is a convenient *abbreviation* we introduce in the logic. In terms of the basic bit-manipulation function FETCH21, FETCH_ABBR(ram,preg) stands for

```
EL
4
(BITS12
 (WORD12
  (V
   (SEG
    (20,31)
    (BITS32
     (FETCH21 ram(WORD21(V(CONS F(BITS20 preg)))))))))))
```

That is, the new instruction is fetched from the address in the 20-bit program
counter (with the twenty-first bit false to indicate main rather than peripheral
memory); it is converted into a list of thirty-two boolean values; the upper 12-bit
segment (the instruction code) is extracted; its value is considered as a 12-bit string;
this is converted to a list of twelve booleans; and the fourth element of the list, the
comparison indicator, is selected.

This whole expression (and most subsequent expressions) are pretty-printed for
ease of parsing (rather than economy of space!)[9].

Occasionally, the 12-bit instruction field (usually represented by the variable inst)
and the other fields are dealt with separately from the 20-bit address. In such cases
we write

```
EL 4(BITS12 inst)

WORD4(V(SEG(0,3)(BITS12 inst)))

WORD3(V(SEG(5,7)(BITS12 inst)))

WORD2(V(SEG(8,9)(BITS12 inst)))

WORD2(V(SEG(10,11)(BITS12 inst)))
```

to denote the comparison indicator, the function selection field, the destination
selection field, the memory indicator and the register selector, respectively.

4.4   Proof in HOL

As an example of HOL features mentioned thus far, consider (in a purely formal
way) the simple HOL definition of the block model function which computes the
source register for a computation. Given the a-, x-, y- and p- registers as arguments,
and the 2-bit source register indicator, the function REGSELECT returns the appropriate
register (negated):

```
|- REGSELECT(areg,xreg,yreg,preg,rsel) =
   (let rf = VAL2 rsel in
    ((rf = 0) => NOT32 areg |
     (rf = 1) => NOT32 xreg |
      (rf = 2) => NOT32 yreg |
       (rf = 3) => NOT32(WORD32(VAL20 preg)) | ARB)))))
```

The turnstile indicates either an axiom, a definition or a theorem (here, a definition).
Terms (if any) appearing to the *left* of the turnstile represent assumptions upon
which the fact to the right depends – as mentioned earlier. Constructs of the logic
used include the *let*-construct and the conditional. The constant ARB serves as the
final *else*-value of the conditional (which is intended never to be returned, since rsel
is a 2-bit word). The HOL bit-string constants used in the definition are: VAL2, NOT32,
WORD32 and VAL20.

---

[9]Expressions of this sort can sometimes be logically simplified, but this expression is one that
actually arises in the analysis, so it is left in the form required.

The basic rules of inference in HOL take the form of ML functions which (roughly speaking) map theorems (and sometimes various parameters as well) to theorems. More elaborate patterns of inference can be constructed from the basic inference rules of the logic by user-designed ML functions. The validity of the compound inferences is preserved by the type system of ML in which *only* inference rules may return theorems as results.

In our example, we can make the definition of REGSELECT more computationally useful by applying a compound inference pattern (an ML function) which expands the *let*-construct into the equivalent lambda expression, and performs the resulting beta-conversions to deduce a new theorem:

```
|- REGSELECT(areg,xreg,yreg,preg,rsel) =
   ((VAL2 rsel = 0) =>
   NOT32 areg |
   ((VAL2 rsel = 1) =>
   NOT32 xreg |
   ((VAL2 rsel = 2) =>
   NOT32 yreg |
   ((VAL2 rsel = 3) => NOT32(WORD32(VAL20 preg)) | ARB))))
```

Applying the inference pattern actually invokes 520 primitive HOL inference steps (as we have written it) but only requires the user to apply one function to one definition in order to generate the 520-step proof. The pattern applies to *any* definition written using the *let*-construct. This relieves the user of constructing a 520-step chain of substitutions of equals for equals, and so on (which no one would want to bother with) whilst still assuring that an actual proof is done. (For perspective, the whole block model correctness proof comprises about seven million primitive inferences[10].)

Much of the Viper block proof consists in successive unfoldings of known facts on specific values. For example, the new form of the definition of REGSELECT can be applied to particular values, and the resulting theorem then simplified. We can prove for example that

```
|- REGSELECT(areg,xreg,yreg,preg,#00) = NOT32 areg
```

by applying a compound inference rule which instantiates the definition to a specific value and then simplifies, using axioms and previously established facts about bit-strings and so on. This particular unfolding consists in 89 primitive inferences. (In a case argument based on the various possible values of rsel, the theorem above is used to simplify one of the cases.)

Proofs based on unfolding are an example of a particularly simple and unsophisticated use of HOL: they procede in a *forward* direction, starting without a fixed idea of the result, but applying known procedures. In the block model case, the method allows us to symbolically evaluate the block model on each of the possible instruction types, and so give a complete description of the model's behaviour. In contrast, one often starts with a conjecture – i.e. with an end result in mind – and constructs a proof *backwards* by engaging proof strategies to produce successively simpler subgoals. As a result of this process, a forward proof is then constructed.

---

[10]A much shorter proof could be constructed in this particular case, but the figure of 520 arises from the generality of the rule used, which relies on the full power of left-to-right recursive rewriting to expand *let*-expressions (250 steps), and then looks for possible beta-conversions at any depth in the resulting expression. It seems desirable, since many definitions of varying complexity have to be treated similarly, to trade computational resources for user effort in this way. However, the apparent length of proofs can be greatly and sometimes misleadingly inflated by use of such general procedures.

5   The Plan for Verification

The original RSRE plan was to establish the correctness of Viper in a series of decreasingly abstract stages.

The top level specification, as mentioned in the introduction, is just a transition function specifying how an abstract state (representing the memory and the visible flags and registers) changes as Viper executes each of the possible instruction types. It thus embodies an operational semantics of the instruction set. The notion of time is implicit in the notion of next state; there are no time variables or clocks. The basic time unit is quite coarse: namely, the execution of a whole instruction schema.

The next (more concrete) level was called the major state level. At the major state level, an instruction is executed by a sequence of events corresponding to major states of the model and representing the phases of processing of an instruction schema. Each event could affect the visible registers of the high level state or any of several internal registers. (These internal registers were still part of an abstract model of Viper, and did not necessarily correspond to parts of the actual Viper chip.) The next event was determined according to the current event, the visible state, and the internal state; some events were recognizably terminal and some were initial, in the sequences. From all of this we extracted a state transition function representing the major state model, which determined a *graph* of events showing all possible major state sequences of Viper. This led to a correctness statement (proved in HOL) of the following form:

> **If** the major state transition function always gives the next internal state based on the current internal state, and **if** the state at the start time is the initial state, and **if** it requires n major state transitions to return for the first time to the initial state, **then** after n transitions the visible part of the internal state agrees with the new state specified by the top level state transition function.

This was proved by analyzing each path through the graph and comparing the results, under the same conditions which force that path to be taken, with the high level results. There was an explicit notion of time at the major state level in the sequencing and accumulation of effects of the various events; several major state transitions simulated the effect of a single state transition at the high level, i.e. simulated the effect of executing a single instruction. The machine proof was based on a paper-and-pencil proof by W. J. Cullyer [8].

The ten (numbered) major states and their associated effects are listed below.

- 0. INDEX: To the temporary register is added the register indicated;

- 1. FETCH: The word indicated by the program counter is got from memory;

- 2. RESET: Various registers are cleared;

- 3. PRECALL: The program counter is stored before a procedure call;

- 4. PERFORM ALU: The appropriate arithmetic-logic operation is done;

- 6. READIO: A peripheral device is read;

- 7. READMEM: The memory is read according to the temporary register;

- 8. STOP: The machine is stopped (because of an error);

- 10. WRITEMEM: The memory is written;

- 11. WRITEIO: A peripheral device is written.

The theorem expressing the correctness of the major state machine, though not easy to prove, did not say very much; merely that the *flow of control* in the major state machine was correct. There was no computation of values at the major state level – that is, additions, comparisons, shifts, and so on – so the essential correctness of Viper was not really addressed; the proof did not require any analysis of the function representing the arithmetic-logic unit, at either level.

Still less abstract than the major state model is the *block level* model [23], whose proof is described in this paper. At the block stage, we begin to approach the functional units and connectivity of the actual circuit, though still in a very abstract way. The block model is a collection of separate units (for example, the arithmetic-logic unit, the decoder, and so on), specified functionally, along with information on their inter-connectedness and timing behaviour, specified pictorially (with labelled schematic diagrams). The combination of the definitions and the pictorial information are given; the block machine's behaviour patterns must be inferred logically from what is supplied. The time scale at the block level is finer again than (but congruent with) the scale at the major state level; each major state transition is implemented by several *minor* state transitions. (There are mechanisms (blocks) to keep track of the major and minor states.) The block model therefore determines another graph, in which each major node consists of several minor event nodes. Before the graph is analyzed, though, it first has to be *derived* from the given information (see Section 6.2). (At the major state level the graph transitions were given explicitly in the definition of the major states.)

At the block level, the behaviour of the arithmetic-logic unit *is* analyzed, so we come closer to the essence of the microprocessor. This means that there is no single major state called 'PERFORM' corresponding to the execution of an arithmetic-logic operation (as at the major state level) but rather a collection of distinct major states for the various arithmetic-logic operations of Viper.

The original plan was to state and prove the equivalence of the high level specification and the major state model on the visible state; and then of the major state model and the block model on the major state; and finally to make the connection by a some sort of transitivity argument. The first equivalence, as mentioned, has been proved completely, and is described in [5]. The second turned out to be awkward (for technical reasons at least). In fact, by using the methods established in the first proof, a direct comparison of the high level and the block level turns out to be straightforward, and this is what is ultimately wanted anyway. The proof has not been completed up to the equivalence (it would not have been practical with present tools and resources), but only to the point of a complete logical description (analysis) of the graph of the block level. This in itself is very valuable because it can be returned to the designers who can then decide whether the results are as predicated and hoped. Further, it reveals what would be required to complete the proof.

The major state proof is thus not necessary for the block proof, and is not used in this paper, but only mentioned in passing. The block level correctness statement, had it been proved, would have had a similar form to the correctness statement at the major state level, except that it would have had the added complication of assumptions carried along with each case. These assumptions limit the correctness of the block machine to certain normal cases in which, for example, no resetting signals come from the outside world during the time taken to process an instruction. This is discussed in Section 8.3.

6   Representing the Problem

6.1   The High Level

The high level or visible state of Viper

(ram,preg,areg,xreg,yreg,bflag,stop)

is represented in HOL as an object with the type

mem21_32#word20#word32#word32#word32#bool#bool

where ram represents the memory (of 32-bit words addressed by 21-bit words); preg the program counter; areg, xreg and yreg the registers; bflag a general-purpose boolean flag; and stop the flag indicating an error requiring Viper to stop. The high level specification gives state changes for: illegal instructions (which stop the machine), no-ops, arithmetic-logic operations involving comparisons (whose the results are put in the bflag), other arithmetic-logical operations whose results go into one of the three registers, arithmetic-logical operations whose results go into the program counter (i.e. jumps), write instructions, and procedure calls. The complete definition is supplied in the Appendix, and explanations can be found in [23]. For purposes of this paper all that is really required is the idea that a visible state (comprising the seven components) is transformed directly by the high level function into a new state upon inspection of the current instruction. The current instruction resides in the memory component of the state, ram, and is pointed to by the program counter component of the state, preg.

An example of a state transformation (which is treated at length in this paper, at the block level) is an addition ALU operation with overflow detection, whose results are placed in one of the three registers *besides* the program counter. In that case, the high level specifies a state in which (i) the program counter is incremented; (ii) the sum of the register source indicated and the memory source is computed and placed in the destination register indicated; and (iii) the value of the stopping flag (i.e. whether the sum overflows) is computed. We abbreviate the memory source as MEM_ABBR(ram,preg) and the register source REGSELECT_ABBR(areg,xreg,yreg,preg,ram)[11].

Results are computed at the high level by a sub-function (called ALU'), the high level function representing the arithmetic-logic unit[12]. The function ALU' maps the

---

[11]The memory abbreviation requires just the program counter and memory as parameters, as explained above. The register abbreviation needs in addition the four registers from amongst which it selects.

[12]In the appendix, and the RSRE formulation, this function has no prime mark, but we add the prime here to distinguish it from the block level function which has the same general purpose but an entirely different type; it is unfortunate that the same name was originally used.

function field, the memory selector field, the destination field, the register source, the memory source and the boolean flag to three values: a 32-bit computed value, a possible new value for the boolean flag, and a possible new value for the stopping flag. An example of an ALU non-comparison function is addition with overflow detection. In this case the ALU' function returns as the 32-bit computed value

```
ADD32(REGSELECT_ABBR(areg,xreg,yreg,preg,ram),MEM_ABBR(ram,preg))
```

where the addition function ADD32 (on *any* two 32-bit words r and m) returns

```
TRIM34T032(WORD34((VAL33(SIGNEXT r)) + (VAL33(SIGNEXT m))))
```

Here, SIGNEXT represents sign-extension:

```
|- SIGNEXT w = WORD33(V(CONS(EL 31(BITS32 w))(BITS32 w)))
```

and TRIM34T032 simply drops the top two bits of the 34-bit sign-extended sum:

```
|- TRIM34T032 w = WORD32(V(TL(TL(BITS34 w))))
```

For the stopping flag, the function ALU' returns:

```
~(EL
  32
  (BITS34
   (WORD34
    ((VAL33(SIGNEXT(REGSELECT_ABBR(areg,xreg,yreg,preg,ram)))) +
     (VAL33(SIGNEXT(MEM_ABBR(ram,preg)))))))) =
  EL
  31
  (BITS34
   (WORD34
    ((VAL33(SIGNEXT(REGSELECT_ABBR(areg,xreg,yreg,preg,ram)))) +
     (VAL33(SIGNEXT(MEM_ABBR(ram,preg))))))))))
```

This represents the overflow condition of the sum; the sum spans bits 0 to 33, and the test concerns bits 31 and 32[13]. When all definitions are unfolded in this example the sum of the register and memory sources works out to be:

```
WORD32
(V
 (TL
  (TL
   (BITS34
    (WORD34
     ((VAL33(SIGNEXT(REGSELECT_ABBR(areg,xreg,yreg,preg,ram)))) +
      (VAL33(SIGNEXT(MEM_ABBR(ram,preg)))))))))))
```

That is, the sign-extended sources are added to form a 34-bit word whose top two bits are subsequently dropped. The stopping flag works out to be:

```
~(EL
  32
  (BITS34
   (WORD34
    ((VAL33(SIGNEXT(REGSELECT_ABBR(areg,xreg,yreg,preg,ram)))) +
     (VAL33(SIGNEXT(MEM_ABBR(ram,preg)))))))) =
  EL
  31
  (BITS34
   (WORD34
    ((VAL33(SIGNEXT(REGSELECT_ABBR(areg,xreg,yreg,preg,ram)))) +
     (VAL33(SIGNEXT(MEM_ABBR(ram,preg))))))))))
```

---

[13]This is a consequence of the high level definition; *why* the expression represents an overflow is another question.

(The boolean flag is not affected.) These expressions, of course, can be further unfolded according to the definition of SIGNEXT.

Another (quite straightforward) ALU operation is a procedure call. The main computed value in that case is the jump address, which is placed in the program counter:

```
WORD20(V(SEG(0,19)(BITS32(MEM_ABBR(ram,preg)))))
```

That is, the address part of the memory source is the address to which to jump. (There is no register source required for a procedure call.)

To illustrate the effect of a comparison type ALU operation on the high level state, we consider first the equality test on the source and memory registers. In this case, the result of the comparison is placed in the boolean flag, and it is simply

```
REGSELECT_ABBR(areg,xreg,yreg,preg,ram) = MEM_ABBR(ram,preg)
```

For the less-than test (i.e. whether the value of the register source is less than the value of the memory source), the result of the comparison is

```
EL
32
(BITS34
 (WORD34
  ((VAL33(SIGNEXT(REGSELECT_ABBR(areg,xreg,yreg,preg,ram)))) +
   ((VAL33(SIGNEXT(MEM_ABBR(ram,preg))) = 0) => 0 |
    (VAL33(NOT33(SIGNEXT(MEM_ABBR(ram,preg)))))) + 1))))
```

That is, the value of the sign-extended source is added to 0 if the value of the sign-extended memory is 0, and to the incremented value of the negated sign-extended memory value otherwise; and the thirty-second bit of that 34-bit sum is the result. (Given the equality and the less-than tests, all other desired comparisons can be constructed.)

Finally, we consider a simple non-ALU operation: an illegal-instruction stop. Various instuction schemata are not legal – for example, any one which tries to use the spare function fields 13, 14 or 15. Likewise, any one in which the program counter cannot be incremented without overflowing its twenty bits is illegal. For all such illegal instructions the new state specified at the high level will have all other registers unchanged but the stop flag set to true.

In all of these examples, the results are *logically inferred* in HOL from the definitions of the high level function and its sub-functions. This is achieved through a process of gradual unfolding of definitions in each case, using the particular case assumptions to simplify resulting expressions. Exactly the same process of unfolding is carried out at the block level, except that there the representing function is not provided explicitly as at the top level, but has instead to be extracted from a partially pictorial representation before the unfolding process can be undertaken.

The examples of this section are compared to the block level results in Section 9. The first example (addition with overflow detection) is the main example of this report.

It should be stressed that the high level specification of Viper is incomplete as regards the actual Viper chip. The specification applies *only* to the fetch-decode-execute cycle, and ignors capabilities such as resetting, timing-out because of memory faults, pausing, single-stepping, externally forced errors, and so on – all of which are possible in the actual circuit, and all of which are represented in the block model. This means that the scope of any relation between the high level specification and the block model must be limited to the fetch-decode-execute behaviour of Viper. Thus, because of the minimality of the high level model, any correctness property one states is bound to be rather unrealistic. It would be interesting, though beyond this report, to explore better high level specifications.

6.2 The Block Level

At the block level there are ten functional units, or blocks. These are described and explained in more detail in [23], but in summary their names and general purposes are as follows:

- MEMORY_BLOCK: Models the external memory, ram, in a simplified way;

- EXTERNAL_BLOCK: Interfaces internal lines with the external world;

- DATAREG: Computes the data registers: the visible areg, xreg, yreg or preg (whichever is the destination), as well as three internal registers: the address (addr), instruction (inst) and temporary (treg) registers;

- TIMING_COMB: Generates the timing (i.e. sequencing) values within each major state;

- MINOR: Keeps track of the current minor state;

- MAJOR: Keeps track of the current major state;

- TIMEOUT_BLOCK: Provides a 64-cycle timeout facility[14];

- DECODER: Generates the control values for the whole block based on the instruction code;

- FSELECT_COMB: Extracts the function selector field from the instruction code;

- ALU_COMB: Performs the arithmetic and logical operations to yield a 32-bit data result as well as a 9-bit word encoding the stopping conditions associated with the computation; and

- BANDSTOP_BLOCK: Computes two boolean flags: the bflag of the visible state and a stop flag (not quite the same as in the visible state).

Each of these blocks is given a functional description in [23]; the HOL versions appear in their entirety in the Appendix. Their inter-connectedness is conveyed in Figure 1, below, which is a compendium (laid out slightly differently) of several diagrams from [23], to which the names of the internal lines count, ram, areg, xreg, yreg and preg have been added by the author. Boxes marked 'LATCH' or 'L' are latches, which cause a delay of one clock cycle (one time unit). These memory elements can be thought of as registers. Purely combinatorial blocks and sub-blocks (i.e. those without latches) are shown with the suffix '_COMB'. Patterns of feedback are indicated in the figure. The nine lines to or from the external world appear at the top of the figure (prefaced by 'e_').

---

[14]This is not relevant to the present analysis because of the simplified memory model used.

To save space, the *types* of the values on the lines are not shown in the diagram, but it can be assumed that they are known. For example, the type of the value on the inst line at any time is word12.

The parts of the Viper block model referred to are explained locally and partially as the need arises, but no attempt is made to present a coherent overall explanation. For that the reader must refer to [23].

6.3 An Example

The nature of the block model and its representation in HOL can be illustrated by studying *one* of the blocks. We choose the DATAREG block (which can be found within Figure 1).

By way of motivation, the function of the whole DATAREG block is firstly to remember the four data values of the high level state: areg, xreg, yreg and preg. It is also to remember three internal values: a 32-bit temporary value called treg, a 12-bit value inst for holding the instruction code part of an instruction, and a 20-bit value addr for holding the address part of an instruction. Besides containing these seven registers, the DATAREG block computes new values for them. It outputs *one* of the four visible registers (whichever is indicated by the rsel line from the decoder block) on rbar.

The new register values are computed on the basis of the old register values, as well as five incoming lines. The informal specification tells us the 'meanings' of these and their relations to the external lines[15]. The five incoming lines are: extdatabar, a 32-bit input from the external interface which, though this is not deducible from the figure, comes from the memory block and represents incoming data (e_data_in); aoutbar, the 32-bit data output of the arithmetic-logic unit; clkenb, a line from the decoder controlling when and which registers of DATAREG are to be loaded with new values; and finally, two boolean timing values, regstrobe and strobeaddr, from the timing block, which respectively indicate the stability of an operation, and when the address latch should be loaded.

The main sub-block of the DATAREG block is called REGISTERS (represented by REGISTERS_COMB in Figure 1). That block is described in HOL by a function called REGISTERS which has the following HOL type:

```
word32#(word32#(word7#(bool#(bool#(word32#(word32#(word32#(word20#(word32#(word20#word12)))))))))) ->
word32#(word32#(word32#(word20#(word32#(word20#word12)))))
```

This type, by conventions of binding power (see Section 4) can be abbreviated as

```
word32#word32#word7#bool#bool#word32#word32#word32#word20#word32#word20#word12 ->
word32#word32#word32#word20#word32#word20#word12
```

REGISTERS has the following definition in HOL:

---

[15]This information is represented only in English prose in [23]. None of it is actually necessary for the formal proof.

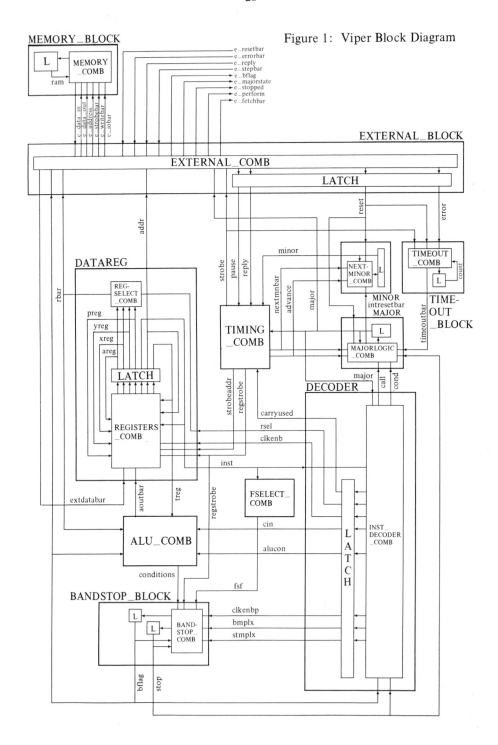

Figure 1: Viper Block Diagram

```
|- REGISTERS
   (extdatabar,aoutbar,clkenb,regstrobe,strobeaddr,areg,xreg,yreg,preg,treg,addr,inst) =
   (let extdata = NOT32 extdatabar in
    let aoutbar_ls20 = WORD20(V(SEG(0,19)(BITS32 aoutbar))) in
    let clkenba = NOT(EL 0(BITS7 clkenb)) in
    let clkenbp = NOT(EL 1(BITS7 clkenb)) in
    let clkenbx = NOT(EL 2(BITS7 clkenb)) in
    let clkenby = NOT(EL 3(BITS7 clkenb)) in
    let clkenbinst = NOT(EL 4(BITS7 clkenb)) in
    let clkenbt = NOT(EL 5(BITS7 clkenb)) in
    let tmplxcon = EL 6(BITS7 clkenb) in
    let new_areg = CLOCK_REGA(aoutbar,areg,clkenba,regstrobe) in
    let new_xreg = CLOCK_REGX(aoutbar,xreg,clkenbx,regstrobe) in
    let new_yreg = CLOCK_REGY(aoutbar,yreg,clkenby,regstrobe) in
    let new_preg = CLOCK_REGP(aoutbar_ls20,preg,clkenbp,regstrobe) in
    let new_inst = CLOCK_INST(extdata,inst,clkenbinst,regstrobe) in
    let new_treg = CLOCK_REGT(aoutbar_ls20,extdata,treg,clkenbt,regstrobe,tmplxcon,clkenbinst) in
    let new_addr = CLOCK_ADDR(preg,treg,addr,clkenbinst,strobeaddr) in
    new_areg,new_xreg,new_yreg,new_preg,new_treg,new_addr,new_inst)
```

Here, the various sub-functions will have been previously defined in HOL. For example, the function CLOCK_REGA, which computes the new value of the a-register, has the HOL type

```
word32#word32#bool#bool->word32
```

and is defined to choose between the ALU_COMB block's (negated) data output and the existing value in the a-register on the basis of two booleans values: clkenba, corresponding to the least significant bit of the 7-bit control value clkenb; and regstrobe, from the timing block.

```
|- CLOCK_REGA(aoutbar,areg,clkenba,regstrobe) = ((clkenba /\ regstrobe) => NOT32 aoutbar | areg)
```

Note that the variable names such as new_areg, new_inst and so on, in the definition of REGISTERS, are purely suggestive of their intended meanings. That is, when the *let*-construct is simplified (see Section 4.4), the seventh output of REGISTERS is

```
CLOCK_INST(NOT32 extdatabar,inst,~EL 4(BITS7 clkenb),regstrobe)
```

which gives no indication that the output represents the new value of the inst register (since the function name 'CLOCK_INST' is also mnemonically chosen and does not formally convey a meaning).

The other function of the DATAREG block is REGSELECT. Its HOL definition (as seen earlier) is

```
|- REGSELECT(areg,xreg,yreg,preg,rsel) =
   (let rf = VAL2 rsel in
    ((rf = 0) => NOT32 areg |
     ((rf = 1) => NOT32 xreg |
      ((rf = 2) => NOT32 yreg |
       ((rf = 3) => NOT32(WORD32(VAL20 preg)) | ARB)))))
```

(See Section 4 for the form without *let*'s.) This function takes rsel (the indicator from the decoder for selecting a destination register from amongst the visible data registers) and the four possible destination registers, and returns the appropriate destination register.

So far, three kinds of information appear to be present in Figure 1 that are not present in the functional definitions of REGISTERS and REGSELECT alone:

1. The internal latch arrangements of the DATAREG block (i.e. the fact that the seven outputs of REGISTERS_COMB are latched)

2. The relation of the inputs and outputs of the DATAREG block to the inputs and outputs of other blocks (e.g. the fact that the input aoutbar comes from the ALU_COMB block and the output inst goes to the DECODER block).

3. The feedback patterns within the DATAREG block (e.g. the fact that the inst value is fed back from the latch to REGISTERS_COMB) so that the input to REGISTERS_COMB on that line at one time was the output *from* REGISTERS_COMB on that line at the previous time

Figure 1 supplies all of this information pictorially, but is not a formal specification in the sense we need for logical analysis. (In any case, it does not supply the internal definitions of the various blocks.) Regarding the second item, we could guess, for example, from the *form* of the definition of the HOL function defining the ALU_COMB block –

```
|- ALU(rbar,treg,cin,bflag,alucon) =
 (  .
    .
    .
    let aoutbar = ... in
    let conditions = ... in
    aoutbar,conditions)
```

– that the output 'aoutbar' of the ALU_COMB block is identical to the input to DATAREG called by the same name; but of course, that is not a formal specification either, since in both definitions the variable 'aoutbar' is bound, and therefore could in either case could be replaced by a different variable.

Some information present in the RSRE informal descriptions is present neither in Figure 1 *nor* in the formal definitions – for example, as we mentioned above, the connection of the extdatabar line to the e_data_in line. These identities, stated in informal prose in [23], obviously assist one's understanding of Viper, but are not formally a part of its definition, nor indeed are they necessary for the proof. More essential ambiguities are discussed in the next section.

In any case, our goal now is to produce a formal expression representing the whole DATAREG block including its latches, feedback patterns and relations to other blocks.

6.3.1 Representing the Whole Block with the Latch

The function REGISTERS was defined to apply to objects (of various type) representing the values on lines; it took inputs such as 32-bit words and boolean values and returned outputs representing seven registers. To describe the behaviour of the whole DATAREG block, however, we have to introduce the notion of *time* so that we can talk about the latch and its behaviour. To do this, we introduce the notion of a *signal*. Signals are functions from times (represented by natural numbers) to values of appropriate type. Thus, while the block's input extdatabar has type word32, the signal called (by convention) extdatabar_sig is a function with type num->word32. We write extdatabar_sig n to denote the *value* of the function extdatabar_sig at time n.

As a first step towards a *function* representing the DATAREG block, we define a new *relation* called REGISTERS_COMB (the suffix '_COMB', as mentioned earlier, suggesting the purely combinational part of the block)[16].

---

[16]From this point on, the development is our own, based on the RSRE definitions and corresponding pictures. The use of relations is a standard hardware verification method.

The relation REGISTERS_COMB holds between the input and output signals corresponding to the input and output values of the function REGISTERS, and it describes the unit's behaviour in time. The output signals of REGISTERS_COMB are given the arbitrary names out1 to out7. The types of these signals can be inferred from the type of REGISTERS: they evaluate respectively to values of type: word32, word32, word32, word20, word32, word20 and word12. The relation REGISTERS_COMB is defined as follows:

```
|- REGISTERS_COMB
    (extdatabar_sig,aoutbar_sig,clkenb_sig,regstrobe_sig,strobeaddr_sig,
     areg_sig,xreg_sig,yreg_sig,preg_sig,treg_sig,addr_sig,inst_sig,
     out1_sig,out2_sig,out3_sig,out4_sig,out5_sig,out6_sig,out7_sig) =
    (!n.
     out1_sig n,out2_sig n,out3_sig n,out4_sig n,out5_sig n,out6_sig n,out7_sig n =
     REGISTERS
     (extdatabar_sig n,aoutbar_sig n,clkenb_sig n,regstrobe_sig n,strobeaddr_sig n,
      areg_sig n,xreg_sig n,yreg_sig n,preg_sig n,treg_sig n,addr_sig n,inst_sig n))
```

This means: the relation REGISTERS_COMB holds over its input and output signal arguments if and only if at every time, the output values of the function REGISTERS at that time are the result of applying REGISTERS to the input values at that time. (The relation is purely combinational because, as yet, there is no advance in time.) Similarly, for the function REGSELECT, we define a relation

```
|- REGSELECT_COMB(areg_sig,xreg_sig,yreg_sig,preg_sig,rsel_sig,rbar_sig) =
    (!n. rbar_sig n = REGSELECT(areg_sig n,xreg_sig n,yreg_sig n,preg_sig n,rsel_sig n))
```

where the rbar_sig signal returns an object of type word32.

In the block model, latches introduce the only delays. To describe a latch formally in HOL – a 32-bit latch, for example – we introduce a relation LATCH32 between two signals each of type num->word32, and we express the fact that the first signal is latched to the second by asserting that

```
|- !in out. LATCH32(in,out) = (!n. out(n+1) = in n)
```

This means that the output at the next time (one time unit later) is *always* equal to the input at the current time.

Thus far, everything has been based on the functional definitions provided by RSRE. We now turn to the pictures provided. The DATAREG block is shown in Figure 2 below. This figure is just an enlarged extract from Figure 1, to which we have added the internal line names out1 to out7. The labelling of the internals lines, though essential to a correct representation, has to be guessed; the discussion below analyzes this problem in detail.

An Aside on Combining Pictures and Text

Considering Figure 2 before the labels out1 to out7 are added, it is clear that the seven output signals of REGISTERS_COMB are each latched. (The latch block is assumed to be a pictorial abbreviation for seven separate latches. Also, 'out1' abbreviates 'out1_sig', and so on.) Four of the output signals of the latch(es) are used only internally (they are fed back to REGISTERS_COMB). The other three are fed back *and* are also (named and typed) outputs of the whole DATAREG block. These latter three outputs must be uniquely identified with outputs of the textual REGISTERS_COMB; since they represent lines to other blocks we need to know how REGISTERS_COMB computes their values. Together the three must include inst, addr and treg, since the picture shows those three names on its outputs to other blocks. The only question is *which* outputs to associate with which names.

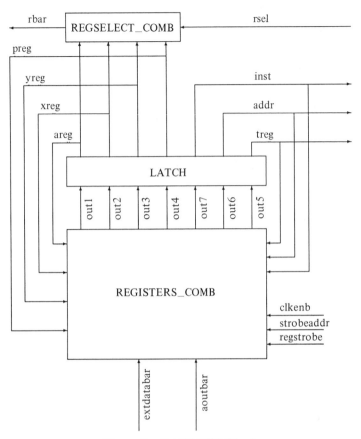

Figure 2: DATAREG Block

In this case, the issue is helpfully resolved by the types of the three signals: out7 has type word12, so it must correspond to inst_sig, whose type we *know* to be word12 from Figure 1 (which, had space pemitted, would have included the types of inter-block lines); and so on. Thus we get the apparently 'reversed ordering' of out5 and out7 in Figure 2.

The point of all this is that the implicit semantics of pictures carry no ordering information; they just tell us that REGISTERS_COMB has nineteen arguments altogether: twelve signals corresponding to the twelve input values plus the seven output values of the function REGISTERS. The textual representation, by its nature, *must* order arguments: out7 is the *seventh* output of REGISTERS and out7_sig is the *nineteenth* argument of REGISTERS_COMB. However, there is no formal indication in the text that inst is intended to be the seventh output value of REGISTERS[17]. Without the fortuitous aid of the distinctness of the types of out5, out6 and out7, these three outputs could be paired off in any fashion with the signals inst, addr and treg. The pairing *does* in

---

[17]The original RSRE text is annotated between lines with comments to this effect, but that, of course, is not a formal indication either.

fact matter, because it could result in the seventh argument being computed by the function CLOCK_ADDR or CLOCK_REGT (see appendix for full definitions) rather than by CLOCK_INST, which would be a serious mis-representation of the 'intended' design. Since the intention is not formally reported, this demonstrates how a picture and its text together may not necessarily determine a correct representation of the designers' intentions. Notational tricks may allow us to make intelligent guesses, but there is no way to be *certain* that a derived representation is correct with respect to intention. This is an important point which is also mentioned in Section 2 and Section 10.

As for the other four outputs of the latch, type information only enables the p-register line to be identified (with out4); the other three are pictorially interchangeable as long as we capture the intended computations by pairing out1 with areg, out2 with xreg and out3 with yreg. Again, nothing supplied formally specifies the intended pairings. In drawing the picture we can choose arbitrarily, since the picture conveys no ordering information. (**This ends the Aside.**)

Next, a consequence of the definition of REGISTERS_COMB is inferred which gives a more convenient form of the definition for reasoning purposes. The consequence requires the introduction of new HOL functions called SECOND, THIRD, and so on, for selecting elements of ordered tuples. (FIRST already exists in HOL as FST, and SEVENTH' differs from SEVENTH in choosing the seventh and final rather the seventh of eight or more elements of a tuple.) From the definition of REGISTERS_COMB, it follows from simple properties of pairing that

```
|- REGISTERS_COMB
    (extdatabar_sig,aoutbar_sig,clkenb_sig,regstrobe_sig,strobeaddr_sig,
     areg_sig,xreg_sig,yreg_sig,preg_sig,treg_sig,addr_sig,inst_sig,
     out1_sig,out2_sig,out3_sig,out4_sig,out5_sig,out6_sig,out7_sig) =
    (!n.
      (out1_sig n =
      FST
      (REGISTERS
        (extdatabar_sig n,aoutbar_sig n,clkenb_sig n,regstrobe_sig n,strobeaddr_sig n,
          areg_sig n,xreg_sig n,yreg_sig n,preg_sig n,treg_sig n,addr_sig n,inst_sig n))) /\
      (out2_sig n =
      SECOND
      (REGISTERS
        (extdatabar_sig n,aoutbar_sig n,clkenb_sig n,regstrobe_sig n,strobeaddr_sig n,
          areg_sig n,xreg_sig n,yreg_sig n,preg_sig n,treg_sig n,addr_sig n,inst_sig n))) /\
      (out3_sig n = ...) /\
      (out4_sig n = ...) /\
      (out5_sig n = ...) /\
      (out6_sig n = ...) /\
      (out7_sig n =
      SEVENTH'
      (REGISTERS
        (extdatabar_sig n,aoutbar_sig n,clkenb_sig n,regstrobe_sig n,strobeaddr_sig n,
          areg_sig n,xreg_sig n,yreg_sig n,preg_sig n,treg_sig n,addr_sig n,inst_sig n))))
```

That is, the *relation* holds over the signals shown if and only if at all times the *function* takes inputs and gives outputs as shown. Note that there is still no advance in time; i.e., this is still a fact only about a combinational part of the DATAREG block.

To specify the whole DATAREG block, which *does* involve an advance in time, we might first try to describe a relation (called DATAREG) in terms of nine already-defined relations:

```
|- DATAREG
   (extdatabar_sig,aoutbar_sig,clkenb_sig,regstrobe_sig,strobeaddr_sig,rsel_sig,rbar_sig,
    treg_sig,addr_sig,inst_sig) =
   (?out1_sig out2_sig out3_sig out4_sig out5_sig out6_sig out7_sig
    areg_sig xreg_sig yreg_sig preg_sig.
    REGISTERS_COMB
    (extdatabar_sig,aoutbar_sig,clkenb_sig,regstrobe_sig,strobeaddr_sig,
     areg_sig,xreg_sig,yreg_sig,preg_sig,treg_sig,addr_sig,inst_sig,
     out1_sig,out2_sig,out3_sig,out4_sig,out5_sig,out6_sig,out7_sig) /\
    LATCH32(out1_sig,areg_sig) /\
    LATCH32(out2_sig,xreg_sig) /\
    LATCH32(out3_sig,yreg_sig) /\
    LATCH20(out4_sig,preg_sig) /\
    LATCH32(out5_sig,treg_sig) /\
    LATCH20(out6_sig,addr_sig) /\
    LATCH12(out7_sig,inst_sig) /\
    REGSELECT_COMB(areg_sig,xreg_sig,yreg_sig,preg_sig,rsel_sig,rbar_sig))
```

That is, the new relation DATAREG would hold of the ten input/output signals of the whole block if and only if there could be found eleven internal signals (appropriately typed) such that all nine relations held as shown – the two computation relations as well as the seven latch relations.

This in itself is perfectly correct, but it turns out not to be useful. The ultimate aim is to 'solve for' the internal variables (i.e. to hide them), and hence derive a relation holding over just the visible state (external variables) of the block. To do this (at least with existing HOL tools), each such variable must have an associated equation of suitable form. For example, for out1_sig n we know (from the consequence of the definition of REGISTERS_COMB) that

```
out1_sig n =
FST
(REGISTERS
 (extdatabar_sig n,aoutbar_sig n,clkenb_sig n,regstrobe_sig n,strobeaddr_sig n,
  areg_sig n,xreg_sig n,yreg_sig n,preg_sig n,treg_sig n,addr_sig n,inst_sig n))
```

which provides a way to solve for out1_sig – by replacing it with another expression not involving it. However, in the case of areg_sig, the latch relations tell us only that

```
|- LATCH32(out1_sig,areg_sig) = (!n. areg_sig(n+1) = out1_sig n)
```

which is not adequate. The syntactic reason it is not adequate is that we have no equation for areg_sig at time n with which to expand; only at time n+1. The intuitive reason it is not adequate is that the a-register (as well as the p-, x- and y-registers) is really a memory element; it 'remembers' its previous values, and so is genuinely part of the *overall state* of the block machine; it should (indeed must) be given explicitly in the overall block level state, and in the overall state of DATAREG. We therefore describe the whole DATAREG block as a relation by writing:

```
|- DATAREG
   (extdatabar_sig,aoutbar_sig,clkenb_sig,regstrobe_sig,strobeaddr_sig,rsel_sig,
    areg_sig,xreg_sig,yreg_sig,preg_sig,
    rbar_sig,
    treg_sig,addr_sig,inst_sig) =
   (?out1_sig out2_sig out3_sig out4_sig out5_sig out6_sig out7_sig.
    REGISTERS_COMB
    (extdatabar_sig,aoutbar_sig,clkenb_sig,regstrobe_sig,strobeaddr_sig,
     areg_sig,xreg_sig,yreg_sig,preg_sig,treg_sig,addr_sig,inst_sig,
     out1_sig,out2_sig,out3_sig,out4_sig,out5_sig,out6_sig,out7_sig) /\
    LATCH32(out1_sig,areg_sig) /\
    LATCH32(out2_sig,xreg_sig) /\
    LATCH32(out3_sig,yreg_sig) /\
    LATCH20(out4_sig,preg_sig) /\
    LATCH32(out5_sig,treg_sig) /\
    LATCH20(out6_sig,addr_sig) /\
    LATCH12(out7_sig,inst_sig) /\
    REGSELECT_COMB(areg_sig,xreg_sig,yreg_sig,preg_sig,rsel_sig,rbar_sig))
```

Here, the relation DATAREG has four more explicit signal parameters than in the first attempt; otherwise it is the same. This definition (a mutually recursive set of equations) can be repeatedly unfolded, with replacements made where possible based on (i) the consequence of the definition of REGISTERS_COMB, and (ii) the definitions of the nine relations, to solve for the seven internal variables. This is a standard technique, commonly used in modelling hardware in HOL. A general ML procedure encodes the inference patterm in HOL. It gives the following result, asserting that the DATAREG relation holds if and only if eight facts hold respectively about the eight state variables (four registers and four outputs) of the block:

```
|- DATAREG
   (extdatabar_sig,aoutbar_sig,clkenb_sig,regstrobe_sig,strobeaddr_sig,rsel_sig,
   areg_sig,xreg_sig,yreg_sig,preg_sig,
   rbar_sig,
   treg_sig,addr_sig,inst_sig) =
   (!n.
     areg_sig(n+1) =
     FST
     (REGISTERS
      (extdatabar_sig n,aoutbar_sig n,clkenb_sig n,regstrobe_sig n,strobeaddr_sig n,
       areg_sig n,xreg_sig n,yreg_sig n,preg_sig n,treg_sig n,addr_sig n,inst_sig n))) /\
   (!n.
     xreg_sig(n+1) =
     SECOND
     (REGISTERS
      (extdatabar_sig n,aoutbar_sig n,clkenb_sig n,regstrobe_sig n,strobeaddr_sig n,
       areg_sig n,xreg_sig n,yreg_sig n,preg_sig n,treg_sig n,addr_sig n,inst_sig n))) /\
   (!n.
     yreg_sig(n+1) = ...) /\
   (!n.
     preg_sig(n+1) = ...) /\
   (!n.
     treg_sig(n+1) = ...) /\
   (!n.
     addr_sig(n+1) = ...) /\
   (!n.
     inst_sig(n+1) =
     SEVENTH'
     (REGISTERS
      (extdatabar_sig n,aoutbar_sig n,clkenb_sig n,regstrobe_sig n,strobeaddr_sig n,
       areg_sig n,xreg_sig n,yreg_sig n,preg_sig n,treg_sig n,addr_sig n,inst_sig n))) /\
   (!n.
     rbar_sig n =
     REGSELECT(areg_sig n,xreg_sig n,yreg_sig n,preg_sig n,rsel_sig n))
```

Note that the constructed relations REGISTERS_COMB and REGSELECT_COMB have disappeared in the end result, and the expression for the DATAREG relation is now in terms of the original functions REGISTERS and REGSELECT. The constructed relations were only devices for deriving the expression for the DATAREG relation. Ultimately, DATAREG itself will turn out to be just a device in deriving the expression representing the whole block model.

We now have expressions for the four outputs of the DATAREG block, as well as for the four internal state values (the a-, x-, y- and p- registers) of the block. These expressions do, finally, reflect the time delay caused by the latch: in each conjunct but the last, the function REGISTERS is applied to a time-n input to yield a time-$(n+1)$ output. In the last conjunct, the block output rbar_sig is not latched, so its expression does not reflect an advance in time. These eight expressions, conjoined, characterize the relation DATAREG.

The other blocks are treated similarly, with relations defined as intermediate devices where required. Some blocks have latches and some do not. The BANDSTOP block, like DATAREG, does – two internal lines are hidden. BANDSTOP_BLOCK has six inputs and two outputs. The inputs, in order, are: two control signals (stmplx_sig and bmplx_sig) from the decoder, to control the stopping flag and the boolean flag, respectively; the 4-bit function selector fsf_sig (projected out from the 12-bit instruction signal); a timing signal regstrobe_sig, from the timing block; another control signal (clkenbp_sig) from the decoder, indicating whether the program counter is the chosen destination (in which case values with more than 20 significant bits must be caught so that the stopping flag is set); and a 9-bit signal (conditions_sig) from the arithmetic-logic unit, coding information about the computed result and the selected destination. The outputs, both latched, are just the boolean flag signal and the stopping flag signal. The main function of the block is the given function BANDSTOP. (SND, below, is the existing HOL function for projecting the second element of a pair.)

```
|- BANDSTOP_BLOCK
   (stmplx_sig,bmplx_sig,fsf_sig,regstrobe_sig,clkenbp_sig,conditions_sig,stop_sig,bflag_sig) =
   (!n. stop_sig(n+1) =
      FST
      (BANDSTOP
       (stmplx_sig n,bmplx_sig n,fsf_sig n,regstrobe_sig n,clkenbp_sig n,conditions_sig n,
        stop_sig n,bflag_sig n))) /\
   (!n. bflag_sig(n+1) =
      SND
      (BANDSTOP
       (stmplx_sig n,bmplx_sig n,fsf_sig n,regstrobe_sig n,clkenbp_sig n,conditions_sig n,
        stop_sig n,bflag_sig n)))
```

The DECODER block takes the (12-bit) instruction signal, the (coded) major state signal, the boolean flag signal and the stopping flag signal. It has ten output signals, all latched except two: an unlatched indicator (cond_sig) of the next major state when there is a choice of next major state; and an unlatched indicator (call_sig) that a procedure call is being processed. The latched signals are: a control signal (alucon_sig) to the arithmetic-logic unit, carrying information on the operation to be performed; bmplx_sig and stmplx_sig as above; an indicator (carryused_sig) of a carry, to the timing block; a boolean-valued signal (cin_sig) of the carry-in (which doubles as the least significant bit in some left shifts); clkenbp_sig as above; and clkenb_sig and rsel_sig as in Section 6.3. The main function of the block is INST_DECODER.

```
|- DECODER
   (inst_sig,major_sig,bflag_sig,stop_sig,cond_sig,call_sig,alucon_sig,bmplx_sig,
    carryused_sig,cin_sig,clkenb_sig,clkenbp_sig,rsel_sig,stmplx_sig) =
   (!n. bmplx_sig(n+1) = FST(INST_DECODER(inst_sig n,major_sig n,bflag_sig n,stop_sig n))) /\
   (!n. stmplx_sig(n+1) = SECOND(INST_DECODER(inst_sig n,major_sig n,bflag_sig n,stop_sig n))) /\
   (!n. rsel_sig(n+1) = THIRD(INST_DECODER(inst_sig n,major_sig n,bflag_sig n,stop_sig n))) /\
   (!n. cin_sig(n+1) = FOURTH(INST_DECODER(inst_sig n,major_sig n,bflag_sig n,stop_sig n))) /\
   (!n. alucon_sig(n+1) = FIFTH(INST_DECODER(inst_sig n,major_sig n,bflag_sig n,stop_sig n))) /\
   (!n. carryused_sig(n+1) = SIXTH(INST_DECODER(inst_sig n,major_sig n,bflag_sig n,stop_sig n))) /\
   (!n. clkenb_sig(n+1) = SEVENTH(INST_DECODER(inst_sig n,major_sig n,bflag_sig n,stop_sig n))) /\
   (!n. clkenbp_sig(n+1) = EIGHTH(INST_DECODER(inst_sig n,major_sig n,bflag_sig n,stop_sig n))) /\
   (!n. cond_sig n = NINTH(INST_DECODER(inst_sig n,major_sig n,bflag_sig n,stop_sig n))) /\
   (!n. call_sig n = TENTH(INST_DECODER(inst_sig n,major_sig n,bflag_sig n,stop_sig n)))
```

Likewise, the MAJOR and MINOR blocks have a latch each. MAJOR takes the stopping signal; call_sig as above; an indication (timeoutbar_sig) of a time-out; two timing signals (nextmnbar_sig and advance_sig); a signal indicating a reset (reset_sig); and cond_sig as above. The latched (main) output is the major state signal, major_sig. The unlatched output (intresetbar_sig) is an indication, sent to MINOR, of an error (so that MINOR can stop counting the sub-states of the major state and re-initialize). The main function of MAJOR is MAJORLOGIC.

```
|- MAJOR
   (stop_sig,call_sig,timeoutbar_sig,nextmnbar_sig,advance_sig,reset_sig,cond_sig,
    major_sig,intresetbar_sig) =
   (!n. major_sig(n+1) =
      FST
      (MAJORLOGIC
       (stop_sig n,call_sig n,timeoutbar_sig n,nextmnbar_sig n,advance_sig n,reset_sig n,
        cond_sig n,major_sig n))) /\
   (!n. intresetbar_sig n =
      SND
      (MAJORLOGIC
       (stop_sig n,call_sig n,timeoutbar_sig n,nextmnbar_sig n,advance_sig n,reset_sig n,
        cond_sig n,major_sig n)))
```

MINOR takes the same two timing signals (nextmnbar_sig and advance_sig), and reset_sig and intresetbar_sig as above. It has only one output (latched) – the minor state signal. The main function of the block is NEXTMINOR.

```
|- MINOR(nextmnbar_sig,advance_sig,reset_sig,intresetbar_sig,minor_sig) =
   (!n. minor_sig(n+1) =
        NEXTMINOR(nextmnbar_sig n,advance_sig n,reset_sig n,intresetbar_sig n,minor_sig n))
```

The external interface, EXTERNAL_BLOCK, takes five signals from the internal part of the machine, one from the memory, and the four signals from the outside world; it returns one unlatched signal to the internal machine, ten unlatched signals to either the memory or the outside world, and four latched signals to the internal machine. The only sub-function is EXTERNAL, which computes the data signal, extdatabar_sig, to the circuit; the four signals to the latch; and the ten signals *not* to the internal circuit.

```
|- EXTERNAL_BLOCK
    (rbar_sig,addr_sig,bflag_sig,strobe_sig,major_sig,e_data_in_sig,
     e_resetbar_sig,e_errorbar_sig,e_stepbar_sig,e_reply_sig,extdatabar_sig,
     e_bflag_sig,e_fetchbar_sig,e_iobar_sig,e_perform_sig,e_data_out_sig,e_address_sig,
     e_stopped_sig,e_majorstate_sig,e_strobebar_sig,e_writebar_sig,
     error_sig,pause_sig,reply_sig,reset_sig) =
    (!n. extdatabar_sig n =
         FST
         (EXTERNAL
          (rbar_sig n,addr_sig n,bflag_sig n,strobe_sig n,major_sig n,e_data_in_sig n,
           e_resetbar_sig n,e_errorbar_sig n,e_stepbar_sig n,e_reply_sig n))) /\
    (!n. reset_sig(n+1) =
         SECOND
         (EXTERNAL
          (rbar_sig n,addr_sig n,bflag_sig n,strobe_sig n,major_sig n,e_data_in_sig n,
           e_resetbar_sig n,e_errorbar_sig n,e_stepbar_sig n,e_reply_sig n))) /\
    (!n. error_sig(n+1) = ...) /\
    (!n. pause_sig(n+1) = ...) /\
    (!n. reply_sig(n+1) = ...) /\
    (!n. e_data_out_sig n = ...) /\
    (!n. e_address_sig n = ...) /\
    (!n. e_bflag_sig n = ...) /\
    (!n. e_majorstate_sig n = ...) /\
    (!n. e_strobebar_sig n = ...) /\
    (!n. e_stopped_sig n = ...) /\
    (!n. e_perform_sig n = ...) /\
    (!n. e_fetchbar_sig n = ...) /\
    (!n. e_iobar_sig n = ...) /\
    (!n. e_writebar_sig n =
     FIFTEENTH
     (EXTERNAL
      (rbar_sig n,addr_sig n,bflag_sig n,strobe_sig n,major_sig n,e_data_in_sig n,
       e_resetbar_sig n,e_errorbar_sig n,e_stepbar_sig n,e_reply_sig n)))
```

The memory, MEMORY_BLOCK, takes the five signals from the external interface and returns one signal to it: (e_data_in_sig). There is one latched signal, ram_sig, purely internal to the memory, representing the state of the memory. The only sub-function is MEMORY, which computes a new memory and data output based on the old memory and the inputs.

```
|- MEMORY_BLOCK
    (e_data_out_sig,e_address_sig,e_iobar_sig,e_writebar_sig,e_strobebar_sig,ram_sig,
     e_data_in_sig) =
    (!n.
     ram_sig(n+1) =
     FST
     (MEMORY
      (ram_sig n,
       e_data_out_sig n,e_address_sig n,e_iobar_sig n,e_writebar_sig n,e_strobebar_sig n))) /\
    (!n.
     e_data_in_sig n =
     SND
     (MEMORY
      (ram_sig n,
       e_data_out_sig n,e_address_sig n,e_iobar_sig n,e_writebar_sig n,e_strobebar_sig n)))
```

Blocks without latches are simpler. For example, as Figure 1 indicates, ALU_COMB has five inputs (two of them *from* DATAREG) and two outputs (one of them *to* DATAREG). The inputs are: rbar_sig, the data signal to the external interface; treg_sig from DATAREG; cin_sig as above; the boolean flag; and alucon_sig as above. The outputs are: the computed data output aoutbar_sig to DATAREG (as above); and a 9-bit word, conditions_sig, encoding certain facts, to BANDSTOP, as above. Without latches there are no internal lines to hide. The main function of the block is ALU.

```
|- ALU_COMB
   (rbar_sig,treg_sig,cin_sig,bflag_sig,alucon_sig,aoutbar_sig,conditions_sig) =
   (!n.
     (aoutbar_sig n = FST(ALU(rbar_sig n,treg_sig n,cin_sig n,bflag_sig n,alucon_sig n))) /\
     (conditions_sig n = SND(ALU(rbar_sig n,treg_sig n,cin_sig n,bflag_sig n,alucon_sig n))))
```

Similarly, the TIMING block has no latches. In order, it takes major_sig and minor_sig as above; two signals, pause_sig and reply_sig, from the external interface; and carryused_sig as above. It outputs five timing signals to various blocks: advance_sig, nextmnbar_sig, regstrobe_sig, strobe_sig and strobeaddr_sig. (See Section 7.2 for the development of these and [23] for further explanations.) The main function of the block is TIMING.

```
|- TIMING_COMB
   (major_sig,minor_sig,pause_sig,reply_sig,carryused_sig,advance_sig,nextmnbar_sig,
    regstrobe_sig,strobe_sig,strobeaddr_sig) =
   (!n.
     (advance_sig n =
      FST(TIMING(major_sig n,minor_sig n,pause_sig n,reply_sig n,carryused_sig n))) /\
     (nextmnbar_sig n =
      SECOND(TIMING(major_sig n,minor_sig n,pause_sig n,reply_sig n,carryused_sig n))) /\
     (regstrobe_sig n =
      THIRD(TIMING(major_sig n,minor_sig n,pause_sig n,reply_sig n,carryused_sig n))) /\
     (strobe_sig n =
      FOURTH(TIMING(major_sig n,minor_sig n,pause_sig n,reply_sig n,carryused_sig n))) /\
     (strobeaddr_sig n =
      FIFTH'(TIMING(major_sig n,minor_sig n,pause_sig n,reply_sig n,carryused_sig n))))
```

The other blocks are not necessary for this exposition; we just point out that the model includes exactly two other purely internal registers which like areg and so on must be included as state variables. The first is ram, the 32-bit memory value in MEMORY_BLOCK, and the second is count in TIMEOUT_BLOCK, a 6-bit counter which at its maximum value indicates that the machine should be timed out (due to a memory failure).

To summarize this section, the DATAREG block was used as an example of the method of deriving a formal expression (a relation) to describe a whole block which is composed of sub-blocks and latches. We began with what was given: the functional definitions of the two sub-blocks REGISTERS and REGSELECT, as well as the picture shown in Figure 2, without names on the internal lines. The notion of signals – functions which produce values given times – was introduced. We discussed why it was necessary to identify internal lines with arguments of REGISTERS, and how these pairings were not fully determined by the combination of picture and text, thus introducing the possibility of incorrectly representing the design of the machine. We then extracted the notion of the *state* of the whole DATAREG block, and used that to infer a description of the whole block in terms of state signals only, with internal lines concealed. All of the blocks in the model are treated similarly. In subsequent sections, the very same method is used to describe the entire block diagram by a relation, taking the ten blocks as the basic units.

6.3.2 Joining Blocks

The ultimate aim in this section is to derive a logical expression representing the whole block model, both its pictorial and the textual parts. The ten block are first represented as relations, just as we have represented DATAREG. They are then combined using exactly the same method as we used to combine the sub-units of the DATAREG block itself. Internal lines which are not state (latch output) values are hidden when possible, as the out_i were hidden. The units of the block model can be combined in any groupings or all at once. Just as an example, suppose we combine just two: DATAREG and ALU_COMB. (These are both discussed in the previous section.)

To represent the combination, we define a relation, called, say DATAREG_ALU, which applies to the lines to and from DATAREG, and to the lines to and from ALU_COMB. (See Figure 1 for a picture of this combination.) Expanding the resulting expression, it is possible to solve for two of the lines internal to the *combined* block: rbar_sig and aoutbar_sig. However, treg_sig is latched in the combined block. This gives expressions for the seven registers (including treg_sig) and for the ALU_COMB output conditions_sig. The resulting a-register expression, for example, is now

```
(!n.
  areg_sig(n+1) =
  FST
  (REGISTERS
  (extdatabar_sig n,
    FST
    (ALU
    (REGSELECT(areg_sig n,xreg_sig n,yreg_sig n,preg_sig n,rsel_sig n),
      treg_sig n,cin_sig n,bflag_sig n,alucon_sig n)),
    clkenb_sig n,regstrobe_sig n,strobeaddr_sig n,
    areg_sig n,xreg_sig n,yreg_sig n,preg_sig n,treg_sig n,addr_sig n,inst_sig n)))
```

while conditions_sig has the expression

```
(!n.
  conditions_sig n =
  SND
  (ALU
  (REGSELECT(areg_sig n,xreg_sig n,yreg_sig n,preg_sig n,rsel_sig n),
    treg_sig n,cin_sig n,bflag_sig n,alucon_sig n)))
```

The expression for the second argument of REGISTERS, in the a-register expression, can be compared with the one which appears in the theorem for DATAREG alone (Section 6.3): by now 'aoutbar_sig' in the earlier expression has expanded into an expression involving the function ALU. In addition, an expression (also in terms of ALU) for conditions has emerged.

All ten blocks of the block model are combined, exactly as DATAREG and ALU_COMB have just been combined in the example. When this is achieved, the expanded expression for the whole relation (called EXTERNAL_AND_BLOCK_AND_MEM, for the block with the external interface and the memory included) is seven pages long; it is the beginning of a series of many lengthy expressions in the analysis of the block model[18]. EXTERNAL_AND_BLOCK_AND_MEM holds over four inputs from the outside world (e_resetbar, e_errorbar, e_stepbar and e_reply) as well as thirty-one state variables. (Those names prefixed by 'e_' are lines to/from memory or to/from the outside world. Of the thirty-one state variables, six, namely e_data_in, e_data_out, e_address, e_strobebar, e_writebar and e_iobar, represent lines to/from memory, which are unlatched, and twenty-five are latched variables internal to the block.) It gives expressions for the thirty-one state variables as well as for the five outputs (e_bflag, e_majorstate, e_stopped, e_perform and e_fetchbar) of

---

[18]The seven pages are formatted and pretty-printed in standard HOL output style

the block model. (The forty lines can all be found in Figure 1; some were introduced in Section 6.3. See [23] for further explanations of the lines.) All other lines shown in Figure 1 are hidden.

```
|- EXTERNAL_AND_BLOCK_AND_MEM
   (e_data_out_sig,e_address_sig,e_strobebar_sig,e_writebar_sig,e_iobar_sig,e_data_in_sig,
    ram_sig,bflag_sig,
    addr_sig,major_sig,pause_sig,reply_sig,reset_sig,error_sig,minor_sig,count_sig,
    alucon_sig,bmplx_sig,carryused_sig,cin_sig,clkenb_sig,clkenbp_sig,rsel_sig,stmplx_sig,
    stop_sig,areg_sig,xreg_sig,yreg_sig,preg_sig,treg_sig,inst_sig,
    e_resetbar_sig,e_errorbar_sig,e_stepbar_sig,e_reply_sig,e_bflag_sig,e_fetchbar_sig,
    e_perform_sig,e_stopped_sig,e_majorstate_sig) =
   (!n.
     areg_sig(n+1) =
     FST
     (REGISTERS
      (FST
       (EXTERNAL
        (REGSELECT
         (areg_sig n,xreg_sig n,yreg_sig n,preg_sig n,rsel_sig n),
         addr_sig n,bflag_sig n,
         FOURTH
         (TIMING
          (major_sig n,minor_sig n,pause_sig n,reply_sig n,carryused_sig n),
          major_sig n,e_data_in_sig n,e_resetbar_sig n,e_errorbar_sig n,
          e_stepbar_sig n,e_reply_sig n)),
        FST
        (ALU
         (REGSELECT
          (areg_sig n,xreg_sig n,yreg_sig n,preg_sig n,rsel_sig n),
          treg_sig n,cin_sig n,bflag_sig n,alucon_sig n)),
         clkenb_sig n,
         THIRD
         (TIMING
          (major_sig n,minor_sig n,pause_sig n,reply_sig n,carryused_sig n)),
         FIFTH'
         (TIMING
          (major_sig n,minor_sig n,pause_sig n,reply_sig n,carryused_sig n)),
          areg_sig n,xreg_sig n,yreg_sig n,preg_sig n,treg_sig n,addr_sig n,inst_sig n))) /\
      .
      .
      .
```

Comparing this expression for the a-register with that in Section 6.3 gives some idea of the evolution of the relation EXTERNAL_AND_BLOCK_AND_MEM, and some idea of how the inter-connectedness of the block model is expressed. There is a similar expression, of course, for each of the state variables and outputs. For brevity, we abbreviate each of these expressions by defining a constant to stand for it. For example, we define

```
AREG_ABBR(treg,areg,xreg,yreg,preg,rsel,addr,bflag,major,minor,pause,reply,carryused,
         e_data_in,e_resetbar,e_errorbar,e_stepbar,e_reply,cin,alucon,clkenb,inst)
```

to stand for the final a-register expression. AREG_ABBR terms can then be unfolded or folded as required. It is helpful with very long expressions to shorten them for as long as possible in the development.

In a similar way, expressions are derived and abbreviations defined for all thirty-one registers and five outputs. Some of the key register expressions are shown below: for bflag, stop, major, minor, alucon and carryused. As mentioned before, the 1-bit carryused value, sent from the decoder to the timing block, indicates a carry, and the 7-bit alucon value, from the decoder to the ALU carries information on the ALU operation to be performed. As mentioned in Section 6.3.1, the main sub-functions of the MAJOR, MINOR and DECODER blocks, respectively, are MAJORLOGIC, NEXTMINOR and INST_DECODER.

```
|- BFLAG_ABBR
   (stop,stmplx,bmplx,inst,major,minor,pause,reply,carryused,clkenbp,areg,xreg,yreg,
    preg,rsel,treg,cin,bflag,alucon) =
   SND
   (BANDSTOP
    (stmplx,bmplx,FSELECT inst,
     THIRD(TIMING(major,minor,pause,reply,carryused)),clkenbp,
     SND(ALU(REGSELECT(areg,xreg,yreg,preg,rsel),treg,cin,bflag,alucon)),stop,bflag))
```

```
|- STOP_ABBR
   (bflag,stmplx,bmplx,inst,major,minor,pause,reply,carryused,clkenbp,areg,xreg,yreg,
   preg,rsel,treg,cin,alucon,stop) =
FST
(BANDSTOP
 (stmplx,bmplx,FSELECT inst,
   THIRD(TIMING(major,minor,pause,reply,carryused)),clkenbp,
   SND(ALU(REGSELECT(areg,xreg,yreg,preg,rsel),treg,cin,bflag,alucon)),stop,bflag))

|- MAJOR_ABBR
   (major,stop,inst,bflag,count,reset,error,minor,pause,reply,carryused) =
FST
(MAJORLOGIC
 (stop,TENTH(INST_DECODER(inst,major,bflag,stop)),
   SND
   (TIMEOUT
    (count,reset,error,
      FOURTH(TIMING(major,minor,pause,reply,carryused)))),
   SECOND(TIMING(major,minor,pause,reply,carryused)),
   FST(TIMING(major,minor,pause,reply,carryused)),reset,
   NINTH(INST_DECODER(inst,major,bflag,stop)),major))

|- MINOR_ABBR
   (minor,major,pause,reply,carryused,reset,stop,inst,bflag,count,error) =
NEXTMINOR
(SECOND(TIMING(major,minor,pause,reply,carryused)),
 FST(TIMING(major,minor,pause,reply,carryused)),reset,
 SND
 (MAJORLOGIC
  (stop,TENTH(INST_DECODER(inst,major,bflag,stop)),
    SND
    (TIMEOUT
     (count,reset,error,
       FOURTH(TIMING(major,minor,pause,reply,carryused)))),
    SECOND(TIMING(major,minor,pause,reply,carryused)),
    FST(TIMING(major,minor,pause,reply,carryused)),reset,
    NINTH(INST_DECODER(inst,major,bflag,stop)),major)),minor)

|- ALUCON_ABBR(inst,major,bflag,stop) = FIFTH(INST_DECODER(inst,major,bflag,stop))

|- CARRYUSED_ABBR(inst,major,bflag,stop) = SIXTH(INST_DECODER(inst,major,bflag,stop))
```

These and the other thirty-five expressions help us to derive what we ultimately want: a *function* respresenting the block model. The function (called WHOLE_BLOCK_NEXT because it computes the next state of the block) applies to the thirty-one state variables and the four inputs. It returns new values for the thirty-one state variables and also for the five outputs to the outside world. The form of the function is:

```
|- WHOLE_BLOCK_NEXT
   (areg,xreg,yreg,preg,treg,addr,inst,error,pause,reply,reset,
   e_iobar,e_data_out,e_address,e_strobebar,e_writebar,ram,e_data_in,
   minor,major,bflag,stop,count,alucon,bmplx,carryused,cin,clkenb,clkenbp,rsel,stmplx,
   e_resetbar,e_errorbar,e_stepbar,e_reply) =
   let e_bflag = ... in let e_majorstate = ... in let e_stopped = ... in
   let e_perform = ... in let e_fetchbar = ... in
   let areg' = AREG_ABBR
               (treg,areg,xreg,yreg,preg,rsel,addr,bflag,major,minor,pause,reply,carryused,
                 e_data_in,e_resetbar,e_errorbar,e_stepbar,e_reply,cin,alucon,clkenb,inst) in
   let xreg' = ... in let yreg' = ... in let preg' = ... in let treg' = ... in
   let addr' = ... in let inst' = ... in
   let error' = ... in let pause' = ... in let reply' = ... in let reset' = ... in
   let e_iobar' = ... in let e_data_out' = ... in let e_address' = ... in
   let e_strobebar' = ... in let e_writebar' = ... in let ram' = ... in let e_data_in' = ... in
   let minor' = MINOR_ABBR
                (minor,major,pause,reply,carryused,reset,stop,inst,bflag,count,error) in
   let major' = MAJOR_ABBR
                (major,stop,inst,bflag,count,reset,error,minor,pause,reply,carryused) in
   let bflag' = BFLAG_ABBR
                (stop,stmplx,bmplx,inst,major,minor,pause,reply,carryused,clkenbp,
                 areg,xreg,yreg,preg,rsel,treg,cin,bflag,alucon) in
   let stop' = STOP_ABBR
                (bflag,stmplx,bmplx,inst,major,minor,pause,reply,carryused,clkenbp,
                 areg,xreg,yreg,preg,rsel,treg,cin,alucon,stop) in
   let count' = ... in let bmplx' = ... in
   let alucon' = ALUCON_ABBR(inst,major,bflag,stop) in
   let carryused' = CARRYUSED_ABBR(inst,major,bflag,stop) in
   let cin' = ... in let clkenb' = ... in let clkenbp' = ... in
   let rsel' = ... in let stmplx' = ... in
   areg',xreg',yreg',preg',treg',addr',inst',error',pause',reply',reset',
   e_iobar',e_data_out',e_address',e_strobebar',e_writebar',
   ram',e_data_in',minor',major',bflag',stop',count',alucon',bmplx',carryused',cin',
   clkenb',clkenbp',rsel',stmplx',
   e_bflag,e_majorstate,e_stopped,e_perform,e_fetchbar
```

Since the function was constructed from the text of another expression (that of the relation EXTERNAL_AND_BLOCK_AND_MEM), it remains to confirm that the function correctly represents the relation. That is, we prove:

```
|- (!n.
    areg_sig(n+1),xreg_sig(n+1),yreg_sig(n+1),preg_sig(n+1),treg_sig(n+1),
    addr_sig(n+1),inst_sig(n+1),
    error_sig(n+1),pause_sig(n+1),reply_sig(n+1),reset_sig(n+1),
    e_iobar_sig n,e_data_out_sig n,e_address_sig n,e_strobebar_sig n,
    e_writebar_sig n,
    ram_sig(n+1),e_data_in_sig n,
    minor_sig(n+1),major_sig(n+1),bflag_sig(n+1),stop_sig(n+1),count_sig(n+1),
    alucon_sig(n+1),bmplx_sig(n+1),carryused_sig(n+1),cin_sig(n+1),
    clkenb_sig(n+1),clkenbp_sig(n+1),rsel_sig(n+1),stmplx_sig(n+1),
    e_bflag_sig n,e_majorstate_sig n,e_stopped_sig n,e_perform_sig n,e_fetchbar_sig n =
  WHOLE_BLOCK_NEXT
    (areg_sig n,xreg_sig n,yreg_sig n,preg_sig n,treg_sig n,addr_sig n,inst_sig n,
     error_sig n,pause_sig n,reply_sig n,reset_sig n,
     e_iobar_sig n,e_data_out_sig n,e_address_sig n,e_strobebar_sig n,e_writebar_sig n,
     ram_sig n,e_data_in_sig n,
     minor_sig n,major_sig n,bflag_sig n,stop_sig n,count_sig n,alucon_sig n,
     bmplx_sig n,carryused_sig n,cin_sig n,clkenb_sig n,clkenbp_sig n,rsel_sig n,stmplx_sig n,
     e_resetbar_sig n,e_errorbar_sig n,e_stepbar_sig n,e_reply_sig n)) =

  EXTERNAL_AND_BLOCK_AND_MEM
    (e_data_out_sig,e_address_sig,e_strobebar_sig,e_writebar_sig,e_iobar_sig,e_data_in_sig,
     ram_sig,bflag_sig,addr_sig,major_sig,
     pause_sig,reply_sig,reset_sig,error_sig,
     minor_sig,count_sig,alucon_sig,bmplx_sig,carryused_sig,cin_sig,clkenb_sig,clkenbp_sig,
     rsel_sig,stmplx_sig,
     stop_sig,areg_sig,xreg_sig,yreg_sig,preg_sig,treg_sig,inst_sig,
     e_resetbar_sig,e_errorbar_sig,e_stepbar_sig,e_reply_sig,e_bflag_sig,e_fetchbar_sig,
     e_perform_sig,e_stopped_sig,e_majorstate_sig)
```

This means that the function WHOLE_BLOCK_NEXT maps time-n values of states and inputs to the values of states and outputs as shown[19] (that is, it sequences values correctly in time) **if and only if** the relation EXTERNAL_AND_BLOCK_AND_MEM holds over input, state and output *signals*. The theorem follows more-or-less immediately from the definitions of the function and the relation, and from properties of pairing. That in turn implies that each of the state and output values can be computed by the (abbreviated) expression derived for it:

Sequencing Theorem

```
|- (!n.
    areg_sig(n+1),xreg_sig(n+1),yreg_sig(n+1),preg_sig(n+1),treg_sig(n+1),
    addr_sig(n+1),inst_sig(n+1),
    error_sig(n+1),pause_sig(n+1),reply_sig(n+1),reset_sig(n+1),
    e_iobar_sig n,e_data_out_sig n,e_address_sig n,e_strobebar_sig n,e_writebar_sig n,
    ram_sig(n+1),e_data_in_sig n,
    minor_sig(n+1),major_sig(n+1),bflag_sig(n+1),stop_sig(n+1),count_sig(n+1),
    alucon_sig(n+1),bmplx_sig(n+1),carryused_sig(n+1),cin_sig(n+1),
    clkenb_sig(n+1),clkenbp_sig(n+1),rsel_sig(n+1),stmplx_sig(n+1),
    e_bflag_sig n,e_majorstate_sig n,e_stopped_sig n,e_perform_sig n,e_fetchbar_sig n =
  WHOLE_BLOCK_NEXT
    (areg_sig n,xreg_sig n,yreg_sig n,preg_sig n,treg_sig n,addr_sig n,inst_sig n,
     error_sig n,pause_sig n,reply_sig n,reset_sig n,
     e_iobar_sig n,e_data_out_sig n,e_address_sig n,e_strobebar_sig n,e_writebar_sig n,
     ram_sig n,e_data_in_sig n,
     minor_sig n,major_sig n,bflag_sig n,stop_sig n,count_sig n,alucon_sig n,bmplx_sig n,
     carryused_sig n,cin_sig n,clkenb_sig n,clkenbp_sig n,rsel_sig n,stmplx_sig n,
     e_resetbar_sig n,e_errorbar_sig n,e_stepbar_sig n,e_reply_sig n))

  ==>

  (areg_sig(n+1) =
  AREG_ABBR
    (treg_sig n,areg_sig n,xreg_sig n,yreg_sig n,preg_sig n,rsel_sig n,addr_sig n,bflag_sig n,
     major_sig n,minor_sig n,pause_sig n,reply_sig n,carryused_sig n,
     e_data_in_sig n,e_resetbar_sig n,e_errorbar_sig n,e_stepbar_sig n,e_reply_sig n,
     cin_sig n,alucon_sig n,clkenb_sig n,inst_sig n) /\
```

---

[19]The twenty-five internal state variables are latched so appear at the 'next' time; the five outputs and six lines to/from memory are not latched so appear at the initial time.

```
(xreg_sig(n+1) = ... /\ (yreg_sig(n+1) = ... /\ (preg_sig(n+1) = ... /\
(treg_sig(n+1) = ... /\ (addr_sig(n+1) = ... /\ (inst_sig(n+1) = ... /\
(error_sig(n+1) = ... /\ (pause_sig(n+1) = ... /\
(reply_sig(n+1) = ... /\ (reset_sig(n+1) = ... /\
(e_iobar_sig n = ... /\ (e_data_out_sig n = ... /\ (e_address_sig n = ... /\
(e_strobebar_sig n = ... /\ (e_writebar_sig n = ... /\ (ram_sig(n+1) = ... /\
(e_data_in_sig n = ... /\
(minor_sig(n+1) =
MINOR_ABBR
(minor,major,pause,reply,carryused,reset,stop,inst,bflag,count,error) /\
(major_sig(n+1) =
MAJOR_ABBR
(major,stop,inst,bflag,count,reset,error,minor,pause,reply,carryused) /\
(bflag_sig(n+1) =
BFLAG_ABBR
(stop,stmplx,bmplx,inst,major,minor,pause,reply,carryused,clkenbp,
 areg,xreg,yreg,preg,rsel,treg,cin,bflag,alucon) /\
(stop_sig(n+1) =
STOP_ABBR
(bflag,stmplx,bmplx,inst,major,minor,pause,reply,carryused,clkenbp,
 areg,xreg,yreg,preg,rsel,treg,cin,alucon,stop) /\
(count_sig(n+1) = ... /\
(alucon_sig(n+1) = ALUCON_ABBR(inst,major,bflag,stop) in
(bmplx_sig(n+1) = ... /\
(carryused_sig(n+1) = CARRYUSED_ABBR(inst,major,bflag,stop) /\
(cin_sig(n+1) = ... /\ (clkenb_sig(n+1) = ... /\ (clkenbp_sig(n+1) = ... /\
(rsel_sig(n+1) = ... /\ (stmplx_sig(n+1) = ... /\
(e_bflag_sig n = ... /\ (e_majorstate_sig n = ... /\ (e_stopped_sig n = ... /\
(e_perform_sig n = ... /\ (e_fetchbar_sig n = )...)
```

The function WHOLE_BLOCK_NEXT (and in particular the property above) is the essential and basic tool for analyzing the block model of Viper. Its usefulness lies in the fact that since it *is* a function, we can give it inputs and it will compute outputs. These outputs, for latched values, represent the values on lines one time unit later. By repeatedly applying the function, it is possible to describe the behaviour of the block machine during and after the processing of Viper's various instruction types. The rest of this paper describes the analysis of the block model based on the functional expression.

7   Using the Representation: The Minor State Transitions

The analysis of the Viper block machine forms a complex layered structure. Simple facts about the functions (such as DATAREG) representing individual blocks, and about their sub-functions and sub-sub-functions (such as REGISTERS and CLOCK_REGA) support more complex facts about the values on lines (such as the value of areg) at given times. In particular, they support facts about the values of the major and minor lines. This pair of values characterizes the progress of the block machine in its execution of an instruction schema. They are particularly important registers in the analysis because as the function WHOLE_BLOCK_NEXT (representing the block model) is computed successively through major and minor values, results accumulate in various registers, thereby giving a complete description of the behaviour of the block machine – which is the goal. As suggested in the introduction, this stage of the analysis can be viewed at one level as a symbolic execution of the block machine, but supported in addition by the security of formal proof; each state transition is logically *inferred*, and not just computed.

The structure of the analysis is reflected in a hierarchy of HOL theories. So far, for example, we have indicated some of what is contained in the HOL theory about the DATAREG block. Each layer of the subsequent analysis is similarly reflected in an HOL theory. The analysis itself rests on a hierarchy of theories containing lemmas.

One example will suffice for all the cases: we assume that the major state is fixed and equal to 4. That indicates the performance of an ALU operation, specified in detail by the current instruction[20]. As for which ALU operation is chosen in the example, it is not necessary at first to be specific.

7.1 Lemmas about the Thirty-Six Lines

The first step in the analysis is to characterize the minor state transitions of the block machine. That means evaluating the **Sequencing Theorem** (Section 6.3.2) up to seven times, assuming in turn that the 3-bit minor state counter holds the values 0, 1, 2, and so on[21], while the 4-bit major state remains fixed at 4. Ultimately, these evaluations will permit the composition of the minor state transformations, and so yield the (provably correct) cumulative effect of the *single* transition through the major state 4 (the performance of an ALU operation). The single transition is effected by a recursive procedure which 'runs through' the minor transitions in sequence, accumulating effects on state and output values. (A similar sort of symbolic execution is considered by J. Joyce in [18]).

To infer the effects on the block state of the various minor state transitions, we infer the value of WHOLE_BLOCK_NEXT at these states. This requires deducing the values on thirty-six lines, each described by an abbreviation such as AREG_ABBR. We therefore prove thirty-six sets of lemmas, dubbed '**Class 2**' lemmas because they depend on one further layer of more basic lemmas. For the a-register, for example, we have the following theorems (recalling that terms to the left of the turnstiles are assumptions on which the theorems depend):

```
VAL3 minor = 0, VAL4 major = 4
|- AREG_ABBR(areg,xreg,yreg,preg,rsel,addr,bflag,major,minor,pause,reply,carryused,
             e_data_in,e_resetbar,e_errorbar,e_stepbar,e_reply,treg,cin,alucon,clkenb,inst) =
   areg

VAL3 minor = 1, VAL4 major = 4
|- AREG_ABBR(areg,xreg,yreg,preg,rsel,addr,bflag,major,minor,pause,reply,carryused,
             e_data_in,e_resetbar,e_errorbar,e_stepbar,e_reply,treg,cin,alucon,clkenb,inst) =
   areg

VAL3 minor = 2, VAL4 major = 4
|- AREG_ABBR(areg,xreg,yreg,preg,rsel,addr,bflag,major,minor,pause,reply,carryused,e_data_in,
             e_resetbar,e_errorbar,e_stepbar,e_reply,treg,cin,alucon,clkenb,inst) =
   areg

VAL3 minor = 3, VAL4 major = 4
|- AREG_ABBR(areg,xreg,yreg,preg,rsel,addr,bflag,major,minor,pause,reply,carryused,
             e_data_in,e_resetbar,e_errorbar,e_stepbar,e_reply,treg,cin,alucon,clkenb,inst) =
   ((~EL 0(BITS7 clkenb) /\ ~carryused) =>
   NOT32
   (FST
    (ALU
     (REGSELECT(areg,xreg,yreg,preg,rsel),treg,cin,bflag,alucon))) |
   areg)

VAL3 minor = 4, VAL4 major = 4
|- AREG_ABBR(areg,xreg,yreg,preg,rsel,addr,bflag,major,minor,pause,reply,carryused,
             e_data_in,e_resetbar,e_errorbar,e_stepbar,e_reply,treg,cin,alucon,clkenb,inst) =
   ((~EL 0(BITS7 clkenb) /\ carryused) =>
   NOT32
   (FST
    (ALU
     (REGSELECT(areg,xreg,yreg,preg,rsel),treg,cin,bflag,alucon))) |
   areg)
```

---

[20]In a sequence of major state transitions, an ALU operation would only happen after the major state had been 1, i.e. sometime after a FETCH operation producing the current instruction. In fact, each sequence of major states begins with 1.

[21]Likewise, each sequence of minor states begins with 0. This and the previous footnote are discussed later.

```
VAL3 minor = 5, VAL4 major = 4
|- AREG_ABBR(areg,xreg,yreg,preg,rsel,addr,bflag,major,minor,pause,reply,carryused,
             e_data_in,e_resetbar,e_errorbar,e_stepbar,e_reply,treg,cin,alucon,clkenb,inst) =
   areg

VAL3 minor = 6, VAL4 major = 4
|- AREG_ABBR(areg,xreg,yreg,preg,rsel,addr,bflag,major,minor,pause,reply,carryused,
             e_data_in,e_resetbar,e_errorbar,e_stepbar,e_reply,treg,cin,alucon,clkenb,inst) =
   areg

VAL3 minor = 7, VAL4 major = 4
|- AREG_ABBR(areg,xreg,yreg,preg,rsel,addr,bflag,major,minor,pause,reply,carryused,
             e_data_in,e_resetbar,e_errorbar,e_stepbar,e_reply,treg,cin,alucon,clkenb,inst) =
   areg
```

The fortunes of the a-register at each of the eight minor states are now explicit: changes can only occur in the fourth and fifth cycles (minor states 3 and 4), during which, depending on two lines from the decoder, the accumulator *may* take on the (negated) data result (i.e. the first output) of the ALU operation. The relevant information is (i) the least significant bit of the control line clkenb, (the bit which controls the loading of the a-register in particular), and (ii) the carry indicator carryused.

The eight theorems above are proved by an ML procedure which takes the definition of AREG_ABBR, unfolds it on the given values, applies relevant **Class 1** lemmas, and simplifies. This is an example for *forward proof* in HOL (see Section 4); a definition is unfolded and simplified, using lemmas and logical identities, to yield a result which was not necessarily foreseen. All of the many lemmas mentioned or suggested in this section are proved in a similar forward manner.

The form in which the lemmas are shown turns out to be a convenient one: that is, equations with assumptions carried along in the background, rather than implications with equational consequents. With assumptions in the background, substitutions (such as '4' for 'VAL4 major') are made directly, and the assumptions are propagated 'behind the scenes'. In the implicative form (|- ... /\ (VAL4 major = 4) => ...), the lemmas would not be useable until the assumptions were dismissed and the substitution(s) made.

Similarly, we prove sets of **Class2** theorems giving the values of the other thirty lines and the five outputs, each at the ten major and eight minor states. As mentioned, the cumulative results for the registers major and minor are particularly important in the operation of the block machine. These results depend on two lines from the external interface: (i) reset, to MINOR, MAJOR and TIMEOUT_BLOCK, which indicates that the machine is to be reset (have its registers cleared, etc), and (ii) error, to TIMEOUT_BLOCK only, which indicates that the machine is to be stopped because of a forced error. The results also depend on the 6-bit internal register, count, which indicates a time-out (due to memory failure) when it reaches the value 64. For example, while the major state is fixed at 4 (PERFORM ALU), we prove:

```
VAL3 minor = 0, VAL4 major = 4
|- MAJOR_ABBR
   (major,stop,inst,bflag,count,reset,error,minor,pause,reply,carryused) =
   (reset => #0010 | (((count = #111111) \/ error) => #1000 | #0100))

VAL3 minor = 0, VAL4 major = 4
|- MINOR_ABBR
   (minor,major,pause,reply,carryused,reset,stop,inst,bflag,count,error) =
   ((reset \/ (count = #111111) \/ error) => #000 | #001)
```

That is, for the minor state 0 and major state 4, unless there is a reset, timeout or error, the major state does not change from 4, and the minor state advances to 1. At the same time, it is inferred that the two boolean flags do not change under any circumstances on this minor cycle:

```
VAL3 minor = 0, VAL4 major = 4
|- BFLAG_ABBR
   (stop,stmplx,bmplx,inst,major,minor,pause,reply,carryused,clkenbp,
    areg,xreg,yreg,preg,rsel,treg,cin,bflag,alucon) =
   bflag

VAL3 minor = 0, VAL4 major = 4
|- STOP_ABBR
   (bflag,stmplx,bmplx,inst,major,minor,pause,reply,carryused,clkenbp,
    areg,xreg,yreg,preg,rsel,treg,cin,alucon,stop) =
   stop
```

The 7-bit control word alucon is sent from the decoder to the ALU, and determines the behaviour of the ALU during its operation. It is computed by DECODE_PERFORM, one of the main sub-functions of the decoder, which takes in the various fields of the instruction and puts out the control values of the DECODER block. The projection function GET_AL selects from the various outputs of DECODE_PERFORM the 7-bit word alucon sent to the ALU.) The alucon register value is inferred (independently of the minor state) to be

```
VAL4 major = 4
|- ALUCON_ABBR(inst,major,bflag,stop) =
   (EL 4(BITS12 inst) => #0011001 |
    GET_AL
    (DECODE_PERFORM
     (WORD4(V(SEG(0,3)(BITS12 inst))),
      WORD3(V(SEG(5,7)(BITS12 inst))),
      WORD2(V(SEG(8,9)(BITS12 inst))),
      WORD2(V(SEG(10,11)(BITS12 inst))),bflag)))
```

That is, there is one 7-bit alucon value for *all* comparison ALU operations (as indicated by bit 4 of the instruction code); while for non-comparisons, DECODE_PERFORM computes the correct control values to be output by the decoder block, and GET_AL selects from these the appropriate alucon value for the ALU operation indicated. Finally, we prove

```
VAL4 major = 4
|- CARRYUSED_ABBR(inst,major,bflag,stop) =
   (EL 4(BITS12 inst) => T |
    ~EL
     5
     (BITS7
      (GET_AL
       (DECODE_PERFORM
        (WORD4(V(SEG(0,3)(BITS12 inst))),
         WORD3(V(SEG(5,7)(BITS12 inst))),
         WORD2(V(SEG(8,9)(BITS12 inst))),
         WORD2(V(SEG(10,11)(BITS12 inst))),bflag)))))
```

meaning that for all comparisons, the carryused value is T; while for non-comparisons it is given by bit 5 of the alucon code.

7.2 Lemmas about Sub-Blocks

The lemmas that are required for proving the **Class 2** theorems exemplified above concern the various functions and sub-functions making up the Viper definition, and they depend on no further layers of lemmas, but simply on the original definitions. We call these 'Class 1' lemmas. To see what is required, consider again the expression for the a-register (Section 6.3.2). The first set of **Class 1** lemmas required to unfold this expression concern the function TIMING. We prove:

```
VAL3 minor = 0, VAL4 major = 4
|- TIMING(major,minor,pause,reply,carryused) = T,T,F,F,F

VAL3 minor = 1, VAL4 major = 4
|- TIMING(major,minor,pause,reply,carryused) = T,T,F,F,F

VAL3 minor = 2, VAL4 major = 4
|- TIMING(major,minor,pause,reply,carryused) = T,T,F,F,F

VAL3 minor = 3, VAL4 major = 4
|- TIMING(major,minor,pause,reply,carryused) = T,T,~carryused,F,F

VAL3 minor = 4, VAL4 major = 4
|- TIMING(major,minor,pause,reply,carryused) = T,carryused,carryused,F,F

VAL3 minor = 5, VAL4 major = 4
|- TIMING(major,minor,pause,reply,carryused) = T,~carryused,F,F,F

VAL3 minor = 6, VAL4 major = 4
|- TIMING(major,minor,pause,reply,carryused) = T,T,F,F,F

VAL3 minor = 7, VAL4 major = 4
|- TIMING(major,minor,pause,reply,carryused) = T,T,F,F,F
```

These theorems give the complete timing pattern for the major state 4 (PERFORM ALU). They are proved simply by unfolding the definition of TIMING under the various assumptions and simplifying. Along with the definitions of THIRD, FOURTH and FIFTH', the timing theorems enable the AREG_ABBR expressions to be reduced to a form with definite values for the TIMING sub-expressions. For example, assuming respectively that minor is 0, 3 and 4, we infer:

```
VAL3 minor = 0, VAL4 major = 4
|- AREG_ABBR
    (areg,xreg,yreg,preg,rsel,addr,bflag,major,minor,pause,reply,carryused,
     e_data_in,e_resetbar,e_errorbar,e_stepbar,e_reply,treg,cin,alucon,clkenb,inst) =
   FST
   (REGISTERS
    (FST
     (EXTERNAL
      (REGSELECT(areg,xreg,yreg,preg,rsel),addr,bflag,F,major,
       e_data_in,e_resetbar,e_errorbar,e_stepbar,e_reply)),
     FST
     (ALU(REGSELECT(areg,xreg,yreg,preg,rsel),treg,cin,bflag,alucon)),
     clkenb,F,F,areg,xreg,yreg,preg,treg,addr,inst))

VAL3 minor = 3, VAL4 major = 4
|- AREG_ABBR
    (areg,xreg,yreg,preg,rsel,addr,bflag,major,minor,pause,reply,carryused,
     e_data_in,e_resetbar,e_errorbar,e_stepbar,e_reply,treg,cin,alucon,clkenb,inst) =
   FST
   (REGISTERS
    (FST
     (EXTERNAL
      (REGSELECT(areg,xreg,yreg,preg,rsel),addr,bflag,F,major,
       e_data_in,e_resetbar,e_errorbar,e_stepbar,e_reply)),
     FST
     (ALU(REGSELECT(areg,xreg,yreg,preg,rsel),treg,cin,bflag,alucon)),
     clkenb,~carryused,F,areg,xreg,yreg,preg,treg,addr,inst))

VAL3 minor = 4, VAL4 major = 4
|- AREG_ABBR
    (areg,xreg,yreg,preg,rsel,addr,bflag,major,minor,pause,reply,carryused,
     e_data_in,e_resetbar,e_errorbar,e_stepbar,e_reply,treg,cin,alucon,clkenb,inst) =
   FST
   (REGISTERS
    (FST
     (EXTERNAL
      (REGSELECT(areg,xreg,yreg,preg,rsel),addr,bflag,F,major,
       e_data_in,e_resetbar,e_errorbar,e_stepbar,e_reply)),
     FST
     (ALU(REGSELECT(areg,xreg,yreg,preg,rsel),treg,cin,bflag,alucon)),
     clkenb,carryused,F,areg,xreg,yreg,preg,treg,addr,inst))
```

(The other five theorems of the set have the same conclusion as the first one above.) This reveals that we next need some further **Class 1** lemmas about the function REGISTERS. The first one simplifies the definition when the fifth argument to REGISTERS is false – this applies when the minor state is 3 or 4. (The fifth argument, corresponding to the variable 'strobeaddr' in the original definition, Section 6.3, indicates to DATAREG when the address register, 'addr', should be loaded.) In that case, new values of the seven registers may be chosen, depending on the decoder's control line clkenb and on the timing line regstrobe:

```
|- REGISTERS
   (extdatabar,aoutbar,clkenb,regstrobe,F,areg,xreg,yreg,preg,treg,addr,inst) =
   (((~EL 0(BITS7 clkenb) /\ regstrobe) => NOT32 aoutbar | areg),
   (((~EL 2(BITS7 clkenb) /\ regstrobe) => NOT32 aoutbar | xreg),
   (((~EL 3(BITS7 clkenb) /\ regstrobe) => NOT32 aoutbar | yreg),
   (((~EL 1(BITS7 clkenb) /\ regstrobe) =>
   NOT20(WORD20(V(SEG(0,19)(BITS32 aoutbar)))) | preg),
   (((~(~EL 5(BITS7 clkenb) /\ regstrobe)) => treg |
   ((~EL 6(BITS7 clkenb) /\ EL 4(BITS7 clkenb)) => NOT32 extdatabar |
   ((EL 6(BITS7 clkenb) /\ EL 4(BITS7 clkenb)) =>
   WORD32(VAL20(NOT20(WORD20(V(SEG(0,19)(BITS32 aoutbar)))))) |
   ((~EL 6(BITS7 clkenb) /\ ~EL 4(BITS7 clkenb)) =>
   WORD32(V(SEG(0,19)(BITS32(NOT32 extdatabar)))) | ARB)))),
   addr,
   (((~EL 4(BITS7 clkenb) /\ regstrobe) =>
   WORD12(V(SEG(20,31)(BITS32(NOT32 extdatabar)))) | inst)
```

This says that the a-, x-, and y-registers and the program counter may each take on the (negated) value computed by the ALU when it is the indicated destination register. The temporary register may take on various values because it has various uses in ALU operations. The address-register does not change in this case, and the instruction register might or might not take on the top twelve bits of the data input from memory.

To simplify the **Class 2** theorem for the minor = 0 case, we prove a theorem which further simplifies the above result by adding the assumption that the fourth input is *also* false. (The fourth argument, corresponding to the variable 'regstrobe' in the original definition, indicates that the results of certain ALU operations are stable.) In this case none of the seven registers is changed:

```
|- REGISTERS(extdatabar,aoutbar,clkenb,F,F,areg,xreg,yreg,preg,treg,addr,inst) =
   areg,xreg,yreg,preg,treg,addr,inst
```

To prove the first of these two lemmas requires an ML procedure which uses the definitions of REGISTER's sub-functions (CLOCK_REGA and so on) to simplify the definition of REGISTERS under the appropriate assumptions. The second is a simple logical consequence.

Using the lemmas gives us simplified values for the three sample a-register expressions, which in the context of specific values of clkenb and carryused (during an actual evaluation sequence of the block machine) will give specific a-register values:

```
VAL3 minor = 0, VAL4 major = 4
|- AREG_ABBR
   (areg,xreg,yreg,preg,rsel,addr,bflag,major,minor,pause,reply,
   carryused,e_data_in,e_resetbar,e_errorbar,e_stepbar,e_reply,
   treg,cin,alucon,clkenb,inst) =
   areg

VAL3 minor = 3, VAL4 major = 4
|- AREG_ABBR
   (areg,xreg,yreg,preg,rsel,addr,bflag,major,minor,pause,reply,carryused,
   e_data_in,e_resetbar,e_errorbar,e_stepbar,e_reply,treg,cin,alucon,clkenb,inst) =
   (((~EL 0(BITS7 clkenb) /\ ~carryused) =>
   NOT32
   (FST
   (ALU
   (REGSELECT(areg,xreg,yreg,preg,rsel),treg,cin,bflag,alucon))) |
   areg)

VAL3 minor = 4, VAL4 major = 4
|- AREG_ABBR
   (areg,xreg,yreg,preg,rsel,addr,bflag,major,minor,pause,reply,carryused,
   e_data_in,e_resetbar,e_errorbar,e_stepbar,e_reply,treg,cin,alucon,clkenb,inst) =
   (((~EL 0(BITS7 clkenb) /\ carryused) =>
   NOT32
   (FST
   (ALU
   (REGSELECT(areg,xreg,yreg,preg,rsel),treg,cin,bflag,alucon))) |
   areg)
```

To put the **Class 1** and **Class 2** lemmas mentioned so far into context, it should be obvious that there are a very large number of them. **Class 1** theorems about the timing patterns have to be inferred for *each* of the ten major states at *each* of eight minor states; the REGISTERS theorems have to be proved for true, false and unknown values of the two key arguments; in some cases, lemmas are also required to simplify the EXTERNAL subexpressions; and analagous theorems have to be proved about *each* of the ten blocks at key specific values. **Class 2** theorems analogous to the theorems about the a-register have to be proved for all of the thirty-one state values and the five outputs, each of these at all combinations of the ten major and eight minor states.

None of the many derivations is difficult, and all can be done easily (and largely automatically) by a set of simple general-purpose ML procedures. The ML procedures required become clear as the unfolding of the block model progresses and facts required become clear. The main point of interest is really the sheer *number* of facts required and the total size of the proofs in time and space (see Section 10).

7.3 The Progression of the Registers

Once the hiererchy of theorems is developed to the point of the **Class 2** lemmas about the register expressions, it is a simple matter to successively substitute into the **Sequencing Theorem** the values for the abbreviation expressions at the eight minor states. By doing this (and using a slightly clever unfolding procedure for recursive unfoldings) a set of eight theorems is proved, all of them of the form sketched below. The theorem for the case minor_sig n = 0, for example, with some of the main register values, is shown below[22]:

```
|- (VAL3(minor_sig n) = 0) ==>
   (VAL4(major_sig n) = 4) ==>
   (!n.
     areg_sig(n+1),xreg_sig(n+1),yreg_sig(n+1),preg_sig(n+1),
     treg_sig(n+1),addr_sig(n+1),inst_sig(n+1),
     error_sig(n+1),pause_sig(n+1),reply_sig(n+1),reset_sig(n+1),
     e_iobar_sig n,e_data_out_sig n,e_address_sig n,e_strobebar_sig n,e_writebar_sig n,
     ram_sig(n+1),e_data_in_sig n,
     minor_sig(n+1),major_sig(n+1),bflag_sig(n+1),stop_sig(n+1),
     count_sig(n+1),alucon_sig(n+1),bmplx_sig(n+1),carryused_sig(n+1),
     cin_sig(n+1),clkenb_sig(n+1),clkenbp_sig(n+1),rsel_sig(n+1),stmplx_sig(n+1),
     e_bflag_sig n,e_majorstate_sig n,e_stopped_sig n,e_perform_sig n,e_fetchbar_sig n =
   WHOLE_BLOCK_NEXT
     (areg_sig n,xreg_sig n,yreg_sig n,preg_sig n,treg_sig n,addr_sig n,inst_sig n,
     error_sig n,pause_sig n,reply_sig n,reset_sig n,
     e_iobar_sig n,e_data_out_sig n,e_address_sig n,e_strobebar_sig n,e_writebar_sig n,
     ram_sig n,e_data_in_sig n,
     minor_sig n,major_sig n,bflag_sig n,stop_sig n,count_sig n,alucon_sig n,
     bmplx_sig n,carryused_sig n,cin_sig n,clkenb_sig n,clkenbp_sig n,
     rsel_sig n,stmplx_sig n,
     e_resetbar_sig n,e_errorbar_sig n,e_stepbar_sig n,e_reply_sig n)) ==>

   (areg_sig(n+1) = areg_sig n) /\
         .
         .
         .
   (minor_sig(n+1) =
   ((reset_sig n \/ (count_sig n = #111111) \/ error_sig n) => #000 | #001)) /\

   (major_sig(n+1) =
   (reset_sig n => #0010 |
    (((count_sig n = #111111) \/ error_sig n) => #1000 | #0100)))  /\
         .
         .
   (bflag_sig(n+1) = bflag_sig n) /\
```

---

[22]The register value lemmas are instantiated, for this purpose, to signals at times, with n the initial time

```
(stop_sig(n+1) = stop_sig n) /\
             .
             .

(alucon_sig(n+1) =
 (EL 4(BITS12(inst_sig n)) => #0011001 |
  GET_AL
  (DECODE_PERFORM
   (WORD4(V(SEG(0,3)(BITS12(inst_sig n)))),
    WORD3(V(SEG(5,7)(BITS12(inst_sig n)))),
    WORD2(V(SEG(8,9)(BITS12(inst_sig n)))),
    WORD2(V(SEG(10,11)(BITS12(inst_sig n)))),bflag_sig n)))) /\
             .
             .

(carryused_sig(n+1) =
 (EL 4(BITS12(inst_sig n)) => T |
  ~EL
  5
  (BITS7
   (GET_AL
    (DECODE_PERFORM
     (WORD4(V(SEG(0,3)(BITS12(inst_sig n)))),
      WORD3(V(SEG(5,7)(BITS12(inst_sig n)))),
      WORD2(V(SEG(8,9)(BITS12(inst_sig n)))),
      WORD2(V(SEG(10,11)(BITS12(inst_sig n)))),bflag_sig n)))))) /\
             .
             .
```

This asserts that if the minor signal at some time is 0, the a-register signal at
the *next* time is unchanged; and the other signals at the next time change or stay
the same as shown. As mentioned, the major and minor values are key factors in
determining the progress of the block machine in time. The areg and bflag values
are typical of block state components which are also components of the high level
state[23]. The alucon and carryused values are typical of the block level control values
(not reflected in the high level state) which determine the subsequent behaviour of
the block model.

By continuing to substitute as above for minor values 1 to 7, the value of each
of the registers can be deduced at each of the minor states. That is, we prove
a sequence of theorems analogous to the one above for all eight possible values
of minor_sig. These contain, for example, the sequence of values of the a-register.
Inspection reveals that the a-register shows no change except when the minor state
is 3 or 4. This follows from the register value lemmas concerning the a-register
(simplified as shown). The fourth theorem in the new sequence tells us that

```
|- (VAL3(minor_sig n) = 3) ==>
   (VAL4(major_sig n) = 4) ==>
             .
             .
   (areg_sig(n+1) =
    ((~EL 0(BITS7(clkenb_sig n)) /\ ~carryused_sig n) =>
     NOT32
     (FST
      (ALU
       (REGSELECT
        (areg_sig n,xreg_sig n,yreg_sig n,preg_sig n,rsel_sig n),
         treg_sig n,cin_sig n,bflag_sig n,alucon_sig n))) |
     areg_sig n)) /\ ...
```

and the fifth tells us

```
|- (VAL3(minor_sig n) = 4) ==>
   (VAL4(major_sig n) = 4) ==>
        .
        .
        .
   (areg_sig(n+1) =
    ((~EL O(BITS7(clkenb_sig n)) /\ carryused_sig n) =>
     NOT32
     (FST
      (ALU
       (REGSELECT
        (areg_sig n,xreg_sig n,yreg_sig n,preg_sig n,rsel_sig n),
         treg_sig n,cin_sig n,bflag_sig n,alucon_sig n))) |
      areg_sig n)) /\ ...
```

and the a-register does not change subsequently. (The pattern for the x-, y-, p-
and t-registers is similar.) Meanwhile (by analogous inferences based on analogous
lemmas), the successive values of the major and minor registers at each of the minor
states are as follows:

```
|- (VAL3(minor_sig n) = 0) ==> (VAL4(major_sig n) = 4) ==> ... ==> ... /\
   (minor_sig(n+1) =
    ((reset_sig n \/ (count_sig n = #111111) \/ error_sig n) => #000 | #001)) /\
   (major_sig(n+1) =
    (reset_sig n => #0010 | (((count_sig n = #111111) \/ error_sig n) => #1000 | #0100)))) /\ ...

|- (VAL3(minor_sig n) = 1) ==> (VAL4(major_sig n) = 4) ==> ... ==> ... /\
   (minor_sig(n+1) =
    ((reset_sig n \/ (count_sig n = #111111) \/ error_sig n) => #000 | #010)) /\
   (major_sig(n+1) =
    (reset_sig n => #0010 | (((count_sig n = #111111) \/ error_sig n) => #1000 | #0100)))) /\ ...

|- (VAL3(minor_sig n) = 2) ==> (VAL4(major_sig n) = 4) ==> ... ==> ... /\
   (minor_sig(n+1) =
    ((reset_sig n \/ (count_sig n = #111111) \/ error_sig n) => #000 | #011)) /\
   (major_sig(n+1) =
    (reset_sig n => #0010 | (((count_sig n = #111111) \/ error_sig n) => #1000 | #0100)))) /\ ...

|- (VAL3(minor_sig n) = 3) ==> (VAL4(major_sig n) = 4) ==> ... ==> ... /\
   (minor_sig(n+1) =
    ((reset_sig n \/ (count_sig n = #111111) \/ error_sig n) => #000 | #100)) /\
   (major_sig(n+1) =
    (reset_sig n =>  #0010 | (((count_sig n = #111111) \/ error_sig n) => #1000 | #0100)))) /\ ...

|- (VAL3(minor_sig n) = 4) ==> (VAL4(major_sig n) = 4) ==> ... ==> ... /\
   (minor_sig(n+1) =
    ((reset_sig n \/ ((count_sig n = #111111) \/ error_sig n) \/
     ~carryused_sig n) => #000 | #101)) /\
   (major_sig(n+1) =
    (reset_sig n => #0010 |
     (((count_sig n = #111111) \/ error_sig n) => #1000 |
      ((~carryused_sig n) => (stop_sig n => #1000 | #0001) | #0100)))) /\ ...

|- (VAL3(minor_sig n) = 5) ==> (VAL4(major_sig n) = 4) ==> ... ==> ... /\
   (minor_sig(n+1) =
    ((reset_sig n \/ ((count_sig n = #111111) \/ error_sig n) \/
     carryused_sig n) => #000 | #110)) /\
   (major_sig(n+1) =
    (reset_sig n => #0010 |
     (((count_sig n = #111111) \/ error_sig n) => #1000 |
      (carryused_sig n => (stop_sig n => #1000 | #0001) | #0100)))) /\ ...

|- (VAL3(minor_sig n) = 6) ==> (VAL4(major_sig n) = 4) ==> ... ==> ... /\
   (minor_sig(n+1) =
    ((reset_sig n \/ (count_sig n = #111111) \/ error_sig n) => #000 | #111)) /\
   (major_sig(n+1) =
    (reset_sig n => #0010 | (((count_sig n = #111111) \/ error_sig n) => #1000 | #0100)))) /\ ...

|- (VAL3(minor_sig n) = 7) ==> (VAL4(major_sig n) = 4) ==> ... ==> ... /\
   (minor_sig(n+1) = #000) /\
   (major_sig(n+1) =
    (reset_sig n => #0010 | (((count_sig n = #111111) \/ error_sig n) => #1000 | #0100)))) /\ ...
```

As can be seen, the advancement of major and minor depends on the absence of signals
for stopping, resetting or timing-out the machine.

The two boolean flags change when the minor state is 3 and again when it is 4
(and at no other minor states). For the b-flag value we have

```
|- (VAL3(minor_sig n) = 3) ==> (VAL4(major_sig n) = 4) ==>
                      .
                      .
                      .
    (bflag_sig(n+1) =
     SND
     (BANDSTOP
      (stmplx_sig n,bmplx_sig n,FSELECT(inst_sig n),~carryused_sig n,clkenbp_sig n,
       SND
       (ALU
        (REGSELECT
         (areg_sig n,xreg_sig n,yreg_sig n,preg_sig n,rsel_sig n),
         treg_sig n,cin_sig n,bflag_sig n,alucon_sig n)),stop_sig n,bflag_sig n))) /\ ...
|- (VAL3(minor_sig n) = 4) ==> (VAL4(major_sig n) = 4) ==>
                      .
                      .
                      .
    (bflag_sig(n+1) =
     SND
     (BANDSTOP
      (stmplx_sig n,bmplx_sig n,FSELECT(inst_sig n),carryused_sig n,clkenbp_sig n,
       SND
       (ALU
        (REGSELECT
         (areg_sig n,xreg_sig n,yreg_sig n,preg_sig n,rsel_sig n),
         treg_sig n,cin_sig n,bflag_sig n,alucon_sig n)),stop_sig n,bflag_sig n))) /\ ...
```

The stop values are similar but depend obviously on the first BANDSTOP output rather than the second. Both stop and bflag also depend on the second ALU output – the 9-bit word (conditions, in Figure 1) which codes the conditions on the source register and on the computed value needed to check for errors.

The carryused and alucon signals do not depend on the minor state, just the major state, so they do not change throughout the minor transitions.

7.4 The Individual Minor State Transitions

In order to compose the minor state transitions to yield their *cumulative* effects, certain assumptions must be made. First, in order to advance the major and minor states (i.e. to resolve their expressions to particular values), some assumptions must be made about certain *internal* lines of the Viper block model at the starting time (time n). For example, it is necessary to assume that

```
(reset_sig n = F) /\
(error_sig n = F) /\
((count_sig n = #111111) = F)
```

That is, the block model cannot be 'run' (starting with a minor state 0 and a major state 4) if the reset or error value is initially set, or if the timout-counter is initially at its maximum value. Of these three, only the count signal is genuinely internal to the block model. The reset and error values stem from external inputs (though, as mentioned earlier, this is not deducible from Figure 1 or any definitions), so that at every time after time n, these two can be expressed in terms of external values at the previous time. In particular, as the transitions are composed, reset and error take on the values ~e_resetbar and ~e_errorbar respectively[24]. To propagate the correct chain of subsequent values for e_resetbar and e_errorbar, *their* initial values must also be assumed[25]:

```
(e_resetbar_sig n = T) /\
(e_errorbar_sig n = T)
```

---

[24]The fact that these connections can be deduced is the reason that they do not have to be indicated in the formal specifications of the blocks, nor drawn in Figure 1.

[25]This is consistent with the RSRE informal explanation (Annex A) of [23] in which these two values are "normally held high".

To evaluate the destination register expressions (e.g. the a-register expression on the minor cycles in which it *does* change), it is necessary to unfold the function ALU. The definition has the form

```
|- ALU(rbar,treg,cin,bflag,alucon) =
  (    .
       .
       .
    let a_c = BITOP(...,treg,bflag,...,cin,alucon) in
    let aout = GET_AOUT a_c in
    let aoutbar = NOT32 aout in
       .
       .
       .
    aoutbar,conditions)
```

BITOP uses the 7-bit control word alucon (from the decoder), whose thirteen (as it happens) permitted values characterize the operation to be performed. BITOP returns the 32-bit computed value (negated), and a boolean value (used for various purposes); GET_AOUT returns the computed value. (See the Appendix for the full definitions.) The value in the alucon register remains the same throughout the minor state sequence – as we have seen:

```
(EL 4(BITS12(inst_sig n)) => #0011001 |
 GET_AL
 (DECODE_PERFORM
  (WORD4(V(SEG(0,3)(BITS12(inst_sig n)))),
   WORD3(V(SEG(5,7)(BITS12(inst_sig n)))),
   WORD2(V(SEG(8,9)(BITS12(inst_sig n)))),
   WORD2(V(SEG(10,11)(BITS12(inst_sig n)))),bflag_sig n))))
```

To resolve the alucon expression, EL 4(BITS12(inst_sig n)) (the fifth bit of the initial instruction register value) must be fixed. That is, it must be known whether the instruction indicates a comparison operation, in which case the alucon value #0011001 happens to be the appropriate code; or whether it is not a comparison, in which case the appropriate code signal is computed by the function DECODE_PERFORM of the decoder. That in turn depends on knowing at least the 4-bit function code of the current instruction, i.e. the value of WORD4(V(SEG(0,3)(BITS12(inst_sig n)))).

Suppose, for example, that the instruction is *not* a comparison, and that the function field indicates addition with detection of overflows. According to the RSRE informal explanations in [23], that case would be indicated by the following assumptions on the current instruction:

```
(EL 4(BITS12(inst_sig n)) = F) /\
(WORD4(V(SEG(0,3)(BITS12(inst_sig n)))) = #0101)
```

For this case, a simple **Class 1** lemma is proved (by the usual unfolding of definitions – see the Appendix for details of DECODE_PERFORM) stating that

```
|- DECODE_PERFORM(#0101,dsf,msf,rsf,bflag) = ...,...,...,F,#0001001,DSFPRIM dsf
```

where DSFPRIM is a sub-function of the decoder which returns (another) 7-bit code, characterizing the destination register. The ALU signal in this case (picked out by the function GET_AL) is #0001001 throughout the sequence. The assumptions also obviously fix the carryused register throughout the sequence to be T (it is fixed by the destination code). Finally, for similar reasons, we also know that

```
cin_sig(n+1) = F /\
clkenb_sig(n+1) = DSFPRIM(WORD3(V(SEG(5,7)(BITS12(inst_sig n)))))
```

throughout, where `clkenb` controls the loading of registers, and `cin` is used for carries and shifts, according to context.

With a specific `alucon` signal (and the others) and the five assumptions about the initial state, it is now possible to deduce a simpler sequence of theorems describing the minor state transitions. For example:

```
|- (VAL3(minor_sig n) = 0) ==>
   (VAL4(major_sig n) = 4) ==>
   (WORD4(V(SEG(0,3)(BITS12(inst_sig n)))) = #0101) ==>
   (EL 4(BITS12(inst_sig n)) = F) ==>
   (e_resetbar_sig n = T) ==>
   (e_errorbar_sig n = T) ==>
   (error_sig n = F) ==>
   (reset_sig n = F) ==>
   ((count_sig n = #111111) = F) ==>
          .
          .
          .
   (minor_sig(n+1) = #001) /\
   (major_sig(n+1) = #0100) /\
          .
          .
          .
```

The subsequent values of the `minor` and `major` registers through the sequence are shown below.

```
(minor_sig(n+1) = #010) /\
(major_sig(n+1) = #0100) /\

(minor_sig(n+1) = #011) /\
(major_sig(n+1) = #0100) /\

(minor_sig(n+1) = #100) /\
(major_sig(n+1) = #0100) /\

(minor_sig(n+1) = #101) /\
(major_sig(n+1) = #0100) /\

(minor_sig(n+1) = #000) /\
(major_sig(n+1) = (stop_sig n => #1000 | #0001)) /\

(minor_sig(n+1) = #111) /\
(major_sig(n+1) = #0100) /\

(minor_sig(n+1) = #000) /\
(major_sig(n+1) = #0100) /\
```

The minor state will therefore advance by increments of 1 until the sixth cycle, at which time it is comes round to 0 again, ready to begin another major cycle. The major state, meanwhile, will stay fixed at 4 until the same sixth cycle, at which time it will then (normally) procede to 1 (FETCH, the beginning of a new fetch-decode-execute loop), unless the stopping flag forces it to 8, the stopped state. (The final two theorems therefore turn out to be superfluous.)

If the minor states had *not* progressed in increments of one, returing to 0 again at some point before or at the last transition, or if the major state had not changed to 1 (FETCH) or 8 (STOP) at the same point, as intended in the informal specification in [23], then an error would have been indicated – either in the block design of Viper *or* in the HOL representation of the block design. The source of the error would have had to have been traced back heuristically through the analysis, possibly to the design. All that would be certain were that (modulo the correctness of the implementation of HOL) the error could not lie in the proof; see Section 4 for a discussion of the security of HOL proofs. In fact, all of the transitions do work correctly in the Viper block machine.

In conclusion, a full account of the individual transitions has now been inferred (illustrated for the seven typical registers). With the pattern of minor and major sequences established, we are finally in a position to compose together the individual minor state transitions.

7.5 Composing the Minor State Transitions

The sequence of theorems describing the individual minor state transitions through the major state 4 are now used to infer the *cumulative* effects on the various registers of the sequence of minor state transitions. The initial effects are described by the simplified theorem for the minor state 0 (and major state 4)[26]:

```
|- (VAL3(minor_sig n) = 0) ==>
   (VAL4(major_sig n) = 4) ==>
   (WORD4(V(SEG(0,3)(BITS12(inst_sig n))))) = #0101) ==>
   (EL 4(BITS12(inst_sig n)) = F) ==>
   (e_resetbar_sig n = T) ==>
   (e_errorbar_sig n = T) ==>
   (error_sig n = F) ==>
   (reset_sig n = F) ==>
   ((count_sig n = #111111) = F) ==>

   (areg_sig(n+1) = areg_sig n) /\
   (minor_sig(n+1) = #001) /\
   (major_sig(n+1) = #0100) /\
   (bflag_sig(n+1) = bflag_sig n) /\
   (stop_sig(n+1) = stop_sig n) /\
   (alucon_sig(n+1) = #0001001) /\
   (carryused_sig(n+1) = T) /\
   (clkenb_sig(n+1) = DSFPRIM(WORD3(V(SEG(5,7)(BITS12(inst_sig n)))))) /\
   (cin_sig(n+1) = F) /\
        .
        .
        .
```

At each subsequent stage until the minor state first returns to 0, the *next* cumulative result in the sequence is inferred from the *current* cumulative theorem. The inference procedure is recursive:

1. For whatever value from 1 to 7 the minor state of the current cumulative theorem is, the appropriate minor state transition theorem is selected, giving time $n+1$ values in terms of time n values, for each n. For example, when the minor state is 4, the corresponding state transition theorem tells us the next states in terms of the current – for example:

```
(areg_sig(n+1) =
 ((~EL 0(BITS7(clkenb_sig n)) /\ carryused_sig n) =>
    NOT32
    (FST
     (ALU
      (REGSELECT
       (areg_sig n,xreg_sig n,yreg_sig n,preg_sig n,rsel_sig n),
        treg_sig n,cin_sig n,bflag_sig n,alucon_sig n))) |
    areg_sig n)) /\

(minor_sig(n+1) = ((~carryused_sig n) => #000 | #101))
```

At this stage (i.e. time $n+4$) the current *cumulative* theorem so far happens to assert that

```
(areg_sig(n+4) = areg_sig n) /\
(inst_sig(n+4) = inst_sig n) /\
(minor_sig(n+4) = #100) /\
(major_sig(n+4) = #0100) /\
(bflag_sig(n+4) = bflag_sig n) /\
(stop_sig(n+4) = stop_sig n) /\
(carryused_sig(n+4) = T) /\
(clkenb_sig(n+4) = DSFPRIM(WORD3(V(SEG(5,7)(BITS12(inst_sig n)))))) /\
(cin_sig(n+4) = F) /\
(rsel_sig(n+4) = WORD2(V(SEG(10,11)(BITS12(inst_sig n))))) /\
(bmplx_sig(n+4) = #000) /\
(clkenbp_sig(n+4) = ~EL 1(BITS7(DSFPRIM(WORD3(V(SEG(5,7)(BITS12(inst_sig n)))))))) /\
(stmplx_sig(n+4) = #011) /\
(alucon_sig(n+4) = #0001001) /\ ...
```

---

[26]In this respect it would be impossible to do an analysis of the block model without some indication that the minor state 0 was distinguished in this way. This indication is in RSRE's English description rather than in the specification itself.

It gives the time-$n+4$ values in terms of the initial values.

2. The appropriate transition theorem is then instantiated to the time of the current theorem, giving a new transition theorem. In the example n is instantiated to $n+4$, giving a transition theorem useful in the present circumstances: it expresses the time $n+5$ values in terms of the known time $n+4$ values. Among other things, the new transition theorem asserts that

```
(areg_sig(n+5) =
((~EL 0(BITS7(clkenb_sig(n+4)))) /\ carryused_sig(n+4)) =>
NOT32
(FST
 (ALU
  (REGSELECT
   (areg_sig(n+4),xreg_sig(n+4),yreg_sig(n+4),preg_sig(n+4),rsel_sig(n+4)),
   treg_sig(n+4),cin_sig(n+4),bflag_sig(n+4),alucon_sig(n+4)))) |
 areg_sig(n+4))) /\

(minor_sig(n+5) = ((~carryused_sig(n+4)) => #000 | #101)) /\

(major_sig(n+5) =
((~carryused_sig(n+4)) => (stop_sig(n+4) => #1000 | #0001) | #0100)) /\

(stop_sig(n+5) =
FST
(BANDSTOP
 (stmplx_sig(n+4),bmplx_sig(n+4),FSELECT(inst_sig(n+4)),carryused_sig(n+4),
 clkenbp_sig(n+4),
 SND
 (ALU
  (REGSELECT
   (areg_sig(n+4),xreg_sig(n+4),yreg_sig(n+4),preg_sig(n+4),rsel_sig(n+4)),
   treg_sig(n+4),cin_sig(n+4),bflag_sig(n+4),alucon_sig(n+4))),
 stop_sig(n+4),bflag_sig(n+4)))) /\ ...
```

3. The last step consists in substituting the known values given by the cumulative theorem so far into the new transition theorem. In the example, the known time $n+4$ values are substituted for the time-$n+4$ terms (this step is justified by the rule of substitution of equals for equals.) The substitution puts the new cumulative theorem entirely in terms of the time-n values. Thus, for example, the new cumulative theorem asserts:

```
(areg_sig(n+5) =
((~EL 0(BITS7(DSFPRIM(WORD3(V(SEG(5,7)(BITS12(inst_sig n)))))))) =>
NOT32
(FST
 (ALU
  (REGSELECT
   (areg_sig n,xreg_sig n,yreg_sig n,preg_sig n,WORD2(V(SEG(10,11)(BITS12(inst_sig n))))),
   treg_sig n,F,bflag_sig n,#0001001))) |
 areg_sig n)) /\

(minor_sig(n+5) = #101) /\

(major_sig(n+5) = #0100) /\

(stop_sig(n+5) =
FST
(BANDSTOP
 (#011,#000,FSELECT(inst_sig n),T,
 ~EL 1(BITS7(DSFPRIM(WORD3(V(SEG(5,7)(BITS12(inst_sig n))))))),
 SND
 (ALU
  (REGSELECT
   (areg_sig n,xreg_sig n,yreg_sig n,preg_sig n,WORD2(V(SEG(10,11)(BITS12(inst_sig n))))),
   treg_sig n,F,bflag_sig n,#0001001)),
 stop_sig n,bflag_sig n)))
```

so that now the a-register, if it is the desination register indicated, takes on the 32-bit result computed by the ALU. The stopping value on this cycle depends on the other computed result – the 9-bit word encoding the knowledge about error conditions. (In this case the stopping expression will evaluate to τ if an addition overflow has occurred.)

Here, a new set of **Class 2** lemmas comes into play; the a-register expression, for example, requires a lemma giving the actual computed value (i.e. the first of the two outputs) of the ALU on the values shown. That lemma in fact implies that

```
areg_sig(n+5) =
((~EL 0(BITS7(DSFPRIM(WORD3(V(SEG(5,7)(BITS12(inst_sig n))))))))) =>
 WORD32
 (V
  (SEG
   (0,31)
   (BITS33
    (WORD33
     ((VAL32
       (NOT32
        (REGSELECT
         (areg_sig n,xreg_sig n,yreg_sig n,preg_sig n,
          WORD2(V(SEG(10,11)(BITS12(inst_sig n)))))))) +
       (VAL32(treg_sig n))))))) |
  areg_sig n)
```

Likewise, lemmas for the stopping expression imply that[27]

```
stop_sig(n+5) =
~EL 1(BITS7(DSFPRIM(WORD3(V(SEG(5,7)(BITS12(inst_sig n))))))) /\
~(V
  (SEG
   (20,31)
   (BITS32
    (WORD32
     (V
      (SEG
       (0,31)
       (BITS33
        (WORD33
         ((VAL32
           (NOT32
            (REGSELECT
             (areg_sig n,xreg_sig n,yreg_sig n,preg_sig n,
              WORD2(V(SEG(10,11)(BITS12(inst_sig n)))))))) +
           (VAL32(treg_sig n))))))))) =
  0) \/

(EL
 31
 (BITS32
  (NOT32
   (REGSELECT
    (areg_sig n,xreg_sig n,yreg_sig n,preg_sig n,
     WORD2(V(SEG(10,11)(BITS12(inst_sig n))))))))) =
 EL 31(BITS32(treg_sig n))) /\
~(EL
   31
   (BITS32
    (NOT32
     (REGSELECT
      (areg_sig n,xreg_sig n,yreg_sig n,preg_sig n,
       WORD2(V(SEG(10,11)(BITS12(inst_sig n))))))))) =
   EL
   31
   (BITS32
    (WORD32
     (V
      (SEG
       (0,31)
       (BITS33
        (WORD33
         ((VAL32
           (NOT32
            (REGSELECT
             (areg_sig n,xreg_sig n,yreg_sig n,preg_sig n,
              WORD2(V(SEG(10,11)(BITS12(inst_sig n)))))))) +
           (VAL32(treg_sig n)))))))))
```

---

[27]The accumulation of values produces a rather large term – it could be subsequently abbreviated if desired.

The a-register result means that if the a-register is the intended destination, then the value initially in the temporary register is added to the (inverse of) the value initially in the source (a-) register to give a 33-bit sum, of which the thirty-third bit is simply ignored (by taking the 32-bit lower segment). If the a-register is not the intended destination it remains unchanged from its initial value.

The stopping value will be true if either of two conditions hold. The first disjunct tests whether the intended destination is the (20-bit) program counter in the case that the sum to be stored there actually uses more than twenty bits. The second disjunct compares the top bit of the source register to the top bits of both the temporary register and the sum, in order to detect addition overflow[28].

At time $n+6$, the minor state becomes 0 again, ready to start a new cycle; and neither the a-register nor the stop flag change. The major state changes in the last cycle from 4, which it has been thus far, to either 8 (STOP) or 1 (FETCH), depending on the stop flag. In this way the final cumulative theorem represents the processing of the single major state 4 (PERFORM ALU) and specifies the *next* major state to be processed.

In the instantiation phase of the recursive accumulation procedure, new assumptions may be generated. In the example, the old cumulative theorem's assumptions included

```
(error_sig n = F) /\
((count_sig n = #111111) = F) /\
(reset_sig n = F) /\

(e_resetbar_sig n = T) /\
(e_resetbar_sig(n+1) = T) /\
(e_resetbar_sig(n+2) = T) /\
(e_resetbar_sig(n+3) = T) /\

(e_errorbar_sig n = T) /\
(e_errorbar_sig(n+1) = T) /\
(e_errorbar_sig(n+2) = T) /\
(e_errorbar_sig(n+3) = T)
```

while the new cumulative theorem's assumptions add to those

```
(e_resetbar_sig(n+4) = T) /\
(e_errorbar_sig(n+4) = T)
```

In the *final* cumulative theorem, `e_errorbar` and `e_resetbar` are assumed to hold steady up to time $n+5$.

The cycle of steps which produces the final cumulative theorem is implemented in HOL by a recursive ML procedure. The procedure produces the next cumulative theorem at each recursive call, while at the same time monitoring the progress of the minor state. The first time after the start that the minor value reaches 0 again, the procedure stops and produces the current cumulative theorem as its final result. In this way, the entire chain of inferences culminating in the final cumulative result is produced automatically – and yet fully formally (once the lemmas are proved). This is an illustration of the power of the linked programming language and the logic in LCF-type systems.

7.6   Lemmas for the Composed Transitions

Stepping experimentally through the procedure, it becomes clear what **Class 2** (and hence **Class 1**) lemmas are required en route. These lemmas make it clearer how

---

[28]This takes some thinking about.

the sum and overflow check evolve in the chain of cumulative results. For example, inferring the time-n+5 value of the a-register shown earlier requires a lemma about the function ALU (on specific values). Namely, we need to know the first output (the computed value) of ALU for the add case:

```
|- FST
   (ALU
    (REGSELECT(areg,xreg,yreg,preg,WORD2(V(SEG(10,11)(BITS12 inst)))),treg,F,bflag,#0001001)) =
   NOT32
   (WORD32
    (V
     (SEG
      (0,31)
      (BITS33
       (WORD33
        ((VAL32
          (NOT32
           (REGSELECT
            (areg,xreg,yreg,preg,WORD2(V(SEG(10,11)(BITS12 inst)))))))) +
         (VAL32 treg)))))))
```

Clearly, this specifies the sum of the selected register and the temporary register. It is proved as usual in a forward manner, by unfolding and simplifying, and using various further **Class 1** lemmas.

Similarly, two lemmas produce the time-n+5 stopping value. The first concerns the *second* output of the application of the function ALU to some specific values. The lemma specifies the 9-bit error conditions code for addition:

```
|- SND
   (ALU
    (REGSELECT(areg,xreg,yreg,preg,WORD2(V(SEG(10,11)(BITS12 inst)))),
     treg,F,bflag,#0001001)) =
   WORD9
   (V
    [EL
     31
     (BITS32
      (WORD32
       (V
        (SEG
         (0,31)
         (BITS33
          (WORD33
           ((VAL32
             (NOT32
              (REGSELECT
               (areg,xreg,yreg,preg,WORD2(V(SEG(10,11)(BITS12 inst)))))))) +
            (VAL32 treg))))))));
     ...;
     ...;

     EL
     31
     (BITS32
      (NOT32
       (REGSELECT
        (areg,xreg,yreg,preg,WORD2(V(SEG(10,11)(BITS12 inst)))))));
     EL 31(BITS32 treg);
     ...;
     ~(V
       (SEG
        (20,31)
        (BITS32
         (WORD32
          (V
           (SEG
            (0,31)
            (BITS33
             (WORD33
              ((VAL32
                (NOT32
                 (REGSELECT
                  (areg,xreg,yreg,preg,WORD2(V(SEG(10,11)(BITS12 inst)))))))) +
               (VAL32 treg)))))))) =
      0);
     ...;
     ...])
```

In the 9-bit word produced, bit 8 is the top bit of the sum; bit 5 is the top bit of the source register; bit 4 is the top bit of the temporary register; and bit 2 indicates whether the top 12 bits of the sum are used (it is false if they are).

The second lemma concerns the first output of the function BANDSTOP when applied to certain values; this gives the stopping value:

```
|- FST
   (BANDSTOP
    (#011,#000,fsf,T,clkenbp,
     SND
     (ALU
      (REGSELECT(areg,xreg,yreg,preg,WORD2(V(SEG(10,11)(BITS12 inst)))),
       treg,F,bflag,#0001001)),stop,bflag)) =
   clkenbp /\
   EL 2(BITS9
        (SND
         (ALU
          (REGSELECT(areg,xreg,yreg,preg,WORD2(V(SEG(10,11)(BITS12 inst)))),
           treg,F,bflag,#0001001)))) \/

   (EL 5(BITS9
         (SND
          (ALU
           (REGSELECT(areg,xreg,yreg,preg,WORD2(V(SEG(10,11)(BITS12 inst)))),
            treg,F,bflag,#0001001)))) =
    EL 4(BITS9
         (SND
          (ALU
           (REGSELECT(areg,xreg,yreg,preg,WORD2(V(SEG(10,11)(BITS12 inst)))),
            treg,F,bflag,#0001001))))) /\
   ~(EL 5(BITS9
          (SND
           (ALU
            (REGSELECT
             (areg,xreg,yreg,preg,WORD2(V(SEG(10,11)(BITS12 inst)))),treg,F,bflag,#0001001)))) =
    EL 8(BITS9
         (SND
          (ALU
           (REGSELECT
            (areg,xreg,yreg,preg,WORD2(V(SEG(10,11)(BITS12 inst)))),treg,F,bflag,#0001001)))))
```

The first disjunct codes the information that the intended destination of the computed result is the program counter (clkenb indicates that) *and* that the top twelve bits of the computed 32-bit ALU result are used. This means that the machine must stop, since the computed result will not fit into the 20-bit p-register. The second disjunct is the overflow condition for addition, which if true must also cause the machine to stop.

These **Class 2** lemmas depend on still simpler **Class 1** lemmas about the functions ALU and BANDSTOP, on specific values. They follow from the basic Viper block definitions by subsitution of the specific values (i.e. unfolding and simplifying again). There are naturally many more lemmas required, of a similar form, for other possible ALU operations, etc. Although all involve different sets of control signals and so on, the ML procedures that generate the proofs are fairly uniform. The required lemmas are all identified as illustrated – by interactively deriving the minor state transitions (in sequence one-by-one), accumulating results, and analyzing the resulting expressions for sub-expressions whose evaluation is essential for producing the *next* theorem in the sequence. This process requires human intelligence to identify the relevant sub-expressions, as well as machine assistance to carry out the numerous and massive inferences. In theory, more of the process could be automated, but that is a research problem of its own, on the boundaries of artificial intelligence.

7.7 Remarks

It is worth stressing at this point that *all* of the proofs in this section are produced by forward reasoning, i.e. straightforward unfoldings of the basic block model definitions or previous lemmas. None of the results were planned or foreseen; we have simply deduced some consequences of the block representation function we derived, and hence analyzed it. The results can usefully be compared with the intended results, or just checked for plausibility; but in any case, the intentions were in no way taken into account in the analysis. The only prior knowledge required for the analysis thus far was the fact that minor cycles always begin at 0.

The minor state analysis proceeds in a similar way for each of the other arithmetic-logic major states, and for all of the non-ALU major states. There are thirty-two ALU operations in all: sixteen comparisons, four arithmetic operations, five logical operations, two read operations, four shifts, and finally, procedure calls. (For purposes of analysis, forty-one ALU operations are considered later on in the analysis, because breaking some up into cases depending on whether the p-register is the destination simplifies matters.) There are nine non-ALU major states, as listed in Section 5.

It is also worth emphasizing the large total size of the cumulative-result theorems; for example, the one used in the example was nine pages long (when pretty-printed in full), and it is only one average-sized theorem among many similar ones. These depend on very large numbers of lemmas of the type illustrated.

Briefly, as this is needed later, the (six) minor transitions comprising the FETCH operation produce a final cumulative theorem asserting that in fetching an instruction, the program counter is (in a roundabout way) incremented:

```
(preg_sig(n+6) =
NOT20
(WORD20
 (V
  (SEG
   (0,19)
   (BITS32
    (NOT32
     (WORD32
      (V
       (SEG
        (0,31)
        (BITS33(WORD33((VAL32(WORD32(VAL20(preg_sig n)))) + 1)))))))))))))
```

It also tells us that the temporary register takes on the address field of the new, fetched instruction:

```
(treg_sig(n+6) =
WORD32
(V
 (SEG
  (0,19)
  (BITS32
   (FETCH21(ram_sig n)(WORD21(V(CONS F(BITS20(preg_sig n)))))))))))
```

Likewise, the inst signal in the end holds the (12-bit) instruction code of the new instruction. The stop signal becomes true *only* if incrementing the program counter makes it more than twenty significant bits long. The major signal can hold a wide variety of values, determined by analysis of the fields of the new fetched instruction. This makes sense, since any other of the major states may follow a fetch operation. The conditional expression representing this analysis is quite complex because of the number of combinations possible. The a-register does not change (nor do the others, aside from the program counter).

The assumptions generated in the process of accumulating minor state effects through the FETCH node are: the external signals e_errorbar_sig, e_resetbar_sig, e_reply_sig and e_stepbar_sig are true from time n to time n+5. Also, the internal signals reset_sig, error_sig and pause_sig are false at time n, while reply_sig is true at time n, and count_sig is not 63 at time n.

The number of minor transitions required varies over the major states. They range in length from one cycle (STOP) to three cycles (READMEM) to four cycles (PRECALL, RESET, WRITEMEM, WRITEIO and READIO) to five cycles (PERFORM ALU operations involving right shifts, calls, reads, and logic operations) to six cycles (FETCH, INDEX, and PERFORM ALU operations involving left shifts, adds, subtracts and comparisons).

To summarize, we have shown through a process of inference how the block definitions (including the definition of the minor state block) together with the information in Figure 1, and the knowledge that minor cycles start at minor state 0, determine the minor state transitions which comprise the major states of the Viper block machine. The minor state transitions give information about the ultimate values on the thirty-one lines of the block model (including the line indicating the next major state), as well as the values on the five lines to the outside world. These values reflect the block state resulting from the execution of each of the major states, such as fetching an instruction or performing an addition operation.

The values determined for the major line in particular put us finally in a position to analyze the *major* state transitions of the block model.

8 Using the Representation: The Major State Transitions

In Section 7 we have shown how the logical analysis of the minor state transitions is carried out. A bonus of that analysis is the information on how the major states follow each other; each minor state transition through a particular major state yields an expression for selecting the next major state. Given adequate case-assumptions to resolve these expressions into definite values, sequences of major states can be composed in the same way that sequences of minor states were composed. This is a much simpler process than composing the minor states; for one thing, there are fewer lemmas needed. The first problem, therefore, is to work out what case assumptions are required.

8.1 The Major State Transition Conditions

In Section 7 we briefly sketched the effects of executing a FETCH instruction: a fresh instruction is produced from the memory according to the program counter and its fields are placed in the appropriate registers of the block machine. One of the results of the FETCH operation, appearing in the major register, is a long, complex expression specifying the next major state. The expression is a conditional in which *any* of the ten major states (including FETCH itself, if the instruction in question is a no-op) can follow the FETCH state. The conditional expression can be helpfully shortened by introducing an abbreviation (as introduced in Section 4.3):

```
|- FETCH_ABBR(ram,preg) =
   BITS12
   (WORD12
    (V
     (SEG(20,31)(BITS32(FETCH21 ram(WORD21(V(CONS F(BITS20 preg))))))))))
```

which denotes the list of booleans representing the twelve bits of the instruction code of the new instruction.

In selecting the next state after FETCH, the conditional expression branches according to the various fields of the 12-bit instruction. It will choose state 4 (PERFORM ALU), for example, under several different sets of circumstances, corresponding to the several possible ALU operations. To characterize the choice of major state 4, for example, we define a predicate (called c4) which holds (over certain components of the visible state) if and only if PERFORM ALU is the major state following the FETCH. This predicate is constructed by analyzing the conditional expression (derived from the block specifications) for changing major states; that is, it follows *entirely* from the block definitions, and is neither invented for the purpose nor derived from the documentation of Viper. The predicate works out as follows[29]:

```
|- c4(ram,preg,areg,xreg,yreg,bflag) =
    (V
      (SEG
        (20,31)
        (BITS32
          (WORD32
            (V(SEG(0,31)(BITS33(WORD33((VAL32(WORD32(VAL20 preg))) + 1)))))))) =
      0) /\

    (EL 4(FETCH_ABBR(ram,preg)) /\
    (VAL2(WORD2(V(SEG(8,9)(FETCH_ABBR(ram,preg)))))) = 0) \/

    ~EL 4(FETCH_ABBR(ram,preg)) /\
    (VAL3(WORD3(V(SEG(5,7)(FETCH_ABBR(ram,preg)))))) = 5) /\
    ~bflag /\
    (VAL2(WORD2(V(SEG(8,9)(FETCH_ABBR(ram,preg)))))) = 0) /\
    ((VAL4(WORD4(V(SEG(0,3)(FETCH_ABBR(ram,preg)))))) = 3) \/
     (VAL4(WORD4(V(SEG(0,3)(FETCH_ABBR(ram,preg)))))) = 5) \/
     (VAL4(WORD4(V(SEG(0,3)(FETCH_ABBR(ram,preg)))))) = 7)) \/

    ~EL 4(FETCH_ABBR(ram,preg)) /\
    (VAL3(WORD3(V(SEG(5,7)(FETCH_ABBR(ram,preg)))))) = 4) /\
    bflag /\
    (VAL2(WORD2(V(SEG(8,9)(FETCH_ABBR(ram,preg)))))) = 0) /\
    ((VAL4(WORD4(V(SEG(0,3)(FETCH_ABBR(ram,preg)))))) = 3) \/
     (VAL4(WORD4(V(SEG(0,3)(FETCH_ABBR(ram,preg)))))) = 5) \/
     (VAL4(WORD4(V(SEG(0,3)(FETCH_ABBR(ram,preg)))))) = 7)) \/

    ~EL 4(FETCH_ABBR(ram,preg)) /\
    (VAL3(WORD3(V(SEG(5,7)(FETCH_ABBR(ram,preg)))))) = 3) /\
    (VAL2(WORD2(V(SEG(8,9)(FETCH_ABBR(ram,preg)))))) = 0) /\
    ((VAL4(WORD4(V(SEG(0,3)(FETCH_ABBR(ram,preg)))))) = 3) \/
     (VAL4(WORD4(V(SEG(0,3)(FETCH_ABBR(ram,preg)))))) = 5) \/
     (VAL4(WORD4(V(SEG(0,3)(FETCH_ABBR(ram,preg)))))) = 7) \/

    ~EL 4(FETCH_ABBR(ram,preg)) /\
    (WORD4(V(SEG(0,3)(FETCH_ABBR(ram,preg)))) = #1100) /\
    ~((VAL3(WORD3(V(SEG(5,7)(FETCH_ABBR(ram,preg)))))) = 3) \/
      (VAL3(WORD3(V(SEG(5,7)(FETCH_ABBR(ram,preg)))))) = 4) \/
      (VAL3(WORD3(V(SEG(5,7)(FETCH_ABBR(ram,preg)))))) = 5)) /\
    ~(VAL3(WORD3(V(SEG(5,7)(FETCH_ABBR(ram,preg)))))) = 6) /\
    ~(VAL3(WORD3(V(SEG(5,7)(FETCH_ABBR(ram,preg)))))) = 7) \/

    ~EL 4(FETCH_ABBR(ram,preg)) /\
    ~((VAL3(WORD3(V(SEG(5,7)(FETCH_ABBR(ram,preg)))))) = 3) \/
      (VAL3(WORD3(V(SEG(5,7)(FETCH_ABBR(ram,preg)))))) = 4) \/
      (VAL3(WORD3(V(SEG(5,7)(FETCH_ABBR(ram,preg)))))) = 5)) /\
    ~(VAL3(WORD3(V(SEG(5,7)(FETCH_ABBR(ram,preg)))))) = 7) /\
    ~(VAL3(WORD3(V(SEG(5,7)(FETCH_ABBR(ram,preg)))))) = 6) /\
    (VAL2(WORD2(V(SEG(8,9)(FETCH_ABBR(ram,preg)))))) = 0) /\
    ~((VAL4(WORD4(V(SEG(0,3)(FETCH_ABBR(ram,preg)))))) = 1) \/
      (VAL4(WORD4(V(SEG(0,3)(FETCH_ABBR(ram,preg)))))) = 13) \/
      (VAL4(WORD4(V(SEG(0,3)(FETCH_ABBR(ram,preg)))))) = 14) \/
      (VAL4(WORD4(V(SEG(0,3)(FETCH_ABBR(ram,preg)))))) = 15)) /\
    ~(WORD4(V(SEG(0,3)(FETCH_ABBR(ram,preg)))) = #1100)))
```

That is, if the incremented program counter has not spread into its top twelve bits (since it must be usable as a 20-bit address), then there are six choices. (See [7] and

---

[29]The predicate could be just defined over the memory and program counter, but the other arguments are for uniformity with other predicates.

[23] for precise details of the codings.) The instruction can indicate a comparison using a literal source; or it can indicate a non-comparison. If the latter, there are five choices:

1. The instruction can indicate the program counter as the destination *if* the boolean flag is false, *with* a false boolean flag, and with a literal source – and a no-op if the boolean flag is true. In that case, the operation indicated can only be a memory-read, an addition with overflow detection, or a subtraction with overflow detection – and no other ALU operation. (The function field values 3, 5 and 7 indicate the operations listed, respectively.)

2. It can be the same as above but with the program counter as the destination *if* the boolean flag is true, with a true boolean flag – and a no-op if it is false.

3. It can be as above but with the program counter as the unconditional destination, and no restriction on the boolean flag.

4. It can indicate the a-, x- or y-register as the destination for a shift operation. (The function field value 12 indicates a shift.)

5. It can indicate the a-, x- or y-register as destination, a literal source, and *any* of the arithmetic-logic operations *except* procedure calls (function field value 1), shifts (function field value 12) or the dis-allowed (spare) operations (function field values 13, 14 and 15).

Similarly, we define condition c_2 to hold if the INDEX operation follows FETCH, c_3 to hold if PRECALL follows FETCH, c_5 if READIO is next, c_6 if READMEM is next, c_7 if STOP is next, c_8 if WRITEMEM is next, c_9 if WRITEIO is next, and c_{10} if FETCH follows immediately after FETCH. All of the condition definitions follow immediately from analysis of the major state expression produced by a FETCH operation.

It is first of all necessary to confirm that the nine conditions cover all logical possibilities; otherwise the conditional expression for choosing would be defective, and we would be able to describe an instruction type which the block machine could not handle. Since the conditional followed from the block representation, this would indicate that *it* was in some way flawed – either the representation itself or the basic definitions and Figure 1. We therefore prove:

```
|- c2(ram,preg,areg,xreg,yreg,bflag) \/
   c3(ram,preg,areg,xreg,yreg,bflag) \/
   c4(ram,preg,areg,xreg,yreg,bflag) \/
   c5(ram,preg,areg,xreg,yreg,bflag) \/
   c6(ram,preg,areg,xreg,yreg,bflag) \/
   c7(ram,preg,areg,xreg,yreg,bflag) \/
   c8(ram,preg,areg,xreg,yreg,bflag) \/
   c9(ram,preg,areg,xreg,yreg,bflag) \/
   c10(ram,preg,areg,xreg,yreg,bflag)
```

The proof of this fact is very long and messy because there are so many cases to consider. Because the condition definitions themselves are so long, this fact is most easily proved in an abstracted form (with various assumptions on the abstract variables), then instantiated, and the assumptions proved and dismissed. This is the first example so far of *goal-oriented* proof; we know to start with what we want to prove, but not necessarily the method of proof. The theorem is achieved by applying to the goal (the candidate theorem above) a strategy based on case-analysis. By choosing cases in a clever order, we can minimize the number of cases considered, though it is still very large.

In any case, we now have a predicate c4 which holds exactly when major state 4 follows major state 1 (in any of several ways). The example case of the previous section, however, is more specific: it is the major state 4 in which the comparison field holds the value F and the function selector field holds the word #0101, indicating addition with overflow detection. That is so far consistent with *four* of the ways in which c4 can hold. To simplify the presentation, we further assume that the program counter is *not* the destination (i.e. the destination field is not 3, 4 or 5); otherwise we would have to check whether the computed result exceeded twenty significant bits. With that assumption, there is only one remaining way in which c4 can hold. Therefore we define a more specific predicate to describe the transition to our *particular* major state 4 – i.e. we define a *sub-condition* of c4 called c4_F_5 (to reflect the comparison and function selector fields. We would write c4_F_345_5 for the sub-condition in which the destination *were* 3, 4 or 5). The sub-condition is:

```
|- c4_F_5(ram,preg,areg,xreg,yreg,bflag) =
  (V
    (SEG
      (20,31)
      (BITS32
        (WORD32
        (V(SEG(0,31)(BITS33(WORD33((VAL32(WORD32(VAL20 preg))) + 1)))))))) =
    0) /\
  ~EL 4(FETCH_ABBR(ram,preg)) /\
  ~((VAL3(WORD3(V(SEG(5,7)(FETCH_ABBR(ram,preg))))) = 3) \/
    (VAL3(WORD3(V(SEG(5,7)(FETCH_ABBR(ram,preg))))) = 4) \/
    (VAL3(WORD3(V(SEG(5,7)(FETCH_ABBR(ram,preg))))) = 5)) /\
  ~(VAL3(WORD3(V(SEG(5,7)(FETCH_ABBR(ram,preg))))) = 7) /\
  ~(VAL3(WORD3(V(SEG(5,7)(FETCH_ABBR(ram,preg))))) = 6) /\
  (VAL2(WORD2(V(SEG(8,9)(FETCH_ABBR(ram,preg))))) = 0) /\
  (WORD4(V(SEG(0,3)(FETCH_ABBR(ram,preg)))) = #0101)
```

There are in all thirty-four sub-conditions of c4, where for each sub-condition the function selector and comparison indicator are given definite values, and the question of whether the program counter is the destination is definitely resolved. All of the sub-conditions are named in the same way that c4_F_5 was named; in the shift case (where the function selector is 12), analysis of the conditional expression reveals that the 2-bit memory selection field must be specified also, because it doubles there as an indicator of the various kinds of shifts; hence the name 'c4_F_12_0', and so on. We prove (much as before) that these thirty-four sub-conditions cover all logical possibilities *within* c4:

```
|- c4(ram,preg,areg,xreg,yreg,bflag) =

  c4_T_0(ram,preg,areg,xreg,yreg,bflag) \/
  c4_T_1(ram,preg,areg,xreg,yreg,bflag) \/
  c4_T_2(ram,preg,areg,xreg,yreg,bflag) \/
  c4_T_3(ram,preg,areg,xreg,yreg,bflag) \/
  c4_T_4(ram,preg,areg,xreg,yreg,bflag) \/
  c4_T_5(ram,preg,areg,xreg,yreg,bflag) \/
  c4_T_6(ram,preg,areg,xreg,yreg,bflag) \/
  c4_T_7(ram,preg,areg,xreg,yreg,bflag) \/
  c4_T_8(ram,preg,areg,xreg,yreg,bflag) \/
  c4_T_9(ram,preg,areg,xreg,yreg,bflag) \/
  c4_T_10(ram,preg,areg,xreg,yreg,bflag) \/
  c4_T_11(ram,preg,areg,xreg,yreg,bflag) \/
  c4_T_12(ram,preg,areg,xreg,yreg,bflag) \/
  c4_T_13(ram,preg,areg,xreg,yreg,bflag) \/
  c4_T_14(ram,preg,areg,xreg,yreg,bflag) \/
  c4_T_15(ram,preg,areg,xreg,yreg,bflag) \/

  c4_F_12_0(ram,preg,areg,xreg,yreg,bflag) \/
  c4_F_12_1(ram,preg,areg,xreg,yreg,bflag) \/
  c4_F_12_2(ram,preg,areg,xreg,yreg,bflag) \/
  c4_F_12_3(ram,preg,areg,xreg,yreg,bflag) \/

  c4_F_345_3(ram,preg,areg,xreg,yreg,bflag) \/
```

```
c4_F_345_5(ram,preg,areg,xreg,yreg,bflag) \/
c4_F_345_7(ram,preg,areg,xreg,yreg,bflag) \/

c4_F_0(ram,preg,areg,xreg,yreg,bflag) \/
c4_F_2(ram,preg,areg,xreg,yreg,bflag) \/
c4_F_3(ram,preg,areg,xreg,yreg,bflag) \/
c4_F_4(ram,preg,areg,xreg,yreg,bflag) \/
c4_F_5(ram,preg,areg,xreg,yreg,bflag) \/
c4_F_6(ram,preg,areg,xreg,yreg,bflag) \/
c4_F_7(ram,preg,areg,xreg,yreg,bflag) \/
c4_F_8(ram,preg,areg,xreg,yreg,bflag) \/
c4_F_9(ram,preg,areg,xreg,yreg,bflag) \/
c4_F_10(ram,preg,areg,xreg,yreg,bflag) \/
c4_F_11(ram,preg,areg,xreg,yreg,bflag)
```

The proof, as before, considers a very large number of cases, represented initially in an abstracted way. The first sixteen sub-conditions shown are comparison operations; the next four are shifts; the next three are non-comparisons with the program counter as destination; and the final eleven are non-comparisons with one of the other registers as destination. (Note that among the latter eleven, no sub-condition has a function field value of 1; 1 is reserved for procedure calls, which are considered as a separate major operation, not a variety of PERFORM ALU operation; and 13, 14 and 15 are dis-allowed as values.)

These conditions define thirty-four major states of the PERFORM ALU type, where initially (and in [5]) we had considered just one. Similarly, the PERFORM ALU states that follows a READMEM state or an INDEX *and* a READMEM state (in sequence) also branch into multiple states (twenty-nine each); these are elaborations of the corresponding PERFORM ALU operations in which the source address is either computed or looked up in memory and then computed, rather than just being taken literally.

The complete set of minor state transition theorems tells us that besides the FETCH state, other major states also offer a choice of next state. For example, the INDEX state can be followed by any of the four read or write operations (or by STOP). Also, as in the add-overflow example, the PERFORM ALU state sometimes specifies a choice between STOP or FETCH as the next state. (Had we considered the case in which the function selector were 4, indicating addition *without* overflow checks, there would have been no choice of next states.) In general, STOP-FETCH choices are governed by conditions which we call c17. In the add-overflow case, we define c17_F_5, based on the complicated stopping expression produced in that case (see Section 7.5), in the context of having previously fetched an instruction. (See also Section 9.) The condition holds if the machine does *not* stop. The corresponding condition is just the negation, so proving that all possibilities are covered is easy.

8.2   The Major State Tree

In the analysis of the minor states, it was never the case that at a given minor state there was a choice of next minor state; the sequences were linear. This is not true of major states; the analysis of the way major states follow each other leads to a representation of the major state transitions of the block machine as a tree (or more accurately, a graph). Each possible sequence of major states (i.e. each path through the graph) is represented in Figure 3, below. The bolder lines in the figure are path *schemata*, e.g. the collective thirty-or-so paths which follow the pattern of major states shown and contain the PERFORM ALU state. Of these, five or so in each path schema admit the possibility of stopping – e.g. the add-overflow case. The stopping cases are also represented collectively by a bolder line.

For each separate path (not path schema), of which there are one hundred and twenty-two, we can now consider composing the relevant minor state transition theorems to infer cumulative *major* path results. This is done by exactly the same method which yielded the cumulative minor state results.

8.3   Major State Transitions

The example case is continued; all one hundred twenty-two major state sequences are treated similarly.

The final cumulative *minor* state theorems for two nodes are considered: FETCH and ADD OVERFLOW. A recursive procedure, just as before, is used to compose them. The first major state in any sequence is always FETCH, so the procedure uses the minor state cumulative theorem for FETCH as its initial theorem[30].

It is assumed that condition `c4_F_5` holds. We then procede as before, using the case assumption where possible, and stopping the first time the major state becomes 1 again, indicating a fresh FETCH cycle. The possibility of the state becoming 8 (STOP) must also be dealt with; the minor state cumulative theorem for the STOP major state (not shown) asserts that at the final time (time $n+1$)

```
major_sig(n+1) = (reset_sig n => #0010 | #1000)
```

so that the next state may be 2 (RESET) or 8 (STOP), depending on the `reset` signal. The minor state cumulative theorem for the RESET state asserts that at the final time (time $n+4$)

```
major_sig(n+4) = #0001
```

This means that the block machine will remain in the stop state indefinitely, until and unless the `reset` line (connected to the external line `e_resetbar`, as we have seen) becomes set, the next time after which it will move to the RESET state and thence to FETCH again. Since a possibly infinite path cannot conveniently be analyzed, the analyis stops at the point at which the STOP state is reached. In any case, since the high level and major state specifications of Viper do not take resetting into account at all, there would be no point in trying to find a method, using an inductive argument, say, that allowed us to trace all paths back to FETCH.

Suppose that `c_17_F_5` holds (the no-stop case). The recursive ML procedure is applied to the initial theorem (the cumulative minor state theorem for FETCH). The procedure then infers the sequence of cumulative theorems for each major state, culminating in the one whose major state points back to FETCH again. In this example, the final theorem is the next one after FETCH: the add-overflow theorem.

To shorten the expressions as results accumulate over major paths, we abbreviated the recurring source and memory sub-expressions:

---

[30] This is another instance in which the informal description of the Viper block machine is required for the anaylsis; there is nothing in the specifications to suggest that FETCH is special in this way – though of course it would be a sensible guess.

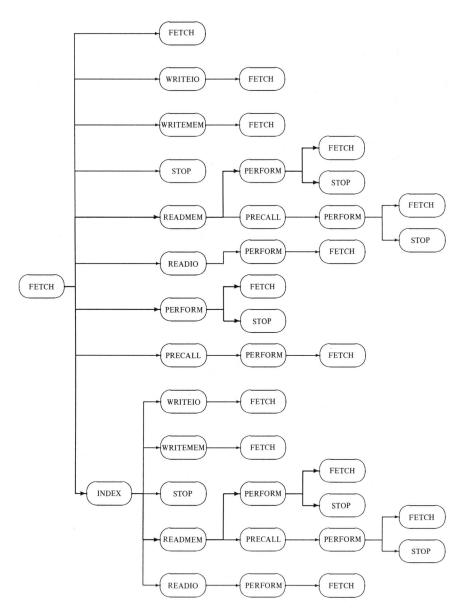

Figure 3: Viper Block Major State Tree Schema

```
|- REGSELECT_ABBR(areg,xreg,yreg,preg,ram) =
   NOT32
   (REGSELECT
    (areg,xreg,yreg,
     NOT20
     (WORD20
      (V
       (SEG
        (0,19)
        (BITS32
         (NOT32
          (WORD32
           (V
            (SEG(0,31)(BITS33(WORD33((VAL32(WORD32(VAL20 preg))) + 1)))))))))))),
      WORD2(V(SEG(10,11)(FETCH_ABBR(ram,preg))))))

  |- MEM_ABBR(ram,preg) =
     WORD32
     (V(SEG(0,19)(BITS32(FETCH21 ram(WORD21(V(CONS F(BITS20 preg)))))))))
```

At the end, the a-register's final value is

```
areg_sig(n+12) =
((~EL
  0
  (BITS7
   (DSFPRIM(WORD3(V(SEG(5,7)(FETCH_ABBR(ram_sig n,preg_sig n))))))))) =>
WORD32
(V
 (SEG
  (0,31)
  (BITS33
   (WORD33
    ((VAL32
      (REGSELECT_ABBR
       (areg_sig n,xreg_sig n,yreg_sig n,preg_sig n,ram_sig n))) +
     (VAL32(MEM_ABBR(ram_sig n,preg_sig n)))))))) |
areg_sig n)
```

Compared with the result for the addition operation in isolation (Section 7.5), this naturally has the same form; but the value taken from the temporary register (t-register) is the word obtained from memory; the destination field found in the instruction register is the destination field of the looked-up word, as so on – as one would expect.

In the corresponding major sequence *with* overflow (i.e. assuming that c17_F_5 does *not* hold), the final value of the a-register, and so on, is the same as above – the addition is still performed. However, the stopping value becomes true in this case, and the next major state is 8 (STOP).

As in the minor state transitions, assumptions accumulate when we compose the theorems. Here, assumption sets from the two major states are produced. As a result, we have to have assumed, by the final theorem in the sequence, that e_errorbar_sig and e_resetbar_sig are true from the initial time through time n+11; that e_reply_sig and e_stepbar_sig are true from the initial time to time n+5; and the former assumptions about the initial state. That is, some signals have to persist through both major states while others only have to persist only until time n+5, through the fetch operation. The rest only have to be true initially.

8.4 Conclusions about the Major State Transitions

We apply the same method to all one hundred twenty-two major state paths. The procedure is similar for all, though in each case there are small variations which make it difficult at first to design a uniform procedure. The total bulk of the theorems can be imagined; the final result in this relatively simple example is a six-page theorem (pretty-printed), and the complications caused by the occurrence of a READMEM (or an INDEX followed by a READMEM) operation before the addition produce very much longer cumulative results.

Once all the major paths through the graph shown in Figure 3 are derived, the formal analysis is provably complete; there is then a complete, proved picture of the block machine's behaviour on all instruction classes (and an assurance that no instruction classes have been overlooked). This in itself is useful because it could be given to the original designers for inspection; just perusing the results could reveal problems. Furthermore, the proved results could be used as the basis for a simulator.

For the non-ALU sequences, the results are not very complicated and they *appear* to be as intended. Some of the arithmetic-logic paths are also apparently correct. Others, in particular the additions, subtractions and comparisons, are neither obviously correct nor incorrect, and require further study. So far, there do not seem to be any definitely incorrect results, but obviously, since the formal analysis ends at this point, there very well could be. For that reason, a great deal of care should be taken in describing the Viper block model as being 'verified'; it has to date only been *analyzed* as described in this section, and inspected (as described in the next section). Section 2 expands on this point.

The rest of the analyis is completely speculative: we explore what facts and equipment would be required to complete the proof up to the point of equivalence with the high level.

9 Speculation on the Rest of the Proof

In Section 6.1, the nature of the state transformations at the specification level was illustrated. We have so far deduced the behaviour of the block machine on every instruction schema, and proved that none was omitted. To complete the proof of correctness of the Viper block model with respect to the high level specification, it would be necessary to compare the high level and block level results for every traversal of the major state tree of the block machine – that is, for over one hundred cases – where each such traversal corresponds to the processing of a single instruction schema. To be able to compare the two sets of results, the logical conditions determining a particular traversal would first have to be related to the conditional choices of the high level specification; then the high level definition would have to be fully unfolded in every case; and finally, the results at the two levels would have to be proved equivalent in some appropriate sense. None of these things has been done, for several reasons:

1. The amount of work involved in proving this number of cases would be considerable, even though there is a certain amount of common effort.

2. Both (i) relating the block's major state path predicates to the high level conditional choices and (ii) relating the block results to the high level results involve a great deal of intricate reasoning about complex bit manipulations. At the time this research was done, there was not enough infrastructure in HOL to support sophisticated reasoning about bits and words; the only method would have been to carry along a large (and possibly inconsistent) set of assumptions about bit-strings. T. Melham [21] at Cambridge is currently

developing a definitional framework for bit-string reasoning in HOL. However, it would still remain to build within that theory a library of theorems adequate to support the complex bit-manipulations involved in the equivalence proof.

3. The results at the two levels are quite disparate in the more difficult cases (particularly in cases of ALU operations which are not shifts). Up to the present point in the analysis, the definitions were capable of being unfolded without deep understanding of Viper's design or methods of computation. However, relating the block results to the high level results could not be approached in the same naive way; it would require extensive co-operation between the verifier and the Viper design team.

Nevertheless, it is possible, and quite useful, to speculate on what would be involved in completing the equivalence proofs.

The general form of the ultimate correctness statement was given in Section 5. Further details of the form can be found in [5]; this applied to a major state machine defined directly, but the form of the statement for the block machine would be very similar. The only forseeable complication is that in the block proof, there are the sets of accumulated assumptions attached to each path result, different in each case, which would probably complicate the task of tying the cases together.

The equivalence proofs are necessarily goal-oriented, unlike most of the proofs described above. That is, it is known what the results are for corresponding cases at the two levels, so one begins with the goal and tries to infer the equivalence by applying proof strategies. Goal-oriented proofs are generally more difficult than the rather deterministic unfolding-style foward proofs seen thus far; they require more insight into the problem on the part of the user.

As mentioned, a large body of facts about bit-strings are required to relate block level and high level terms. These are often phrased rather differently – for example, high level expressions are in terms of 32-bit fetched instructions, while block level expressions are in terms of segments of 12-bit instruction codes placed in various internal registers. Because the facts required are not very profound, we assume that they will eventually form part of a theory of bit-strings, and ignore them by simply equating the register and memory source expressions at the two levels. As in Section 5, the two expressions are abbreviated respectively as REGSELECT_ABBR(areg,xreg,yreg,preg,ram) and MEM_ABBR(ram,preg).

Unfolding the high level definition would be straightforward; it is unfolded in a forward way exactly as the block definitions were unfolded. The really difficult part of the equivalence is not relating the predicates to the conditional choices, but relating the results at the two levels. For example, in the addition-overflow case (where the destination is not the program counter) we have claimed (Section 6.1) that the sum specified at the high level by the function ALU' is

```
WORD32
(V
 (TL
  (TL
   (BITS34
    (WORD34
     ((VAL33(SIGNEXT(REGSELECT_ABBR(areg,xreg,yreg,preg,ram)))) +
      (VAL33(SIGNEXT(MEM_ABBR(ram,preg)))))))))))))
```

where sign extension (SIGNEXT) was defined as the construction of a 33-bit word from a 32-bit word by copying the top bit:

```
|- !w. SIGNEXT w = WORD33(V(CONS(EL 31(BITS32 w))(BITS32 w)))
```

The values of the two sign-extended words are added to form a 34-bit word from which the top two bits are then dropped. On the other hand, the sum we have laboriously inferred at the block level is

```
WORD32
(V
 (SEG
  (0,31)
  (BITS33
   (WORD33
    ((VAL32(REGSELECT_ABBR(areg,xreg,yreg,preg,ram))) +
    (VAL32(MEM_ABBR(ram,preg))))))))
```

which involves adding the values of the two registers to form a 33-bit word and dropping the top (thirty-second) bit. One can convince oneself that the two sums are equivalent, but again, the infrastructure to do this formally in HOL is at present rather limited. Further, the difficulty in this case, and more so in some others, stems from the different methods of computation used at the two levels; here, one level is in terms of a 34-bit sign-extension addition algorithm, while the other is an abstract addition with a carry into the thirty-third bit which is then ignored. The difference is clearer in the case of the overflow expressions; the new overflow condition (the stopping flag) at the high level was claimed to be

```
~(EL
  32
  (BITS34
   (WORD34
    ((VAL33(SIGNEXT(REGSELECT_ABBR(areg,xreg,yreg,preg,ram)))) +
    (VAL33(SIGNEXT(MEM_ABBR(ram,preg))))))) =
  EL
  31
  (BITS34
   (WORD34
    ((VAL33(SIGNEXT(REGSELECT_ABBR(areg,xreg,yreg,preg,ram)))) +
    (VAL33(SIGNEXT(MEM_ABBR(ram,preg))))))))
```

That is, overflow depended on the equality of the thirty-second and thirty-first bits of the 34-bit sum (comprising bits 0 to 33) of the values of the 33-bit sign-extended registers. At the block level the corresponding overflow expression is inferred to be

```
(EL 31(BITS32(REGSELECT_ABBR(areg,xreg,yreg,preg,ram))) =
EL 31(BITS32(MEM_ABBR(ram,preg)))) /\
~(EL 31(BITS32(REGSELECT_ABBR(areg,xreg,yreg,preg,ram))) =
  EL
  31
  (BITS32
   (WORD32
    (V
     (SEG
      (0,31)
      (BITS33
       (WORD33
        ((VAL32(REGSELECT_ABBR(areg,xreg,yreg,preg,ram))) +
        (VAL32(MEM_ABBR(ram,preg)))))))))))
```

Here, the overflow depends on the top bits of the two original registers as well as on the penultimate bit of the 33-bit sum of the values in the two original registers. It is difficult to relate these two expressions even informally, aside from problems of reasoning about bit-strings.

In a typical comparison case, we have claimed that the equality of two registers, at the high level, is tested simply by

```
REGSELECT_ABBR(areg,xreg,yreg,preg,ram) = MEM_ABBR(ram,preg)
```

At the low level, it happens that the expression inferred is

```
(V
 (SEG
  (0,15)
  (BITS32
   (WORD32
    (V
     (SEG
      (0,31)
      (BITS33
       (WORD33
        ((VAL32(REGSELECT_ABBR(areg,xreg,yreg,preg,ram))) +
         ((VAL32(NOT32(MEM_ABBR(ram,preg)))) + 1))))))))) =
0) /\
(V
 (SEG
  (16,19)
  (BITS32
   (WORD32
    (V
     (SEG
      (0,31)
      (BITS33
       (WORD33
        ((VAL32(REGSELECT_ABBR(areg,xreg,yreg,preg,ram))) +
         ((VAL32(NOT32(MEM_ABBR(ram,preg)))) + 1))))))))) =
0) /\
(V
 (SEG
  (20,31)
  (BITS32
   (WORD32
    (V
     (SEG
      (0,31)
      (BITS33
       (WORD33
        ((VAL32(REGSELECT_ABBR(areg,xreg,yreg,preg,ram))) +
         ((VAL32(NOT32(MEM_ABBR(ram,preg)))) + 1))))))))) =
0)
```

Here, the register and memory source are combined by negating the memory and adding one to its value, then adding that to the value of the register source. The thirty-third bit of that sum is dropped, and the 32-bit result is partitioned into three segments, each of which is tested for equality to 0. Again, it takes some thought to convince oneself that the two levels are equivalent; and only the designers of Viper could explain the computational significance of the block level partition.

For the less-than comparison (source less than memory) the top level result was

```
EL
32
(BITS34
 (WORD34
  ((VAL33(SIGNEXT(REGSELECT_ABBR(areg,xreg,yreg,preg,ram)))) +
   ((VAL33(SIGNEXT(MEM_ABBR(ram,preg)))) = 0) => 0 |
   (VAL33(NOT33(SIGNEXT(MEM_ABBR(ram,preg)))))) + 1))))
```

while at the block level what we derive is

```
~(EL 31(BITS32(REGSELECT_ABBR(areg,xreg,yreg,preg,ram))) =
 EL
 31
 (BITS32
  (WORD32
   (V
    (SEG
     (0,31)
     (BITS33
      (WORD33
       ((VAL32(REGSELECT_ABBR(areg,xreg,yreg,preg,ram))) +
        ((VAL32(NOT32(MEM_ABBR(ram,preg)))) + 1))))))) /\
~EL 31(BITS32(MEM_ABBR(ram,preg))) \/
EL 31(BITS32(REGSELECT_ABBR(areg,xreg,yreg,preg,ram))) /\
EL
31
(BITS32
 (WORD32
  (V
   (SEG
    (0,31)
    (BITS33
     (WORD33
      ((VAL32(REGSELECT_ABBR(areg,xreg,yreg,preg,ram))) +
       ((VAL32(NOT32(MEM_ABBR(ram,preg)))) + 1)))))))
```

Here, the top bits of the original registers and the top bits of the same sum as before are the determining factors. This is a case in which it is not all easy to see the connection between levels, and not easy to explain the differing methods of computation.

Finally, the procedure call case was said to give the following jump address at the high level:

```
WORD20(V(SEG(0,19)(BITS32(MEM_ABBR(ram,preg)))))
```

At the block level the address is inferred to be:

```
NOT20(WORD20(V(SEG(0,19)(BITS32(NOT32(MEM_ABBR(ram,preg)))))))
```

which is clearly equivalent (once we have proved suitable lemmas about inversion).

The expressions resulting from indirect addressing, or worse, offset addressing, are particularly long and cumbersome, though the ideas should be the same, for both, as in the corresponding simple cases.

In summary, some of the pairs of high and block level expressions are not related in an obvious way, while others require only simple lemmas to be shown equal. The main difficulty rests in understanding and relating the different methods of computation at the two levels.

A minor oddity is that at the block level, two store operations are performed when one would be expected. The high level memory value resulting from this operation is simply

```
STORE21
(WORD21
 (V
  (CONS
   F
   (BITS20(WORD20(V(SEG(0,19)(BITS32(MEM_ABBR(ram,preg)))))))))))
 (REGSELECT_ABBR(areg,xreg,yreg,preg,ram))
 ram
```

whereas at the low level the new memory value is

```
STORE21
(WORD21
 (V
  (CONS
   F
   (BITS20(WORD20(V(SEG(0,19)(BITS32(MEM_ABBR(ram,preg)))))))))))
 (REGSELECT_ABBR(areg,xreg,yreg,preg,ram))
 (STORE21
  (WORD21
   (V
    (CONS
     F
     (BITS20(WORD20(V(SEG(0,19)(BITS32(MEM_ABBR(ram,preg)))))))))))
   (REGSELECT_ABBR(areg,xreg,yreg,preg,ram))
   ram)
```

The double store is clearly equivalent to a single store, and there is apparently a reason for this construction[31].

A slightly more puzzling discrepancy is that the stop flag component of the high level (visible) state is, in exactly one case, not equivalent to the block level stop flag. Because of this, the two flags are not identifiable, although one might well expect them to be. The instruction schema on which they differ is the illegal instruction. The high level function specifies that in any of several illegal instruction types, the new state will have all other registers unchanged but the stop flag set to true

---

[31]C. Pygott, private communication.

(see Section 6.1). At the block level, on the other hand, the new value of the stopping flag is determined exclusively by whether the program counter can be incremented without exceeding twenty bits – and any other illegal conditions are ignored. Because of this fact, the two stopping flags cannot be identified (even though in all other instances they correspond). To circumvent the problem (which seems unintentional), we have to relate the stopping flag at the high level to whether `major is 8`, at the block level – i.e. to whether the major state is STOP. This does give a pair of corresponding expressions, but it is rather unaesthetic.

Since the equivalence proof can be explored but cannot reasonably be completed at the present time, the next logical step was for the designers to carefully examine the block results. With suitable abbreviations these are quite readable, and the designers were able to report that the block machine results bore at least a roughly plausible relation to the specified results. This is of course no substitute for a proof, but is a good sign. Nevertheless, it remains possible that a completed formal proof might reveal subtle errors not obvious from an informal scan of the results by the designers. Thus it is really too soon to claim that the Viper block model has been verified, and, as discussed in Section 2, a great deal of care should be taken in making that sort of claim.

10 Lessons and Conclusions

In this section we amplify some of the points made in the Introduction.

This report has described the partially completed proof of correctness of the block model of the Viper microprocessor. The block model is a register-transfer level model of Viper, comprising functional descriptions of the computational units such as the ALU, together with pictorial descriptions of the flow of data and control signals, over time, between units. Viper's correctness is analyzed relative to a higher level specification given in terms of abstract state transformations. The block model was intended to assist in Viper's circuit design phase. Both Viper and the block model were developed at RSRE, by W. J. Cullyer, C. Pygott and J. Kershaw. The idea of structuring the proof according to the tree of execution paths is due to W. J. Cullyer.

The project thus far has achieved the following:

- We have extracted a fully formal representation of the block machine based on the original block definitions and diagrams. This has been done in the HOL theorem proving system, in which functions (in the mathematical sense) are regarded as first-class objects of the logic. The block machine itself is expressed as a function, so that one can *infer* its value when applied to arguments. The representation determines the set of possible execution sequences of the machine.

- Through a process of inference consisting in recursively instantiating the theorems expressing the results of individual major state traversals, we have deduced the cumulative effects of each of the block model's major state sequences. Each sequence in this sense represents the execution of a single instruction type. Each major state traversal is achieved in turn by a sequence of minor events itself having a cumulative effect. Therefore the analysis is hierarchically structured according to major and minor sequences. The individual major and minor state traversal theorems themselves rely on a very large number of lemmas about the basic blocks comprising the block model. These lemmas are all proved by straightforward unfolding of definitions[32].

In effect, we have symbolically evaluated the machine through each execution path, but with the additional security of having proved that the results follow from the definitions. As a side-effect, this provides a secure basis for building a simulator.

Assumptions about certain incoming and initial signals limit the cases considered; this is necessary in part because the high level specification is itself incomplete as regards certain inputs to the block model (e.g. externally forced resets).

- We prove further that all possible instruction types have been considered, making the description complete as well as correct.

The fourth and concluding phase of the project, about which we have only speculated, would involve proving a congruence between the block results and the high level results for corresponding instruction types. This phase is impractical to complete at the moment, for several reasons. One reason is that the number of cases and the size of some of the resulting expressions is dauntingly large (for what would appear to be dwindling research interest). A second reason is the current lack of a well-developed HOL infrastructure for supporting advanced reasoning about bit-string manipulations. (T. Melham at Cambridge ([21]) is working on developing an appropriate HOL theory, but it still remains to build within this theory a library of useful theorems leading up to the sort of theorem needed in the equivalence statement – see Section 9.) Finally, it became clear that relating the two disparate descriptions would involve close interaction between verification team and design team, and could not be done as independently as the proof thus far has been.

The success of the proof thus far must be qualified by three caveats, all of which have been discussed in detail in Section 2. Firstly, as just mentioned, the high level specification itself is incomplete as regards Viper as a whole; it describes only the fetch-decode-execute cycle of the machine. Proving that a design satisfies an incomplete specification still leaves plenty of scope for an unacceptable implementation of the design. Secondly, the problem of establishing that the formal representation of the block is itself accurate has no real solution. For example, in the present proof, an incorrect representation was originally derived by cross-naming two block outputs of the same type. The description was subsequently used to generate plausible block results (in most cases) until the error happened to be noticed. Thirdly, the block level itself is still very abstract. Between the block level and the manufactured chip there remain, for example, the gate level and the electronic level – at which problems seem at least as likely to occur. Furthermore, a physical chip cannot be verified in a mathematical sense. Any model is inaccurate to some extent; a 'verified' safety-critical chip can still go wrong and cause loss of life and environment.

As has been described at length, all of the analytical concepts of the proof stem from the functional representation we derived of the block machine. These include

---

[32]By 'unfolding' we mean the application of lambda-expressions to values, followed by beta-reduction.

the state of the block machine, the shape of the major state tree, and all of the path predicates. Almost the whole of the proof (with the exception of the proofs that no instruction types have been omitted) consisted in unfolding definitions on typical values. This is not the only mode in which HOL can be used, but it is the easiest, as it requires the user to follow rather than lead the analysis. Indeed, the proof thus far requires minimal understanding of the operation of Viper and its design rationale; all that is really essential is to understand that there are major and minor cycles, and that the major start with FETCH (1), while the minor start with 0. The fourth phase, however, in which the block results are shown to implement the specified results, is different. There, instead of applying routine transformations to known facts to yield unforeseen results, conjectures (goals) are reduced to subgoals by application of carefully planned strategies. These strategies have to embody some understanding of the reason for the implementation, or the proof drifts into useless subgoals.

A brief account of the size of and time required by the proof may be of interest. (It should be borne in mind here that all HOL proofs, like LCF proofs, are actually *performed* to the smallest logical steps; any lemmas used must already have been proved; and no procedures are used to shortcut the process of formal deduction.) The approximate number of primitive inference steps for the various groups of theorems is given below, with the approximate elapsed cpu time on a 12-megabyte Sun-3, and the relevant section of this report (if the issue was discussed).

- Deriving the representation: 95,000 steps (2 hours) (Section 6)

- Lemmas about particular block values for specific minor and/or major states: 5,130,000 steps (50 hours) (Section 7.6)

- Theorems about the transitions through minor states: 210,000 steps (15 hours) (Section 7)

- Proof that all cases have been covered: 160,000 steps (14 hours) (Section 8.1)

- Lemmas about major path predicates: 530,000 steps (7 hours)

- Theorems about the transitions through major states: 128,000 steps (10 hours) (Section 8)

- Projected theorems expressing equivalence to the high level, provided with appropriate lemmas about bit-strings: 900,000 steps (14 hours) (Section 9)

- Total without equivalence proof: 6,253,000 steps (98 hours)

- Total with equivalence proof: about 7,153,000 steps (about 112 hours)

This effort took one person working full-time eleven months to complete. It can be observed that the greatest bulk of the proof by far is in the lemmas which describe the behaviour of the individual blocks on specific major and minor values. These lemmas are all very simple unfoldings; there are just a large number of them, since the block results all have to be inferred for most possible major and minor values. This is encouraging, since that phase involved relatively little time or effort on the part of the user. The most user time was spent in tracing the one-hundred or so execution paths, but again that was a fairly routine process. The most difficult (creative) part of the proof effort was deriving the logical representation of the block machine.

These observations suggest that for a trained user it is not difficult to verify computer designs; it requires more patience perhaps than cleverness. This holds despite the absence of sophisticated proof-editors, friendly interfaces, helpful error messages, or support for proof structuring (beyond a simple subgoal stack). However, it is true that unfolding one hundred-odd cases was tedious. Although the cases were quite similar to one another (especially the sets of corresponding ALU cases using different means of addressing), each seemed to involve some complication which prevented previous strategies from working without some minor change. Generalizing the strategies would be a useful exercise.

The representation methods used in the block proof were all standard in 'Cambridge style' hardware verification circles; the author cannot claim any original contribution. Likewise, the proof strategies deployed (as mentioned) were simple ones – not particularly interesting or original in themselves. Only the size of the proof and the realistic nature of the problem are unique (and probably also the length of some of the theorems).

In terms of the original plan [8] to prove that the block machine implements major state machine (which was already proved [5] to implement the high level specification), it appeared to be more sensible to make the link directly from the block level to the high level. We list some of the issues related to this decision:

- The use of the intermediate level seemed unnecessary, for a start; since equivalence with the top level was the ultimate goal anyway, the need for some sort of transitivity argument was avoided by a direct connection.

- The block and major state levels, though both culminated in major state transition trees, produced rather differently structured trees with different path predicates; the tree and conditions at the block level stem directly from the block representation function, whereas the major state level was designed directly to implement the top level specification. This meant that the local contexts of the trees were difficult to relate. (At the high level, however, the conditional expression branches directly into a state transformation for each possible operation, so there is no problem about local contexts in a tree.)

- The expressions derived at the major state level were never fully unfolded (in particular, ALU expressions were not unfolded) because the property proved did not involve any computation. That is, in the previous major state tree there was only one FETCH-PERFORM path rather than one for each possible ALU operation, so the major state level missed the essence of the block level: Viper's computational behaviour. This further complicated the use of the major state machine as an intermediate level, as the previous major state results would have to be much further unfolded to be useful.

- Finally, the block level path predicates only emerge from a consideration of the block level major paths, because of the accumulation of results. That is, block level conditions such as $c17\_F\_5$ (Section 8.1) require the whole FETCH-PERFORM sequences to be considered, so that the use of the intermediate level does not seem to get round that stage of analysis (as hoped).

These problems raise the more general issue of using intermediate levels at all, and if so, how they are designed. In a layered approach, it may be difficult to predict, in advance of analyzing the low level in detail, what are the useful concepts at intermediate stages, and what are useful bridging levels. In this case it *seemed* a good idea to isolate the flow of control correctness at the major state level, and the computational problem at the block level – but in fact the block level involved *both* control and computation, and the two were intertwined, so the separation did not work to assist the proof. This is a subject deserving further thought and research. (J. Joyce at Cambridge is also working on this problem [19].) In any case, though not engaged directly, the major state proof was certainly a useful preliminary exercise for the block proof; it established the methodology on a more manageable scale.

Another conclusion, in considering the block proof as a feasibility study in realistically sized proofs, is the performance of the HOL system. HOL does not have a particularly good implementation; it was laced togther from old LCF code which itself has complex historical origins. It is therefore amazing and very encouraging that the system performed so well on a very large proof – orders of magnitude larger than was originally intended when the first LCF system was implemented in the 1970's. This can only be attributed to the successful design principles devised R. Milner, which shine through the layers of implementations and patches. Indeed the only bottleneck in the performance of the block proof was a shortage of disc-space, not the performance of the system or its capacity for handling (parsing, retreiving, manipulating, etc) massive expressions or for managing huge intermediate proof structures.

The massive expressions that comprise the proof arise mostly through the accumulation of effects through layers of processing, as well as through the absence of logical equipment for simplifying expressions about bits and bit-strings. Apart from that, it would have been better (in retrospect) to have abbreviated the longer recurring sub-expressions as much as possible, for readability and economy; some abbreviation was used, as noted in the paper, but on the whole it was difficult to predict intelligent abbreviations in advance, and after the fact, a bit tedious to re-do the proof for that reason alone.

Re-doing proofs is another issue worth mentioning; at one stage our representation was actually incorrect (involving crossed lines, as mentioned) and it was of course necessary to re-do the proof up to that point. (In fact, this happened a second time because of a typographical error by the author.) The re-doing was facilitated by having preserved the ML code which generated the proof, and simply re-running it on corrected definitions. In fact, since proofs are *performed* rather than produced as sequences of (millions of) steps, the ML code generating the proof is the only tangible end product of a proof effort, and indeed the only 'abstract' description of the proof that one ever has. Two facilities would be desirable in HOL, though both raise philosophical/research problems. One would be a genuinely abstract description of the structure of a proof – more abstract than the sequence of tactics or inference patterns that generate the proof. The other would be a way of tracing 'material dependence' of theorems, so that when it were necessary to go back and change the problem statement, there were a way of knowing which areas of the proof might be or definitely could not be affected. As it stands, the entire proof has to be redone in such cases, which means replacing documentation, and so on.

Many readers will want a simple answer to the question: is Viper correct at the block level? The major state proof [5] was wholly completed, and errors *were* found in the design of the major state model. It may therefore come as a disappointment that the present proof has not been carried far enough for a definite answer to be given; it amounts to an *analysis* of the block machine under all circumstances covered by the specification, but is not a proof that the block model meets the specification. On the positive side, it is now possible for the designers to (i) make a visual inspection of the complete description of the block model's behaviour, and (ii) use the description to simulate the block model in a secure way. These factors may go some way to answering the question. Aside from some confusion about the stopping behaviour of the machine (Section 9), which in any case can be circumvented as described, there does not so far appear to be any problem – that is the best answer we can give.

Finally, the author would like to express two hopes: first, that Viper chips are used with an intelligent understanding of the limited sense in which a 'verified chip' is 'safe' from failure; and second, that Viper chips find uses in constructive applications (civil aviation, medical computing, railway signalling, and so on) as well as in destructive ones.

11   Acknowledgements

Thanks to Mike Gordon who assisted enormously and advised throughout the project, and who helped with this report; and to the rest of the Hardware Verification Group at Cambridge University Computer Laboratory. Thanks also to Clive Pygott and (particularly) to John Cullyer, of RSRE, for their support and encouragement, and for their confidence in, and pioneering efforts in promoting, formal methods. The work described here was supported by a grant from RSRE under Research Agreement 2029/205(RSRE). This report first appeared as a University of Cambridge Computer Laboratory Technical Report: No. 134, May 1988.

References

[1] H. G. Barrow, VERIFY: A Program for Proving Correctness of Digital Hardware Designs, Artificial Intelligence Vol. 24, 1984

[2] A. Camilieri, M. Gordon and T. Melham, Hardware Verification using Higher-Order Logic, Proceedings of the IFIP WG 10.2 Working Conference: From H.D.L. Descriptions to Guaranteed Correct Circuit Designs, Grenoble, September 1986, ed. D. Borrione, North-Holland, Amsterdam, 1987

[3] A. Church, A Formulation of the Simple Theory of Types, Journal of Symbolic Logic 5, 1940

[4] A. Cohn and M. Gordon, A Mechanized Proof of Correctness of a Simple Counter, University of Cambridge, Computer Laboratory, Tech. Report No. 94, 1986

[5] A. Cohn, A Proof of Correctness of the Viper Microprocessor: the First Level, VLSI Specification, Verification and Synthesis, eds. G. Birtwistle and P.A. Subrahmanyam, Kluwer, 1987; Also University of Cambridge, Computer Laboratory, Tech. Report No. 104

[6] W. J. Cullyer and C. H. Pygott, Hardware Proofs using LCF_LSM and ELLA, RSRE Memo. 3832, Sept. 1985

[7] W. J. Cullyer, Viper Microprocessor: Formal Specification, RSRE Report 85013, Oct. 1985

[8] W. J. Cullyer, Viper — Correspondence between the Specification and the "Major State Machine", RSRE report No. 86004, Jan. 1986

[9] W. J. Cullyer, Implementing Safety-Critical Systems: The Viper Microprocessor, VLSI Specification, Verification and Synthesis, eds. G. Birtwistle and P.A. Subrahmanyam, Kluwer, 1987

[10] C. Gane (Computing Devices Company Ltd.), Computing Devices, Hastings' VIPER-VENOM Project: VIPER in Weapons Stores Management, SafetyNet: Viper Microprocessors in High Integrity Systems, Enq. No. 021, Issue 2, July-August-September 1988, Viper Technologies Ltd., Worcester

[11] J. Cullyer et. al., forthcoming book

[12] M. Gordon, R. Milner and C. P. Wadsworth, Edinburgh LCF, Lecture Notes in Computer Science No. 78, Springer-Verlag, 1979

[13] M. Gordon, Proving a Computer Correct, University of Cambridge, Computer Laboratory, Tech. Report No. 42, 1983

[14] M. Gordon, HOL: A Machine Oriented Formulation of Higher-Order Logic, University of Cambridge, Computer Laboratory, Tech. Report No. 68, 1985

[15] M. Gordon, HOL: A Proof Generating System for Higher-Order Logic, University of Cambridge, Computer Laboratory, Tech. Report No. 103, 1987; Revised version in VLSI Specification, Verification and Synthesis, eds. G. Birtwistle and P.A. Subrahmanyam, Kluwer, 1987

[16] M. P. Halbert (Cambridge Consultants Ltd.), Selfchecking Computer Module Based on the Viper1A Microprocessor, SafetyNet: Viper Microprocessors in High Integrity Systems, Enq. No. 017, Issue 2, July-August-September 1988, Viper Technologies Ltd., Worcester

[17] W. A. Hunt Jr., FM8501: A Verified Microprocessor, University of Texas, Austin, Tech. Report 47, 1985

[18] J. J. Joyce, Formal Verification and Implementation of a Microprocessor, VLSI Specification, Verification and Synthesis, eds. G. Birtwistle and P.A. Subrahmanyam, Kluwer, 1987

[19] J. J. Joyce, private communication

[20] J. Kershaw, Viper: A Microprocessor for Safety-Critical Applications, RSRE Memo. No. 3754, Dec. 1985

[21] T. Melham, private communication

[22] L. Paulson, Interactive Theorem Proving with Cambridge LCF, Cambridge University Press, 1987

[23] C. H. Pygott, Viper: The Electronic Block Model, RSRE Report. No. 86006, July 1986

[24] Viper Microprocessor: Verifiable Integrated Processor for Enhanced Reliability: Development Tools, Charter Technologies Ltd., Publication No. VDT1, Issue 1, Dec. 1987

[25] Application for Admission and Registration Form, Second VIPER Symposium, RSRE, Malvern, 6-7 September, 1988

12 Appendix: The HOL Viper High Level and Block Level Definitions

The following two sections give the complete, corrected RSRE definitions in HOL format of (i) the high level specification and (ii) the blocks of the block model. The author takes full responsibility for any errors which may remain undetected in the typing and translation into HOL format of the original definitions. In each section, the constants (with types) are shown first, followed by the definitions. The block definitions are to be interpreted in conjunction with Figure 1 (Section 6.3). Note that the constants at the two levels are completely distinct; in particular the constant ALU, representing the arithmetic-logic units, occurs at both levels but does not represent the same function in the two cases. This should be regarded as an unfortunate coincidence of names. In the text, the high level function with this name has been distinguished with a prime-mark. The state in the high level definitions uses the variable names

```
(ram:mem21_32,p:word20,a:word32,x:word32,y:word32,b:bool,stop:bool)
```

whereas the visible part of the block level state uses

```
(ram:mem21_32,preg:word20,areg:word32,xreg:word32,yreg:word3,bflag:bool,stop:bool)
```

where p corresponds to preg, and similarly for all the other variables, except that the stop component at the high level does not work out to correspond to the stop component at the block level. (See Section 9.)

The definitions are implicitly universally quantified over all free variables.

12.1 The High Level Specification

12.1.1 The Types

```
VALUE:word32#bool#bool->word32
OFLO:word32#bool#bool->bool
CARRY:word32#bool#bool->bool
SVAL:word32#bool#bool->bool
BVAL:word32#bool#bool->bool
TRIM33TO20:word33->word20
TRIM32TO20:word32->word20
TRIM34TO32:word34->word32
PAD20TO32:word20->word32
SIGNEXT:word32->word33
RIGHT:bool#word32->word32
LEFT:word32#bool->word32
RIGHTARITH:word32->word32
NEG:word33->num
ADD32:word32#word32->word32#bool#bool
SUB32:word32#word32->word32#bool#bool
INCP32:word20->word32
FINDINDEX:word2#word32#word32#word32->word32
DEC:(num # num)#word32->num
R:word32->word2
M:word32->word2
D:word32->word3
C:word32->word1
FF:word32->word4
A:word32->word20
NOOP:word3#word1#bool->bool
COMPARE:word4#word32#word32#bool->bool
REG:word2#word32#word32#word32#word20->word32
INVALID:word32->bool
INSTFETCH:mem21_32 # word20 -> word32"
ALU:word4#word2#word3#word32#word32#bool->word32#bool#bool
WRITE:word3#word1->bool
SPAREFUNC:word3#word1#word4->bool
ILLEGALCALL:word3#word1#word4->bool
ILLEGALPDEST:word3#word1#word4->bool
ILLEGALWRITE:word3#word1#word2->bool
```

```
OFFSET:word2#word20#word32#word32->word32
MEMREAD:mem21_32#word2#word20#word32#bool#bool->word32
MEMWRITE:mem21_32#word32#word2#word20#word32#word32#bool->mem21_32
NILM:word3#word1#word4->bool
OUTPUT:word3#word1->bool
INPUT:word3#word1#word4->bool
NEXT:mem21_32#word20#word32#word32#bool#bool->
      mem21_32#word20#word32#word32#bool#bool
```

12.1.2   The Definitions

```
|- VALUE(result,carry,overflow) = result
|- OFLO(result,carry,overflow) = overflow
|- CARRY(result,carry,overflow) = carry
|- SVAL(result,b,abort) = abort
|- BVAL(result,b,abort) = b
|- TRIM33TO20 w = WORD20(V(SEG(0,19)(BITS33 w)))
|- TRIM32TO20 w = WORD20(V(SEG(0,19)(BITS32 w)))
|- TRIM34TO32 w = WORD32(V(TL(TL(BITS34 w))))
|- PAD20TO32 w = WORD32(VAL20 w)
|- SIGNEXT w = (let bitlist = BITS32 w in WORD33(V(CONS(EL 31 bitlist)bitlist)))
|- RIGHT(b,r) = WORD32(V(CONS b(SEG(1,31)(BITS32 r))))
|- LEFT(r,b) = (let twice = V(TL(BITS32 r)) in
                  (b => WORD32((twice + twice) + 1) | WORD32(twice + twice)))
|- RIGHTARITH r = (let sign = EL 31(BITS32 r) in
                     WORD32(V(CONS sign(SEG(1,31)(BITS32 r)))))
|- NEG m = ((VAL33 m = 0) => 0 | (VAL33(NOT3 m)) + 1)
|- ADD32(r,m) = (let sum = WORD34((VAL33(SIGNEXT r)) + (VAL33(SIGNEXT m))) in
                   let opposite = (EL 31(BITS32 r)) XOR (EL 31(BITS32 m)) in
                   TRIM34TO32 sum,(EL 32(BITS34 sum)) XOR opposite,
                   (EL 32(BITS34 sum)) XOR (EL 31(BITS34 sum)))
|- SUB32(r,m) = (let dif = WORD34((VAL33(SIGNEXT r)) + (NEG(SIGNEXT m))) in
                   let opposite = (EL 31(BITS32 r)) XOR (EL 31(BITS32 m)) in
                   TRIM34TO32 dif,(EL 32(BITS34 dif)) XOR opposite,
                   (EL 32(BITS34 dif)) XOR (EL 31(BITS34 dif)))
|- INCP32 p = VALUE(ADD32(PAD20TO32 p,WORD32 1))
|- FINDINDEX(msf,t,x,y) = (let xindex = VAL2 msf = 2 in
                             (xindex => VALUE(ADD32(t,x)) | VALUE(ADD32(t,y))))
|- DEC((low,high),w) = V(SEG(low,high)(BITS32 w))
|- R w = WORD2(DEC((30,31),w))
|- M w = WORD2(DEC((28,29),w))
|- D w = WORD3(DEC((25,27),w))
|- C w = WORD1(DEC((24,24),w))
|- FF w = WORD4(DEC((20,23),w))
|- A w = WORD20(DEC((0,19),w))
|- NOOP(dsf,csf,b) = (let df = VAL3 dsf in
                        let cf = VAL1 csf in
                        (cf = 0) AND (((df = 5) AND b) OR ((df = 4) AND (NOT b))))
|- COMPARE(fsf,r,m,b) =
   (let op = VAL4 fsf in
    let dif = WORD34((VAL33(SIGNEXT r)) + (NEG(SIGNEXT m))) in
    let equal = r = m in
    let less = EL 32(BITS34 dif) in
    let borrow = (EL 32(BITS34 dif)) XNOR ((EL 31(BITS32 r)) XNOR (EL 31(BITS32 m))) in
    ((op = 0) => less |
     ((op = 1) => NOT less |
      ((op = 2) => equal |
       ((op = 3) => NOT equal |
        ((op = 4) => less OR equal |
         ((op = 5) => NOT(less OR equal) |
          ((op = 6) => borrow |
           ((op = 7) => NOT borrow |
            ((op = 8) => less OR b |
             ((op = 9) => (NOT less) OR b |
              ((op = 10) => equal OR b |
               ((op = 11) => (NOT equal) OR b |
                ((op = 12) => (less OR equal) OR b |
                 ((op = 13) => (NOT(less OR equal)) OR b |
                  ((op = 14) => borrow OR b | (NOT borrow) OR b)))))))))))))))
|- REG(rsf,a,x,y,p) = (let r = VAL2 rsf in
                        ((r = 0) => a | ((r = 1) => x | ((r = 2) => y | PAD20TO32 p))))
|- INVALID value = NOT(value = PAD20TO32(TRIM32TO20 value))
|- INSTFETCH(ram,p) = FETCH21 ram(WORD21(V(CONS F(BITS20 p))))
|- ALU(fsf,msf,dsf,r,m,b) =
   (let ff = VAL4 fsf in
    let mf = VAL2 msf in
    let df = VAL3 dsf in
    let pwrite = (df = 3) OR ((df = 4) OR (df = 5)) in
    ((ff = 0) => (NOT32 m,b,pwrite) |
     ((ff = 1) => (m,b,(NOT pwrite) OR (INVALID m)) |
      ((ff = 2) => (m,b,pwrite) |
       ((ff = 3) => (m,b,pwrite AND (INVALID m)) |
        ((ff = 4) => let sum = ADD32(r,m) in VALUE sum,CARRY sum,pwrite |
         ((ff = 5) => let sum = ADD32(r,m) in
                        VALUE sum,b,(OFLO sum) OR (pwrite AND (INVALID(VALUE sum))) |
          ((ff = 6) => let dif = SUB32(r,m) in VALUE dif,CARRY dif,pwrite |
           ((ff = 7) => let dif = SUB32(r,m) in
                          VALUE dif,b,(OFLO dif) OR (pwrite AND (INVALID(VALUE dif))) |
```

```
                    ((ff = 8) => ((r OR32 m) AND32 (NOT32(r AND32 m)),b,pwrite) |
                     ((ff = 9) => (r AND32 m,b,pwrite) |
                      ((ff = 10) => (NOT32(r OR32 m),b,pwrite) |
                       ((ff = 11) => (r AND32 (NOT32 m),b,pwrite) |
                        ((ff = 12) => ((mf = 0) => (RIGHTARITH r,b,pwrite) |
                                       ((mf = 1) => (RIGHT(b,r),EL 0(BITS32 r),pwrite) |
                                        ((mf = 2) => let double = ADD32(r,r) in
                                                     VALUE double,b,(OFLO double) OR pwrite |
                                                     (LEFT(r,b),EL 31(BITS32 r),pwrite)))) |
                         ((ff = 13) => (r,b,T) |
                          ((ff = 14) => (r,b,T) | (r,b,T))))))))))))))))))
|- WRITE(dsf,csf) = (let df = VAL3 dsf in
                     let cf = VAL1 csf in (cf = 0) AND ((df = 7) OR (df = 6)))
|- SPAREFUNC(dsf,csf,fsf) =
   (let df = VAL3 dsf in
    let cf = VAL1 csf in
    let ff = VAL4 fsf in
    (cf = 0) AND ((NOT((df = 6) OR (df = 7))) AND ((ff = 13) OR ((ff = 14) OR (ff = 15)))))
|- ILLEGALCALL(dsf,csf,fsf) =
   (let df = VAL3 dsf in
    let cf = VAL1 csf in
    let ff = VAL4 fsf in
    (cf = 0) AND ((ff = 1) AND ((df = 0) OR ((df = 1) OR (df = 2)))))
|- ILLEGALPDEST(dsf,csf,fsf) =
   (let df = VAL3 dsf in
    let cf = VAL1 csf in
    let ff = VAL4 fsf in
    (cf = 0) AND (((df = 3) OR ((df = 4) OR (df = 5))) AND
                  (NOT((ff = 1) OR ((ff = 3) OR ((ff = 5) OR (ff = 7)))))))
|- ILLEGALWRITE(dsf,csf,msf) = (let mf = VAL2 msf in (WRITE(dsf,csf)) AND (mf = 0))
|- OFFSET(msf,addr,x,y) =
   (let mf = VAL2 msf in
    let addr32 = PAD20TO32 addr in
    ((mf = 0) => addr32 |
     ((mf = 1) => addr32 |
      ((mf = 2) => VALUE(ADD32(addr32,x)) | VALUE(ADD32(addr32,y))))))
|- MEMREAD(ram,msf,addr,x,y,io,nil) =
   (let m = VAL2 msf in
    (nil => WORD32 0 |
     ((m = 0) => PAD20TO32 addr |
      FETCH21
      ram
      (WORD21(V(CONS io(BITS20(TRIM32TO20(OFFSET(msf,addr,x,y)))))))))))
|- MEMWRITE(ram,source,msf,addr,x,y,io) =
   (let m = VAL2 msf in
    ((m = 0) =>  ram |
     STORE21
     (WORD21(V(CONS io(BITS20(TRIM32TO20(OFFSET(msf,addr,x,y)))))))
     source
     ram))
|- NILM(dsf,csf,fsf) =
   (let df = VAL3 dsf in
    let cf = VAL1 csf in
    let ff = VAL4 fsf in (cf = 0) AND ((NOT((df = 7) OR (df = 6))) AND (ff = 12)))
|- OUTPUT(dsf,csf) = (let df = VAL3 dsf in let cf = VAL1 csf in (cf = 0) AND (df = 6))
|- INPUT(dsf,csf,fsf) =
   (let df = VAL3 dsf in
    let cf = VAL1 csf in
    let ff = VAL4 fsf in (cf = 0) AND ((NOT((df = 7) OR (df = 6))) AND (ff = 2)))
```

12.1.3   The High Level State Transition Function

```
|- NEXT(ram,p,a,x,y,b,stop) =
   (let instbits = BITS32(INSTFETCH(ram,p)) in
    let newp = TRIM32TO20(INCP32 p) in
    let rsf = WORD2(V(SEG(30,31)instbits)) in
    let msf = WORD2(V(SEG(28,29)instbits)) in
    let dsf = WORD3(V(SEG(25,27)instbits)) in
    let csf = WORD1(V(SEG(24,24)instbits)) in
    let fsf = WORD4(V(SEG(20,23)instbits)) in
    let addr = WORD20(V(SEG(0,19)instbits)) in
    let df = VAL3 dsf in
    let cf = VAL1 csf in
    let ff = VAL4 fsf in
    let comp = cf = 1 in
    let call = (cf = 0) AND (ff = 1) in
    let output = OUTPUT(dsf,csf) in
    let input = INPUT(dsf,csf,fsf) in
    let io = output OR input in
    let writeop = WRITE(dsf,csf) in
    let skip = NOOP(dsf,csf,b) in
    let noinc = INVALID(INCP32 p) in
    let illegaladdr = (NOT(NILM(dsf,csf,fsf))) AND
                      ((INVALID(OFFSET(msf,addr,x,y))) AND
                       (NOT skip)) in
    let illegalcl = ILLEGALCALL(dsf,csf,fsf) in
    let illegalsp = SPAREFUNC(dsf,csf,fsf) in
    let illegalonp = ILLEGALPDEST(dsf,csf,fsf) in
```

```
let illegalwr = ILLEGALWRITE(dsf,csf,msf) in
let source = REG(rsf,a,x,y,newp) in
(stop => (ram,p,a,x,y,b,T) |
 ((noinc OR illegaladdr) OR ((illegalcl OR illegalsp) OR (illegalonp OR illegalwr)) =>
  (ram,newp,a,x,y,b,T) |
  (comp => (ram,newp,a,x,y,COMPARE(fsf,source,MEMREAD(ram,msf,addr,x,y,io,F),b),F) |
   (writeop => (MEMWRITE(ram,source,msf,addr,x,y,io),newp,a,x,y,b,F) |
    (skip => (ram,newp,a,x,y,b,F) |
    let m = MEMREAD(ram,msf,addr,x,y,io,NILM(dsf,csf,fsf)) in
    let aluout = ALU(fsf,msf,dsf,source,m,b) in
    ((df = 0) => (ram,newp,VALUE aluout,x,y,BVAL aluout,SVAL aluout) |
     ((df = 1) => (ram,newp,a,VALUE aluout,y,BVAL aluout,SVAL aluout) |
      ((df = 2) => (ram,newp,a,x,VALUE aluout,BVAL aluout,SVAL aluout) |
       (call => (ram,TRIM32TO2O(VALUE aluout),a,x,INCP32 p,BVAL aluout,SVAL aluout) |
        (ram,TRIM32TO2O(VALUE aluout),a,x,y,BVAL aluout,SVAL aluout))))))))))
```

12.2   The Block Definitions

Both the types and the definitions are grouped according to the eleven blocks. The types are as follows:

12.2.1   Minor Block Types

```
INCWORD3:word3->word3
NEXTMINOR:bool#bool#bool#bool#word3->word3
```

12.2.2   Major Block Types

```
ILLEGAL_MAJOR:word4->bool
INTRESET:bool#word4->bool
NEXT_MAJOR:bool#bool#bool#word4#word4->word4
FIND_MAJOR:bool#bool#bool#bool#bool#bool#word4#word4->word4
MAJORLOGIC:bool#bool#bool#bool#bool#bool#word4#word4->word4#bool
```

12.2.3   Timeout Block Types

```
NEXT_COUNT:word6#bool#bool->word6
FIND_TIMEOUT:word6#bool->bool
TIMEOUT:word6#bool#bool#bool->word6#bool
```

12.2.4   Timing Block Types

```
FIND_OPTYPE:word4#bool->num
FIND_STA:num#word3->bool
FIND_HALT:num#word3#bool#bool->bool
FIND_STB:num#word3#bool->bool
FIND_RS:num#word3->bool
FIND_NM:num#word3->bool
TIMING:word4#word3#bool#bool#bool#bool->bool#bool#bool#bool#bool
```

12.2.5   BandStop Block Types

```
STOP_LOGIC:word3#bool#bool#bool#bool#bool#bool#bool#bool->bool
B_COMPARE:word4#bool#bool#bool#bool#bool#bool#bool->bool
B_LOGIC:word3#word4#bool#bool#bool#bool#bool#bool#bool#bool#bool->bool
BANDSTOP:word3#word3#word4#bool#bool#word9#bool#bool->bool#bool
```

12.2.6   Decoder Block Types

```
OSEL:word2->word2
DSFPRIM:word3->word7
FIND_CALL:bool#word4->bool
IO_READ:word4->word4
COND_INDEX:word4#bool#word3#bool->word4
ILLEGAL_OP:word4#word3->bool
COND_JUMP:word4#word2->word4
COND_FETCH:word4#bool#word3#word2#bool#bool->word4
FIND_COND:word4#bool#word3#word4#bool#bool#word4->word4
DECODE_PERFORM:word4#word3#word2#word2#bool->word3#word3#word2#bool#word7#word7
DECODE_MAJOR:word4#bool#word3#word2#word2#bool#word4->word3#word3#word2#bool #word7#word7
GET_BM:word3#word3#word2#bool#word7#word7->word3
GET_SM:word3#word3#word2#bool#word7#word7->word3
GET_RS:word3#word3#word2#bool#word7#word7->word2
GET_CI:word3#word3#word2#bool#word7#word7->bool
```

```
GET_AL:word3#word3#word2#bool#word7#word7->word7
GET_CL:word3#word3#word2#bool#word7#word7->word7
INST_DECODER:word12#word4#bool#bool->word3#word3#word2#bool#word7#bool#word7#bool#word4#bool
```

12.2.7 ALU Block Types

```
ADD32:word32#word32#bool->word33
ADD32BIT:word32#word32#bool->word32#bool
FIND_SR:bool#bool#bool->bool
BITOP:word32#word32#bool#bool#bool#word7->word32#bool
GET_RM31:bool#word7->bool
GET_AOUT:word32#bool->word32
GET_COUT:word32#bool->bool
ALU:word32#word32#bool#bool#word7->word32#word9
```

12.2.8 Datareg Block Types

```
CLOCK_REGA:word32#word32#bool#bool->word32
CLOCK_REGX:word32#word32#bool#bool->word32
CLOCK_REGY:word32#word32#bool#bool->word32
CLOCK_REGP:word20#word20#bool#bool->word20
CLOCK_REGT:word20#word32#word32#bool#bool#bool#bool->word32
CLOCK_INST:word32#word12#bool#bool->word12
CLOCK_ADDR:word20#word32#word20#bool#bool->word20
REGISTERS:word32#word32#word7#bool#bool#bool#word32#word32#word32#word20#word32#word20#word12->
          word32#word32#word32#word20#word32#word20#word12
REGSELECT:word32#word32#word32#word20#word2->word32
```

12.2.9 FSelect Block Types

```
FSELECT:word12->word4
```

12.2.10 External Block Types

```
FIND_EDB:word32#word4->word32
FIND_DO:word32#bool->word32
EXTERNAL:word32#word20#bool#bool#word4#word32#bool#bool#bool#bool->
         word32#bool#bool#bool#bool#word32#word20#bool#word4#bool#bool#bool#bool#bool#bool#bool
```

12.2.11 Memory Block Types

```
MEMORY:mem21_32#word32#word20#bool#bool#bool->mem21_32#word32
```

The block definitions are as follows:

12.2.12 Minor Block

```
|- INCWORD3 w = (let valw = VAL3 w in ((valw = 7) => WORD3 0 | WORD3(valw + 1)))
|- NEXTMINOR(nextmnbar,advance,reset,intresetbar,minor) =
   (let clear = reset OR ((NOT intresetbar) OR (advance AND (NOT nextmnbar))) in
   (clear => WORD3 0 | (advance => INCWORD3 minor | minor)))
```

12.2.13 Major Block

```
|- ILLEGAL_MAJOR major =
   (let majorval = VAL4 major in
    (majorval = 5) OR
    ((majorval = 8) OR
     ((majorval = 9) OR
      ((majorval = 12) OR
       ((majorval = 13) OR ((majorval = 14) OR (majorval = 15)))))))
|- INTRESET(timeoutbar,major) = (ILLEGAL_MAJOR major) OR (NOT timeoutbar)
|- NEXT_MAJOR(stop,call,cond,major) =
   (let majorval = VAL4 major in
    ((majorval = 0) => cond |
     ((majorval = 1) => cond |
      ((majorval = 2) => WORD4 1 |
       ((majorval = 3) => WORD4 4 |
        ((majorval = 4) => (stop => WORD4 8 | WORD4 1) |
         ((majorval = 6) => WORD4 4 |
          ((majorval = 7) => (call => WORD4 3 | WORD4 4) |
           ((majorval = 10) => WORD4 1 |
            ((majorval = 11) => WORD4 1 | ARB)))))))))))
|- FIND_MAJOR(reset,intreset,advance,nextmain,stop,call,cond,major) =
   (reset => WORD4 2 |
```

```
      (intreset => WORD4 8 |
      (advance AND nextmain => NEXT_MAJOR(stop,call,cond,major) | major)))
|- MAJORLOGIC
   (stop,call,timeoutbar,nextmnbar,advance,reset,cond,major) =
   (let intreset = INTRESET(timeoutbar,major) in
    let nextmajor = FIND_MAJOR(reset,intreset,advance,NOT nextmnbar,stop,call,cond,major) in
    nextmajor,NOT intreset)
```

12.2.14 Timeout Block

```
|- NEXT_COUNT(count,reset,strobe) =
   (let countval = VAL6 count in
    (reset OR (NOT strobe) => WORD6 0 | ((countval = 63) => WORD6 0 | WORD6(countval + 1))))
|- FIND_TIMEOUT(count,error) = (count = #111111) OR error
|- TIMEOUT(count,reset,error,strobe) =
   (let nextcount = NEXT_COUNT(count,reset,strobe) in
    let timeoutbar = NOT(FIND_TIMEOUT(count,error)) in nextcount,timeoutbar)
```

12.2.15 Timing Block

```
|- FIND_OPTYPE(major,carryused) =
   (let majorval = VAL4 major in
    ((majorval = 0) => 0 |
     ((majorval = 1) => 3 |
      ((majorval = 2) => 2 |
       ((majorval = 3) => 2 |
        ((majorval = 4) => (carryused => 0 | 1) |
         ((majorval = 6) => 4 |
          ((majorval = 7) => 4 |
           ((majorval = 8) => 5 |
            ((majorval = 10) => 4 | ((majorval = 11) => 4 | ARB)))))))))))
|- FIND_STA(optype,minor) =
   ((optype = 0) => F |
    ((optype = 1) => F |
     ((optype = 2) => F |
      ((optype = 3) => (minor = #001) |
       ((optype = 4) => (minor = #001) | ((optype = 5) => F | ARB))))))
|- FIND_HALT(optype,minor,pause,reply) =
   (let wait = NOT reply in
    ((optype = 0) => F |
     ((optype = 1) => F |
      ((optype = 2) => F |
       ((optype = 3) =>
        ((minor = #001) AND pause) OR ((minor = #011) AND wait) |
         ((optype = 4) => (minor = #011) AND wait |
          ((optype = 5) => F | ARB)))))))
|- FIND_STB(optype,minor) =
   ((optype = 0) => F |
    ((optype = 1) => F |
     ((optype = 2) => F |
      ((optype = 3) =>
       (minor = #010) OR ((minor = #011) OR (minor = #100)) |
        ((optype = 4) => (minor = #010) OR (minor = #011) |
         ((optype = 5) => F | ARB))))))
|- FIND_RS(optype,minor) =
   ((optype = 0) => (minor = #100) |
    ((optype = 1) => (minor = #011) |
     ((optype = 2) => (minor = #011) |
      ((optype = 3) => (minor = #100) |
       ((optype = 4) => (minor = #011) | ((optype = 5) => F | ARB))))))
|- FIND_NM(optype,minor) =
   ((optype = 0) => (minor = #101) |
    ((optype = 1) => (minor = #100) |
     ((optype = 2) => (minor = #011) |
      ((optype = 3) => (minor = #101) |
       ((optype = 4) => (minor = #011) | ((optype = 5) => F | ARB))))))
|- TIMING(major,minor,pause,reply,carryused) =
   (let optype = FIND_OPTYPE(major,carryused) in
    let advance = NOT(FIND_HALT(optype,minor,pause,reply)) in
    let nextmnbar = NOT(FIND_NM(optype,minor)) in
    let regstrobe = FIND_RS(optype,minor) in
    let strobe = FIND_STB(optype,minor) in
    let strobeaddr = FIND_STA(optype,minor) in advance,nextmnbar,regstrobe,strobe,strobeaddr)
```

12.2.16 BandStop Block

```
|- STOP_LOGIC
   (stmplx,regstrobe,clkenbp,nztop12,rm31,r31,r30,aout31,stop) =
   (let stopsel = VAL3 stmplx in
    let illegal_jump = clkenbp AND nztop12 in
    let overflow = (r31 = rm31) AND (NOT(r31 = aout31)) in
    (NOT regstrobe => stop |
     ((stopsel = 0) => F |
      ((stopsel = 1) => illegal_jump |
```

```
                     ((stopsel = 2) => NOT(r31 = r30) |
                     ((stopsel = 3) => illegal_jump OR overflow |
                      ((stopsel = 4) => T |
                       ((stopsel = 5) => nztop12 |
                        ((stopsel = 6) => T | ((stopsel = 7) => T | ARB)))))))))
|- B_COMPARE
   (fsf,nzbot16,zmid4,nztop12,coutbar,rm31,r31,aout31,bflag) =
   (let ff = VAL4 fsf in
   let reqm = (NOT nzbot16) AND (zmid4 AND (NOT nztop12)) in
   let rltm = ((NOT(r31 = aout31)) AND rm31) OR (r31 AND aout31) in
   let rgtm = NOT(rltm OR reqm) in
   ((ff = 0) => rltm |
    ((ff = 1) => NOT rltm |
     ((ff = 2) => reqm |
      ((ff = 3) => NOT reqm |
       ((ff = 4) => NOT rgtm |
        ((ff = 5) => rgtm |
         ((ff = 6) => coutbar |
          ((ff = 7) => NOT coutbar |
           ((ff = 8) => bflag OR rltm |
            ((ff = 9) => bflag OR (NOT rltm) |
             ((ff = 10) => bflag OR reqm |
              ((ff = 11) => bflag OR (NOT reqm) |
               ((ff = 12) => bflag OR (NOT rgtm) |
                ((ff = 13) => bflag OR rgtm |
                 ((ff = 14) => bflag OR coutbar |
                  ((ff = 15) => bflag OR (NOT coutbar) | ARB)))))))))))))))))
|- B_LOGIC
   (bmplx,fsf,regstrobe,nzbot16,zmid4,nztop12,coutbar,rm31,r31,r0,aout31,bflag) =
   (let bsel = VAL3 bmplx in
   (NOT regstrobe => bflag |
    ((bsel = 0) => bflag |
     ((bsel = 1) => r0 |
      ((bsel = 2) => r31 |
       ((bsel = 3) => NOT coutbar |
        ((bsel = 4) => coutbar |
         ((bsel = 5) => F |
          ((bsel = 6) =>
            B_COMPARE
            (fsf,nzbot16,zmid4,nztop12,coutbar,rm31,r31,aout31,bflag) |
           ((bsel = 7) => bflag | ARB)))))))))
|- BANDSTOP
   (stmplx,bmplx,fsf,regstrobe,clkenbp,conditions,stop,bflag) =
   (let nzbot16 = EL 0(BITS9 conditions) in
   let zmid4 = EL 1(BITS9 conditions) in
   let nztop12 = EL 2(BITS9 conditions) in
   let coutbar = EL 3(BITS9 conditions) in
   let rm31 = EL 4(BITS9 conditions) in
   let r31 = EL 5(BITS9 conditions) in
   let r30 = EL 6(BITS9 conditions) in
   let r0 = EL 7(BITS9 conditions) in
   let aout31 = EL 8(BITS9 conditions) in
   let nextstop = STOP_LOGIC(stmplx,regstrobe,clkenbp,nztop12,rm31,r31,r30,aout31,stop) in
   let nextbflag = B_LOGIC
                     (bmplx,fsf,regstrobe,nzbot16,zmid4,nztop12,coutbar,rm31,r31,r0,
                      aout31,bflag) in
   nextstop,nextbflag)
```

12.2.17 Decoder Block

```
|- OSEL msf =
   (let msfval = VAL2 msf in ((msfval = 2) => #01 | ((msfval = 3) => #10 | ARB)))
|- DSFPRIM dsf =
   (let dsfval = VAL3 dsf in
   ((dsfval = 0) => #0111110 |
    ((dsfval = 1) => #0111011 |
     ((dsfval = 2) => #0110111 |
      ((dsfval = 3) => #0111101 |
       ((dsfval = 4) => #0111101 |
        ((dsfval = 5) => #0111101 | ARB)))))))
|- FIND_CALL(csf,fsf) = (NOT csf) AND (fsf = #0001)
|- IO_READ fsf = ((fsf = #0010) => WORD4 6 | WORD4 7)
|- COND_INDEX(fsf,csf,dsf,stop) =
   (let io_read = IO_READ fsf in
   let dsfval = VAL3 dsf in
   (stop => WORD4 8 |
    (csf => WORD4 7 |
     ((dsfval = 7) => WORD4 10 |
      ((dsfval = 6) => WORD4 11 | io_read)))))
|- ILLEGAL_OP(fsf,dsf) =
   (let ff = VAL4 fsf in
   let df = VAL3 dsf in
   let pdest = (df = 3) OR ((df = 4) OR (df = 5)) in
   let legal_on_p = (ff = 1) OR ((ff = 3) OR ((ff = 5) OR (ff = 7))) in
   let illegal_not_p = (ff = 1) OR ((ff = 13) OR ((ff = 14) OR (ff = 15))) in
   ((NOT pdest) AND illegal_not_p) OR (pdest AND (NOT legal_on_p)))
|- COND_JUMP(fsf,msf) =
   (let mf = VAL2 msf in
```

```
            let shift = fsf = #1100 in
            let call = fsf = #0001 in
            let io = fsf = #0010 in
            (shift =>  WORD4 4 |
              ((mf = 0) => (call => WORD4 3 | WORD4 4) |
              ((mf = 1) => (io => WORD4 6 | WORD4 7) | WORD4 0))))
|- COND_FETCH(fsf,csf,dsf,msf,bflag,stop) =
    (let cond_jump = COND_JUMP(fsf,msf) in
      let illegal_op = ILLEGAL_OP(fsf,dsf) in
      let mf = VAL2 msf in
      let df = VAL3 dsf in
      (stop => WORD4 8 |
        (csf =>
          ((mf = 0) => WORD4 4 | ((mf = 1) => WORD4 7 | WORD4 0)) |
          ((df = 7) => ((mf = 0) => WORD4 8 | ((mf = 1) => WORD4 10 | WORD4 0)) |
          ((df = 6) => ((mf = 0) => WORD4 8 | ((mf = 1) => WORD4 11 | WORD4 0)) |
            (illegal_op => WORD4 8 |
              ((df = 5) => (bflag => WORD4 1 | cond_jump) |
              ((df = 4) => (NOT bflag => WORD4 1 | cond_jump) | cond_jump)))))))))
|- FIND_COND(fsf,csf,dsf,msf,bflag,stop,major) =
    ((major = #0000) => COND_INDEX(fsf,csf,dsf,stop) |
      ((major = #0001) => COND_FETCH(fsf,csf,dsf,msf,bflag,stop) | ARB))
|- DECODE_PERFORM(fsf,dsf,msf,rsf,bflag) =
    (let bmp_b = #000 in
      let bmp_lsbr = #001 in
      let bmp_msbr = #010 in
      let bmp_carry = #011 in
      let bmp_borrw = #100 in
      let bmp_0 = #101 in
      let bmp_comp = #110 in
      let smp_0 = #000 in
      let smp_nsf = #001 in
      let smp_shfov = #010 in
      let smp_ovf = #011 in
      let smp_1 = #100 in
      let smp_tail = #101 in
      let rsel_x = #00 in
      let cin_t = T in
      let cin_f = F in
      let cin_x = F in
      let alu_and = #1101001 in
      let alu_rmb = #1111001 in
      let alu_zero = #1100000 in
      let alu_m = #1101000 in
      let alu_comm = #1111000 in
      let alu_r = #1110001 in
      let alu_sr = #1110010 in
      let alu_xor = #0111001 in
      let alu_nor = #0111100 in
      let alu_add = #0001001 in
      let alu_sub = #0011001 in
      let alu_incr = #0000001 in
      let alu_sl = #0000101 in
      let fsfval = VAL4 fsf in
      let msfval = VAL2 msf in
      let dsfprim = DSFPRIM dsf in
      ((fsfval = 0) => (bmp_b,smp_0,rsel_x,cin_x,alu_comm,dsfprim) |
        ((fsfval = 1) =>
          (bmp_b,smp_nsf,rsel_x,cin_x,alu_m,dsfprim) |
          ((fsfval = 2) => (bmp_b,smp_0,rsel_x,cin_x,alu_m,dsfprim) |
          ((fsfval = 3) => (bmp_b,smp_nsf,rsel_x,cin_x,alu_m,dsfprim) |
          ((fsfval = 4) => (bmp_carry,smp_0,rsf,cin_f,alu_add,dsfprim) |
          ((fsfval = 5) => (bmp_b,smp_ovf,rsf,cin_f,alu_add,dsfprim) |
          ((fsfval = 6) => (bmp_borrw,smp_0,rsf,cin_t,alu_sub,dsfprim) |
          ((fsfval = 7) => (bmp_b,smp_ovf,rsf,cin_t,alu_sub,dsfprim) |
          ((fsfval = 8) => (bmp_b,smp_0,rsf,cin_x,alu_xor,dsfprim) |
          ((fsfval = 9) => (bmp_b,smp_0,rsf,cin_x,alu_and,dsfprim) |
          ((fsfval = 10) => (bmp_b,smp_0,rsf,cin_x,alu_nor,dsfprim) |
          ((fsfval = 11) => (bmp_b,smp_0,rsf,cin_x,alu_rmb,dsfprim) |
          ((fsfval = 12) =>
            ((msfval = 0) => (bmp_b,smp_0,rsf,cin_t,alu_sr,dsfprim) |
            ((msfval = 1) => (bmp_lsbr,smp_0,rsf,cin_f,alu_sr,dsfprim) |
            ((msfval = 2) => (bmp_b,smp_shfov,rsf,cin_f,alu_sl,dsfprim) |
            ((msfval = 3) => (bmp_msbr,smp_0,rsf,bflag,alu_sl,dsfprim) |
              ARB)))) |
                ARB)))))))))))))
|- DECODE_MAJOR(fsf,csf,dsf,msf,rsf,bflag,major) =
    (let bmp_b = #000 in
      let bmp_lsbr = #001 in
      let bmp_msbr = #010 in
      let bmp_carry = #011 in
      let bmp_borrw = #100 in
      let bmp_0 = #101 in
      let bmp_comp = #110 in
      let smp_0 = #000 in
      let smp_nsf = #001 in
      let smp_shfov = #010 in
      let smp_ovf = #011 in
      let smp_1 = #100 in
      let smp_tail = #101 in
```

```
        let reg_p = #11 in
        let rsel_x = #00 in
        let cin_t = T in
        let cin_f = F in
        let cin_x = F in
        let alu_and = #1101001 in
        let alu_rmb = #1111001 in
        let alu_zero = #1100000 in
        let alu_m = #1101000 in
        let alu_comm = #1111000 in
        let alu_r = #1110001 in
        let alu_sr = #1110010 in
        let alu_xor = #0111001 in
        let alu_nor = #0111100 in
        let alu_add = #0001001 in
        let alu_sub = #0011001 in
        let alu_incr = #0000001 in
        let alu_sl = #0000101 in
        let dest_t = #1011111 in
        let dest_y = #0110111 in
        let dest_fet = #0001101 in
        let dest_all = #0110000 in
        let dest_te = #0011111 in
        let dest_non = #0111111 in
        let majorval = VAL4 major in
        let osel = OSEL msf in
        ((majorval = 0) => (bmp_b,smp_tail,osel,cin_f,alu_add,dest_t) |
         ((majorval = 1) => (bmp_b,smp_nsf,reg_p,cin_t,alu_incr,dest_fet) |
          ((majorval = 2) => (bmp_0,smp_0,rsel_x,cin_x,alu_zero,dest_all) |
           ((majorval = 3) => (bmp_b,smp_0,reg_p,cin_x,alu_r,dest_y) |
            ((majorval = 6) => (bmp_b,smp_0,rsel_x,cin_x,alu_zero,dest_te) |
             ((majorval = 7) => (bmp_b,smp_0,rsel_x,cin_x,alu_zero,dest_te) |
              ((majorval = 8) => (bmp_b,smp_0,rsel_x,cin_x,alu_zero,dest_non) |
               ((majorval = 10) => (bmp_b,smp_0,rsf,cin_x,alu_zero,dest_non) |
                ((majorval = 11) => (bmp_0,smp_0,rsf,cin_x,alu_zero,dest_non) |
                 ((majorval = 4) => (csf =>
                                   (bmp_comp,smp_0,rsf,cin_t,alu_sub,dest_non) |
                                   DECODE_PERFORM(fsf,dsf,msf,rsf,bflag)) | ARB))))))))))
|- GET_BM(bmplx,stmplx,rsel,cin,alucon,clkenb) = bmplx
|- GET_SM(bmplx,stmplx,rsel,cin,alucon,clkenb) = stmplx
|- GET_RS(bmplx,stmplx,rsel,cin,alucon,clkenb) = rsel
|- GET_CI(bmplx,stmplx,rsel,cin,alucon,clkenb) = cin
|- GET_AL(bmplx,stmplx,rsel,cin,alucon,clkenb) = alucon
|- GET_CL(bmplx,stmplx,rsel,cin,alucon,clkenb) = clkenb
|- INST_DECODER(inst,major,bflag,stop) =
    (let instlist = BITS12 inst in
     let fsf = WORD4(V(SEG(0,3)instlist)) in
     let csf = EL 4 instlist in
     let dsf = WORD3(V(SEG(5,7)instlist)) in
     let msf = WORD2(V(SEG(8,9)instlist)) in
     let rsf = WORD2(V(SEG(10,11)instlist)) in
     let decode_major = DECODE_MAJOR(fsf,csf,dsf,msf,rsf,bflag,major) in
     let bmplx = GET_BM decode_major in
     let stmplx = GET_SM decode_major in
     let rsel = GET_RS decode_major in
     let cin = GET_CI decode_major in
     let alucon = GET_AL decode_major in
     let clkenb = GET_CL decode_major in
     let carryused = NOT(EL 5(BITS7 alucon)) in
     let clkenbp = NOT(EL 1(BITS7 clkenb)) in
     let cond = FIND_COND(fsf,csf,dsf,msf,bflag,stop,major) in
     let call = FIND_CALL(csf,fsf) in
     bmplx,stmplx,rsel,cin,alucon,carryused,clkenb,clkenbp,cond,call)
```

12.2.18 ALU Block

```
|- ADD32(r,t,cin) = (cin => WORD33((VAL32 r) + ((VAL32 t) + 1)) | WORD33((VAL32 r) + (VAL32 t)))
|- ADD32BIT(r,t,cin) =
    (let sum33 = ADD32(r,t,cin) in
     let aout = WORD32(V(SEG(0,31)(BITS33 sum33))) in
     let carry = EL 32(BITS33 sum33) in aout,carry)
|- FIND_SR(cin,bflag,r31) = (cin => r31 | bflag)
|- BITOP(r,t,bflag,r31,cin,op) =
    (let cout_x = F in
     let tbar = NOT32 t in
     let rbar = NOT32 r in
     let sr = FIND_SR(cin,bflag,r31) in
     let r_xor_t = (r AND32 tbar) OR32 (rbar AND32 t) in
     let shift_right = WORD32(V(CONS sr(SEG(1,31)(BITS32 r)))) in
     ((op = #1101001) => (r AND32 t,cout_x) |
      ((op = #1111001) => (r AND32 tbar,cout_x) |
       ((op = #1100000) => (WORD32 0,cout_x) |
        ((op = #1101000) => (t,cout_x) |
         ((op = #1111000) => (tbar,cout_x) |
          ((op = #1110001) => (r,cout_x) |
           ((op = #1110010) => (shift_right,cout_x) |
            ((op = #0111001) => (r_xor_t,cout_x) |
             ((op = #0111100) => (NOT32(r OR32 t),cout_x) |
```

```
                       ((op = #0001101) => ADD32BIT(r,r,cin) |
                        ((op = #0001001) => ADD32BIT(r,t,cin) |
                         ((op = #0011001) => ADD32BIT(r,tbar,cin) |
                          ((op = #0000001) => ADD32BIT(r,WORD32 0,cin) | ARB))))))))))))))
|- GET_RM31(t31,alucon) = ((alucon = #0011001) => NOT t31 |
                          ((alucon = #0001001) => t31 | F))
|- GET_AOUT(aout,cout) = aout
|- GET_COUT(aout,cout) = cout
|- ALU(rbar,treg,cin,bflag,alucon) =
   (let r = NOT32 rbar in
    let t31 = EL 31(BITS32 treg) in
    let rm31 = GET_RM31(t31,alucon) in
    let r0 = EL 0(BITS32 r) in
    let r30 = EL 30(BITS32 r) in
    let r31 = EL 31(BITS32 r) in
    let a_c = BITOP(r,treg,bflag,r31,cin,alucon) in
    let aout = GET_AOUT a_c in
    let aoutbar = NOT32 aout in
    let coutbar = NOT(GET_COUT a_c) in
    let aout31 = EL 31(BITS32 aout) in
    let nzbot16 = NOT(V(SEG(0,15)(BITS32 aout)) = 0) in
    let zmid4 = V(SEG(16,19)(BITS32 aout)) = 0 in
    let nztop12 = NOT(V(SEG(20,31)(BITS32 aout)) = 0) in
    let conditions = WORD9(V[aout31;r0;r30;r31;rm31;coutbar;nztop12;zmid4;nzbot16]) in
    aoutbar,conditions)
```

12.2.19   Datareg Block

```
|- CLOCK_REGA(aoutbar,areg,clkenba,regstrobe) = (clkenba AND regstrobe => NOT32 aoutbar | areg)
|- CLOCK_REGX(aoutbar,xreg,clkenbx,regstrobe) = (clkenbx AND regstrobe => NOT32 aoutbar | xreg)
|- CLOCK_REGY(aoutbar,yreg,clkenby,regstrobe) = (clkenby AND regstrobe => NOT32 aoutbar | yreg)
|- CLOCK_REGP(aoutbar_ls20,preg,clkenbp,regstrobe) =
   (clkenbp AND regstrobe => NOT20 aoutbar_ls20 | preg)
|- CLOCK_REGT
   (aoutbar_ls20,extdata,treg,clkenbt,regstrobe,tmplxcon,clkenbinst) =
   (let extdata_ls20 = WORD32(V(SEG(0,19)(BITS32 extdata))) in
    let aout_tail = WORD32(VAL20(NOT20 aoutbar_ls20)) in
    (NOT(clkenbt AND regstrobe) => treg |
     ((NOT tmplxcon) AND (NOT clkenbinst) => extdata |
      (tmplxcon AND (NOT clkenbinst) => aout_tail |
       ((NOT tmplxcon) AND clkenbinst => extdata_ls20 | ARB)))))
|- CLOCK_INST(extdata,inst,clkenbinst,regstrobe) =
   (let extdata_ms12 = WORD12(V(SEG(20,31)(BITS32 extdata))) in
    (clkenbinst AND regstrobe => extdata_ms12 | inst))
|- CLOCK_ADDR(preg,treg,addr,clkenbinst,strobeaddr) =
   (NOT strobeaddr => addr |
    (clkenbinst => preg | WORD20(V(SEG(0,19)(BITS32 treg)))))
|- REGISTERS
   (extdatabar,aoutbar,clkenb,regstrobe,strobeaddr,areg,xreg,yreg,preg,treg,addr,inst) =
   (let extdata = NOT32 extdatabar in
    let aoutbar_ls20 = WORD20(V(SEG(0,19)(BITS32 aoutbar))) in
    let clkenba = NOT(EL 0(BITS7 clkenb)) in
    let clkenbp = NOT(EL 1(BITS7 clkenb)) in
    let clkenbx = NOT(EL 2(BITS7 clkenb)) in
    let clkenby = NOT(EL 3(BITS7 clkenb)) in
    let clkenbinst = NOT(EL 4(BITS7 clkenb)) in
    let clkenbt = NOT(EL 5(BITS7 clkenb)) in
    let tmplxcon = EL 6(BITS7 clkenb) in
    let new_areg = CLOCK_REGA(aoutbar,areg,clkenba,regstrobe) in
    let new_xreg = CLOCK_REGX(aoutbar,xreg,clkenbx,regstrobe) in
    let new_yreg = CLOCK_REGY(aoutbar,yreg,clkenby,regstrobe) in
    let new_preg = CLOCK_REGP(aoutbar_ls20,preg,clkenbp,regstrobe) in
    let new_inst = CLOCK_INST(extdata,inst,clkenbinst,regstrobe) in
    let new_treg = CLOCK_REGT
                   (aoutbar_ls20,extdata,treg,clkenbt,regstrobe,tmplxcon,clkenbinst) in
    let new_addr = CLOCK_ADDR(preg,treg,addr,clkenbinst,strobeaddr) in
    new_areg,new_xreg,new_yreg,new_preg,new_treg,new_addr,new_inst)
|- REGSELECT(areg,xreg,yreg,preg,rsel) =
   (let rf = VAL2 rsel in
    ((rf = 0) => NOT32 areg |
     ((rf = 1) => NOT32 xreg |
      ((rf = 2) => NOT32 yreg |
       ((rf = 3) => NOT32(WORD32(VAL20 preg)) | ARB)))))
```

12.2.20   FSelect Block

```
|- FSELECT inst = WORD4(V(SEG(0,3)(BITS12 inst)))
```

12.2.21   External Block

```
|- FIND_EDB(e_data_in,major) =
   (let unused_input = WORD32 0 in
    let majorval = VAL4 major in
    ((majorval = 1) => NOT32 e_data_in |
```

```
      ((majorval = 6) => NOT32 e_data_in |
        ((majorval = 7) => NOT32 e_data_in | unused_input))))
|- FIND_DO(rbar,writegate) = (let tri_state = WORD32 0 in (writegate => NOT32 rbar | tri_state))
|- EXTERNAL
    (rbar,addr,bflag,strobe,major,e_data_in,e_resetbar,e_errorbar,e_stepbar,e_reply) =
    (let writegate = (major = #1010) OR (major = #1011) in
     FIND_EDB(e_data_in,major),NOT e_resetbar,NOT e_errorbar,NOT e_stepbar,e_reply,
     FIND_DO(rbar,writegate),addr,bflag,major,NOT strobe,(major = #1000),(major = #0100),
     NOT(major = #0001),NOT((major = #0110) OR (major = #1011)),
     NOT((major = #1010) OR (major = #1011)))
```

12.2.22   Memory Block

```
|- MEMORY(ram,e_data_out,e_address,e_iobar,e_writebar,e_strobebar) =
    (let address = WORD21(V(CONS(NOT e_iobar)(BITS20 e_address))) in
     (e_strobebar => (ram,e_data_out) |
      (e_writebar => (ram,FETCH21 ram address) |
       (STORE21 address e_data_out ram,e_data_out))))
```

Formal Verification of the Sobel Image Processing Chip*

Paliath Narendran

General Electric Company

Corporate Research and Development

Schenectady, NY 12345

Jonathan Stillman

State University of New York at Albany

Albany, NY 12222

Abstract

We describe an approach to hardware verification in the context of our recent success in formally verifying the description of an image processing chip currently under development at Research Triangle Institute. We demonstrate that our approach, which uses an implementation of an equational approach to theorem proving developed by Kapur and Narendran, can be a viable alternative to simulation. In particular, we are able to take advantage of the "recursive" nature of many circuits, such as n-bit adders, and our techniques allow verification of sequential circuits. To the best of our knowledge this is the first time a complex sequential circuit which was not designed with verification specifically in mind has been verified. Finally, we describe the discovery of several design errors in the circuit description, detected during the verification attempt (the actual verification could only take place once these errors were removed), and discuss directions that future work will take.

1 Introduction

Formal verification of hardware involves using theorem-proving techniques to verify that a stated behavioral definition of a circuit is a logical consequence of the structural description of the circuit, i.e., proving that the structure of the circuit forces it to behave as stated. There has been a great deal of recent interest in formal verification as an alternative to exhaustive simulation, since simulation is often not a practical approach for large circuits. Related work can be found in [Ba 84, Go 85], among others. It is hoped that formal verification will be a useful tool in many of the situations where simulation is impractical.

Our approach utilizes an equational approach to theorem-proving that was developed by Kapur and Narendran, as implemented in the Rewrite Rule Laboratory (RRL) developed under NSF grant No. CCR-8408461. Both the structural and

---

*Work partially supported by Contract F33615-85-C-1862, AFWAL, Wright-Patterson Air Force Base, Ohio.

behavioral definitions are specified in a first-order predicate calculus with equality. The approach is refutational in nature: the statements describing the structure are assumed to hold, the behavior is assumed not to hold, and an attempt is made to reach a contradiction. Previous work which is closely related is described in [NaSt 87] and, for combinational circuits, in [ChPC 87]. In addition to the circuit described herein, we have verified a number of the leaf cells of a CMOS bit-serial compiler currently under development at GE CRD. This work will be described in an upcoming report.

Our recent work on formal hardware verification has been part of an effort to develop a designer's workstation in which the designer uses graphical tools to build circuits hierarchically; the resulting information can be used both to generate VHDL programs and to generate the structural information for formal verification of the design. A more complete description of the workstation (IVW) and the current level of integration of verification tools into it can be found in [NaSt 87]. To the best of our knowledge, this is the first time that formal verification tools have been integrated into a design environment.

The Sobel chip has approximately 10,000 transistors, and implements the Sobel edge detection algorithm [VaBK 87,Pr 78], which is commonly used in the image segmentation phase of military image processing systems. The structure of the design was completely specified in VHDL. We were not provided with a formal behavioral description of the chip, but we were able to extract the necessary information from [VaBK 87]. The overall effort took about two man-months.

2 Theoretical Background

The basis of our approach is the Kapur-Narendran method for theorem-proving for first-order predicate calculus. We briefly describe the method, and give a simple example of how it may be applied to hardware verification; for a detailed description of the Kapur-Narendran method, one should consult [KaNa 85]. The method is closely related to methods for term-rewriting; the interested reader is referred to [Bu 83] for an overview of this topic, and to [HuOp 80,KnBe 70] for more detailed accounts. Basically, the Kapur-Narendran method involves taking a set of first-order formulae, performing any necessary Skolemization (see [ChLe 74]) to remove existential quantifiers, then translating the resulting formulae into an equivalent set of polynomials over a boolean ring whose operators are "exclusive-or" and "and" (sometimes denoted by $*$ and $+$, respectively), in which the atoms appearing in the original formulae make up the indeterminates of the polynomial equations. The original set of formulae is satisfiable if and only if the resulting system of polynomials has a solution. It is possible that a contradiction is found in the translation process itself; if not, the polynomials are oriented into rewrite rules and a modified version of the Knuth-Bendix completion procedure is performed on them using the method of critical pairs to compute new rules. One can view the set of polynomials as a basis of an ideal, and the completion procedure as computing the Gröbner basis of the original basis (see [Buc 85,MiYa 86] for details). Such a Gröbner basis has the property that every polynomial in the ideal can be rewritten to 0 using the polynomials in the basis for simplification. It can be shown that a set of first-order formulae is unsatisfiable if and only if the corresponding Gröbner basis contains 1 (where 1 stands for true and 0 stands for false). One should note, however, that if the formulae are simultaneously satisfiable, the Gröbner basis may be infinite, i.e., the completion procedure may not terminate.

To insure termination of rewriting, a partial well-founded ordering can be introduced on monomials; such an ordering is easily extended to polynomials by considering them to be multisets of monomials.

In order to orient the polynomials into rewrite rules, each polynomial can be broken up into two parts as follows: for a polynomial P, let $HD(P)$ be the set of maximal monomials (using the chosen partial ordering) in P; let $TL(P)$ be $P - HD(P)$. With the polynomial P we associate the rule $HD(P) \rightarrow TL(P)$. A polynomial Q can be rewritten using the rule associated with P if and only if there is a monomial m and a substitution θ such that $m * \theta(HD(P)) \subseteq Q$. When this occurs we obtain the polynomial Q' by replacing the occurrence of $m * \theta(HD(P))$ in Q with $m * \theta(TL(P))$, and say that $Q \rightarrow Q'$ by the rule associated with P. The above defines a *rewrite relation* $\rightarrow$ induced by a set of rules, and we denote the reflexive and transitive closure of $\rightarrow$ by $\overset{*}{\rightarrow}$. In addition to the rules obtained from the polynomials, rules are also introduced for idempotence of "and" (for every predicate $P(x_1, \ldots x_n)$, introduce the rule $P(x_1, \ldots x_n) * P(x_1, \ldots x_n) \rightarrow P(x_1, \ldots x_n)$) and nilpotence of "exclusive-or" (add the rule $1 + 1 \rightarrow 0$).

New rules are generated by computing *critical pairs*. This is done by finding, for each pair of rules, a minimal polynomial which can be rewritten by both rules. The critical pair is the pair of polynomials p_1, p_2 resulting from rewriting the minimal superposition using each of the two rules involved. The polynomial $p_1 - p_2$ associated with a critical pair is called an *S-polynomial*. If the polynomial $p_1 - p_2$ can be rewritten to 0 the critical pair is said to be *trivial*, and no new rule is introduced; if not, the resulting polynomial is oriented into a rule and added to the rule set. In general, computing superpositions is somewhat involved; in the interest of brevity, we only illustrate the case when $HD(p_1)$ and $HD(p_2)$ are both monomials. The general case is fully explored in [KaNa 85]. Let $HD(p_1) \rightarrow TL(p_1)$ and $HD(p_2) \rightarrow TL(p_2)$ be two (not necessarily distinct) rules, where $HD(p_i)$ can be partitioned into monomials $g_i f_i, i = 1, 2$, and there exists a most general unifier σ which unifies g_1 and g_2. Let $\sigma(g_1) = \sigma(g_2) = g$. Then the polynomial $p = \sigma(f_1) g \sigma(f_2) = \sigma(HD(p_1))\sigma(f_2) = \sigma(HD(p_2))\sigma(f_1)$ is a superposition of p_1 and p_2. The resulting *S-polynomial* is $\sigma(TL(p_1))\sigma(f_2) - \sigma(TL(p_2))\sigma(f_1)$.

The completion procedure simply involves repeatedly computing critical pairs until either no more rules can be superposed or the rule $1 \rightarrow 0$ is generated, signaling a contradiction (since we are using a refutational approach, this means that the original formula is valid). If the procedure halts without finding a contradiction, the polynomials in the resulting Gröbner basis can be used to find a model for the original formula; remember, however, that there is no guarantee that the procedure will terminate in all cases.

Rather than go into more technical detail, we present the following example to help illustrate the method.

3   An Example

We demonstrate a proof that the Muller C-element (see figure below) indeed functions as a consensus mechanism; in the description of the circuit the variable x denotes an arbitrary point in time and the function f denotes the next point in time. Logical gates are assumed to take 1 time unit to propagate a signal. Thus the formulae describing the structure are

1. `a1(f(x)) equ (in1(x) and in2(x))`

2. `a2(f(x)) equ (in1(x) and out(x))`

3. `a3(f(x)) equ (in2(x) and out(x))`

4. `out(f(x)) equ (a1(x) or a2(x) or a3(x))`

5. `not( all x ((in1(x) equ in2(x)) ⇒`
   `(out(f(f(x))) equ in1(x)) and ((in1(x) xor in2(x)) ⇒`
   `(out(f(f(x))) equ out(x)))))`

The proof proceeds as follows:
Input 1 produces the rule:
[1] `(in1(x) and in2(x)) → a1(f(x))`
Input 2 produces the rule:
[2] `(in1(x) and out(x)) → a2(f(x))`
Input 3 produces the rule:
[3] `(in2(x) and out(x)) → a3(f(x))`
Input 4 produces the rule:
[4] `(a1(x) and a2(x) and a3(x)) →`
`(a1(x) xor a2(x) xor a3(x) xor out(f(x))`
`xor (a1(x) and a2(x)) xor (a1(x) and a3(x)) xor (a2(x) and a3(x)))`
Input 5 reduced by Rules [1], [2], [3], produces the rule:
[5] `out(f(f(s1))) →`
`(true xor a1(f(s1)) xor a2(f(s1)) xor a3(f(s1)))`
Rule [3] superposed itself[1] , reduced by Rule [3], produces the rule:
[6] `(a3(f(x)) and in2(x)) → a3(f(x))`
Rule [3] superposed itself, reduced by Rule [3], produces the rule:
[7] `(a3(f(x)) and out(x)) → a3(f(x))`
Rule [1] superposed with Rule [3], produces the rule:
[8] `(a1(f(x)) and out(x)) → (a3(f(x)) and in1(x))`
Rule [2] superposed itself, reduced by Rule [2], produces the rule:
[9] `(a2(f(x)) and in1(x)) → a2(f(x))`
Rule [8] superposed with Rule [2], reduced by Rule [9], produces the rule:
[10] `(a2(f(x)) and a3(f(x))) →`
`(a1(f(x)) and a2(f(x)))`
Rule [7] superposed with Rule [2], produces the rule:
[11] `(a3(f(x)) and in1(x)) →`
`(a1(f(x)) and a2(f(x)))`
which simplifies [8] to:
[8] `(a1(f(x)) and out(x)) → (a1(f(x)) and a2(f(x)))`
Rule [6] superposed with Rule [1], reduced by Rule [11], produces the rule:
[12] `(a1(f(x)) and a3(f(x))) →`
`(a1(f(x)) and a2(f(x)))`
Rule [11] superposed itself, reduced by Rules [11], [4], [12], [10], produces the rule:

---

[1]A rule can superpose itself using the idempotence property

[13] out(f(f(x))) →
(a1(f(x)) xor a2(f(x)) xor a3(f(x)))
Rule [13] deleted Rule [5], produces the rule:
[14] true → false

This completes the proof, which was derived automatically using RRL; since the negation of the behavior in conjunction with the structure results in a contradiction, it follows that the behavior is correct (note that it may not completely describe the circuit's behavior; we only know that the stated behavior is forced by the circuit).

4   The Circuit Description Model

4.1   Primitive Elements

The specification of the Sobel chip was written in VHDL; the primitive elements (those with no internal components) included the following:

- registers, which were essentially modelled as simple delay elements,

- multiplexors, and

- gates.

These components were also treated as primitive in our verification of the chip.

4.2   Basic Axiomatization

Since the Sobel algorithm is numeric in nature, we needed to introduce a set of axioms to deal with numbers. These axioms are listed below with a brief explanation where necessary.

4.2.1   Axioms For Integers

The axioms introduced for the integers follow:

- a is a constant representing the number '0'.

- x + a = a + x = x; adding '0' to a number doesn't change the number.

- not(eq(s(x),x)); no number is the same as its successor.

- x + s(y) = s(x) + y = s(x + y); adding '1' to a sum is the same as adding '1' to one of the summands.

- tw(x) = x + x; axiom for doubling numbers (tw stands for "twice").

- a * x = x * a = a; any number times '0' is '0'.

- s(x) * y = (x * y) + y, and x * s(y) = (x * y) + x; x * (y + 1) is the same as (x * y) + x, and (x + 1) * y is the same as (x * y) + y.

- texp(a) = s(a), and texp(s(x)) = tw(texp(x)); texp(x) is a function representing 2^x.

- x + minus(x) = minus(x) + x = a; each number has an inverse, its minus such that the number plus its inverse equals '0'.

- minus(minus(x)) = x; the inverse of the inverse yields the original number.

- minus(a) = a; '0' is its own inverse.

- (x + y) + minus(y) = x, and (x + y) + minus(x) = y; subtracting x from x + y yields y, and vice versa.

4.2.2   Bit Strings

In order to facilitate the recursive specification of circuits we chose to model bit strings as infinite arrays of bits, of which we can see an n-bit slice at any given time. Three critical functions are:

- apply(bitstring,x) which returns the xth bit of bitstring

- intv(boolean) which returns the integer associated with a truth value, and

- intval(bitstring,x,y) which returns the integer value of the segment of length y + 1 of bitstring starting at the xth position. We define this function inductively as follows:
  intval(x,y,a) = intv(apply(x,y)); in the case of a single bit, just take the integer associated with the truth value of the bit,
  intval(x,y,s(z)) = intval(x,y,a) + tw(intval(x,s(y),z)); if the bit string is longer than a single bit, then the integer value is the sum of the integer value of the least-significant bit and twice the integer value of the remaining bits to the left. (Alternatively, one could define the inductive part as follows:

  ```
  intval(x,y,s(z)) = intval(x,y,z) +
                     (texp(s(z)) * intv(apply(x,y + s(z)))).
  ```

  The two definitions can be proved equivalent by induction.)

4.2.3   Two's Complement and Absolute Value

The axioms introduced to represent the two's complement of a binary number, and to represent the absolute value of a number are shown below:

- tcval(x,y,s(z)) = intval(x,y,z) + intv(apply(x,y + s(z))) * minus(texp(s(z))); the two's complement is obtained by taking the integer value and subtracting it from the sign bit times $2^{length+1}$.

- eq(apply(x,y + s(z)),e0) -> eq(absval(x,y,s(z)), intval(x,y,z)); if the sign bit of the number is '0', then the absolute value is the number itself.

- eq(apply(x,y + s(z)),e1) -> eq(absval(x,y,s(z)), texp(s(z)) + minus(intval(x,y,z))); if the sign bit is '1', the absolute value of the number is the two's complement of the original number.

4.2.4 Relations

Some axioms were also introduced to represent relations between numbers such as 'greater-than' ($>$) and 'greater-than-or-equal-to' ($\geq$); typical axioms follow:

- (gt(s(x),x)); $x + 1 > x$

- ((gt(x,y) and ge(y,z)) -> gt(x,z))

- (gt(x,y) equ not(ge(y,x)))

- (ge(x,y) equ (eq(x,y) or gt(x,y)))

- (gt(x,y) and ge(u,v)) -> gt(x + u, y + v)

- ((gt(x,y) and gt(x,z)) -> gt(x, dmax(y,z))); dmax stands for "maximum of."

- gt(texp(s(x)),texp(x)); $2^{x+1} > 2^x$

4.3 An Example

We demonstrate a typical input to the theorem prover with a very simple example, a circuit with n input lines which outputs *true* if and only if each of the inputs is *false*, i.e. the circuit tests whether of not the integer value of the input is zero. The input follows:

```
; axioms for integers
a + x = x
x + a = x
x + s(y) = s(x + y)
s(x) + y = s(x + y)
twice(x) = x + x
not(eq(s(x),x))
; axioms for boolean representation.
; e0 stands for 'false', e1 for 'true'.
not(eq(e0,e1))
bar(e0) = e1
bar(e1) = e0
; axioms defining the integer
; value of a bit string
intv(e0) = a
intv(e1) = s(a)
intval(x,y,a) = intv(apply(x,y))
intval(x,y,s(z)) = intval(x,y,a) + twice(intval(x,s(y),z))
eq(apply(x,y),e0) xor eq(apply(x,y),e1)
; definition of the predicate bszero
; when a single bit is considered
bszero(x,y,a) equ eq(e0,apply(x,y))
; negation of theorem (basis)
not(all x all y (bszero(x,y,a) equ (eq(intval(x,y,a),a))))
```

```
; recursive definition of bszero
bszero(x,y,s(z)) equ (eq(e0,apply(x,y)) and bszero(x,s(y),z))
; inductive hypothesis (used
; once the basis is verified)
bszero(x,y,k) equ (eq(intval(x,y,k),a))
; inductive step (negation)
not(all x all y (bszero(x,y,s(k)) equ (eq(intval(x,y,s(k)),a))))
```

Thus the proof that the predicate bszero is true if and only if each of the n input bits are *false* proceeds by verifying the basis case, then, using an inductive hypothesis, verifying the property for an arbitrary $n > 0$. In general, both the predicates defining circuits and the properties of the circuits are much more complex; the proofs of these more complex circuits currently require a fair amount of user guidance.

5 The Sobel Chip

5.1 Overview

The chip under consideration was designed at Research Triangle Institute (RTI) partly as a test case for the VHDL design methodology. It has approximately 10,000 transistors, and implements the Sobel edge detection algorithm [VaBK 87,Pr 78], which is commonly used in the image segmentation phase of military image processing systems. The chip can be broken down into two subsections (see Figure 2). The first is a windowing section which filters the image with four 3×3 windows, generating four values which measure the differences in intensity along the vertical, horizontal, left diagonal, and right diagonal. These values are passed to the second section of the chip, a magnitude and direction processor, which combines the window processor outputs to compute both gradient magnitude and gradient direction

The window processors consist of a network of adders, subtractors, and registers; while their size was fixed at 12 bits, the verification was done inductively and (except for overflow conditions) the algorithm was shown to be correct for arbitrary widths. The verification was hierarchical, first confirming that the adders indeed added, etc. then showing that under the assumption that the components worked, the implementation of the algorithm was correct.

The magnitude-and-direction processor consists of adders, comparators, multiplexors, and registers. Basically, this phase consists of computing the absolute values of the four inputs, comparing these to find the maximum and minimum, then using these two values to compute the gradient direction and gradient magnitude.

The structural specification of the chip was presented to us in the form of VHDL code; although we feel that the extraction of first-order statements from VHDL can be largely automated, the translation was done by hand in this test case.

As we have previously stated, we were not provided with a formal behavioral description of the chip, but were able to get the necessary information from the high-level statements provided in the document [VaBK 87] cited above.

5.2   Verification of the Sobel Chip

As mentioned above, the architecture of the circuit is presented as consisting of
two essentially separate components, the window processor and the magnitude-and-
direction processor. The window processor was decomposed into its 4 windows, each
of which were verified separately. The verification was hierarchical, and the basic
arithmetic operations involved were addition, subtraction and doubling (shifting the
bits leftward toward the higher bits and attaching a '0' at the right end) of bit-strings
represented in two's complement form. The adders and subtractors were relatively
straightforward to verify inductively since the ripple-carry paradigm was used. Some
complexity was introduced, however, by the fact that in general the ripple-carrying
technique does not work correctly for two's complement numbers (the sign bit is
not guaranteed to be correct). The designers of the circuit compensated for this
problem by making sure that the absolute values of the summands were always
small enough relative to the size of the adders so that the adders worked correctly.
Thus, we had to verify the following property:

> if the absolute values of two n-bit integers in two's-complement form
> are less than $2^{(n-2)}$, then adders (as well as subtractors) that use the
> ripple-carrying technique work correctly.

In our notation (see Section 4), this is expressed as the following theorem (for
addition):

```
(all x1 all x2 all z all y all u all xb all xc all zc
   ((bsadd(x1,x2,y,u,xb,xc,z) and
     fulladder(apply(x1,y + s(u)), apply(x2,y + s(u)),
             xc, apply(z,y + s(u)), zc) and
     gt(texp(u), absval(x1,y,s(u))) and
     gt(texp(u), absval(x2,y,s(u)))) =>
   eq(tcval(z,y,s(u)), tcval(x1,y,s(u)) + tcval(x2,y,s(u))))))
```

The final full adder is 'separated out' merely for convenience, since the leftmost
(highest order) bits determine the signs. The complete proof is given in the appendix. (Similar theorems were proved also for subtraction and doubling.)

We also had to keep track of the absolute values of the numbers at every point;
essentially we had to derive upper bounds for them and make sure that the bounds
obtained at every point were low enough for the adders and the subtractors to work.
For instance, we had to prove the following:

> if x3 is the sum of two two's-complement numbers x1 and x2 and the
> signs of x1 and x2 differ, then the absolute value of x3 is no greater than
> the maximum of the absolute values of x1 and x2.

We express this as follows:

```
(all x1 all x2 all x3 all y all z
   ((eq(tcval(x1,y,s(z)) + tcval(x2,y,s(z)), tcval(x3,y,s(z))) and
     eq(apply(x1, y + s(z)), bar(apply(x2, y + s(z))))) =>
    ge(dmax(absval(x1,y,s(z)), absval(x2,y,s(z))),
       absval(x3,y,s(z)))))
```

A list of theorems we needed to prove and some sample proofs are given in the appendix.

For our purposes it was convenient to further decompose the magnitude-and-direction processor into two components (see Figure 3), one which computes absolute values (the "absolute value processor") and one which computes the gradient (the "gradient processor"). Verification of the absolute value processor was conceptually straightforward, since all that had to be done was to verify that the circuit computing the absolute value of a two's complement number worked correctly. The gradient processor receives four integers in the form of their absolute values and the corresponding sign bits and uses these to compute the magnitude and direction of the intensity gradient. We formalize these steps as follows. We take "angle" as a basic entity (or concept); the four basic angles that we deal with are horizontal, vertical, left-diagonal and right-diagonal, and these are represented by the constant symbols ach1, acv1, adl1, and adr1. The functions aval, sign, ortho, code1, and code2 operate on angles, or, in other words, have angles as arguments, where

- aval stands for the absolute value of the 12-bit integer output by the window processor corresponding to the angle,

- sign stands for the sign (i.e. the 12th bit) of the 12-bit integer output by the window processor corresponding to the angle,

- ortho stands for the direction orthogonal (perpendicular) to the given direction, and

- code1 and code2 stand for the first and second bits in the 2-bit codes associated with the four angles.

Of these, we have two arguments for all functions except ortho, the second argument being an integer, signifying that their values vary with time (though we really don't need these for code1 and code2).

The function maxdir stands for "maximum direction" and returns the angle that has the higher absolute value associated with it in accordance with the following axioms:

```
(ge(aval(x,z),aval(y,z)) => eq(maxdir(x,y,z),x))
(not(ge(aval(x,z),aval(y,z))) => eq(maxdir(x,y,z),y))
```

(Note the asymmetry involved: if the absolute values are equal, then the first argument is returned.)

We can now prove that the absolute value of the maximum direction is the maximum of the two absolute values input, i.e.,

```
aval(maxdir(x,y,z),z) == dmax(aval(x,z),aval(y,z)),
```

where the function dmax returns the maximum of two integers.

The circuit was broken down into several smaller components. A typical sub-circuit is the comparator-and-mux combination that computes the maximum and minimum directions of two angles. The circuit consists of a comparator and two muxes that multiplex two 11-bit numbers in accordance with (but differently) the flag raised by the comparator. We specify these as follows:

```
compare(aval(x,z),aval(y,z),u)
mux(aval(y,z),aval(x,z),vmax,u)
mux(aval(x,z),aval(y,z),vmin,u)
```

Now, with the additional assumption that

```
y = ortho(x)
```

we can prove that

```
vmin = aval(ortho(maxdir(x,y,z)),z)
```

and

```
vmax = aval(maxdir(x,y,z),z).
```

6 Error Detection

6.1 Overview

In the process of verifying the design of the Sobel chip, several errors were detected. The two errors that we detected that the design team was unaware of involved the reversal of multiplexor input lines in the direction and magnitude processor. In addition, the proof process pointed out an error in a comparator circuit (Compare_GE) which had already been caught and remedied by the design team, but which had not been corrected in the version of the specification that we were working with. While the multiplexor errors weren't semantically deep, neither were they obvious. In fact, they were not detected until a point was reached in the proof attempt where a consistent statement contrary to what we knew to be true was derived. We conjecture that the most common design errors will be of this form (with the design being basically correct, but having some simple errors which can easily go undetected), especially when a hierarchical approach to design is used.

6.2 A Case Study

We will illustrate one of the errors that we found in the VHDL specification. This error was found in the magnitude-and-direction processor and the specific piece of code where it occurred is shown below:

```
Cmp      : Compare    port (HV_Max,  DRL_Max,  GE3);
Max_Mux  : Mux12_by_2 port (DRL_Max, HV_Max,  Pre_Max,  GE3);
Min_Mux  : Mux12_by_2 port (HV_Min,  DRL_Min, Pre_Perp, GE3);

Max_Hold  : Reg12 port (Pre_Max,  Max,  Clock);
Perp_Hold : Reg12 port (Pre_Perp, Perp, Clock);
```

where Max was to be the maximum value of the four absolute values received by the magnitude-and-direction processor and Perp was to be the value corresponding to the angle orthogonal to the angle with the maximum value. For instance, if the maximum value was in the horizontal direction, then Pre_Perp should be the value in the vertical direction.

Thus we had to prove that

```
(all z
    eq(max(f(f(z))),
        aval(maxdir(maxdir(ach1,acv1,z),maxdir(adr1,adl1,z),z),z)))
```

and that

```
(all z
    eq(perp(f(f(z))),
        aval(ortho(maxdir(maxdir(ach1,acv1,z),
                          maxdir(adr1,adl1,z),z)),
             z))).
```

The structure was specified as

```
compare(hvmax(x),drlmax(x),ge3(x))
mux(drlmax(x),hvmax(x),premax(x),ge3(x))
mux(hvmin(x),drlmin(x),preperp(x),ge3(x))
max(f(x)) == premax(x)
perp(f(x)) == preperp(x)
```

We had already proved that

(a) HV_Max was the maximum value among the values in the horizontal and vertical directions,

(b) DRL_Max the maximum value among the values in the two diagonal directions,

(c) HV_Min was the value in the direction orthogonal to the direction of HV_Max, and

(d) DRL_Min was the value in the direction orthogonal to the direction of DRL_Max.

In our notation,

```
hvmax(f(x))  = aval(maxdir(ach1,acv1,x),x)

hvmin(f(x))  = aval(ortho(maxdir(ach1,acv1,x),x))

drlmax(f(x)) = aval(maxdir(adr1,adl1,x),x)

drlmin(f(x)) = aval(ortho(maxdir(adr1,adl1,x),x))
```

Hence to simplify the proofs, we were trying to prove the intermediate lemma

```
(all x all y all z all u all vp
    (compare(aval(x,z),aval(y,z),u) and
     mux(aval(ortho(x),z),aval(ortho(y),z),vp,u))
    => eq(vp, aval(ortho(maxdir(x,y,z)),z)))
```

and found that we could not prove it unless x and y were the same. By further inspection, we found that the inputs to the muxes should be reversed, i.e. the lemma should be

```
(all x all y all z all u all vp
   (compare(aval(x,z),aval(y,z),u) and
    mux(aval(ortho(y),z),aval(ortho(x),z),vp,u))
    => eq(vp, aval(ortho(maxdir(x,y,z)),z))).
```

This also indicated that the same reversal should take place in the circuit, i.e. the third line should be

```
Min_Mux : Mux12_by_2  port (DRL_Min, HV_Min, Pre_Perp, GE3);
```

and this was confirmed by Jeff Bartlett of RTI.

7   Summary and Research Directions

In verifying the Sobel chip, we have developed a useful approach to hardware verification, especially in that we were able to detect several design errors before fabrication of the chip. The tools we used are at the stage of development where they can be used only by those conversant with theorem proving techniques, however, and a great deal of work needs to be done before they can be made accessible to a larger community. In the following we briefly describe several research ideas which should be pursued in order to improve these tools and make them more readily usable by the design community.

7.1   Semantics

The proof method currently being used is purely syntactic; much more information could potentially be gained by exploiting semantic knowledge. For example, there are often a number of rules introduced which will only be used for simplification, and thus will only play a minor part in the proof. These can be largely ignored, but the current system doesn't 'know' this, and can waste time and space processing them. By keeping the search space as small as possible, many more proofs can be derived automatically, decreasing the need for user proof-guidance.

7.2   Proof Management

As we worked our way through the verification of the Sobel chip it quickly became clear to us that when dealing with a circuit where the verification consists of a large number of proofs of components, a good proof-management system is essential. Such a system would help the user keep track of what has been verified, what hypotheses were introduced, etc. For instance, at a given point in verifying a circuit the user may find it convenient to split a proof into subcases. A proof-manager could keep track of which subcases have been verified and which haven't. Also, it is often helpful to introduce intermediate lemmas in trying to derive a proof; such lemmas will themselves need to be proven at some point, and a proof-manager could keep track of these lemmas, their status, and which of the main proofs depend on them (should one of the lemmas be incorrect, those proofs which used it are invalid).

7.3 Incorporating Decision Procedures

Incorporating decision procedures for classes of formulae which are used frequently would help considerably with many of the proofs we had to derive. A case in point is the cumbersome way we had to deal with simple arithmetic expressions. Some work has been done in this area (see, for instance, [Sho 79,BoMo 85]), but no such procedures have been incorporated into RRL as of this date. In the example given below we illustrate where such a tool might be helpful in deriving a proof.

The example is taken from our verification of the absolute-value element, which takes a two's complement number as input and returns its absolute value The formulae given below were generated from the input to RRL:

```
eq(((intv(apply(s7, s5))
      + (intval(s7, s(s5), s2) + intval(s7, s(s5), s2)))
      + (intv(bar(s3))
      + ((intv(apply(s4, s5))
      + (intval(s4, s(s5), s2) + intval(s4, s(s5), s2)))
      + (((texp(s2) * intv(s6)) + (texp(s2) * intv(s6)))
      + ((texp(s2) * intv(s6)) + (texp(s2) * intv(s6))))))
      ),
      ((texp(s2) + texp(s2)) + (texp(s2) + texp(s2)))))
== false

((intv(s1(s4, s5, s3, s7, s2, s6))
   + intv(s1(s4, s5, s3, s7, s2, s6)))
   + intv(apply(s7, s5)))
==   (intv(bar(apply(s4, s5))) + intv(s3))

(intval(s7, s(s5), s2)
   + (intv(bar(s1(s4, s5, s3, s7, s2, s6)))
   + (intval(s4, s(s5), s2)
   + ((texp(s2) * intv(s6)) + (texp(s2) * intv(s6))))))
== (texp(s2) + texp(s2))
```

Renaming expressions for the sake of clarity, this can be shown to be equivalent to

$$\text{intv}(c1) + 2d_1 + \text{intv}(\neg s3) + \text{intv}(c2)$$
$$+ \ 2d_2 + 4d_3 * \text{intv}(s6) \neq 4d_3$$

$$2 * \text{intv}(c3) + \text{intv}(c1) = \text{intv}(\neg c2) + \text{intv}(s3)$$

$$d_1 + \text{intv}(\neg c3) + d_2 + 2d_3 * \text{intv}(s6) = 2d_3$$

which is unsatisfiable, thus completing the proof. If we had a way of handling integer inequalities, our task (of showing unsatisfiability) would have been easier than it was, since once the boolean values that are arguments to intv are instantiated, we are left with a system of linear integer inequalities. An idea of further research would be to incorporate a decision procedure for this "combined" theory of pseudo-linear integer inequalities, where multiplication is allowed with terms of the form intv(x) (whose integer values are bounded).

7.4   Heuristics

We mentioned earlier that the proof process may not halt if the circuit description is incorrect. At present, the user can only determine that the circuit description is incorrect by his failure to derive a proof of its correctness. Despite the fact that the method described herein encourages a hierarchical approach (in this way many errors will be discovered early in the proof process), it is unlikely that a circuit description will be completely correct on the first attempt at verification. Thus, heuristics for helping a user determine what might be wrong with an incorrect description need to be developed.

References

[Ba 84]   Barrow, H., "VERIFY: A Program for Proving Correctness of Digital Hardware Designs," *Artificial Intelligence*, 24, pp.437-491, 1984.

[BoMo 85] Boyer, R.S., Moore, J.S., "Integrating Decision Procedures into Heuristic Theorem Provers: A Case Study of Linear Arithmetic," Technical Report ICSCA-CMP-44, The University of Texas at Austin, Austin, TX 78712, 1985.

[Buc 85]   Buchberger, B., "Gröbner: An algorithmic method in polynomial ideal theory," in *Multidimensional Systems Theory* (N.K. Bose, ed.), D. Reidel, 1985.

[Bu 83]   Bundy, A., *The Computer Modelling of Human Reasoning*, Academic Press, New York, 1983.

[ChPC 87] Chandrasekhar, M., Privitera, J., Conradt, K., "Application of term rewriting techniques to hardware design verification," in *Proceedings of the 24th Design Automation Conference*, Miami Beach, FL, 1987.

[ChLe 74] Chang, C., Lee, R., *"Symbolic Logic and Mechanical Theorem Proving,"* Academic Press, New York, 1974.

[Go 85]   Gordon, M., "Why Higher-Order Logic is a good formalism for specifying and verifying hardware," Technical Report 77, Computer Laboratory, University of Cambridge, Cambridge, U.K., 1986.

[HsDe 83] Hsiang, J., and Dershowitz, N., "Rewrite methods for clausal and non-clausal theorem proving," in *Proceedings of the 10th EATCS Intl. Colloq. on Automata, Languages, and Programming*, Barcelona, Spain, 1983.

[Hu 85]   Hunt, W., "FM8501: A Verified Microprocessor," Technical Report 47, Institute for Computing Science, Univ. of Texas at Austin, Austin TX, 1985.

[HuOp 80] Huet, G., and Oppen, D., "Equations and rewrite rules: a survey," in *Formal Languages: Perspectives and Open Problems* (R. Book, ed.), Academic Press, New York, 1980.

[KaNa 85] Kapur, D., and Narendran, P., "An equational approach to theorem proving in first-order predicate calculus," in *Proceedings of the 9th Intl. Joint Conference on Artificial Intelligence*, Los Angeles, California, 1985.

[KaSZ 86] Kapur, D., Sivakumar, G., and Zhang, H., "RRL: a rewrite rule laboratory," in *Proceedings of the 8th Conference on Automated Deduction*, Oxford, U.K., 1986.

[KnBe 70] Knuth, D., and Bendix, P., "Simple word problems in universal algebras," in *Computational Problems in Abstract Algebra* (J. Leech, ed.), Pergamon Press, Oxford, 1970, pp. 263-297.

[MiYa 86] Mishra, B., and Yap, C., "Notes on Gröbner Bases," Technical Report # 257, New York University Courant Institute of Mathematical Sciences, New York, N.Y., 1986.

[NaSt 87] Narendran, P., and Stillman, J., "Hardware verification in the Interactive VHDL Workstation," in *VLSI Specification, Verification, and Synthesis*, (G. Birtwistle, P.A. Subrahmanyam, eds.), Kluwer Academic Publishers, Boston, 1988, pp. 235-255.

[Pr 78]   Pratt, W., *"Digital Image Processing,"* John Wiley & Sons, Inc. 1978.

[Sho 79]   Shostak, R., "A Practical Decision Procedure for Arithmetic with Function Symbols," *JACM 26,* 2 (1979), pp. 351-360.

[VaBK 87] Vasanthavada, N., Baker, R., Kanopoulos, N., "A monolithic image edge detection filter," in *Proceedings of the IEEE Custom Integrated Circuits Conference*, 1987.

8 Figures

8.1 Figure 1: The Muller C-Element

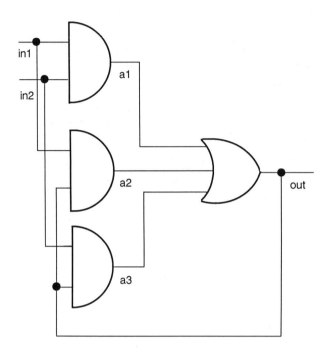

8.2   Figure 2: A View of the Sobel Chip

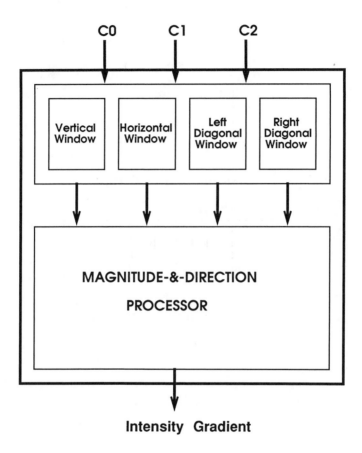

Intensity Gradient

8.3 Figure 3: The Magnitude-and-Direction Processor

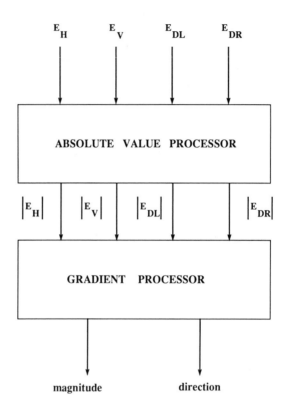

magnitude   direction

9   Appendix

9.1   Specification of the Ripple-Carry Adder

The input file for the ripple-carry adder is shown below:

```
; axioms
a + x == x
x + a == x
x + s(y) == s(x + y)
s(x) + y == s(x + y)
tw(x) == x + x
x * a == a
a * x == a
x * s(y) == (x * y) + x
s(x) * y == (x * y) + y
((x + y) * z) == ((x * z) + (y * z))
texp(a) == s(a)
texp(s(x)) == tw(texp(x))
eq(e0,e1) == false
intv(e0) == a
intv(e1) == s(a)
intval(x,y,a) == intv(apply(x,y))
intval(x,y,s(z)) == intval(x,y,a) + tw(intval(x,s(y),z))

; definition of fulladder; this was proved to follow from
; the gate-level description elsewhere.

fulladder(x,y,zcin,u,vcout) equ
        eq(tw(intv(vcout)) + intv(u),
           intv(x) + (intv(y) + intv(zcin)))

; structural description of the adder (defined inductively)

bsadd(x1,x2,y,a,xb,xc,z) equ
 fulladder(apply(x1,y),apply(x2,y),xb,apply(z,y),xc)


bsadd(x1,x2,y,s(u),xb,xc,z) equ
(exist xb1
 ((fulladder(apply(x1,y),apply(x2,y),xb,apply(z,y),xb1))
  and bsadd(x1,x2,s(y),u,xb1,xc,z)))

; basis part of behavioral description; this part is proved
; (by negating it and deriving a contradiction), then the
; inductive statement is proved to complete the
; proof (see below)
```

```
(all x1 all x2 all y all xb all xc all z
  (bsadd(x1,x2,y,a,xb,xc,z) ->
    eq(intval(z,y,a) + (texp(s(a)) * intv(xc)),
      (intval(x1,y,a) + (intval(x2,y,a) + intv(xb))))))
; inductive part of behavioral description

(all u ((all x1 all x2 all y all xb all xc all z
        (bsadd(x1,x2,y,u,xb,xc,z) ->
          eq(intval(z,y,u) + (texp(s(u)) * intv(xc)),
            (intval(x1,y,u) + (intval(x2,y,u) + intv(xb)))))) ->
      (all x1 all x2 all y all xb all xc all z
      (bsadd(x1,x2,y,s(u),xb,xc,z) ->
        eq(intval(z,y,s(u)) + (texp(s(s(u))) * intv(xc)),
          (intval(x1,y,s(u)) +
          (intval(x2,y,s(u)) + intv(xb))))))))
```

9.1.1 Proof of the Ripple-Carry Adder (basis)

The proof of the basis case is given below:

Input #1, produced:

    [1] (a + x) ---> x

Input #4, produced:

    [4] (s(x) + y) ---> s((x + y))

Input #5, produced:

    [5] tw(x) ---> (x + x)

Input #7, produced:

    [7] (a * x) ---> a

Input #9, produced:

    [9] (s(x) * y) ---> (y + (x * y))

Input #10, produced:

    [10] texp(a) ---> s(a)

Input #11, reduced by Rule [5], produced:

    [11] texp(s(x)) ---> (texp(x) + texp(x))

Input #15, produced:

```
[15] intval(x, y, a) ---> intv(apply(x, y))
```

Input #17, reduced by Rule [5], produced:

```
[17] fulladder(x, y, zcin, u, vcout) --->
       eq((intv(u) + intv(vcout) + intv(vcout)),
       (intv(x) + intv(y) + intv(zcin)))
```

Input #18, reduced by Rule [17], produced:

```
[18] bsadd(x1, x2, y, a, xb, xc, z) --->
       eq((intv(xb) + intv(apply(x1, y)) +
       intv(apply(x2, y))),
       (intv(xc) + intv(xc) + intv(apply(z, y))))
```

Input #19, reduced by Rule [18], produced:

```
[19] (intv(apply(s1, s3)) + (intv(apply(s2, s3))
       + intv(s4))) --->
       ((intv(s5) + intv(s5)) + intv(apply(s6, s3)))
```

Input #19, reduced by Rules [15], [19], [11], [10], [4],
         [9], [1], [7], produced:

```
[20] eq((intv(apply(s6, s3)) + (intv(s5) + intv(s5))),
   ((intv(s5) + intv(s5)) + intv(apply(s6, s3)))) ---> false
```

Rule [ac-op] deleted Rule [20], produced:

```
[24] true ---> false.
```

9.1.2   Proof of the Ripple-Carry Adder (induction)

The inductive case is proved below:

Input #5, produced:

```
[5] tw(x) ---> (x + x)
```

Input #11, reduced by Rule [5], produced:

```
[11] texp(s(x)) ---> (texp(x) + texp(x))
```

Input #15, produced:

[15] intval(x, y, a) ---> intv(apply(x, y))

Input #16, reduced by Rules [15], [5], produced:

```
[16] intval(x, y, s(z)) --->
     (intv(apply(x, y)) +
      intval(x, s(y), z) + intval(x, s(y), z))
```

Input #17, reduced by Rule [5], produced:

```
[17] fulladder(x, y, zcin, u, vcout) --->
     eq((intv(u) + intv(vcout) + intv(vcout)),
       (intv(x) + intv(y) + intv(zcin)))
```

Input #19, reduced by Rule [17], produced:

```
[19] (eq(((intv(apply(x1, y)) +
          (intv(apply(x2, y)) + intv(xb))),
         ((intv(s1(x1, y, x2, xb, z, u, xc))
          + intv(s1(x1, y, x2, xb, z, u, xc)))
          + intv(apply(z, y)))))
     and bsadd(x1, x2, y, s(u), xb, xc, z)
     and bsadd(x1, x2, s(y), u,
                 s1(x1, y, x2, xb, z, u, xc), xc, z)) --->
     bsadd(x1, x2, y, s(u), xb, xc, z)
```

Input #19, reduced by Rule [17], produced:

```
[20] (eq(((intv(apply(x1, y)) +
          (intv(apply(x2, y)) + intv(xb))),
         ((intv(xb1) + intv(xb1)) + intv(apply(z, y)))))
     and bsadd(x1, x2, y, s(u), xb, xc, z)
     and bsadd(x1, x2, s(y), u, xb1, xc, z)) --->
     (eq(((intv(apply(x1, y)) + (intv(apply(x2, y)) +
                                intv(xb))),
         ((intv(xb1) + intv(xb1)) + intv(apply(z, y)))))
     and bsadd(x1, x2, s(y), u, xb1, xc, z))
```

Rule [20] deleted Rule [19], produced:

```
[21] (eq(((intv(apply(x1, y)) + (intv(apply(x2, y)) +
                                intv(xb))),
         ((intv(s1(x1, y, x2, xb, z, u, xc)) +
           intv(s1(x1, y, x2, xb, z, u, xc))) +
          intv(apply(z, y)))))
     and bsadd(x1, x2, s(y), u,
                 s1(x1, y, x2, xb, z, u, xc), xc, z)) --->
     bsadd(x1, x2, y, s(u), xb, xc, z)
```

Input #20, reduced by Rule [11], produced:

```
[22] (eq(((intval(x1, y, s2) + (intval(x2, y, s2) +
                             intv(xb))),
          (intval(z, y, s2) +
          ((texp(s2) + texp(s2)) * intv(xc)))))
      and bsadd(x1, x2, y, s2, xb, xc, z)) --->
      bsadd(x1, x2, y, s2, xb, xc, z)
```

Input #20, produced:

```
[23] bsadd(s3, s4, s5, s(s2), s6, s7, s8) ---> true
```

Input #20, reduced by Rules [16], [11], produced:

```
[24] eq((((intv(apply(s3, s5)) +
           (intval(s3, s(s5), s2) + intval(s3, s(s5), s2)))
       + ((intv(apply(s4, s5)) + (intval(s4, s(s5), s2)
                        + intval(s4, s(s5), s2))) + intv(s6))),
       ((intv(apply(s8, s5)) +
           (intval(s8, s(s5), s2) + intval(s8, s(s5), s2)))
       + (((texp(s2) + texp(s2)) +
           (texp(s2) + texp(s2))) * intv(s7))))) ---> false
```

Rule [21] superposed itself, reduced by Rule [21], produced:

```
[25] (eq(((intv(apply(x1, y)) + (intv(apply(x2, y)) +
                             intv(xb))),
          ((intv(s1(x1, y, x2, xb, z, u, xc)) +
            intv(s1(x1, y, x2, xb, z, u, xc))) +
          intv(apply(z, y)))))
      and bsadd(x1, x2, y, s(u), xb, xc, z)) --->
      bsadd(x1, x2, y, s(u), xb, xc, z)
```

Rule [21] superposed itself, reduced by Rule [21], produced:

```
[26] (bsadd(x1, x2, y, s(u), xb, xc, z)
        and bsadd(x1, x2, s(y), u,
                  s1(x1, y, x2, xb, z, u, xc), xc, z)) --->
      bsadd(x1, x2, y, s(u), xb, xc, z)
```

Rule [23] superposed with Rule [25],
                    reduced by Rule [23], produced:

```
[27] ((intv(s1(s3, s5, s4, s6, s8, s2, s7)) +
        intv(s1(s3, s5, s4, s6, s8, s2, s7))) +
        intv(apply(s8, s5))) --->
```

```
            (intv(apply(s3, s5)) + (intv(apply(s4, s5)) +
                              intv(s6)))
```

Rule [26] superposed with Rule [23],
                    reduced by Rule [23], produced:

```
    [28] bsadd(s3, s4, s(s5), s2,
              s1(s3, s5, s4, s6, s8, s2, s7), s7, s8) --->
        true
```

Rule [22] superposed with Rule [28],
                    reduced by Rule [28], produced:

```
    [30] (intval(s3, s(s5), s2) + (intval(s4, s(s5), s2) +
              intv(s1(s3, s5, s4, s6, s8, s2, s7)))) --->
        (intval(s8, s(s5), s2) + ((texp(s2) * intv(s7)) +
                              (texp(s2) * intv(s7))))
```

Input #21, produced:

```
    [31] ((x + y) * z) ---> ((x * z) + (y * z))
```

Rule [31] deleted Rule [24], reduced by Rule [31], produced:

```
    [33] eq(((intv(apply(s3, s5)) +
              (intval(s3, s(s5), s2) + intval(s3, s(s5), s2)))
            + ((intv(apply(s4, s5)) +
                (intval(s4, s(s5), s2) + intval(s4, s(s5), s2)))
              + intv(s6))),
            ((intv(apply(s8, s5)) + (intval(s8, s(s5), s2) +
                              intval(s8, s(s5), s2))) +
              (((texp(s2) * intv(s7)) + (texp(s2) * intv(s7)))
              + ((texp(s2) * intv(s7)) +
                (texp(s2) * intv(s7))))))) ---> false
```

Rule [ac-op] deleted Rule [27], produced:

```
    [39] (intv(s6) + intv(apply(s3, s5)) +
          intv(apply(s4, s5))) --->
          (intv(apply(s8, s5)) +
          intv(s1(s3, s5, s4, s6, s8, s2, s7)) +
          intv(s1(s3, s5, s4, s6, s8, s2, s7)))
```

Rule [ac-op] deleted Rule [30], produced:

```
    [41] (intv(s1(s3, s5, s4, s6, s8, s2, s7)) +
          intval(s3, s(s5), s2) + intval(s4, s(s5), s2)) --->
          ((texp(s2) * intv(s7)) +
```

```
    (texp(s2) * intv(s7)) + intval(s8, s(s5), s2))
```

Rule [ac-op] deleted Rule [33],
                reduced by Rules [39], [41], produced:

    [44] true ---> false

9.2 Specification of the Adder

The theorem we are trying to prove is:

```
(all x1 all x2 all z all y all u all xc all zc
 ((bsadd(x1,x2,y,u,e0,xc,z) and
    fulladder(apply(x1,y + s(u)), apply(x2,y + s(u)),
                xc, apply(z,y + s(u)), zc) and
   gt(texp(u), absval(x1,y,s(u))) and
   gt(texp(u), absval(x2,y,s(u)))) ->
  eq(tcval(z,y,s(u)), tcval(x1,y,s(u)) + tcval(x2,y,s(u)))))
```

where

(a) bsadd is the predicate used to recursively specify the adder,

(b) tcval specifies the two's-complement value of a segment of a bit string, and

(c) absval specifies the absolute value of a segment of a bit string.

Since the adder has already been proved (see Section 9.1), we use its behavior, namely that

```
(all x1 all x2 all y all u all xb all xc all z
 (bsadd(x1,x2,y,u,xb,xc,z)
  -> eq(intval(z,y,u) + (texp(s(u)) * intv(xc)),
        (intval(x1,y,u) + (intval(x2,y,u) + intv(xb))))))
```

The full adder is assumed to be implemented of majority and exclusive-or gates and is specified by

```
fulladder(x,y,zcin,u,vcout) equ
    (eq(vcout,maj(x,y,zcin)) and
     eq(u, dxor(dxor(x,y),zcin)))
```

Apart from the axiomatizations of integers and the greater-than and greater-than-or-equal-to relations, we also need the following properties of intval:

```
gt(texp(s(z)), intval(x,y,z))
ge(intval(x,y,u),a)
```

Thus the final input to RRL is

```
a + x == x
x + a == x
x + s(y) == s(x + y)
s(x) + y == s(x + y)
tw(x) == x + x
x + minus(x) == a
minus(x) + x == a
minus(minus(x)) == x
(x + y) + minus(y) == x
(x + y) + minus(x) == y
x * a == a
a * x == a
x * s(y) == (x * y) + x
s(x) * y == (x * y) + y
texp(a) == s(a)
texp(s(x)) == tw(texp(x))

intv(e0) == a
intv(e1) == s(a)
maj(x,x,y) == x
maj(x,y,x) == x
maj(y,x,x) == x
dxor(e0,x) == x
dxor(x,e0) == x
dxor(x,x) == e0
eq(e0,e1) == false

((gt(x,y) and ge(y,z)) -> gt(x,z))
(gt(x,y) equ not(ge(y,x)))
(ge(x,y) equ (eq(x,y) or gt(x,y)))
(gt(x,y) and ge(u,v)) -> gt(x + u, y + v)

gt(texp(s(z)), intval(x,y,z))
ge(intval(x,y,u),a)

tcval(x,y,s(z)) ==
 intval(x,y,z) + intv(apply(x,y + s(z))) * minus(texp(s(z)))

eq(apply(x,y + s(z)), e0) ->
          eq(absval(x,y,s(z)), intval(x,y,z))

eq(apply(x,y + s(z)), e1) -> eq(absval(x,y,s(z)),
                              texp(s(z)) + minus(intval(x,y,z)))

(all x1 all x2 all y all u all xb all xc all z
 (bsadd(x1,x2,y,u,xb,xc,z)
  -> eq(intval(z,y,u) + (texp(s(u)) * intv(xc)),
```

```
     (intval(x1,y,u) + (intval(x2,y,u) + intv(xb)))
     )))

fulladder(x,y,zcin,u,vcout) equ
   (eq(vcout,maj(x,y,zcin)) and eq(u, dxor(dxor(x,y),zcin)))

not(all x1 all x2 all z all y all u all xc all zc
    ((bsadd(x1,x2,y,u,e0,xc,z) and
       fulladder(apply(x1,y + s(u)), apply(x2,y + s(u)),
                 xc, apply(z,y + s(u)), zc) and
       gt(texp(u), absval(x1,y,s(u))) and
       gt(texp(u), absval(x2,y,s(u)))) ->
       eq(tcval(z,y,s(u)), tcval(x1,y,s(u)) + tcval(x2,y,s(u)))))
```

The proof is done by splitting into cases in accordance with the values apply(x1, y + s(u)), apply(x2, y + s(u)), and xc can take. Since they are all bits, we have to effectively consider 8 cases. We show a typical proof (that of tcadd in the case where xc = e0, apply(x1, y + s(u)) = e0 and apply(x2, y + s(u)) = e0) below:

Input #2, produced:

```
   [2] (x + a) ---> x
```

Input #3, produced:

```
   [3] (x + s(y)) ---> s((x + y))
```

Input #5, produced:

```
   [5] tw(x) ---> (x + x)
```

Input #11, produced:

```
   [11] intv(e0) ---> a
```

Input #16, produced:

```
   [16] dxor(e0, x) ---> x
```

Input #17, produced:

```
   [17] dxor(x, e0) ---> x
```

Input #21, produced:

```
[21] (a * x) ---> a
```

Input #25, reduced by Rule [5], produced:

```
[25] texp(s(x)) ---> (texp(x) + texp(x))
```

Input #33, reduced by Rules [3], [25], produced:

```
[33] tcval(x, y, s(z)) --->
     (intval(x, y, z) +
     (intv(apply(x, s((y + z)))))
     * minus((texp(z) + texp(z)))))
```

Input #36, reduced by Rule [25], produced:

```
[36] (eq(((intval(x1, y, u) +
          (intval(x2, y, u) + intv(xb))),
          (intval(z, y, u) +
          ((texp(u) + texp(u)) * intv(xc))))
     and bsadd(x1, x2, y, u, xb, xc, z)) --->
     bsadd(x1, x2, y, u, xb, xc, z)
```

Input #37, produced:

```
[37] fulladder(x, y, zcin, u, vcout) --->
        (eq(u, dxor(dxor(x, y), zcin))
     and eq(vcout, maj(x, y, zcin)))
```

Input #38, produced:

```
[38] bsadd(s1, s2, s4, s5, e0, s6, s3) ---> true
```

Input #38, reduced by Rules [37], [3], produced:

```
[39] (eq(s7, maj(apply(s1, s((s4 + s5))),
                 apply(s2, s((s4 + s5))), s6))
     and eq(apply(s3, s((s4 + s5))),
            dxor(dxor(apply(s1, s((s4 + s5))),
                      apply(s2, s((s4 + s5)))), s6))) ---> true
```

Input #38, reduced by Rule [33], produced:

```
[42] eq((((intval(s1, s4, s5) +
          (intv(apply(s1, s((s4 + s5)))))
          * minus((texp(s5) + texp(s5)))))
        + (intval(s2, s4, s5) +
          (intv(apply(s2, s((s4 + s5)))))
          * minus((texp(s5) + texp(s5)))))),
```

```
        (intval(s3, s4, s5) +
        (intv(apply(s3, s((s4 + s5))))
         * minus((texp(s5) + texp(s5)))))) ---> false
```

Rule [39] superposed itself, reduced by Rules
[39], [52], [55], produced:

```
   [43] s7 ---> e0
```

Rule [43] deleted Rule [39], reduced by Rules
[52], [55], [57], produced:

```
   [44] apply(s3, s((s4 + s5))) ---> e0
```

Rule [44] deleted Rule [42], produced:

```
   [45] eq(((intval(s1, s4, s5) +
             (intv(apply(s1, s((s4 + s5))))
            * minus((texp(s5) + texp(s5)))))
          + (intval(s2, s4, s5) +
             (intv(apply(s2, s((s4 + s5))))
             * minus((texp(s5) + texp(s5))))))),
          (intval(s3, s4, s5)
          + (intv(dxor(dxor(apply(s1, s((s4 + s5))),
                            apply(s2, s((s4 + s5)))), s6))
           * minus((texp(s5) + texp(s5)))))) ---> false
```

Rule [36] superposed with Rule [38], reduced by Rules
[11], [2], [38], [52], produced:

```
   [46] (intval(s1, s4, s5) + intval(s2, s4, s5)) --->
        intval(s3, s4, s5)
```

Input #40, produced:

```
   [52] s6 ---> e0
```

Rule [52] deleted Rule [45], reduced by Rule [17], produced:

```
   [54] eq(((intval(s1, s4, s5)
            + (intv(apply(s1, s((s4 + s5))))
              * minus((texp(s5) + texp(s5)))))
          + (intval(s2, s4, s5)
            + (intv(apply(s2, s((s4 + s5))))
                * minus((texp(s5) + texp(s5))))))),
          (intval(s3, s4, s5)
          + (intv(dxor(apply(s1, s((s4 + s5))),
                  apply(s2, s((s4 + s5)))))
```

```
                    * minus((texp(s5) + texp(s5)))))) ---> false
```

Input #41, produced:

```
    [55] apply(s1, s((s4 + s5))) ---> e0
```

Rule [55] deleted Rule [54], reduced by Rules
[11], [21], [2], [55], [16], produced:

```
    [56] eq((intval(s1, s4, s5) +
              (intval(s2, s4, s5)
               + (intv(apply(s2, s((s4 + s5))))
                  * minus((texp(s5) + texp(s5)))))),
             (intval(s3, s4, s5)
              + (intv(apply(s2, s((s4 + s5))))
                 * minus((texp(s5) + texp(s5)))))) ---> false
```

Input #42, produced:

```
    [57] apply(s2, s((s4 + s5))) ---> e0
```

Rule [57] deleted Rule [56], reduced by Rules
[11], [21], [2], [46], [57], produced:

```
    [58] true ---> false
```

9.3   Bounds Theorems

The following theorems were proved about the upper bounds of absolute values of numbers arising from additions and subtractions. As mentioned before, these had to be proved in order to show that none of the intermediate numbers in the window circuits had absolute values greater than 2^{11}, or texp(sc11) in our notation.

```
(all x1 all x2 all x3 all y all z
 ((eq(tcval(x1,y,s(z)) + tcval(x2,y,s(z)),
      tcval(x3,y,s(z))))
  -> ge(absval(x1,y,s(z)) + absval(x2,y,s(z)),
        absval(x3,y,s(z))))))

(all x1 all x2 all x3 all y all z
 ((eq(tcval(x1,y,s(z)) + tcval(x2,y,s(z)),
      tcval(x3,y,s(z)))
   and
   eq(apply(x1, y + s(z)),
      bar(apply(x2, y + s(z)))))
  -> ge(dmax(absval(x1,y,s(z)),absval(x2,y,s(z))),
        absval(x3,y,s(z))))))
```

```
(all x1 all x2 all x3 all y all z
 ((eq(tcval(x1,y,s(z)) + minus(tcval(x2,y,s(z))),
      tcval(x3,y,s(z)))
   and
   eq(apply(x1, y + s(z)),
      apply(x2, y + s(z))))
  -> ge(dmax(absval(x1,y,s(z)), absval(x2,y,s(z))),
        absval(x3,y,s(z)))))

(all x1 all x2 all x3 all y all z
 ((eq(tcval(x1,y,s(z)) + minus(tcval(x2,y,s(z))),
      tcval(x3,y,s(z))))
  -> ge(absval(x1,y,s(z)) + absval(x2,y,s(z)),
        absval(x3,y,s(z)))))
```

9.4   A Sample Proof

We present a typical proof; that of the theorem

```
(all x1 all x2 all x3 all y all z
 ((eq(tcval(x1,y,s(z)) + minus(tcval(x2,y,s(z))),
      tcval(x3,y,s(z)))
   and
   eq(apply(x1, y + s(z)), apply(x2, y + s(z))))
  -> ge(dmax(absval(x1,y,s(z)), absval(x2,y,s(z))),
        absval(x3,y,s(z)))))
```

in the case where the sign bits of both the first and third inputs (x1 and x3) are 0, i.e., both x1 and x3 are positive. (The proofs of the other cases are similar and are omitted from this document.)

Input #2, produced:

    [2] (x + a) ---> x

Input #3, produced:

    [3] (x + s(y)) ---> s((x + y))

Input #5, produced:

    [5] tw(x) ---> (x + x)

Input #7, produced:

    [7] (a * x) ---> a

Input #11, reduced by Rule [5], produced:

```
[11] texp(s(x)) ---> (texp(x) + texp(x))
```

Input #14, produced:

```
[14] minus(minus(x)) ---> x
```

Input #16, produced:

```
[16] ((x + y) + minus(y)) ---> x
```

Input #19, produced:

```
[19] ge(y, x) ---> (true xor gt(x, y))
```

Input #24, reduced by Rule [19], produced:

```
[25] gt(x, dmax(x, y)) ---> false
```

Input #25, reduced by Rule [19], produced:

```
[26] (gt(v, u) and gt(x, y) and gt((x + u), (y + v))) --->
     (gt(x, y) xor (gt(v, u) and gt(x, y)) xor
     (gt(x, y) and gt((x + u), (y + v))))
```

Input #27, reduced by Rule [19], produced:

```
[28] gt(a, intval(x, y, u)) ---> false
```

Input #31, produced:

```
[32] intv(e0) ---> a
```

Input #34, reduced by Rules [3], [11], produced:

```
[35] tcval(x, y, s(z)) ---> (intval(x, y, z) +
     (intv(apply(x, s((y + z)))) * minus((texp(z) + texp(z)))))
```

Input #35, reduced by Rule [3], produced:

```
[36] (eq(e0, apply(x, s((y + z))))
     and eq(absval(x, y, s(z)), intval(x, y, z)))
     ---> eq(e0, apply(x, s((y + z))))
```

Input #37, reduced by Rule [19], produced:

```
[38] gt(absval(s3, s4, s(s5)),
        dmax(absval(s1, s4, s(s5)),
```

```
                    absval(s2, s4, s(s5)))) ---> true
```

Input #37, reduced by Rule [3], produced:

```
  [39] apply(s2, s((s4 + s5))) ---> e0
```

Input #37, reduced by Rules [35], [39], produced:

```
  [40] ((intval(s1, s4, s5)
       + (intv(apply(s1, s((s4 + s5))))
          * minus((texp(s5) + texp(s5)))))
       + minus((intval(s2, s4, s5)
             + (intv(apply(s1, s((s4 + s5))))
                 * minus((texp(s5) + texp(s5)))))))
      ---> (intval(s3, s4, s5)
             + (intv(apply(s3, s((s4 + s5))))
                 * minus((texp(s5) + texp(s5)))))
```

Rule [40] superposed with Rule [16],
                    reduced by Rule [14], produced:

```
  [41] ((intval(s3, s4, s5)
        + (intv(apply(s3, s((s4 + s5))))
           * minus((texp(s5) + texp(s5)))))
       + (intval(s2, s4, s5)
         + (intv(apply(s1, s((s4 + s5))))
            * minus((texp(s5) + texp(s5))))))
      ---> (intval(s1, s4, s5)
             + (intv(apply(s1, s((s4 + s5))))
                 * minus((texp(s5) + texp(s5)))))
```

Rule [26] superposed with Rule [28],
                    reduced by Rules [28], [2], produced:

```
  [49] (gt(x, y) and gt((x + intval(x1, y2, u)), y)) ---> gt(x, y)
```

Input #38, produced:

```
  [50] apply(s1, s((s4 + s5))) ---> e0
```

Rule [50] deleted Rule [41],
        reduced by Rules [32], [7], [2], [50], produced:

```
  [52] ((intval(s3, s4, s5)
        + (intv(apply(s3, s((s4 + s5))))
           * minus((texp(s5) + texp(s5)))))
       + intval(s2, s4, s5))
      ---> intval(s1, s4, s5)
```

Rule [36] superposed with Rule [50],
                    reduced by Rule [50], produced:

   [53] absval(s1, s4, s(s5)) ---> intval(s1, s4, s5)

Rule [53] deleted Rule [38], produced:

   [54] gt(absval(s3, s4, s(s5)),
       dmax(intval(s1, s4, s5),
           absval(s2, s4, s(s5))))) ---> true

Rule [36] superposed with Rule [39],
                    reduced by Rule [39], produced:

   [55] absval(s2, s4, s(s5)) ---> intval(s2, s4, s5)

Rule [55] deleted Rule [54], produced:

   [56] gt(absval(s3, s4, s(s5)),
       dmax(intval(s1, s4, s5), intval(s2, s4, s5))) ---> true

Rule [49] superposed with Rule [52], produced:

   [57] (gt((intval(s3, s4, s5) +
        (intv(apply(s3, s((s4 + s5))))
         * minus((texp(s5) + texp(s5)))))), y)
      and gt(intval(s1, s4, s5), y))
   ---> gt((intval(s3, s4, s5) +
        (intv(apply(s3, s((s4 + s5))))
        * minus((texp(s5) + texp(s5)))))), y)

Input #39, produced:

   [58] apply(s3, s((s4 + s5))) ---> e0

Rule [58] deleted Rule [57], reduced by Rules
[58], [32], [7], [2], produced:

   [60] (gt(intval(s1, s4, s5), y) and
       gt(intval(s3, s4, s5), y)) ---> gt(intval(s3, s4, s5), y)

Rule [36] superposed with Rule [58],
                    reduced by Rule [58], produced:

   [61] absval(s3, s4, s(s5)) ---> intval(s3, s4, s5)

Rule [61] deleted Rule [56], produced:

[62] gt(intval(s3, s4, s5),
         dmax(intval(s1, s4, s5), intval(s2, s4, s5))) ---> true

Rule [62] superposed with Rule [60], reduced by Rules
[25], [62], produced:

[63] true ---> false

Specification-driven Design of Custom Hardware in HOP

Ganesh C. Gopalakrishnan, Richard M. Fujimoto,
Venkatesh Akella, N.S.Mani, and Kevin N. Smith
Dept. of Computer Science, University of Utah
Salt Lake City, Utah 84112, U.S.A

Abstract. *We present a language "Hardware viewed as Objects and Processes" (HOP) for specifying the structure, behavior, and timing of hardware systems. HOP embodies a simple process model for lock-step synchronous processes. Processes may be described both as a black-box and as a collection of interacting sub-processes. The latter can be statically simplified using an algorithm 'PARCOMP'. PARCOMP symbolically simulates a collection of interacting processes. The advantages claimed for HOP include simple semantics, intuitiveness, high expressive power, and numerous provisions to support easily verifiable designs all the way to VLSI layout.*

After introducing HOP, and presenting some of the results obtained from experimenting with the HOP design system, we present the design of a large hardware system (the "Utah Simulation Engine") currently being developed to speed-up distributed discrete event simulation using Time Warp. Issues in the specification driven design of this system are discussed and illustrated using HOP.

1 Introduction

The use of formal specifications for specifying, verifying, manually designing, and automatically synthesizing hardware systems is becoming widespread. Not only are there different formal specification languages, but also there are a number of different *formalisms* in use: Functional Programming [JBB88,She85] Prolog [WFC87], Petri Nets [Chu87], Temporal Logic [BCDM85], various Calculii of Communicating Systems [Mil83,Hen84] Trace Theory [Sne85], Higher Order Logic [CGM86,JB85], Algebraic Specifications and Equational Techniques [GSS87,Sub83,NS88], Synchronized Transition Systems [GGS88], and Path Expressions [ACFM85], to name a few. Enough impressive results have been demonstrated to justify the use of formal specifications for VLSI design. However, as will be discussed momentarily, many problems remain unsolved.

Our research is aimed at solving some of these problems. We have designed a simple hardware description language called HOP (Hardware viewed as Objects and Processes) that embodies solutions to these problems. HOP deals with the structure, behavior, and timing of digital hardware systems. We are also using HOP to specify VLSI circuits that are designed and implemented by various student groups. One such design called the Roll Back Chip (RBC) is described in this paper. Our research thus involves developing HOP as well as using it in many ways for designing real-life hardware.

Our effort to date has given us the following insights, as well as specific results.

Insights

Complex designs evolve in several dimensions. When designing custom computer architectures, many major design decisions are taken over a prolonged duration of time. Some of these decisions are: (i) should a computation be implemented in hardware, firmware, or software? (ii) should a hardware unit be asynchronous or synchronous? (iii) how should the data and control interactions be organized? The problem of effective communication among the various hardware designers of a project (and also between the same designer, over many days!) is often as severe as reported in [Bro75].

There is no satisfactory design language that specifies a finished hardware architecture completely and formally, or accomodates the evolving nature of the design of a custom computer architecture and helps facilitate communication among designers. Although some impressive work has been reported, (e.g. [Bae86]), most works either focus on control flow and ignore data flow, or ignore resource utilization, or ignore complications such as interrupts, exception handling, etc.

What we require is a simple and semantically well understood hardware description language (HDL) that can be used to capture the *structural*, *behavioral*, and *temporal* aspects of an evolving hardware+software design. In addition, a VLSI design system (an assemblage of tools) that supports the simulation and validation of descriptions in such a language is also inevitable [Seq87].

It has been reported that the complete formal verification of even extremely simple ICs is really hard [Coh88]. More importantly, impressive results with theorem provers have almost always been exhibited by persons who played a major role in developing the theorem prover (and hence knew its innards)—not by end-users. Until this situation changes dramatically, custom architectures would at best partially proven correct for *certain* critical safety properties. Therefore formal specification will be used to a large extent for its *indirect benefits*—better understanding, better communication among hardware designers and system software writers, better testing, the ability to measure what exactly has been tested, etc.

Given this, it is unquestionable that well designed HDLs, that help the designers *think effectively* using high-level abstractions and thereby intuitively establish the correctness of designs, have a major role to play. To paraphrase Stoy [Sto77, Page 7], "well designed HDLs of the above nature will help write specifications that will more probably be correct because we will less likely have forgotten about the crucial little exception to some general rule that applies in our particular case."

Finally, computations are often done in custom hardware for gaining speed. Thus, lack of throughput is also a design error. Hardware Verification does not yet address performance issues.

Specific Results

1. We have designed a version of HOP that can describe *lockstep synchronous processes*. Specific instances of such processes are: (i) synchronous hardware systems; (ii) hardware systems studied under the unit-delay timing model. (Note: In this paper the language "HOP" refers to the lockstep synchronous version of HOP.)

2. We have described the semantics of HOP both operationally as well as through linear-time temporal logic.

3. We have discovered several new techniques for simulating and debugging (via static analysis) specifications. These procedures have been implemented, and experimental results are presented.

4. An implementation the HOP VLSI design system is in progress, with some parts already operational.

5. We have designed a number of VLSI chips, and have also specified them in HOP.

Organization of the Paper

Section 2 presents the HOP language. Section 3 presents the semantics of HOP. Section 4 presents the HOP design system, and some results obtained to date. Section 5 presents the specification-driven design of the RBC. Section 6 presents the detailed design of one of the submodules of the RBC. Section 7 presents our concluding remarks, and an outline of the planned extensions to HOP. The Appendix describes the algorithms PARCOMP and PARCOMP-DC.

Related Work

The HOP *language* is inspired in many ways by the work of Milner [Mil82] and Milne [Mil83]. The SBL language designed by the first author under Smith and Srivas also [GSS87,Gop86] influenced HOP. Our work is different from related work in these respects: (i) we have focussed on lock-step synchronous timing model,thus obtaining results useful for synchronous hardware; (ii) we model value communications using *data queries* and *data assertions* (as opposed to existing ways of value communication in CSP or CCS); doing so has several advantages, to be discussed; (iii) our work addresses a number of practical design issues in a formal framework.

Broadly speaking, there are two kinds of design automation researchers: (a) Those who take one formalism (such as lazy functional languages or HOL), "piously" believe in it, and "get the most mileage out of it", at the risk of making some practically unrealistic assumptions; (b) those who don't carry such pious beliefs, but treat all formalisms with equal interest, use them as "tools", assimilate group (a)'s results and render them more practical. We belong more in group (b) than in group (a).

131

```
ABSPROC <ModuleName> [<formal params pertaining to sizes & types>]
CONST <list of constants of the same value>
TYPE  <list of type identifiers of the same type>
PORT  <list of ports of the same type>
CLOCK <a clock agent and the ports imported from it>
EVENT <events and their encodings in terms of port values>
PROTOCOL <a list of process definitions>
DEFUN    <a list of function definitions>
END <ModuleName>
```

Figure 1: The Skeleton of an Absproc Specification

```
REALPROC <ModuleName> [<formal params pertaining to sizes & types>]
CONST <list of constants of the same value>
TYPE  <list of type identifiers of the same type>
PORT  <the external ports of the module being defined>
SUBPROCESS <instantiations of prev. defined abs/real/vec processes>
CONNECT   <the set of interconnections among the subprocesses>
END <ModuleName>
```

Figure 2: The Skeleton of a Realproc Specification

2   The HOP Language

The basic unit of specification in HOP is an *module*. The external attributes of a module are:

- Zero or more uni- or bi-directional *data ports*;
- Zero or more uni-directional *events*;
- An external protocol specification.

A module specified as a black box is called an *absproc*, standing for *abstract process*. The skeleton of an ABSPROC is shown in figure 1. A module specified as a network of subprocesses is called a *realproc*, the skeleton of which appears in figure 2. (Note: For ease of parsing, currently we use a lisp-like syntax for HOP; we have

```
VECPROC <ModuleName> [<formal params pertaining to sizes & types>]
CONST <list of constants of the same value>
TYPE  <list of type identifiers of the same type>
PORT  <the external ports of the module being defined>
SUBPROCESS <instantiations of prev. defined abs/real/vec processes>
DIMENSIONS <the SIZES of each dimensions of regularity>
CONNECT <interconnections betn. subprocesses, via recurrence eqns.>
END <ModuleName>
```

Figure 3: The Skeleton of a Vecproc Specification

hand edited *almost* all syntactic descriptions in this paper to an easier-to-understand higher-level syntax.)

Since topologically regular realprocs (*e.g.* single and two-dimensional arrays of modules) occur very frequently in practice we identify a sub-category of realprocs called *vecprocs* (figure 3). Vecprocs in HOP may best be regarded as "arhythmic arrays"—geometrically regular arrays in which computations aren't regular, or rhythmic, as in systolic arrays. Previous work involving regular arrays ([She84,She85], [Pat85], [MH85]) has dealt more with systolic arrays than with arythmic arrays. We will show that having the special category of Vecprocs is useful in many ways.

A realproc is built using one or more absprocs by connecting some of the ports and events of the absprocs, by composing the external protocols of the absprocs, and by internalizing (hiding) some of the events and ports of the absprocs. A syntactically sugered notation (**DATANODE** and **EVENTNODE**) mitigates the burden of specifying the *renaming* and *hiding* ([Mil80]) information for large systems. A vecproc is essentially built in the same fashion; however a notation based on recurrence relations is provided for easily specifing the regular placement of modules as well as regular interconnections among them.

An algorithm called PARCOMP ("Parallel Composition") has been designed. It takes as input a realproc or a vecproc and produces as output an absproc. It works by symbolically simulating all possible interactions between the subprocesses of a realproc or vecproc. PACOMP implements the operational rules of HOP presented in section 3.

The absproc inferred by PARCOMP captures, via symbolic expressions, the the behavior of the realproc or vecproc for all possible starting states of the submodules, and for all external inputs. The text of the inferred absproc can be manually studied to see if the system behaves as understood by the designer. Thus, PARCOMP greatly facilitates the understanding of the *collective behavior* of a collection of synchronous systems.

PARCOMP, as well as its planned uses, are similar to the work reported in [HK87], and to the idea of *constructive simulation* reported in [Mil85]. However our work is done for a much higher level language that includes user-defined abstract data types. Our algorithm embodies useful static checks of timing protocols. Our algorithm capitalizes on the structural information (specifically, knowledge about events that are completely hidden within a module) to save on computation time. This is accomplished thus (explained in detail later): "states reachable via transitions labeled by unsynchronized and hidden events are never visited, and consequently the search-space is pruned." Further, we have developed a version of PARCOMP called PARCOMP-DC that can exploit the regularity of vecprocs using a *divide-and-conquer* technique. The complexity analysis of PARCOMP-DC shows that it could often be faster. Finally, PARCOMP can save the time of simulation; we can perform a "pre simulation" of the tester and the testee using PARCOMP, and run the resultant process. These computational-effort saving measures are believed to be new.

Due to the availability of PARCOMP, it is helpful to think of HOP realprocs and vecprocs as having only absprocs as their submodules (*i.e.* if the submodules are themselves realprocs or vecprocs, one could infer equivalent absprocs using PARCOMP).

We now examine the specification of an absproc in detail.

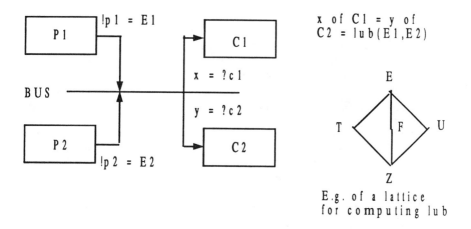

Figure 4: Use of Data Assertions and Queries for Value Communication

2.1 Specifying an Absproc

An absproc is specified by its ports, its events, and its protocol.

2.1.1 Ports and Value Communication

The mechanism of synchronized communication as used in [Mil83] does not accurately model the value communication in hardware systems. As an example, consider figure 4 which depicts a system consisting of two producer processes $P1$ and $P2$ that can communicate with two consumer processes $C1$ and $C2$ over a bus. In this system, it is perfectly acceptable to have a query without a simultaneous assertion, or vice versa. It is even permissible to have two simultaneously active data assertions (say, with compatible "strengths" [Bry84]) on the bus.

In HOP, value communication is performed through a mechanism called *data assertions and queries*. A data assertion, written as !p=E , binds an individual variable p representing the *output port* to the value E at the time the data assertion is made. In general, data assertions are of the form !p=E until e, where e is a future event, where the until operator has the same meaning as the until operator of temporal logic. (Events are discussed shortly.) The lack of an assertion can be modeled by the assertion !p=Z, where Z denotes high impedance.

A data query, written as x=?q, binds x to the value bound to the *input port* q at the time the query is made. Multiple data assertions (as in bus connections) end up asserting the *least upper bound* (LUB) of the asserted values on the port. For handling multiple data assertions, the type of values communicable via ports in HOP must be organized into a strength lattice [Bry84]. For example the *bit type* of HOP includes the weakest value Z (*high-impedance*), truth values T and F, an unknown value U, and the most dominant value E, *error*. T,F, and U are incomparable amongst themselves and lie in-between Z and E.

The mechanism of data assertions and queries is satisfactory for modeling bidirectionality and bussing at the architectural level where "sufficient time is given for combinational paths to settle". Pass transistors can be modeled as idealized switches, ignoring the threshold drop (as in HOL[CGM86]). Busses can be semantically modeled as *logical variables* [Lin85] (as pointed out to us by the author of the cited paper). By having two processes interaction mechanisms (*events* and *data assertions*) we have essentially *separated synchronization from communication.*

Advantages of Data Assertions and Queries

We now show through an example that this separation is advantageous for hardware modeling. Consider a counter with two commands *reset* and *up* that are triggered via events with the same names. After the counter has been subject to the *reset* event and until it is subject to the *up* event, it asserts a data of 0 on its output port. The process that is responsible for the *reset* and the *up* events can, after it has applied the *reset* event (but before it has applied the *up* event) safely assume that the output will be well-defined (and equal to zero) and sample this output as many times as it wishes, without any participation of the counter. In contrast, if value communication is bundled up with rendezvous—as is the case with CSP, CCS, and Circal, the counter would have to actually rendezvous, causing the counter process to make progress in its computation. The writer of the counter process thus has to anticipate all possible places where such rendezvous are possible, and make provisions for them in the specification. Our experience is that this renders hardware specifications unnatural and more complicated. In contrast, with data assertions, once the counter has asserted 0 on its output, it has "discharged all its duties".

One may complain that the separation of communication from synchronization is error-prone. In our experience, this is not true. We have written specifications of systems using data assertions as well as traditional synchronization-plus-communication constructs; the former proved to be much more elegant for modeling value communications hardware. The *spirit* in which this extension to communication mechanisms was made, is similar to the extension made by Martin to CSP to include *Probes* [Mar85]. Both these mechanisms show that concurrency constructs developed for software may not be the best possible ones for hardware modeling.

Data assertions and queries have well-understood and simple semantics. They can model "thru connections" (to be discussed shortly). Interactions such as shown in figure 4 are too awkward to model using existing "CCS/CSP like" languages.

2.1.2 Thru Connections

Now we turn our attention to an important (but hitherto neglected) class of ports called *through connections*. Through connections are wires that pass through a module with or without touching an internal node. It is not satisfactory to model a through connection as a node resident outside the module because by doing so the correspondence between the layout and the high-level specification is not maintained at module boundaries. Maintaining this correspondence would simplify the implementation of a VLSI design system considerably, as has been our experience so far in our use HOP to model Path Programmable Logic (see section 6). In addition, we believe that mixed-mode simulation, layout synthesis, formal verification etc. would be greatly facilitated by maintaining identical interfaces for modules at all levels.

Information regarding through connections is used while compiling a realproc and a vecproc, as well as during PARCOMP. Often thru connections result in bus structures that are embedded within modules. These effects of thru connections are handled appropriately.

2.1.3 Events

Events are of two kinds: input, and output. An input event **e** (written **Ie**) denotes a *condition* that a module senses via wires. An output event **e** (written **Oe**) denotes a *condition* that a module generates via wires. Most modules have, at every point in time, a set of events GE ("good events") that would steer the module into well defined modes of activity. Modules also have, at every point in time, a set of events BE ("bad events") for which they do not have any useful behavior defined. We call the GEs at every point in time as the "synchronization points" of the module.

Events help in making implicit synchronization points explicit. For illustration, consider a clocked synchronous system supporting multiple operations. In traditional designs of synchronous systems, the completion of an operation is not explicitly notified, but is *tacitly* assumed after the elapse of a certain interval of time from the start of the operation. However this approach is worse than hard-wiring literal constants in programs leading to programs that are hard to debug or modify. A better approach would be to encourage the writers of module specifications to "highlight" these synchronization points by introducing *events*. These events may be thought of as being implemented by fictitious control and status wires.

Events have a *conceptual reality* even at very early stages of the design; however they attain *implementational reality* (e.g. "should an event be represented in unary, or in binary?", etc.) only much later. The latter decision is influenced by the nature of the controller, and this is typically decided much later in a design life-cycle.

Some of the advantages of using events are:

1. It becomes possible to statically check for sequencing errors. We show some examples in section 4.

2. It highlights the allowed modes of usage of a module. Hardware specifications must not merely attempt to model hardware as it is; rather they must model hardware as *it is expected to be used*. Hardware systems have astronomically more useless combinations of inputs (as well as *sequences* of combinations of inputs) than useful ones.

3. As digital designs evolve, the events that were originally thought to represent fictitious control wires may be implemented as combinations of control signals and clocks. Combinational logic necessary to decode these combinations and raise the corresponding input event will be tacitly assumed, and not modeled explicitly. This is of advantage on two occasions: (i) when these encodings haven't been decided; (ii) in later stages of a design, when these encodings would be excess baggage to carry around.

4. Event connections between modules is achieved via *renaming*. The actual implementation of renaming is through combinational logic that translates a condition in one module to a condition in another. This could pave the way for the synthesis of "glue logic" that connect modules. This connection between a language operator (*renaming*) and its hardware interpretation (*glue logic*) is pleasant.

2.1.4 Data Path States

In the specification of an absproc, the data path state of the system being specified can be modeled using an appropriate high-level ADT. In our experience, (and as illustrated by the Roll Back Chip (section 5), the use of ADTs having simple definitions can make *reference specifications* far more *reliable* and *easier to understand.*

The introduction of new abstract data types into HOP is greatly facilitated by using an underlying object oriented language called FROBS [Mue87]. The class mechanism of FROBS is used to implement *generic types* (such as the class of stacks over various sizes and element types). Each class acts as a repository of the attributes of the type concerned. Subtyping is realized through class inheritance. The creation of an instance of a generic type amounts to creating an instance of a class. The values of a type themselves are implemented as "(type-descriptor . lisp-data-structure)" pairs. (Overloadable) class methods implement the data type operations. These decisions have made HOP's ADT library very well organized.

2.1.5 The Timing Model

Time is a way to order events. In HOP, processes are lockstep synchronous. Therefore the time of every process advances at the same rate, and thus the event ordering we have can be described via three relations: *simultaneous, before,* and *after.* A HOP specification may or may not refer to a central clock depending on whether it models a clocked synchronous system or a unit-delay combinational system. Currently we do not have the ability to model some subsystems at the unit-delay combinational level, and the remaining subsystems at the clocked level. We hope to add this capability later on, by specifying clock periods to be fixed integral multiples of unit-delays (an idea proposed in [ISD88]).

In later versions of HOP, we will provide a "clock library", i.e. an expandable library of various clocking schemes. Each entry in this library would specify a clock generator of a certain kind; for instance there would be a *two-phase* clock generator in this library.

2.1.6 An Example of an Absproc: A Pipelined Memory

Consider memory module MEM which has an address input port ?addr, a data input ?din port, and a data output port !dout. It can, in its "quiescent state", entertain events Inop, Iwrite, and Iread, each of which implement the commands nop (no op), write, and read. MEM is pipelined thus: the delivery of the result of a read request is overlapped with waiting for the next command. Operation write as well as operation nop (no operation) aren't pipelined.

Let us study figure 5. The header declares two size parameters. The PORT section declares the I/O ports. The EVENT section defines three events, and equates them to "To Be Defined" (TBD). Thus, the designer of MEM doesn't yet care about the encodings of the control inputs as well as clocks (if any). He/she assumes that Iwrite, Iread, and Inop are three control wires coming in.

Consider the PROTOCOL section. This section can always be depicted as shown in figure 6. This is because HOP processes are finitely representable processes (that is, they have a finite-state control skeleton, and this control skeleton can be annotated ("decorated") with data path state changes and port value assertions.) These annotations are done in a purely functional notation. The functional notation improves

```
-- This is a comment.
ABSPROC MEM [ address_size, data_size : int ] -- Note-0

TYPE
 addressType = 0 .. address_size - 1
 dataType    = 0 .. data_size - 1
 memoryType  = array[addressType] of dataType

PORT
 ?din, !dout : array [data_size] of bit
 ?ain : array [address_size] of bit

EVENT
 Imnop, Iread, Iwrite = TBD

PROTOCOL

      MEM  [ms : memoryType] <=
              Imnop -> MEM [ms]
            | Iwrite, va=?addr, vd=?din -> MEM [write(ms,va,vd)]
            | Iread, va=?addr -> MEM1[ms, va]      --^-- Note-1

      MEM1 [ms : memoryType, oa : addressType] <=
              Imnop, !dout=read(ms,oa) -> MEM [ms]
            | Iwrite, na=?addr, vd=?din,
                    !dout=read(ms,oa) -> MEM [write(ms,na,vd)]
            | Iread, na=?addr, !dout=read(ms,oa) -> MEM1[ms, na]
DEFUN

 write :: m : memoryType, a: addressType, d:dataType -> m1 : memoryType
              IF (> addr memSize)
                  (print "Illegal memory address")
                  (error-obj memType)                -- Note-2
              ELSE (update-vector memType m a d) -- Note-3

 read :: m : memoryType, a: addressType -> d : dataType
              IF (> addr memSize)
                  (print "Illegal memory address")
                  (error-obj int)                    -- Note-2
              ELSE (index-vector memType m a)      -- Note-3

END MEM
-- Note-0 : Upper and Lower Cases are Treated the Same in HOP.
-- Note-1 : write (defined in DEFUN) computes the new data path state.
-- Note-2 : error-obj is supported for memoryType by our ADT library
-- Note-3 : index-vector and update-vector supported by memoryType
--          which is defined in ADT Library.
```

Figure 5: Specifications of a Memory

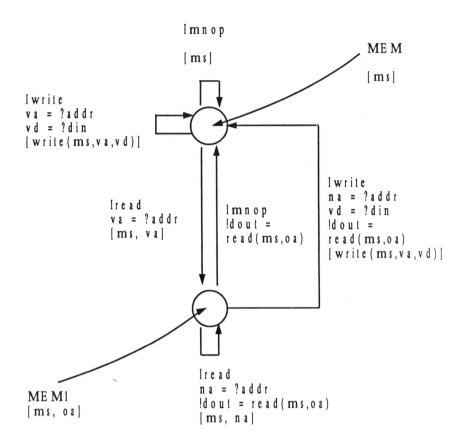

Figure 6: Depiction of the PROTOCOL Specification of MEM

the readability and conciseness of specifications considerably.

The functional expressions *used* in the PROTOCOL section are *defined* in the DE-FUN section and/or in the ADT library. Since the ADT library is implemented using object oriented techniques (our technique: "generic types are classes"), functions are overloaded and dispatched correctly. Besides, subtyping is available for free through class inheritance. The data types support both immutable and mutable constructors. We are currently implementing the *in situ evaluation* technique [GS88] to use mutable constructors whenever possible, while preserving the referential transparency of HOP functional expressions.

Let us study the text of the PROTOCOL section. This section is also depicted in figure 6. In this figure, we have *annotated* the transitions with current events, data queries and assertions, and the *next* data path state; the next data path state is shown only if it is different from the *current* data path state. Process MEM begins in control state MEM and in datapath state ms. It offers a choice of three events, Imnop, Iwrite, and Iwrite. If none of these events is asserted from outside, the behavior of MEM is undefined. Event Imnop (realized by the unasserted combination of the read and write controls) causes MEM to go back to its top control state; Event Iwrite when asserted from outside must be accompanied by data assertions va on the ?addr bus, and vd on the data bus ?din. It causes MEM to go back to the control state MEM; however its datapath state changes to write(ms,va,vd). Event Iread must be accompanied by a data assertion va on port ?addr. The next control state attained is MEM1, and the next data path state is a pair [ms,va].

In control state MEM1, process MEM1 is in data path state [ms,oa]. It again offers the choice of three events. However note that while waiting here, the data assertion !dout=read(ms,oa) is made (this is the pipelining effect). This assertion corresponds to the result of the *previously* requested read. A Iwrite or Imnop takes MEM1 back to MEM; however while reads keep coming, MEM1 goes back to MEM1.

In this specification, the user can model datapath states using an abstract data type of his/her choice. The unit of time is unspecified. If this memory were to be used in a clocked system, the events Iwrite, Iread, etc. would be generated at the appropriate clock phases. Thus, details such as multiphase clocking would be described in the EVENT section of an ABSPROC by replacing the "TBD"s by boolean expressions involving input control wires and clocks.

We assume that Inop is a special event that is asserted if none of the other events are asserted. Such an event exists in most modules, and should be defined to be the "unasserted combination of control+clock inputs".

2.2   Specifying Realprocs and Vecprocs

A realproc specifies a system's realization. As an example let us use the memory unit in figure 5 to build a stack using an absproc CTR to implement the stack pointer and a controller SCTL to control the stack. The design of the stack would be specified by writing a realproc specification, as shown in figure 9. This specification captures the schematic shown in figure 8. Let us now discuss the sections that are important to highlight the roles played by a Realproc.

In the PORT and EVENT sections, the *external* ports and events of the realproc are declared. All other ports and events are assumed to be *internal*, and hence hidden from the outside world.

In the SUBPROCESS section of a Realproc, previously specified abs/real/vec processes are instantiated to the required sizes as well as types. For example we could now instantiate a generic stack to be a stack over bytes. The subprocesses themselves are described in figure 7. We present only the PROTOCOL section of the subprocesses. In the CONNECT section, interconnections between ports as well as events among the submodules, and between the submodules and the external ports/events of the stack are specified. Semantically, connections are treated as *renamings*, in the style of [Mil80]. That is, connected entities are renamed to common names that are unique.

Let us look at the first two lines of the DATANODE subsection of the CONNECT section. (The remainder of the realproc is similar.) The node that connects ?cdo of MEM and !cdo of CTR is hidden. The ?din port of MEM connects to ?din of the stack.

Given the above stack realproc specification and given the specifications for CTR and SCTL shown in figure 7, we can use PARCOMP to infer the equivalent absproc specification STACK shown in figure 10. (Again, only the PROTOCOL section of the inferred process is shown.) This description was obtained automatically, using our implementation of PARCOMP. Inferring the behavior of the stack takes less than ten seconds of elapsed time running on an HP-Bobcat running compiled HP Common Lisp.

The inferred PROTOCOL specification asserts that the STACK system offers a choice of events Ireset, Ipush, Itop, Ipop, and Inop.

Let us study Itop. After asserting this event, the external world (say, the "tester process" of the stack) has to idle for one tick. No event is entertained by the stack (signified by the absence of any input events following Itop), as it is internally busy. During the second tick, it asserts the data value $read(ms, cs)$ on the !dout port. This symbolic expression confirms that the stack would output the correct result on port !dout following the *top* command. Finally, the STACK[cs,ms] process continues to behave like STACK[cs,ms] itself, meaning that the STACK process did not suffer any state changes.

Now let us study the *push* operation. The external world is expected to supply the item to be pushed *two ticks* after it applied the Opush trigger that matched with the Ipush event. If this value were vd, then the future behavior of STACK would be like that of STACK[add1(cs),write(ms,add1(cs),vd)]. This symbolic expression shows that the *push* operation was implemented correctly. This is because the counter state has advanced from cs to add1(cs), and the memory state has advanced from ms to write(ms,add1(cs),vd). Informally, the stack pointer was incremented, and the memory location pointed to by the new stack pointer was written with vd.

The other operations are similarly correct. (Note: While doing the reset, the initial stack pointer value has to be fed from outside via ?cdi.)

2.3 Links to Formal Verification

PARCOMP can be used to simplify the task of verifying hardware realizations. Suppose that the designer had written an independent ABSPROC specification for the stack, as shown in figure 11. Here, the operator ~> signifies "indefinite delay"— the designer doesn't know the exact timings. The designer also uses an Abstract Data Type "stackType" to model the data path state. We can then prove that STKREQ and STACK are indistinguishable with respect to the set of commands they can perform and the results they would deliver. We omit further details.

```
--------------------- An up/down Counter ----------------------

CTR [cs] <=    Icnop, !cdo=cs    -> CTR [cs]
             | Iload, vdin=?cdi -> CTR [vdin]
             | Iup,  !cdo=cs    -> CTR [add1(cs)]
             | Idown, !cdo=cs    -> CTR [sub1(cs)]

--------------------- A stack controller ----------------------

SCTL <=    Isnop,  Omnop, Ocnop  -> SCTL
         | Ireset, Omnop, Ocnop  -> Oload, Omnop -> SCTL
         | Ipush,  Omnop, Ocnop  -> Oup, Omnop   -> Owrite, Ocnop
           -> SCTL
         | Ipop,   Omnop, Ocnop  -> Odown, Omnop -> SCTL
         | Itop,   Omnop, Ocnop  -> Oread, Ocnop -> Omnop, Ocnop
           -> SCTL

-- Note: the ''nop'' events have to be specified at present.
-- Could be specified as defaults later.
```

Figure 7: Specifications of the Submodules of the Stack

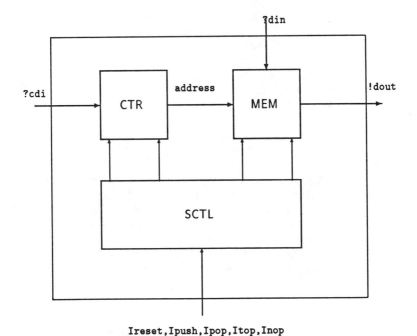

Figure 8: Schematic of the Realproc of a Stack

```
REALPROC stack [<various size & type parameters>]
PORT
  ?cdi, ?din, !dout : <suitable types>
EVENT
  Ireset, Ipush, Ipop, Itop, Inop = TBD
SUBPROCESS -- Note-4
  MEM : mem [<actual size parameters>]
  CTR : ctr [<actual size parameters>]
  SCTL : sctl

CONNECT
  DATANODE
  -- Note-1
  HIDDEN CONNECTS ((MEM ?cdo) (CTR !cdo))
  ?din CONNECTS ((MEM ?din))
  ?cdi CONNECTS ((CTR ?cdi))
  !dout CONNECTS ((MEM !dout))

  EVENTNODE
  -- Notes-2,3
  HIDDEN CONNECTS ((MEM Imnop) (SCTL Omnop))
  HIDDEN CONNECTS ((MEM Iread) (SCTL Oread))
  HIDDEN CONNECTS ((MEM Iwrite) (SCTL Owrite))
  HIDDEN CONNECTS ((CTR Icnop) (SCTL Ocnop))
  HIDDEN CONNECTS ((CTR Iload) (SCTL Oload))
  HIDDEN CONNECTS ((CTR Iup) (SCTL Oup))
  HIDDEN CONNECTS ((CTR Idown) (SCTL Odown))

  Ipush CONNECTS ((SCTL  Ipush))
  Ireset CONNECTS ((SCTL  Ireset))
  Ipop CONNECTS ((SCTL  Ipop))
  Itop CONNECTS ((SCTL  Itop))
  Inop CONNECTS ((SCTL  Isnop))

END stack
--Note-1: Each line of form <extport>/<hidden> CONNECTS <ports>
--Note-2: Each line of form <extevent>/<hidden> CONNECTS <events>
--Note-3: Currently we have to specify even ''obvious defaults''.
-- Later such defaults (such as unasserted values of events etc.)
-- will be automatically provided.
--Note-4: In general module instance names and module type names
-- are different. Here they are the same. E.g. SCTL and sctl.
```

Figure 9: Realproc of a Stack

```
PROTOCOL
STACK [cs,ms] <=
     Ireset -> di = ?cdi ->  STACK [di,ms]
   | Ipush  -> Oidle -> vd=?din ->  STACK [add1(cs), write(ms,add1(cs),vd)]
   | Itop   -> Oidle -> !dout=read(ms,cs) ->  STACK [cs,ms]
   | Ipop   -> Oidle ->  STACK [sub1(cs), ms]
   | Inop   ->  STACK [cs,ms]
```

Figure 10: Absproc Automatically Inferred from stkreal using PARCOMP

```
STKREQ [dps : stackType] <=
   Ireset ~> Ofree -> STKREQ[reset(dps)]
 | Ipush  ~> Idata_avail, vdin = ?din ~> Ofree -> STKREQ[push(dps,vdin)]
 | Ipop   ~> Ofree -> STKREQ[pop(dps)]
 | Itop   ~> Otop_avail, !dout=top(dps) ~> Ofree -> STKREQ[dps]
 | Isdef  ~> Ofree -> STKREQ[dps]
```

Figure 11: Requirements Specification for a Stack

3 Semantics of HOP

3.1 An Operational Semantics for HOP

In this section, we provide an operational semantics for HOP, using many of the conventions presented by Plotkin [Plo81] for writing operational definitions. In addition to describing HOP unambiguously, these rules form the basis for implementing design tools based on HOP. For instance, PARCOMP is written by following these operational rules. Towards the end of this section, we also briefly touch upon the subject of viewing HOP specifications as Temporal Logic formulae or as Higher-Order Logic specifications.

The operational meaning of a HOP process is its *transition relation* $\overset{ca}{\rightarrow} = Proc \times act \times Proc$ where the domain of actions for a process is *act* and that of processes is *Proc*. This relation is defined via structural induction using the notation $\frac{ante}{conse}$ where *ante* is an already defined HOP process (the "antecedent"), and *conse* (the "consequent") introduces the next syntactic category of processes that has not been defined so far.

3.1.1 Action Product

Action product captures how simultaneous actions (events and data actions) interact.

An input event Ie represents a logical condition that is awaited (at some time) by a module. An output event Oe represents the assertion of a logical condition at a particular time instant. Event product, written $e1, e2$ captures how two simultaneous events interact.

Data actions have only one simplification rule defined for them by action product: when two different data assertions $!p = E_1$ and $!p = E_2$ are made, the resultant value on the port $!p$ is defined by the function $lub(E_1, E_2)$. The *lub* function computes the least upper bound of its two arguments over a value lattice. (See figure 4 for an

$$
\begin{aligned}
Ie, Ie &\Rightarrow Ie & (1)\\
Ie, Oe &\Rightarrow Oe & (2)\\
Oe, Oe &\Rightarrow Oe & (3)\\
Oidle, e &\Rightarrow e & (4)\\
!p = E_1, !p = E_2 &\Rightarrow !p = bus(E_1, E_2) & (5)
\end{aligned}
$$

Figure 12: Definition of *Action Product* in HOP

example.) A complete definition of the action product operator is given in figure 12.

3.1.2 Definition of the Transition Relation $\overset{ca}{\rightarrow}$

In this section, we define the transition relation by structural induction. Before these definitions are applied to a realproc or a vecproc, all the port and event names in their submodules are assumed to be renamed so as to be distinct. Also every compound action used in a definition is assumed to have been reduced to an irreducible form by repeated applications of the action product operator ','.

Process STOP

STOP is the simplest of HOP processes. It has a null transition relation; *i.e.* it always remains halted.

A *finite process* is defined to be one that will become STOP in a finite number of steps. A finite process does not usually represent any practically useful hardware system. Therefore if PARCOMP results in a finite process starting from non-finite processes, there is room for suspicion that there are sequencing errors in the system. When none of the *input* events in the branches of a CHOICE process P are synchronized, and when these input events are all hidden, process P is turned into a finite process. This can happen (for example) due to the erroneous sequencing of control inputs.

Sequential Processes

Action: $(ca \rightarrow P) \overset{ca}{\longrightarrow} P$

If P is a process, $ca \rightarrow P$ is a process that first performs the compound action ca and then behaves like P. Sequential Processes are a special case of *deterministic choices* where there is exactly one choice available.

Deterministic Choice

Det-choice: $(|_i \ ca_i \rightarrow P_i) \overset{ca_i}{\longrightarrow} P_i$

A process $P =|_i ca_i \rightarrow P_i$, where i ranges over an index set I is one that offers a *deterministic choice* consisting of the compound actions ca_i during its first computational step. If choice c_M is accepted, P continues to behave like P_M.

If I has more than one element, then there must be an input event e_i present in each ca_i. Since the e_is govern the selection of one of the alternatives of the choices, the e_is must have pairwise mutually exclusive definitions for their control encodings.

Adding Actions To Initials

If P is a process, $ca1, P$ is a process which adds $ca1$ to the initials of P.

$$\text{Add-to-initials:} \quad \frac{P \xrightarrow{ca} P'}{ca1, P \xrightarrow{ca1,ca} P'}$$

Hiding

"Hiding an event e" is a shorthand for saying that both Ie and Oe are hidden from a process. Rule *Hiding-sync* considers the hiding of Oe. Oe is replaced by $Oidle$.

$$\text{Hiding-sync} \quad \frac{P \xrightarrow{ca} P'}{\text{Hide } e \text{ in } P \xrightarrow{ca[Oidle/Oe]} \text{Hide } e \text{ in } P'}$$

The notation "[new/old]" is used to mean that "new" replaces "old".

Hiding Ie from a process prevents it from synchronizing on this event. This can be captured by *pruning* those branches of the synchronization tree that are labeled by Ie:

$$\text{Hiding-unsync} \quad \frac{P \xrightarrow{ca1} P', \ P \xrightarrow{ca2} P'', \ e \in ca1}{(\text{Hide } e \text{ in } P) \xrightarrow{ca2} (\text{Hide } e \text{ in} P'')}$$

Hiding a data output port removes data assertions made on that port from the current compound-action of the process. This would affect those processes that perform a data query from a connected port at the same time:

$$\text{Hiding-dout} \quad \frac{P \xrightarrow{ca,!p=E} P'}{\text{Hide } p \text{ in } P \xrightarrow{ca} \text{Hide } p \text{ in } P'}$$

Hiding a data input port causes those variables that would have been bound by a data query on this port to remain unbound:

$$\text{Hiding-din} \quad \frac{P \xrightarrow{ca,x=?p} P'}{\text{Hide } p \text{ in } P \xrightarrow{ca} \text{Hide } p \text{ in } P' \text{with } x \text{ free in } P'}$$

Renaming

Processes are made to interact with each other either via events or via data actions (da) on ports by renaming their individual event and port names to common names:

$$\text{Renaming-e} \quad \frac{P \xrightarrow{e} P'}{\text{Rename } e \text{ to } e1 \text{ in } P \xrightarrow{e1} \text{Rename } e \text{ to } e1 \text{ in} P'}$$

$$\text{Renaming-port} \quad \frac{P \xrightarrow{da} P', \ da \text{ uses } p}{\text{Rename } p \text{ to } p1 \text{ in } P \xrightarrow{da[p1/p]} \text{Rename } p \text{ to } p1 \text{ in} P'}$$

Parallel Composition

The parallel composition operator $\|$ models the process of realizing a system by putting together several sub-processes, and permitting their interaction through events and ports that are connected.

$$Parcomp \ \frac{P \xrightarrow{ca1} P', \ Q \xrightarrow{ca2} Q'}{(P\|Q) \xrightarrow{ca1,ca2} (P'\|Q')}$$

After performing parallel composition according to the above rule, we may simplify the result by using the following rule (if applicable). This rule captures the effect of value communication:

$$Value \ Communication \ During \ Parallel \ Composition \ \frac{P \xrightarrow{(x=?p),(!p=E),ca} P'}{P \xrightarrow{(!p=E),ca} P' \ [E/x]}$$

Conditionals

HOP processes are usually defined as process schemas $P[dps]$, where for each value of dps we have one specific process. dps usually represents the data path state of the process. We have the notion of *conditional processes* in HOP that allows us to specify the behavior of a process based on its dps variable. Thus we may define a process P as:

$$P[dps] \Leftarrow if \ p(dps) \ then \ P1[f(dps)] \ else \ P2[g(dps)].$$

After reducing the predicate application $p(dps)$ to $true$ or $false$, one of the following rules would apply:

$$Conditional \ \frac{P1 \xrightarrow{ca} P'}{(\text{if } true \text{ then } P1 \text{ else } P2) \xrightarrow{ca} P'} \quad ; \quad \frac{P2 \xrightarrow{ca} P'}{(\text{if } false \text{ then } P1 \text{ else } P2) \xrightarrow{ca} P'}$$

Recursion

A collection of one or more processes may be defined recursively. Since only tail-recursion is allowed, recursion can be modeled as iteration.

Indefinite Delay

The constructs introduced so far are all deterministic. However, nondeterminism seems to be unavoidable at really high-levels of specification when specifying an ordering between every event in the system can be tedious, or may not be possible. Besides this can also over-constrain the specification, thus leading to inefficient hardware designs.

We have begun using an "indefinite delay" operator for writing a priori specifications. The phrase "$\rightsquigarrow e$" stands for: "Delay indefinitely until e occurs." Its definition is as follows:

$$P1 \quad \Leftarrow \quad ca1 \rightsquigarrow e, ca2 \rightarrow R1$$
$$is \ equivalent \ to$$
$$P1 \quad \Leftarrow \quad ca1 \rightarrow Q1$$
$$Q1 \quad \Leftarrow \quad not(e) \rightarrow Q1$$
$$| \ e, ca2 \rightarrow R1$$

(In the *syntactic* rules of the action product operation, a definition $not(Ie), not(Oe) \Rightarrow not(Oe)$ should be introduced. As we will show below, events can be *semantically* interpreted as propositional temporal logic variables.)

An indefinite delay before producing an output event is an instance of nondeterministic behavior. Currently we use $\rightsquigarrow$ only in writing requirements specification for a module. The corresponding implementation of the module may have any specific delay at all, corresponding to every use of the indefinite delay operator in the requirements specification.

We are working on extensions of HOP to include concurrency-related constructs, as briefly discussed in section 7.

Some Relationships of HOP with Other Languages

It is possible to view HOP as stylized formulae in either Temporal Logic or in HOL. For instance the specification in figure 13 can be modeled in temporal logic as shown in figure 14, or in Higher Order Logic [CGM86] as shown in figure 15. In the Tem-

```
P [s] <=    Ie1 -> !dout = 55 -> P [f(s)]
          | Ie2, x=?din -> Q [g(s,x)]
```

Figure 13: An Example HOP Specification

$$P(s) \equiv \Box((Ie1 \supset \bigcirc((!dout = 55) \land \bigcirc P(f(s)))))$$
$$\land (Ie2 \supset (x =?din \land \bigcirc Q(g(s,x)))))$$
$$\land (not(Ie1) \land not(Ie2)) \supset \Box ERROR).$$

Figure 14: Temporal Logic Equivalent of the Example HOP Specification

$$P(s,t) \equiv Ie1 \supset (!dout(t+1) = 55) \land P(f(s),t+2)$$
$$\land \ Ie2 \supset (x(t) =?din(t) \land Q(g(s,x(t)),t+1))$$
$$\land \ (not(Ie1) \land not(Ie2)) \supset \forall k \ ERROR(t+k)$$

Figure 15: HOL Equivalent of the Example HOP Specification

poral Logic specification, we treat port names $?din$ and $!din$ as individual variables. Renaming and hiding are modeled in an obvious way. The effect of simultaneous data assertions and queries on a bus can be handled by first computing the LUB of the asserted values (over the value-lattice of the data items asserted), and then binding this LUB to the variables involved in all the queries on this bus.

In the HOL description, ports are treated as streams, and explicit quantification over time is used. Hiding is modeled in HOL using existential quantification, as described by Gordon in [CGM86].

One benefit of using pragmatically oriented HDLs that have a clean semantics (like HOP), as opposed to directly using universal functional/relational calculii, is simplicity. HOP processes may be viewed as a collection of communicating automatons.

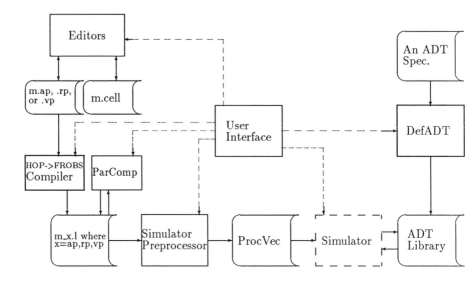

Figure 16: Data Flow Diagram of the HOP Design System

The operational semantics provided in this section define the rules of communication, and they may be understood syntactically. Milner [Mil82] and Plotkin [Plo81] have extolled the virtues of this approach.

Another major benefit of using HDLs is the following. Useful "idioms"—commonly occurring patterns in HDL descriptions—can be identified by trying out a large number of examples. Then we can identify a *subset* of Temporal Logic (or another formalism) that matches these idioms. The advantages of identifying such subsets of (*inherently undecidable*) theories is obvious—we can make a focussed attack on the problem of verification and testing of hardware.

4 The HOP Design System

Figure 16 illustrates the data flow diagram of the HOP design system that is currently under development. Subsystems for which prototypes currently do not exist are shown in dashed boxes. Rectangular boxes indicate functional units, and boxes with curved sides indicate intermediate storage units. Dotted lines show the flow of control, and solid lines show the flow of data.

Input specifications are entered through text editors. File name extensions .ap, .rp, and .vp refer to absproc, realproc, and vecproc. Cell specifications are entered using the PPL layout editor called Tiler [JS86]. HOP specifications are compiled into FROBS representations using the HOP→FROBS compiler. The algorithm PAR-COMP can now be applied on realprocs and vecprocs (presently implemented only

```
TESTER <=  Oreset -> !cot = 0
                 -> Opush  -> Oidle -> !dot = 1
                 -> Opush  -> Oidle -> !dot = 2
                 -> Opop   -> Oidle
                 -> Otop   -> Oidle -> topval = ?dit, !result = topval
                 -> TESTER
```

Figure 17: Description of the Stack Tester Process

for realprocs). PARCOMP infers functionally equivalent absproc specifications from realproc and vecproc specifications. The inferred behavior will be much faster to simulate. The simulator preprocessor compiles the FROBS database into a form suitable for the simulator (under development). A data type definition mechanism has been implemented using FROBS [MLG]. During simulation, the simulator will be called upon to evaluate functional expressions that compute new datapath states as well as output port values. These will be achieved by invoking the operations defined on the various data types.

We now list specific results obtained to date. Details have been omitted.

- The stack realproc was subject to PARCOMP. The result is shown in figure 10. Considerable pruning of the state space was obtained by capitalizing on the *event hiding* information. See appendix 8.1 for details of PARCOMP.

- We deliberately introduced mistakes into the stack controller. Here is a specific experiment: do not generate the Oread event after synchronizing on event Itop, in process SCTL. PARCOMP is able to detect this as an error.

  This is possible because at this point in time, the MEM process offers a choice of input events, none of which is asserted by any other process. Thus, the behavior of the MEM process, and hence the stack process beyond this point is undefined. See figure 19 which shows the STACK process entering a state called STOP. A STOP control state in a process is indicative of a design error, because a hardware system's behavior must be defined for every time instant.

- We composed the stack with a *tester process*, and deduced the behavior of the "tester+testee" system using PARCOMP. The particular tester we used, and the results we obtained are shown in figure 18. We now explain this result in some depth.

Figure 17 shows the stack tester designed by the designer to test the stack for the following sequence of operations: apply reset; then push 1 into the stack; then push 2; then pop the stack, and finally observe the top of the stack and return it through the !result port. In order for this experiment to succeed, the design of the stack has to be correct, and more importantly, the timing protocol followed by the tester in using the stack should also be correct! The result of PARCOMP shown in figure 18 shows that both are correct in this case.

As a specific example, the final result delivered has been inferred completely symbolically to be:

!result =

```
STACK+TESTER [MS, CS] <= Oidle -> Oidle -> STACK+TESTER-1 [MS, 0]

STACK+TESTER-1 [MS, CS] <= Oidle -> Oidle -> STACK+TESTER-2 [MS, (Add1 0)]

STACK+TESTER-2 [MS, CS] <= Oidle -> !dot = 1
      -> STACK+TESTER-3 [(Write MS (Add1 0) 1), (Add1 0)]

STACK+TESTER-3 [MS, CS] <=
      Oidle -> STACK+TESTER-4 [(Write MS (Add1 0) 1), (Add1 (Add1 0))]

STACK+TESTER-4 [MS, CS] <= Oidle -> !dot = 2
      -> STACK+TESTER-5 [(write (Write MS (Add1 0) 1) (Add1 (Add1 0)) 2),
                         (Add1 (Add1 0))]

STACK+TESTER-5 [MS, CS] <= Oidle -> Oidle
      -> STACK+TESTER-6 [(write (Write MS (Add1 0) 1) (Add1 (Add1 0)) 2),
                         (Sub1 (Add1 (Add1 0)))]

STACK+TESTER-6 [MS, CS] <= Oidle -> Oidle ->
  !result =
  (READ
   (WRITE (WRITE MS (Add1 0) 1) (Add1 (Add1 0)) 2)
   (Sub1 (Add1 (Add1 0)))
  )
                  -> STACK+TESTER
```

Figure 18: Inferred Behavior of the Stack Interacting with the Tester

```
STACK [MS,cs]
 <= Itop -> STOP [MS,CS]
   | Ipop  ... same as before ...
   | Ipush ... same as before ...
   | Isnop ... same as before ...

** Protocol Error in Input Specifications **
```

Figure 19: Inferred Behavior of the Stack using an Erroneous SCTL

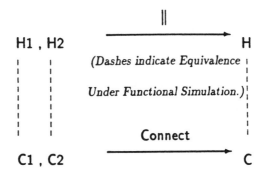

Figure 20: HOP Provides a Compositional Model

```
(READ
 (WRITE (WRITE MS (ADD1 0) 1) (ADD1 (ADD1 0)) 2)
 (SUB1 (ADD1 (ADD1 0))))
)
```

From this inferred behavior it is clear that all value communications that would occur between the various modules have been "compiled away"—i.e. statically determined via symbolic simulation, and are represented as function calls. Since function calls are much cheaper than maintaining and propagating node values, simulation can be greatly speeded-up.

Further, it can be noted that many of the functional expressions contained in the inferred behavior can be simplified using simple rewrite rules. For example, Add1 and Sub1 cancel.

This optimization needs to be implemented in our system. Since FROBS comes with a forward-chaining inference engine, and since the ADT library is implemented using FROBS, the implementation of the expression simplifier will be very modular in the HOP design system.

5 From HOP Specifications to VLSI Layout

In this section, we present a VLSI design technique that is supported by HOP. In this approach, the circuit/layout level is realized using PPL (path programmable logic). (The use of a high-level language for PPL has previously been studied by [SR85].)

PPL is a cellular grid-based design methodology. The layout is accomplished by tiling a plane with predefined cells, called the PPL cells. Until recently, one disadvantage of PPL was the limited set of predefined cells; however the capability for users to add (compact) custom cells is planned. Considering this, HOP seems to be an attractive PPL cell library specification language.

Figure 20 illustrates the correspondence we will aim for between the PPL and HOP levels; if this correspondence were to be established in some rigorous way, it could pave the way for the development of provably correct circuits from verified HOP specifications. PARCOMP could then serve as a "high-level circuit extractor."

For illustration we show how a D flip-flop fed through a tri-state driver can be described both in PPL and HOP. Figure 21 shows the PPL circuit schematic of the

152

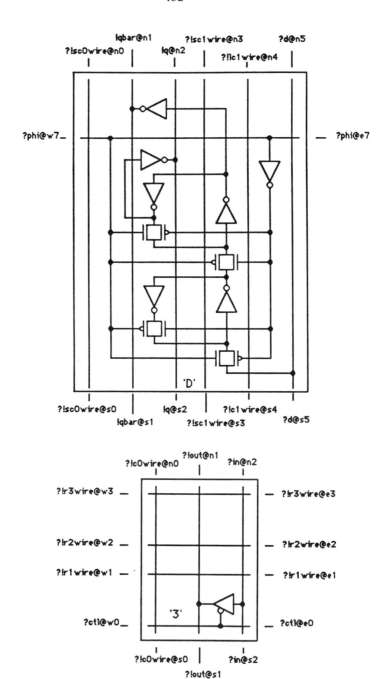

Figure 21: The Circuit Schematic of a 'D' cell Connected to a '3' cell

```
ABSPROC dff[] -- a D flip flop
PORT
   ?phi@e7, ?phi@w7, ?d@s5, ?d@n5, !q@n2, !q@s2, !qbar@n1, !qbar@s1,
   ?!sc0wire@n0, ?!sc0wire@s0, ?!sc1wire@n3, ?!sc1wire@n3,
   ?!c1wire@n4,  ?!c1wire@s4 : bit
CLOCK singlephase ?phi
EVENT
  Iload = (?phi)
  Ihold = (bit-not ?phi)
PROTOCOL
  dff[dps1,dps2] <=
           Iload, vd=?d, !q = dps2, !qbar = not(dps2) -> dff[not(vd), dps2 ]
         | Ihold, !q = not(dps1), !qbar = dps1 -> dff[dps1, not(dps1)]
END dff


----------------------------------------------------------------------


ABSPROC 3 [] -- a tristate driver
PORT
   ?in@s2, ?in@n2, ?!out@s1, ?!out@n1, ?ctl@wo, ?ctl@e0,
   ?!r2wire@e2, ?!r2wire@w2, ?!r3wire@e3, ?!r3wire@w3,
   ?!r4wire@e4, ?!r4wire@w4, ?!c0wire@n0, ?!c0wire@s0 : bit

PROTOCOL
  !out (if ?ctl ?in Z)

END 3
```

Figure 22: HOP Specifications of the 'D' and '3' Cells

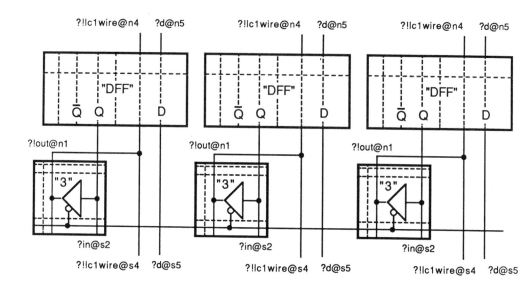

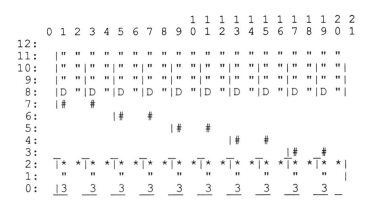

Figure 23: An Array of D-Flip Flops and Tri-State Drivers in PPL

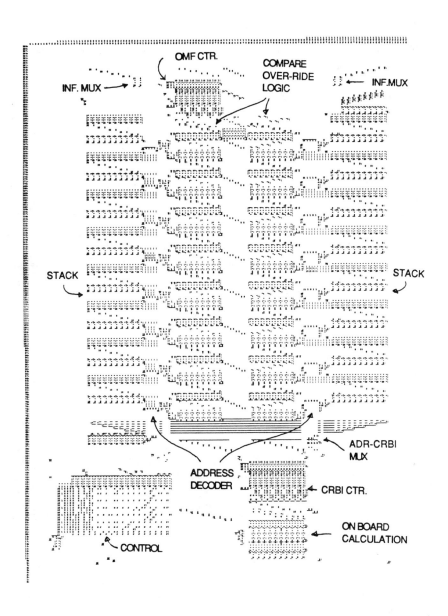

Figure 24: The Layout of the RBHC Chip in PPL

"3" cell (a tristate driver) and the "D" cell (a D flip-flop). Shown in figure 22 are the HOP descriptions for these cells. The HOP description gives a physical description of the location of ports along with a behavioral description of the cell. For example, the port labeled "?ctl@w0" in the '3' cell corresponds to the an input port named "ctl" located at the west side of the zero'th row location. All of the port numbers are relative to the lower left hand corner of the cell. In PPL the connections between ports are implied based on their placement in the circuit plane.

Figure 23 shows a PPL circuit as viewed when tiling a circuit using PPL tools. Such an ASCII file is all that is needed to generate the CIF code and have the chip fabricated (of course there are no pads shown in this figure, but they could be easily placed). This figure is an excerpt from the design of a large chip called the Rollback History Chip (RBHC) which we have conducted. The corresponding schematic is also shown in the same figure. PARCOMP could be run on the realproc generated from the PPL description, thus giving a new behavioral and physical description of the circuit. PARCOMP-DC (section 8.2) also seems promising since PPL systems often possess high degrees of geometrical regularity. As can be seen, layout composition is paralleled by PARCOMP at the HOP level. This process could be repeated to any level of design. Simulation can then be performed on the realproc with much improved speed. Verification can also be performed by the designer.

To illustrate that this technique can be used even for non-trivial designs, we present the complete layout of the RBHC in figure 24.

6   The Design of the Roll Back Chip (RBC)

In this section we present a case study of specification driven design. The example considered is called the Roll Back Chip, which forms part of the Utah Simulation Engine. Our presentation will be qualitative in nature, and is intended to highlight the issues that arise during the design of a large hardware system. We present four levels of refinement of the RBC.

Introduction to the RBC

Simulation plays an important role in the study of systems. For instance, a proposed computer architecture must be evaluated through simulation before it is built; an assembly-line simulated before it is put into operation. Such simulations are usually conducted using *discrete event simulation* techniques. Processes in the real-world are modeled as logical processes in the simulator. These processes communicate using time-stamped messages. A (fairly strong) sufficient condition for correctness in discrete event simulation is that messages be processed in non-decreasing time-stamp order, thus preserving data dependency relations.

Due to the decreasing cost of computers, there has been growing interest in speeding up simulation by spreading the logical processes of a simulation in the various nodes of a multiprocessor. A central global clock cannot be used if the simulation is asynchronous, forcing one to use independent, local clocks. Unfortunately many strategies that have been proposed for doing so are fraught with problems such as the proliferation of null messages, the risk of running into a deadlock, etc. [Fuj88]. Time-Warp is a promising alternative that doesn't suffer from the problems due to null messages or deadlock. To support time-warp, the following capabilities are necessary:

- It must be possible to take a snapshot of a process's data segment at any instance of time, and save this snapshot.
- It must be possible to restore a saved snapshot; by restoring a snapshot we mean throwing away the current data segment and reverting to a previously saved data segment.

Doing the above operations in software has proven to be extremely expensive. On the other hand it has also been observed that if the above operations could be made negligibly expensive, then time-warp would be one of the best possible approaches to distributed discrete event simulation.

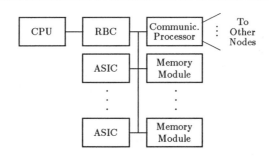

Figure 25: Configuration for Each Node of the Utah Simulation Engine

At the University of Utah, a hardware architecture called the USE (Utah Simulation Engine) has been under development for the past year for speeding up distributed discrete event simulation using time-warp. The Roll Back Chip (RBC) is a key component in USE. A node of the USE is shown in figure 25.

In this section we present the development of the RBC to date. The vesion of RBC presented in this paper corresponds to that presented in [FTG88]. Chronologically, the development of the RBC progressed along the following lines:

- We studied the problem at hand and captured the behavior of the intended architecture in pseudo-code;
- We specified the early design of the RBC in HOP. In this specification we used a data type similar to an ordinary stack to model the data path state of the RBC (discussed in section 6.1). This reference specification has proven to be valuable in many ways (to be discussed).
- We identified heuristics for speeding-up the initial design specification by identifying "clever data structures and algorithms"—i.e. by adopting problem-specific heuristics;
- For each such heuristic, we informally argued why each optimization strategy adopted is sound;
- We evaluated the architecture by writing a simulation of the chip in the C programming language. (The implementation of the HOP simulator isn't complete yet.)
- We coded the original reference specification in C and compared the behavior of the detailed description of the RBC against the reference specification via simulation. Both systems were subject to over a million operations, and the

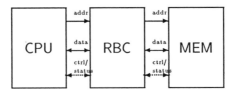

Figure 26: The Architectural Context of the RBC Chip

---

results produced were compared. (A form of "bisimulation".) Though not as effective as formal verification, we did discover a few bugs via this process. This supports the observation that formal reference specification directed simulation is useful in practice.

- A key component of the Roll Back Chip, the "Roll Back Cache" (RBCache) was designed, laid out, and simulated using the Path Programmable Logic (PPL) Tools [NM]. A complete specification of the TLB in HOP is currently being written. Currently another major component, the "Roll Back History Unit" has been fully designed and simulated in parts, and is being laid out in PPL.

The purpose of the remainder of this section is to reenact the development of the RBC (Parnas calls this "faking a rational design process" [PC86]). It has been our experience (hopefully shared by many of you) that as opposed to the development of relatively well-known architectures such as microprocessors, etc. the development of large, non-conventional VLSI architectures involves several iterations of the design, adjusting it to newly discovered opportunities as well as newly discovered hindrances almost on a day-to-day basis. Although one hopes that such changes be discovered as well as performed automatically on a formal specification of the system without human involvement, a more tangible hope seems to be that formal specifications be used in the following roles:

- Provide a notation for the digital system architect that is *formal, easily amenable to verification*, and *practically oriented*;
- Support the recording of "hunches" leading to optimizations as *invariants, or lemmas* to be proved subsequently; The LP theorem prover [GG88] can be used for reasoning about such equational formulae. (See also [GGS88].)
- Support the evaluation of performance (speed etc.) of the design using simulation studies; conduct only informal design verification at this stage, as the proposed design may not be selected owing to poor performance;
- Conduct a formal verification of the entire system when the final architecture has been selected.

Some of the levels of refinement of the RBC system are now examined.

6.1 The First Level: System RM1

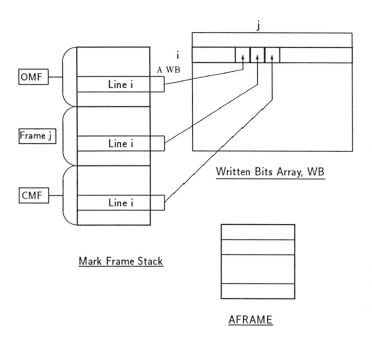

Figure 27: The First Two Levels: Systems RM1 and RM2

One way to understand the RBC is as a memory management unit situated between the CPU and the memory, as shown in figure 26. The behavior of the RBC plus a memory module from the point of view of the CPU is now described (Note: the name "RM1" corresponds to RBC plus Memory, level 1.)

- A program P running in the above system uses a data segment ranging from address *Amin* to *Amax*. This is called the *version controlled memory* of which snapshots can be taken and restored.
- At any time P can *read* from the data segment from address A (operation $Read(A)$).
- At any time program P can *write* into the data segment at address A the data d.
- Periodically, program P issues a request to the 'RM1' system to take a version ("snap-shot") of the data segment contents; P then continues with the computation; this operation is called $Mark()$.
- At some future time, P encounters an error and wants to revert to an earlier version of the data segment state, effectively restarting the computation at some time in the past. This operation is called *Rollback*.
- At any time P can discard very old versions of checkpointed state that are no longer needed. This operation is called *Advance*.

Figure 27 shows a data structure similar to an ordinary stack that was used to model the above requirements specification. (Please ignore the WB array and the Aframe for now.) Specifically, the data type used is:

```
type RM1type =
  RECORD
    STK : array [nframes] of array [nlines] of <data,wb>
    CMF, OMF : int
  END
```

The data type consists of a stack STK and two pointers CMF and OMF. STK contains **nframes** frames, each of which is as large as the data segment. STK is used to hold the multiple snapshots of the data segment. CMF points to the current "mark frame", or the current data segment. OMF points to the oldest mark frame in use. Each frame of the stack holds **nlines** "lines" of data. A line is a group of bytes, and is similar to the lines in a cache. Actually pairs **<data,wb>** are maintained. The **wb** stands for "written bit", and is set only if the line has been written into. A **wb=0** corresponds to a "hole"; *e.g.* if location i of a frame F has a zero written bit, then location i was not written into while F was the "current" mark frame.

(For simplicity assume that the fields of the record correspond to variables with the same names as the fields.) The operations of the RBC are:

read(a) finds the least f starting from $f = CMF$ such that STK[f][a].wb is set; it then returns the data item STK[f][a].data. Frame f in this example corresponds to the *most recent version* (MRV) of the data at line address a. Hence we will call such a frame the "MRV frame".

write(a, d) updates STK[CMF][a].data with d, and also sets STK[CMF][a].wb.

mark(k) increments CMF by k.

rollback(k) decrements CMF by k, and if CMF falls behind OMF, an error is reported.

advance does nothing at all—we are not interested in garbage collecting the conceptual stack!

6.2 The Second Level: System RM2

We move towards practical reality by making the following changes (see figure 27):

- We detach the written bits from the stack and pool them into a matrix;
- We treat the stack as a circular buffer; the data area is shown side-by-side with the WB array;
- We make a provision for garbage collection by introducing the frame called Aframe. The idea behind garbage collection is simple. When the OMF frame is no longer needed (this happens when the *Global Virtual Time*(GVT) [Jef85] exceeds the the time-stamp of the OMF frame), we may free-up the OMF frame *provided* we are not throwing away a legal version of any data. The steps in archiving are the following:

```
Suppose the GVT advances to frame OMF+K. Then
For every line l that doesn't have a set written bit in frame OMF+k
   determine the MRV of the data looking back from OMF+K;
   copy this MRV data to Aframe[l]
end For;
Free up frames from OMF to OMF+k.
```

All the operations in level RM2 are implemented as in level RM1 except for the MRV computation. In RM2, if no set written bits exist for a line l between positions CMF and OMF, then the MRV is read from Aframe[l].

From the point of view of the user, all the levels of the RM system are best regarded as abstract data types with *constructors write, mark, rollback*, and *advance*, and the only observer *read*. Hence the verification condition for *functional correctness* can be stated using *data type induction* [GHM78]. For example, the verification condition going from level RM1 to level RM2 is:

For all constructor operations *op* defined at level RM1 (op_{RM1}) and RM2 (op_{RM2}) and their arguments *oparg*

$$\frac{read_{RM1}(dps_{RM1}, l) = read_{RM2}(dps_{RM2}, l)}{read_{RM1}(op_{RM1}(dps_{RM1}, opargs), l) = read_{RM2}(op_{RM2}(dps_{RM2}, opargs), l)}.$$

Each level following RM2 has to be similarly verified with respect to RM1.

For an earlier design of the RBC chip, we did prove some of these invariants by hand by unfolding the consequent part (the portion below the line) using the definition of *read* and *op* at the lower level, performing a case analysis on it, and showing that in each of these cases the consequent either reduces to an instance of the antecedent or follows from the antecedent. In a few cases, proof by contradiction proved to be much easier.

Our current focus is in formulating verification criteria, and understanding what exactly verification means in the context of complex timings.

6.3 The Third Level: System RM3

Obviously, searching down the stack for every *read* operation is inefficient. Hence we introduce a cache unit to cache the data in the MRV frame (see figure 28). The management of this cache is quite different from traditional cache units because we have to guarantee that whenever there is a cache hit, we do return the MRV data item. This also means that when a roll-back or advance occurs, suitable invalidations of the cache entries are performed.

At level RM3 we have three processes, LRU, RBCache, and CTRLR communicating via their interface protocols. The process CTRLR may not realized as a stand-alone unit; it simply represents the portion of the microcode devoted to the management of level RM3.

It is possible to model the system state as a three tuple of the states of these modules and specify the RBC operations as mappings on these states. A more useful approach—the one we plan to use—would be to model these as stand-alone processes, compose them using PARCOMP, and then compare the process inferred by PARCOMP with a process representation of level RM2. While we have begun working in this direction, we do not have a satisfactory solution yet. The questions we would like to answer are: (i) how best to write a prior specifications in a manner that permits systematic refinement to more detailed timing? (ii) how to discover optimizations such as pipelining in this process? (ii) What forms of nondeterminism are useful at the architectural level of specification?

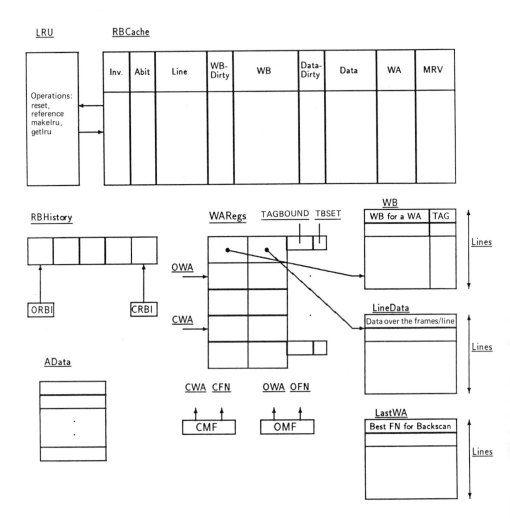

Figure 28: The Levels RM3, RM4 and RM5

6.4   The Fourth Level: System RM4

When a cache miss occurs in the RBCache, the MRV entry has to be retrieved from the memory. A backward search beginning at CMF searching for the first set written bit has to be performed. Is it possible to keep a good estimate of the point from which to begin the backwards scan? The answer is 'yes', and incorporating this optimization leads to model RM4.

In this scheme (shown in figure 28), a frame full of pointers (one for each line) is maintained in the array LastWa. The contents of LastWa are updated *whenever a cache entry is evicted*. Our simulation studies show that the use of LastWa cuts down the average distance of backscan for the MRV dramatically.

6.5   The Fifth Level: System RM5

When a Rollback occurs, not only must we invalidate entries in the cache that correspond to frames that have been "rolled over", but we must also clear the written bits in the memory corresponding to these "rolled over" frames. This is because we are effectively forgetting the existence of these frames. However, industriously clearing the written bits isn't wise. We have developed a lazy approach to clearing the written bits. In figure 28 the addition to our "data structures" for achieving this optimization is shown. The mechanism is called the "Roll Back History Unit", or the RBhistory unit.

Section 5 presents the design of the RBH unit.

Section Summary

The current design of the RBC chip is still at least five more such refinement steps away! The five steps shown so far make it amply clear to us that digital system architects routinely think of data abstractions and are aware of optimization invariants. However they don't seem to document these thoughts in a formal specification language. Here is where specification driven design approaches have a definite contribution to make. Also as we mentioned in section 1, lack of performance is also a design error. (Interestingly, the USE system could be used to conduct performance studies of digital architectures—even of itself—quite rapidly!)

7   Summary

- We have developed a hardware specification language HOP, and are using it for the specification driven design of a large custom architecture. This cooperative research is expected to lead to a language well balanced in formal details as well as pragmatic ones.
- One of the key results so far has been the recognition of the importance of data and structural abstraction mechanisms in real-world design. HOP seems to be satisfactory in these respects.
- Constructs for modeling truly concurrent processes (fork/join, barrier synchronization, mutual exclusion of resources, etc.) are missing in the present version of HOP. We are designing a successor to HOP called "HOP-CP" (CP for concurrent processes) that includes these constructs. The present version of HOP

would be truly compatible with HOP-CP. Specifications for the components of USE will be written in HOP-CP.

- A library based design approach where certified circuit/layout "tiles" are stored along with high-level specifications is believed to be an effective way to translate correctness proofs at the high-level into correctness assurances regarding the circuit/layout. Our experiments with PPL substantiate this observation.

Acknowledgements

The first author acknowledges the support of his research by the National Science Foundation under Grant No.MIP-8710874. The second author acknowledges the support of his research by ONR, under contract number 00014-87K-0184.

8 Appendix

8.1 A Specification of PARCOMP

$\boxed{\text{Input:}}$ An expression Hide HS in $\parallel \{P_i[\overline{X_i}], ..., C_j[\overline{X_j}], ...\}$ for $i \in \{1..m\}$, $j \in \{1..n\}$. C_j are conditional processes of the form
$C_j[\overline{X_j}] =$ if q_j then $T_j[g_j(\overline{X_j})]$else $F_j[h_j(\overline{X_j})]$ and P_i are non-conditional processes of the form
$P_i[\overline{X_i}] = y_i : initials_i \to R_i(y_i)$;

Each P_i offers a set of initial choices $initials_i$ and for each choice y_i that is offered, the future behavior of P_i is $R_i(y_i)$. HS is the *Hidden Set*, the set of events and ports hidden from the parallel composition.

$\boxed{\text{Output:}}$ A behaviorally identical process $P[\overline{X_i}, ..., \overline{X_j}, ...]$.

$\boxed{\text{Method:}}$ A *done-list* is maintained for each parallel composition $\parallel \{P_i[\overline{X_i}], ...\}$ that has already been computed. Upon getting a call for performing parallel composition, the *done-list* is first consulted.

- If the requested parallel composition is in the *done-list*, return. Else enter it in the *done-list* and proceed as follows.

- Combine all conditional processes into one conditional process C. Combining two conditional processes is done as follows:

$$C_1[\overline{X_1}] = \text{if } q_1 \text{ then } T_1[g_1(\overline{X_1})] \text{ else } F_1[h_1(\overline{X_1})]$$

$$C_2[\overline{X_2}] = \text{if } q_2 \text{ then } T_2[g_2(\overline{X_2})] \text{ else } F_2[h_2(\overline{X_2})]$$

$$
\begin{aligned}
C_1[\overline{X_1}] \parallel C_2[\overline{X_2}] \quad = \quad & \text{if } (q_1 \wedge q_2) \text{ then } T_1[g_1(\overline{X_1})] \parallel T_2[g_2(\overline{X_2})] \\
& \text{else if } (q_1 \wedge not(q_2)) \text{ then } T_1[g_1(\overline{X_1})] \parallel F_2[h_2(\overline{X_2})] \\
& else \ ...etc. \ (all \ four \ combinations)
\end{aligned}
$$

- Now we are left with the task of computing Hide HS in $\parallel \{P_i[\overline{X_i}], ..., C\}$. Let C be of the form

$$\text{if } q_1 \text{ then } C_1[g_1(\overline{X_1})]\text{else if } q_2 \text{ then } C_2[g_2(\overline{X_2})]etc.$$

$\| \{P_i[\overline{X_i}], ..., \ C\}$ reduces to a conditional process with q_i as the conditions. This conditional has in it parallel compositions of the form $\| \{P_i[\overline{X_i}], ..., \ C_i\}$. that is (recursively) computed. Eventually we are faced with composing non-conditional processes in parallel. We take this up next.

- Consider $\| \{P_i[\overline{X_i}], ...\}$. Let each P_i be

$$P_i[\overline{X_i}] = ca_i^1 \to R_i^1[f_i^1(\overline{X_i})]$$
$$| \quad ca_i^2 \to R_i^2[f_i^2(\overline{X_i})]$$
$$| \quad ...$$
$$| \quad ca_i^{n_i} \to R_i^{n_i}[f_i^{n_i}(\overline{X_i})]$$

- Generate tuples

$$T =< ca_1^{x_1}, \ ca_2^{x_2}, \ ...ca_m^{x_m} >$$

i.e. a tuple of the x_1th initial compound action offered by P_1, the x_2th initial compound action offered by P_2, etc. This tuple T is assumed to be the irreducible form arrived at after applying the action product rules of figure 12. According to the rule for parallel composition Parcomp all such tuples would become the initial choices of the resultant process. Following such choices, the resultant process would continue to behave like $\| \{R_1^{x_1}[f_1^{x_1}(\overline{X_1})]R_2^{x_2}[f_2^{x_2}(\overline{X_2})], ...\}$. However using the hiding information HS, we can prune many of these choices. In particular,

- those tuples T that contain *unsynchronized* events Ie that belong to HS are dropped, and the corresponding arm of the synchronization tree is pruned;
- those tuples T that contain Oe that belong to HS are replaced via the substitution $T[Oidle/Oe]$.

- In computing

$$\| \{R_1^{x_1}[f_1^{x_1}(\overline{X_1})], R_2^{x_2}[f_2^{x_2}(\overline{X_2})], ...\},$$

the bindings generated by taking action products of the members of T are taken into account. □

8.2 A Divide-and-Conquer PARCOMP, PARCOMP-DC

Consider the array A shown in figure 29. It consists of a collection of modules M connected in a regular interconnection pattern. For simplicity of explanation, assume a nearest-neighbor connection that is regular in both the dimensions.

Consider the problem of computing $PARCOMP(A)$; *i.e.* the composition of all the Ms constituting A. $PARCOMP$ is both commutative and associative. Hence, we can split A into two halves, say A_T standing for "the top of A" and A_B, standing for "the bottom of A", and assert:

$$PARCOMP(A) = PARCOMP(PARCOMP(A_T), \ PARCOMP(A_B)).$$

Since A_T and A_B differ only in the names of their external ports, we need compute only $PARCOMP(A_T)$. $PARCOMP(A_B)$ can be obtained from this, by renaming the ports of A_T to the corresponding ports of A_B.

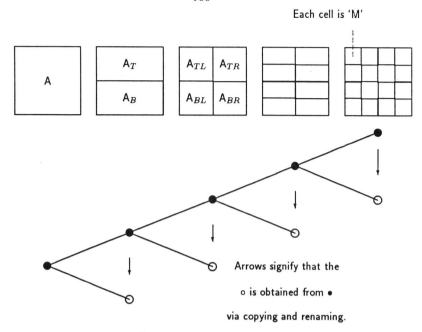

Figure 29: Divide and Conquer PARCOMP

This division process can be carried down to the leaf cells, as depicted in figure 29.

PARCOMP-DC is often more efficient than PARCOMP. Let us make an approximate cost analysis. The worst-case time complexity of PARCOMP is primarily dependent on the number of control states that we have in a process diagram. Specifically, it can be equal to the cross-product of the number of control states in each of the processes. Suppose for simplicity that array A is square, and has N modules of type M, M has C control states in it, and that N be a power of 2. Then

$$cost\_parcomp(A) = O(C^N)$$

because we may, in the worst-case, end-up taking a full cross-product of the process diagrams of the N modules.

Suppose that the modules formed during the division process of PARCOMP-DC are $M, ..., A_{TL}, A_T, A$. Let $ncs(M)$ denote the number of control states in a module M. Further let $C\_copying$ denote the cost of copying the process descriptions (see figure 29). Then

$$cost\_parcomp\_dc(A) = O(ncs(M)^2 + ... + ncs(A_{TL})^2 + ncs(A_T)^2 + ncs(A)^2 + C\_copying).$$

The above sum has $log_2(N)$ terms. Let D be the *root mean square* (RMS) value of the number of control states in $M, ..., A_{TL}, A_T, A$. Let the cost of copying and applying renamings to a process description not exceed the number of control states in it. (This is the case for our data structures that represent processes.) The number of queries and assertions in a compound action is assumed to be bounded by a constant. Then,

$$cost\_parcomp\_dc(A) = O(\log_2(N) \times (D^2 + D^2)) = O(\log_2(N) \times D^2).$$

Firstly we note that D does not tend to increase as the size of the modules grow. This is a fact of practical systems because when designing a module using several submodules, only very few of the astronomically large number of sequences of the submodule operations are actually used. Hence the number of control states in a module is often vastly smaller than what it could be. (Consider for example the total number of possible microprograms for a typical datapath .vs. the number of microroutines that are actually ever used!) Thus if D is close to C and if M is large, then there is a significant payoff by using PARCOMP-DC.

In conclusion, the following approach is suggested for handling arhythmic arrays:

- Perform PARCOMP of two modules of the array;
- Study the inferred behavior and see if it is verifiable manually or through exhaustive simulation.
- The behavior inferred by PARCOMP (or PARCOMP-DC) will have complex if-then-else functions. Construct tabular functions corresponding to these.
- Use these tabular functions for efficient simulation.
- Try to perform formal verification of the whole array by setting up an induction.

References

[ACFM85] T.S. Anantharaman, E.M. Clarke, M.J. Foster, and B. Mishra. Compiling Path Expressions into VLSI Circuits. In *Proceedings of the 12th Symposium on Principles of Programming Languages*, ACM, January 1985.

[Bae86] Jean-Loup Baer. Modelling Architectural Features With Petri Nets. In *Petri Nets: Applications and Relationships to Other Models of Concurrency*, pages 258–275, Springer Verlag, September 1986. LNCS 255.

[BCDM85] M. Browne, Edmund Clarke, D. Dill, and B. Mishra. Automatic Verification of Sequential Circuits using Temporal Logic. In *Proceedings of the Seventh International Conference on Computer Hardware Description Languages*, pages 98–113, North-Holland, 1985.

[Bro75] Frederick P. Brooks. *The Mythical Man-month*. Addison-Wesley, 1975.

[Bry84] Randall E. Bryant. A Switch Level Model and Simulator for MOS Digital Systems. *IEEE Transactions on Computer*, C-33:160–177, February 1984.

[CGM86] Albert Camilleri, Michael C. Gordon, and Tom Melham. Hardware Specification and Verification using Higher Order Logic. In *Processings of the IFIP WG 10.2 Working Conference on "From HDL Descriptions to Guaranteed Correct Circuit Designs", Grenoble, August 1986*, North-Holland, 1986.

[Chu87] Tam-Anh Chu. Synthesis of Self-timed VLSI Circuits from Graph-theoretic Specifications. In *International Workshop on Petri Nets and Performance Models, Madison, Wisconsin*, August 1987. See also MIT VLSI Memo no.87-410, September 1987, with the same title.

[Coh88] Avra Cohn. Correctness Properties of the Viper Block Model: The Second Level. In *1988 Banff Workshop on Hardware Verification (this volume)*, Springer Verlag, 1988.

[FTG88]   Richard Fujimoto, Jya-Jang Tsai, and Ganesh Gopalakrishnan. Design and Performance of Special Purpose Hardware for Time Warp. In *15th Annual International Symposium on Computer Architecture, Honolulu*, pages 401–408, 1988.

[Fuj88]   R. M. Fujimoto. Performance Measurements of Distributed Simulation Programs. *1988 Society for Computer Simulation Multiconference*, feb 1988.

[GG88]   Stephen J. Garland and John Guttag. Inductive Methods for Reasoning About Abstract Data Types. In *15th ACM Conference on Principles of Programming Languages*, January 1988. San Diego, CA, January 13-15.; This article describes the theory behind the Larch theorem prover (LP).

[GGS88]   Stephen Garland, John Guttag, and Jorgen Staunstrup. Verification of VLSI circuits using LP. In George Milne, editor, *1988 Glasgow Workshop (IFIP WG 10.2) on Hardware Verification*, 1988.

[GHM78]   John V. Guttag, Ellis Horowitz, and David R. Musser. Abstract Data Types and Software Validation. *Communications of the ACM*, 21(12):1048–1064, December 1978.

[Gop86]   Ganesh C. Gopalakrishnan. *From Algebraic Specifications to Correct VLSI Systems*. PhD thesis, Dept. of Computer Science, State University of New York, December 1986. (Also Tech. Report UU-CS-86-117 of Univ. of Utah).

[GS88]   Ganesh C. Gopalakrishnan and Mandayam K. Srivas. Implementing Functional Programs Using Mutable Abstract Data Types. *Information Processing Letters*, 26(6):277–286, January 1988.

[GSS87]   Ganesh C. Gopalakrishnan, Mandayam K. Srivas, and David R. Smith. From Algebraic Specifications to Correct VLSI Circuits. In D.Borrione, editor, *From HDL Descriptions to Guaranted Correct Circuit Designs*, pages 197–225, North-Holland, 1987. (Proc of the IFIP WG 10.2 Working Conference with the same title.).

[Hen84]   Matthew Hennessy. *Proving Systolic Systems Correct*. Technical Report CSR-162-84, Department of Computer Science, University of Edinburg, June 1984.

[HK87]   Richard H. Lathrop Robert J. Hall and Robert S. Kirk. Functional Abstraction from Structure in VLSI Simulation Models. In *Proc. 24st Design Automation Conference*, pages 822–828, 1987.

[ISD88]   I.S.Dhingra. Formal Verification of a Design Style. In Graham Birtwistle and P.A.Subrahmanyam, editors, *VLSI Specification, Verification and Synthesis*, pages 293–321, Kluwer Academic Publishers, Boston, 1988. ISBN-0-89838-246-7.

[JB85]   Jeffrey Joyce and Graham Birtwistle. *Proving a Computer Correct in Higher Order Logic*. Technical Report 85/208/21, Dept. of Computer Science, Univ. of Calgary, August 1985.

[JBB88]   Stephen Johnson, B. Bose, and C. Boyer. A Tactical Framework for Hardware Design. In Graham Birtwistle and P.A.Subrahmanyam, editors, *VLSI Specification, Verification and Synthesis*, pages 349–383, Kluwer Academic Publishers, Boston, 1988. ISBN-0-89838-246-7.

[Jef85]   D. R. Jefferson. Virtual Time. *ACM Transactions on Programming Languages and Systems*, 7(3):404–425, July 1985.

[JS86]   Steve Jacobs and Kent Smith. TILER User's Guide. 1986. User's Manual Available from the Univ. of Utah, Dept. of Computer Science VLSI Group.

[Lin85]   Gary Lindstrom. Functional Programming and the Logical Variable. In *Proceedings of the 12th ACM Symposium on Principles of Programming Languages*, pages 266–280, January 1985.

[Mar85]   Alain J. Martin. The Probe: An Addition to Communication Primitives. *Information Processing Letters*, 20(3):125–130, April 1985. An Erratum related to this article appeared in the August 1985 issue of the Info. Proc. Letters.

[MH85]   M.Lam and H.T.Kung. A Transformational Approach to Systolic System Design. *IEEE Computer*, 18(2), 1985.

[Mil80]   Robin Milner. *A Calculus of Communicating Systems*. Springer-Verlag, 1980. LNCS 92.

[Mil82]   Robin Milner. *Calculii for Synchrony and Asynchrony*. Technical Report CSR-104-82, Univ. of Edinburg, 1982. Internal Report.

[Mil83]   George J. Milne. CIRCAL: A calculus for circuit description. *Integration*, (1):121–160, 1983.

[Mil85]   George J. Milne. Simulation and Verification: Related Techniques for Hardware Analysis. In *Proceedings of the Seventh International Conference on Computer Hardware Description Languages*, pages 404–417, North-Holland, 1985.

[MLG]   John Merk, John Lalonde, and Ganesh Gopalakrishnan. ADTP User's Manual. Requirements Specification and User Manual for the Abstract Data Type definition Package (ADTP), Software Engineering Lab., Spring 1988.

[Mue87]   Eric G. Muehle. *FROBS: A Merger of Two Knowledge Representation Paradigms*. Master's thesis, Dept. of Computer Science, University of Utah, Salt Lake City, UT 84112, December 1987. FROBS Stands for Frames+Objects.

[NM]   Mani Narayana and Surya Mantha. The Design of a TLB for the Roll Back Chip. VLSI Class Project Report, Winter 1988.

[NS88]   P. Narendran and J. Stillman. Hardware Verification in the Interactive VHDL Workstation. In Graham Birtwistle and P.A.Subrahmanyam, editors, *VLSI Specification, Verification and Synthesis*, pages 235–255, Kluwer Academic Publishers, Boston, 1988. ISBN-0-89838-246-7.

[Pat85]   Dorab Patel. nuFP: An Environment for the Multi-level Specification, Analysis and Synthesis of Hardware Algorithms. In *Proceedings of the Functional Programming and Computer Architecture Conference*, Springer-Verlag, LNCS 201, September 1985. Nancy, France.

[PC86]   David Lorge Parnas and Paul C. Clements. A Rational Design Process: How and Why to Fake It. *IEEE Transactions on Software Engineering*, SE-12(2):251–257, February 1986.

[Plo81]   Gordon D. Plotkin. *A Structural Approach to Operational Semantics*. Technical Report DAIMI FN-19, Aarhus University, Denmark, September 1981.

[Seq87]   Carlo H. Sequin. VLSI Design Strategies. In Wolfgang Fichtner and Martin Morf, editors, *VLSI CAD Tools and Applications*, pages 1–16, Kluwer Academic Press, 1987.

[She84]   Mary Sheeran. muFP, a Language for VLSI Design. In *Proceedings of the ACM Symposium on Lisp and Functional Programming*, pages 104–112, 1984.

[She85]   Mary Sheeran. Design of Regular Hardware Structures Using Higher Order Functions. In *Proceedings of the Functional Programming and Computer Architecture Conference*, Springer-Verlag, LNCS 201, September 1985. Nancy, France.

[Sne85]   Jan Snepscheut. *Trace Theory and VLSI Design.* Springer Verlag, 1985. LNCS 200.

[SR85]   Pashupathy A. Subramaniam and Sanjay Rajopadhye. Formal Semantics for a Symbolic IC Design Technique: Examples and Applications. *Integration: The VLSI Journal*, (3):13–32, March 1985.

[Sto77]   Joseph E. Stoy. *Denotational Semantics.* The MIT Press, 1977.

[Sub83]   Pashupathy A. Subramaniam. Overview of a Conceptual and Formal Basis for An Automatable High Level Design Paradigm for Integrated Systems. In *Proceedings of the International Conference for Computer Design and VLSI, Westchester*, pages 647–651, 1983.

[WFC87]   W.F.Clocksin. Logic Programming and Digital Circuit Analysis. *Journal of Logic Programming*, (4):59–82, 1987.

Formal Verification of a Microprocessor Using Equational Techniques

R.C. Sekar

Department of Computer Science

SUNY at Stony Brook

Stony Brook, NY 11794.

M.K. Srivas

Odyssey Research Associates Inc.

301A Harris B. Dates Drive

Ithaca, NY 14850.

Abstract

This paper develops a method for formally proving the correctness of a microprocessor using an equational method. The behavioral and structural specifications of the processor are expressed in a functional language. The asynchronous interaction between the memory and CPU is expressed in the specification by using a nondeterministic function *random*. This makes our specification more direct and natural than previous efforts based on a functional formalism. The previous efforts have had to encode in the specification the anticipated asynchronous interaction between the memory and the CPU.

The statement of correctness is expressed as a sentence in FOPC with equations relating expressions of the specification language as atomic formulae. The verification is carried out by formulating a set of invariants on the reachable states of the processor. The invariants are verified via symbolic evaluation employing appropriate induction and case analysis. The invariant approach makes the verification method modular in that it isolates the parts of the proof which are microcode-dependent from the rest. The method has been applied to verify an abstract version of the processor architecture used in Wirth's Modula II machine LILITH.

1   Introduction

Research into formally proving correctness of software has continued for many years with some notable successes. The correctness of hardware design has received much less attention until recently, but it seems, as indicated by some recent works [13], that progress might be made fairly rapidly. Hardware is in some ways more constrained in its organization than software, which makes many problems more tractable. It seems like we can apply correctness proof techniques to a broad enough range of hardware systems to be useful. In this paper we present a method of formally proving the correctness of a microprocessor design using equational techniques. We apply the method to prove the correctness of an abstract version of a large scale realistic design: the processor of Wirth's Modula II machine LILITH [15].

The motivation for this work arose in the context of the *SBL* VLSI design system [4]. The *SBL* system is built around a high level hardware description language also called *SBL* [14]. *SBL* is a functional language [5] with an underlying equational semantics, and is designed to support hierarchical development of custom VLSI designs. The current implementation of the system provides a symbolic interpretor to simulate *SBL* specifications, and "hooks" [9] to various VLSI CAD tools to help translate the high level design specification into silicon systematically. The present work was initiated as an exercise in using *SBL* to specify and verify a large scale realistic hardware design.

The main objectives of our work are the following. Firstly, to investigate how the external behavior (*abstract specification*) and the internal structural design (*realization specification*) of a microprocessor can be captured using an equational/functional specification method. Secondly, to develop a method of formally verifying the correctness of a microprocessor using "equational proof methods." Let us elaborate on what we mean by this. The current *SBL* simulator is essentially an equational interpretor, i.e., it simplifies expressions treating the specifications as a set of oriented equations (*rewrite rules* [10]). Our goal is to verify the microprocessor by rigorously refining the formal statement of correctness into a tree of lemmas so that every leaf in the tree is an *equationally verifiable* equation. An equation is equationally verifiable if it can be proved by simplifying the two sides to a common expression by the *SBL* interpreter. (In simplifying an equation the interpreter may use any axiom in the specification and/or a lemma which is not an ancestor of the equation.) Refinement of the statement of correctness into the equational assertions is itself carried out manually using a set of sound *proof strategies*. Some of the refinement steps used can be automated by existing equational/rewriting-based theorem provers, such as ORACLE [12], REVE [10] and RRL [8].

One of the difficult parts of specifying and verifying a microprocessor design in a "nontemporal" formalism, such as an equational logic, in which timing is not modeled explicitly, lies in characterizing and reasoning about the asynchronous interaction between the memory and the cpu modules. Previous works either assume ([1], [2]) that the memory is dedicated to the cpu or use an explicit "oracle" [6] in the specification. The oracle is like a timer that encodes the asynchronous interactions between the cpu and the memory. The use of oracles makes the specification unnatural, and the proof more complex because of the need to define an "oracle constructing function" [6].

Our approach permits the memory to interact asynchronously with one or more modules in the design, and yet does not use any oracles in the specification or proof. The memory may take an arbitrary (but finite) amount of time to process competing requests for its access. We assume, however, that the system as a whole is synchronous in the sense that it is governed by a global clock which is used by a centralized controller to synchronize the asynchronous interactions.

In our approach the specification of the processor does not take the form of an interpreter for the microprocessor, unlike in [6]. Rather it defines the effect of executing (*Step*) a single macro instruction cycle at the abstract level and a single micro instruction cycle (*Execute*) at the realization level. The asynchronous nature of the memory is modelled by making use of a nondeterministic random number generating function *random* in the specification of the memory. Because of this our specification is no longer referentially transparent under the substitution rules of equality. We handle this situation by amending our substitution rules by means of a safety condition. The safety condition has the effect of interpreting every occurrence of *random* in an expression to be actually denoting a distinct function symbol which obeys the same (minimal) assumption we make about *random*, namely that it denotes a number less than a finite quantity (*access_time*). Our proof itself is parameterized with respect to *access_time*.

The verification method is based on identifying a condition which must hold whenever the processor is about to start execution of the next macro instruction, and on proving certain *Liveness* and *Safety* properties for every module which interacts with the memory. The structure of our proof is such that it is robust with respect to changes in the microcode of the processor. The parts of the proof that must be redone for a change in the microcode are the proofs of the assertions at the leaves (of the tree of lemmas) which are done (almost) completely automatically. It is also possible to establish the correctness of every macro instruction independently.

The liveness property for a module asserts that every memory request by the module is guaranteed to be processed after a finite time. The safety properties assert certain invariants on the state of the processor. A liveness property asserts that when the system is in a certain state a certain other state is reachable after a finite number of invocations of the *Execute* operation. We use a form of computational induction [11] to transform the proof of the liveness property into a set of equationally verifiable assertions. However, a liveness-like property can be used in the proof of another liveness-like property in a pseudo-equational fashion.

The next section describes the hardware semantics attributed to our specification. Sections 3 and 4 illustrate our approach on the specification and verification of a very simple processor *Simple*. Sections 4 and 5 apply the method to the specification and verification of a more complex architecture, *Mini Lilith*, an abstract version of the Lilith architecture. Section 7 gives a summary of our proof strategies. The last section summarizes the work and compares our work to other microprocessor verification efforts.

2   The Hardware Model

We are interested in describing the behavior of synchronous systems with a global clock. That is, every sequential component in our system (regardless of where it appears in the design hierarchy) is assumed to be clocked by the global clock. We assume that our system is designed so that it can be partitioned into a centralized *controller* and a *data path*. The controller, which is a finite state machine, controls and coordinates interactions between different modules in the data path. Every transition of the controller corresponds to a cycle of the clock and triggers the desired "actions" on zero or more data path modules. Triggering an action on a module causes the state of the module to change which in turn causes the outputs of the module to change.

The behavior of a hardware module is characterized in terms of its *state*, inputs, outputs, and a set of *operations* that the module supports. There are two kinds of operations: *Out* and *Next*. The Next operations are the ones which can be triggered as "actions" by the controller. (In real hardware this is normally done by applying an appropriate *control signal* to the module. The control signal, however, remains implicit in our specifications.) A Next operation is a function of the type $Current\,State \times I \rightarrow New\,State$, where $I \subseteq Inputs$. It defines the new state attained by the module as a result of the action.

The Out operation describes the combinational behavior of the device. It defines the signal values which are present at the output ports of a module when the module is in a stable state and its inputs are stable. Hence, it is defined as a function of the type $Current\ State \times I \to O$, where $I \subseteq Inputs$ and $O \subseteq Outputs$. A module which is purely combinational does not support any next operation, and hence does not need any explicit triggering.

The action denoted by a next operation may take one or more clock cycles to complete. In SBL, which allows a more general timing model than the one used here, a module may support several next-like operations. Each of them may take a different amount of time (definite or indefinite) to complete. This makes the synchronization task of the controller, which is synthesized automatically in the SBL design system, more complex, and also necessitates a timing specification language for describing the timing characteristics of the operations. We avoid such a need here by making a set of implicit timing assumptions about the behavior of our modules which is summarized below.

1. Every module has at most one next and at most one out operation. The behavior of the processor as a whole is described by a single next-like operation: at the abstract specification level, this operation, called *Step*, describes the execution of a single (macro) instruction of the processor; at the realization level (i.e. design specification level), this operation, called *Execute*, describes the behavior of the processor for a single micro cycle. Every submodule used in the data path of the processor is described using a single Out operation, and a single Next operation, which describes its behavior over a single microcycle.

2. All inputs to a next operation must be stable during the clock cycle in which it is triggered.

3. An out operation "takes" zero time. (Note that this means that we assume that our clock cycle is long enough to subsume the longest delay incurred due to combinational blocks.)

3   The Processor Simple

We use a simple processor, *Simple*, to illustrate our specification and verification methods. Fig.1 shows the architecture of the Simple processor. The processor uses

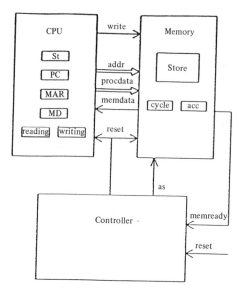

Figure 1

hardwired control. It consists of a memory module and a central processing unit (CPU). The memory is an "intelligent" memory, permitting asynchronous handshake using its *ready* signal. A valid request is presented to the memory by asserting *as*, with the address, data and read/write signals on the appropriate busses. The memory processes the request and responds by asserting *ready* signal after an indeterminate (but finite) amount of time. The CPU views the memory as responding synchronously in the sense that, if a request were presented to the memory in the beginning of a CPU clock cycle, it is serviced before the beginning of the next CPU clock cycle. It is the controller's job to interface these modules. It achieves this by simply inhibiting the control signal that invokes the Next operation on the CPU, whenever the memory is not ready.

3.1 Abstract Specification

The abstract specification describes the programmer's view of the processor. In our description, it supports a single operation, *Step*, which specifies the effect of executing a single instruction. The state of the processor is modelled in a way that the programmer would see it: as a record ⟨PC, STORE⟩, where PC stands for the program counter and STORE stands for the memory. The PC always points to the next instruction in the memory. The abstract specification of Simple is given below. The operation Step takes one input, the *reset* signal. When Step is invoked with reset signal asserted, the PC is reset to zero and the memory is unaffected. If reset is not asserted, it 'executes' the instruction in the memory pointed to by the PC. For the sake of simplicity, the behavior of the processor is specified only for two instructions: Store PC (STPC) and jump (JMP). Each of these instructions occupies two words of memory, the first one holding the instruction and the second one holding the address of the operand. As a result of executing STPC, the contents of PC incremented by two is stored at the address specified. The result of executing JMP instruction is to replace the contents of the PC with the operand.

```
ABS_MODULE Processor

ABS_STATE IS RECORD
                PC: BITVECTOR [16] % program counter
                STORE: memory(2**16) of BITVECTOR [16] % memory is usual memory.
            END
EXTOP
   Step :: CurSt, ?reset ==> newstate
   WHERE {
    newstate =
      COND{ ?reset: <0, STORE>  % PC becomes zero, STORE unchanged.            S1
            ELSE: COND{ (STORE[PC]==JMP):<read(STORE, inc(PC)), STORE>          S2
                        % PC is loaded with the operand, STORE unchanged.
                        (STORE[PC]==STPC):<inc(inc(PC)), write(STORE,
                                     read(STORE, inc(PC)), inc(inc(PC)))>        S3
                        %PC+2 stored at the address specified by operand. }}}
END Processor
```

3.2 Abstract Specification of the CPU

The state of the CPU is modelled as a record which consists of the program counter (PC), memory address register (MAR), memory data out register (MDout), read/write flags (reading and writing) and a component (St) that stores the current state of computation of the processor. The processor can be in one of four states: fetching an instruction, executing a jump instruction or executing the first or second parts of the STPC instruction. The CPU supports two operations: Out and Next. The Out operation, which is a function of the current state only, outputs the address of the memory location to be accessed, i.e., the contents of MAR, a write signal which specifies the intended direction of data transfer, and the contents of the output data register. The Next operation takes two inputs, the reset signal and data input from the memory.

If reset is asserted, the CPU is unconditionally reset, which includes clearing the PC. Otherwise, if the CPU is in fetch state, the data input is interpreted as an opcode and the execution of the corresponding instruction is started. In the succeeding invocations of Next, the instruction execution is completed, after which the CPU would again be in a fetch state, waiting for the next instruction. The abstract specification of the CPU is given in appendix A.

3.3 Abstract Specification of the Memory

The state of the memory is modelled as a record which consists of three components: *Store*, the storage part, and *cycle_num* and *acc_num*, which are used to model the non deterministic response time of the memory. For each memory access, the memory module 'guesses' the time it is going to take to process the request and records it in acc_num. It then counts the time that has passed since the beginning of the current memory access, in cycle_num [1]. Though this design may look artificial, note that we are just giving an abstract specification that encompasses all possible behaviors of any real memory.

The memory also supports two operations, Out and Next. The Out operation takes an address and a write signal as inputs and outputs a ready signal and a data. When the ready signal is asserted by the memory, it signals that the memory has processed the data request and is ready for the next access. In that case, the data output contains the data requested in case of a read cycle (i.e., the write input is negated). The Next operation takes the memory to a new state, in which the contents of the memory may be updated (depending on whether the current access has been serviced and the write input is asserted) and the cycle_num and acc_num are updated. The abstract specification of memory is given in Appendix A.

3.4 Realization Specification

A realization description specifies the internal design of the processor, which in the case of Simple is made up of the data path modules - CPU and memory - and the controller. The purpose of the controller is to gate the signals appropriately between the CPU and the memory, and to 'shield' the asynchronous nature of the memory from the CPU. The CPU and the memory have their own abstract specifications which

---

[1] All units of time are in clock cycles; in this case, it is also equal to the number of invocations of Next on the memory since the controller (as will be seen in the realization specification to follow) triggers Next at every transition.

describe their behavior. The realization specification can be viewed as an abstract functional description of the actions scheduled by the controller.

In a realization specification, the state of the processor is modelled as a record of the states of the submodules used in the design. The Out and Next operations of the module are defined as a composition of the Out and Next operations of the submodules. Note that the composition of the submodule operations implicitly determine the way the signals are going to be routed from one submodule to another. The realization specification of Simple supports one Next-like operation, *Execute*, which takes the single input, the reset signal, and defines the logic used by the controller for coordinating the activities in the data path over a single micro cycle. The Execute operation invokes Out operations on the CPU and memory to fetch their outputs. The outputs are then used to invoke the Next operation on memory, and a Next on the CPU. Next on CPU is invoked only if the memory is ready and the reset input signal is not asserted. (Inhibiting the Next on the CPU corresponds to lengthening the clock pulses to the CPU when the memory is not ready.) The realization specification of Simple is given below:

```
REAL_MODULE Processor

REAL STATE IS  P = <CPU:cpu, MEM:mem>

EXTOP
   Execute:: <CPU:cpu, MEM:mem>, ?reset ==> newstate
   WHERE{
     newstate = LETSEQ{
        <addr, procdata, write> = Out(cpu)                          S4
        <memready, memdata> = Out(mem, write, addr)                 S5
        ncpu = COND{ ?reset OR memready: Next(cpu, memdata, ?reset) S6
                       ELSE: cpu }
        nmem = Next(mem, t, write, addr, procdata, ?reset) }        S7
     IN  <ncpu, nmem> }                                             S8
END Processor
```

Note that the CPU and MEM may not be implemented in hardware exactly as described in their respective abstract specifications. If so, the correctness of these modules should in turn be verified with respect to their realization specifications. This process must be repeated till the design is expressed in terms of a set of primitive *library cells* such as registers, gates, etc. We assume that each of the library cells has been validated with respect to its abstract specification by other methods.

4   Verification of Simple

At this point we introduce some notation to simplify the description of our proofs. In what follows, A and R stand for arbitrary abstract and realization states, E and S for Execute and Step. The unspecified inputs in function applications, as well as the variables that are not quantified, are assumed to be universally quantified. $E^n(R, x)$ and $S^n(A, x)$ are formally defined by the equations:

$$E^0(R, x) = R, \; E^{n+1}(R, x) = E^n(E(R, x), x)^\dagger$$
$$S^0(A, x) = A, \; S^{n+1}(A, x) = S^n(S(A, x), x)^\dagger$$

4.1   Formal Statement of Correctness

We prove the correctness of the implementation of the processor by showing that the realization description *meets* the abstract specification. By this, we mean that there is a mapping between the states of the realization and abstract specifications, and that there is a correspondence between the Step and Execute operations. There are two kinds of abstractions to be considered here.

- *Structural*: given by the function F, this abstracts away the components of the realization state that have no analogues in the abstract state.

- *Temporal*: This bridges the gap arising due to the difference in the time scales over which *Execute* and *Step* describe the behavior of the processor. This is handled by relating a histories of *Step* and *Execute* invocations.

Formally, the correctness condition is stated as:

$$\forall A \, \forall R \, \forall n \, \exists n' \, [F(R) = A] \supset [F(E^{n'}(E(R, t), f)) = S^n(S(A, t), f)]$$

where n and n' are finite integers. This says that for any A and any R that can be mapped onto A, the abstract state attained by any number of invocations of Step (following a reset) on A can be mapped onto by the realization state reached by a finite number of invocations of Execute (following an invocation of reset) on R. The function F is defined by

$$F(R) = \langle \mathbf{PC} : R.\mathsf{cpu.pc}, \mathbf{STORE} : R.\mathsf{mem.store} \rangle \qquad \cdots (*)$$

---

[1]$\dagger$The second argument to E and S is a 1-bit reset input.

4.2 The Verification Approach

The verification approach identifies two levels. The *object level* consists of the specification which defines objects (or states modeled as abstract data types), and (collections of) functions on states. The *metalevel* consists of statements of properties about the objects defined by the specification. A metalevel statement can take one of the following forms. Every variable that appears as an argument to a function symbol in a meta-level statement is assumed to be universally quantified.

1. An equation $t_1 = t_2$, where t_1 and t_2 are object-level

2. An *iterative* statement of the form: $\forall n \; \exists n' \; f^{n'}(R)) = g^n(R)$, where g^n denotes n repeated applications of g.

3. An implication, where the antecedent and the consequent are conjunctions of meta-level statements. (The antecedent may be empty.)

For convenience, we allow the definition of meta-level predicates to serve as abbreviations for meta-level statements. A meta-level predicate definition must be non-recursive and may take parameters which are intended to be instantiated by object-level terms.

The basic method used in the verification is equational simplification, i.e., using the definition of functions to suitably *rewrite* expressions in equations to a common expression. An equation which is verifiable strictly by rewriting in the above fashion is said to be *equationally verifiable*. A meta-level statement which is not an equation or which is not equationally verifiable is verified by refining it into a set of equationally verifiable assertions using a set of sound proof strategies which includes structural induction, case-splitting and a form of iteration induction for proving existentially quantified iterative meta-statements. The proof strategies used will be summarized in section 5.

4.2.1 Dealing with Nondeterminism

As the reader will see shortly, some parts of our specification use a nondeterministic function *random*. Equational simplification (replacement of equals by equals) becomes "unsafe", i.e., no longer preserves equality, in the presence of nondeterministic functions, such as *random*. For example $f(random(m), \; random(m))$ can be rewritten into $g(random(m))$ if we had the equation $f(x, x) = g(x)$ in the specification although the two terms need not denote the same object if *random* is a nondeterministic function.

One way to ensure that a proof based on rewriting is correct in the presence of nondeterministic functions is to make sure that the proof does not rely on two syntactically identical *nondeterministic expressions* being equivalent. (Any expression which consists of one or more nondeterministic function symbol is considered a nondeterministic expression.) We do this by treating distinct occurrences of the same nondeterministic function symbol in an equation to be actually denoting different function symbols governed by a replica of the set of axioms defining the nondeterministic function symbol. In other words, whenever a subterm on either side of an equation is replaced by another term, a check must be made if there is more than one occurrence of the same nondeterministic function symbol. If so they must be appropriately renamed to eliminate all collisions.

Note that determining which of the function symbols defined by our specification is nondeterministic is trivial. The only way in which a defined function can be nondeterministic is if one of the function symbols used in the definition of the defined symbol is nondeterministic. The only nondeterministic function among the primitive symbols used in the specification is *random*. Hence, every function whose specification contains *random* is assumed to be nondeterministic.

4.2.2   Verification Conditions

Lets start with the statement of correctness given above which is obviously not equationally verifiable.

Removing reference to A**:** The antecedent of the statement of correctness, being an equation, can be used to eliminate A from the consequent, yielding the following equivalent statement of correctness:

$$\forall R \,\forall n \,\exists n' \, [F(E^{n'}(E(R,t),f)) = S^n(S(F(R),t),f)]$$

Handling $\forall n$ **:** This is proved by induction on the number of macro instructions, i.e., Step operations, executed after reset. For this, it is necessary to establish an invariant on the realization state, which is true on reset and true 'between' the executions of any two consecutive macro instructions.

Let us define a *meta predicate* $fetching(R)$ that characterizes the real states when the processor is attempting to fetch an instruction from the memory.

$$fetching(R) \;::\; [(R.\mathsf{cpu.st} = fetch) \wedge (R.\mathsf{cpu.reading} = t)$$
$$\wedge (R.\mathsf{cpu.writing} = f) \wedge (R.\mathsf{cpu.mar} = R.\mathsf{cpu.pc})$$

The basis and induction steps of the proof of the above statement of correctness give the following verification conditions to be proved.

$$\forall R\ [(F(E(R,t)) = S(F(R),t)) \wedge fetching(E(R,t))] \qquad \cdots (1)$$

$$\forall R\ \exists n\ fetching(R) \supset [(F(E^n(R,f)) = S(F(R),f)) \wedge fetching(E^n(R,f))] \qquad \cdots (2)$$

4.3   Proof of Verification Condition (1)

The verification condition 1 could be stated in the form of equations and proved by symbolically evaluating the left and right hand sides of the equation to a common expression. Stated as equations, (1) becomes :

$$F(E(R,t)) = S(F(R),t) \qquad \cdots (1a)$$
$$E(R,t).\text{cpu.st} = fetch \qquad \cdots (1b)$$
$$E(R,t).\text{cpu.reading} = t \qquad \cdots (1c)$$
$$E(R,t).\text{cpu.writing} = f \qquad \cdots (1d)$$
$$E(R,t).\text{cpu.mar} = E(R,t).\text{cpu.pc} \qquad \cdots (1e)$$

The rules used in simplification are the equation (*) defining F, and the ones that constitute the realization specification of the processor and the abstract specifications of the processor, CPU and MEM (given in appendix A). The equations used for symbolic evaluation are numbered S1 through S25. The actual proof uses *structural induction* on R, which is of record type, and is given below:

Let $R = \langle \mathbf{CPU} : \langle \mathrm{ST}, \mathrm{PC}, \mathrm{MA}, \mathrm{MD}, \mathrm{RD}, \mathrm{WR} \rangle, \mathbf{MEM} : \langle \mathrm{STORE}, \mathrm{N1}, \mathrm{N2} \rangle \rangle$

$$S(F(R),t) = \langle \mathbf{PC} :0, \mathbf{STORE} : \mathrm{STORE} \rangle - \text{from S1}$$

$F(\ E(\langle \mathbf{CPU} : \langle \mathrm{ST}, \mathrm{PC}, \mathrm{MA}, \mathrm{MD}, \mathrm{RD}, \mathrm{WR} \rangle, \mathbf{MEM} : \langle \mathrm{STORE}, \mathrm{N1}, \mathrm{N2} \rangle \rangle, t)\ )$

$= F(\langle \mathbf{CPU} : NCPU, \mathbf{MEM} : NMEM \rangle)$ by S8

$= F(\langle \mathbf{CPU} : Next(R.CPU, memdata, t),$

$\qquad \mathbf{MEM} : Next(R.MEM, t, write, addr, procdata, t) \rangle),$

$\qquad\qquad$ by S6, S7 and LETSEQ and COND rules, since $reset = t$

$= F(\langle \mathbf{CPU} : \langle fetch, 0, 0, 0, t, f \rangle,$ by S12

$\qquad \mathbf{MEM} : \langle \mathrm{STORE}, 0, \mathrm{AN} \rangle \rangle),$ by S21 , S22, and S24 $\qquad\qquad \cdots (4)$

$= \langle \mathbf{PC} :0, \mathbf{STORE} :\mathsf{STORE}\rangle$ – by defn of F

Thus, $F(E(R, t)) = S(F(R), t)$, and it is clear from (4) that (1b) through (1e) hold.

4.4 Towards a Proof of Verification Condition (2)

The approach used in the case of (1) does not work for condition 2 because of the existential quantification. The existential quantification in (2) captures the fact that different macro instructions may take different number of micro cycles (or Execute's) and that the memory accesses take an indefinite amount of time. In this section, we present techniques to handle this problem.

Let us define a few meta predicates and functions that simplify the discussion in this section:

$$WAIT(R, R') :: [(R'.\mathsf{cpu} = R.\mathsf{cpu}) \wedge (R'.\mathsf{mem.store} = R.\mathsf{mem.store})$$
$$\wedge (R'.\mathsf{mem.acc\_num} = R.\mathsf{mem.acc\_num})]$$

$$RT(R) :: (R.\mathsf{mem.acc\_num} - R.\mathsf{mem.cycle\_num})$$

$$READY(R) :: (RT(R) = 0)$$

To handle the existential quantification, we first need to establish a Liveness Lemma for the memory.

Lemma 1 Liveness property of the Memory
$$\forall R \, \exists n \, [WAIT(R, E^n(R, f)) \wedge READY(E^n(R, f))]$$

This lemma captures the fact that every access to the memory will terminate ultimately, and during this time, essentially no part of the state of the processor changes. To use this lemma in later proofs, we need an equational version of this theorem, as follows:

$$E^{n'}(\langle \mathbf{CPU} :\langle s, pc, a, d, r, w\rangle, \mathbf{MEM} :\langle m, x, y\rangle\rangle, f) =$$
$$E^{n''}(\langle \mathbf{CPU} :\langle s, pc, a, d, r, w\rangle, \mathbf{MEM} :\langle m, y, y\rangle\rangle, f) \text{ such that } n'' \leq n'$$

Since $E^{n'}(R, f) = E^{n''}(E^{n'-n''}(R, f), f)$, with $n = n' - n''$, and from the definition of $WAIT$ and $READY$, it is clear that this equation is just a restatement of the lemma.

Proof: The proof of this lemma is given in Appendix B.

4.5 Proof of Verification Condition (2)

For ease of reference, the verification condition (2) is restated here:

$$\forall R\, \exists n\; fetching(R) \supset [(F(E^n(R,f)) = S(F(R),f)) \wedge fetching(E^n(R,f))] \quad \cdots (2)$$

The strategy used in the proof is the following:

Premise Instantiation: This is the strategy we use for proving an implication. We instantiate R to the most general symbolic constant of the appropriate record type, i.e., add the equation

$$R = \langle \mathbf{CPU} : \langle \mathsf{ST}, \mathsf{PC}, \mathsf{MA}, \mathsf{D}, \mathsf{RD}, \mathsf{WR} \rangle, \mathbf{MEM} : \langle \mathsf{M}, \mathsf{X}, \mathsf{Y} \rangle \rangle$$

Then, we add equations in the antecedent of the implication. For verification condition (2), the hypotheses, after simplification, are the following:

$\mathsf{ST} = fetch$
$\mathsf{RD} = t$
$\mathsf{WR} = f$
$\mathsf{MA} = \mathsf{PC}$

Case Splitting: The proof of correctness of verification condition (2), will depend on the macro instruction that resides in the memory location pointed to by PC. There will be a case for every possible macro instruction. An equation characterizing this case is added to the system of equations as a hypothesis. For example, for the case of JMP instruction, we add the equation

$$read(F(R).\mathsf{store}, F(R).\mathsf{pc}) = JMPINS$$

Simplify: Simplify the right hand side of the verification condition to its (irreducible) normal form. Now, reduce the left hand side by applying liveness lemma and the equations in the specifications till it becomes equal to the reduced right hand side. The liveness lemma is used in reduction if $R.\mathsf{mem.acc\_num} \neq R.\mathsf{mem.cycle\_num}$. If they are equal, the equations defining the Execute operation are used.

When we attempt to prove (2), we are faced with the problem of simplifying $E^n(R,f)$ when n is unknown. Since n is existentially quantified, we will use as many E's as needed till the left hand side can be simplified to the normal form of right hand side. We show the proof of the JMP instruction here. Appendix B gives the proof of the STPC instruction.

JMP - Jump Instruction: $read(\text{STORE}, PC) = JMPINS$

From here onwards, we indicate only the main steps of the proof and skip the intermediates to keep the proof short and concise.

$S(F(R), f) = \langle \mathbf{PC} : read(\text{STORE}, inc(PC)), \mathbf{STORE} : \text{STORE} \rangle$ by S2

$F(E^n(\langle \mathbf{CPU} : \langle FETCH, PC, PC, MD, t, f \rangle,$

$\quad \mathbf{MEM} : \langle \text{STORE}, \text{CN}, \text{AN} \rangle \rangle, f))$, by premise instantiation

$= F(E^{n'}(\langle \mathbf{CPU} : \langle FETCH, PC, PC, MD, t, f \rangle,$

$\quad \mathbf{MEM} : \langle \text{STORE}, \text{AN}, \text{AN} \rangle \rangle, f))$ by liveness lemma

$= F(E^{n'-1}(\langle \mathbf{CPU} : \langle JMP, inc(PC), inc(PC), MD, t, f \rangle, \mathbf{MEM} : \langle \text{STORE}, 0, Z \rangle \rangle, f))$

$\quad$ by S4,S9,S10,S11,S5,S18, S19, S6, S13, S7, S21,S22 and S24

$= F(E^{n''}(\langle \mathbf{CPU} : \langle JMP, inc(PC), inc(PC), MD, t, f \rangle,$

$\quad \mathbf{MEM} : \langle \text{STORE}, Z, Z \rangle \rangle, f))$ by liveness lemma

$= F(E^{n''-1}(\langle \mathbf{CPU} : \langle FETCH, read(\text{STORE}, inc(PC)), read(\text{STORE}, inc(PC)), MD, t, f \rangle,$

$\quad \mathbf{MEM} : \langle \text{STORE}, 0, W \rangle \rangle, f))$

$\quad$ by S4,S9,S10,S11,S5,S18,S19, S6, S15, S7, S21, S22 and S24

$= F(\langle \mathbf{CPU} : \langle FETCH, read(\text{STORE}, inc(PC)), read(\text{STORE}, inc(PC)), MD, t, f \rangle,$

$\quad \mathbf{MEM} : \langle \text{STORE}, 0, W \rangle \rangle)$ putting $n'' - 1 = 0$ $\qquad \cdots (5)$

$= \langle \mathbf{PC} : read(\text{STORE}, inc(PC)), \mathbf{MEM} : \text{STORE} \rangle$ by defn of F

It is clear that $F(E(R, f)) = S(F(R), f)$ and from (5) $fetching(E(R, f))$ holds.

In general, the above proof is the only one that depends on the microprogram. The refinement steps that precede this proof are independent of the microcode. It is possible to automate entirely the part that depends on the micro program. Thus, it is possible to verify many 'different' processors, corresponding to different micro programs, with the same amount of manual effort.

At this point, we note that this process of verification for each instruction individually can be fully automated only in case there are no loops in the controller or the CPU (or, in the micro programmed controller, when there are loops in the micro program). Where the micro program contains loops, (with the number of iterations depending on data), manual assistance is required in order to set up the invariant for the loop.

5 Discussion of the Proof Strategies Used

The techniques made use of in our proofs are discussed here:

Premise Instantiation: This technique is made use of in proving implications. In proving the condition $[P(x)] \supset e_1 = e_2$, where P is a meta-predicate, x is instantiated to the most general symbolic constant of the appropriate type. Now, the set of equations is augmented with the set of equations $\mathcal{H}$ such that $P(x) = t$ is a consequence of $\mathcal{H}$. This technique was first illustrated in section 4.4.

Case Splitting: This technique is often used in proving conditions of the form $\forall x \; P(x)$ when the domain of x is finite. If the possible values x can take are in the set $\{v_1, \ldots, v_k\}$, $\forall x \; P(x)$ is proved by establishing each one of $P(v_1), \ldots, P(v_k)$. For proving $P(v_i)$, we augment the set of equations with $x = v_i$. (In some cases, we need to prove that the set of values x assumes is a finite set for all possible R or A). We make use of this technique in the proofs of most of the lemmas.

Handling $\forall R$: When we have to prove conditions of the form $\forall R \; P(R)$, we use reachability induction on n. The states R of interest to us are the ones that can be reached from the reset state $E(R', t)$ for some R', by a finite number of applications of E. Hence, we need only prove $\forall n \; P(E^n(E(R, t), t), f)$. The condition is established by proving $\forall R \; P(E(R, t))$ and $\forall R \; P(R) \supset P(E(R, f))$. We made use of this technique in the proofs of the Safety lemmas.

Proving Liveness Properties: Theorems of the form $\forall R \; \exists n \; Q(R, E^n(R, f)), n \geq 0$ are proved using this technique. The general method involves identifying:

- A rank function $g : R \to D$
- A well-founded ordering $<$ on D, with a least element ϕ
- A transitive invariant binary predicate on $R : I(R_1, R_2)$

such that the following hold:

1. Invariance: $\phi < g(R) \supset I(R, E(R, f))$
2. Final Condition 1: $[(g(R') = \phi) \wedge I(R, R')] \supset Q(R, R')$
3. Final Condition 2: $g(R) = \phi \supset I(R, R)$
4. Convergence: $\phi < g(R) \supset g(E(R, f)) < g(R)$

The application of this technique is first illustrated in the proof of lemma 1 (Appendix B). In that case, RT corresponds to g, the usual less than ordering on natural numbers for $<$, 0 for ϕ, and $WAIT$ for I. The propositions 1,2 and 3 in the proof in Appendix C correspond directly to the conditions stated above.

Simplification: In proving $e_1 = e_2$, we reduce e_1 and e_2 to a common expression. Being the fundamental technique in equational theorem proving, it needs no further elaboration.

Handling $\exists n$: For proving theorems of the form $\exists n\ P(E^n(R, f))$, we need to show that for some value k of n, $P(E^k(R, f))$ holds. The proof process needs to construct this value of n. In our proofs, we simplify the expression using 'as many E's as required', till the $E^n(R, f)$ gets simplified to $E^{n-l}(R', f)$ such that $P(R')$ holds. At this point, $n - l$ is replaced by 0, completing the proof. The justification for this replacement is that with $n = l$, the condition $P(E^n(R, f))$ holds. This technique is used in the proof of verification condition (2).

6 To a More Complex Architecture: Mini Lilith

We started out trying to verify the correctness of the Lilith processor ([15]). We succeeded in giving a full description of the processor, but a complete proof became far too tedious to carry through in the absence of a proof checker. In our attempt to verify the Lilith, we realized that the most interesting aspects of the architecture from the point of view of verification were the following [2]:

1. Pre-fetching of instructions for a higher degree of parallelism.

2. Buffered writes to the memory.

3. The access time of the memory is non-deterministic but finite, as in the case of Simple.

4. More than one module competes asynchronously for memory access.

To help us investigate the extension of our verification methods to cover the above aspects, we use a carefully designed abstraction of the Lilith machine, called *Mini Lilith*. The description of Mini Lilith is given below. The aspects of Lilith that have been abstracted away can be summarized as follows:

---

[2] Lemmas 2 through 5 (to follow) capture these interesting aspects.

- Interrupts and Interrupt controller

- Lilith uses a 2-port memory with separate ports for data and instruction memory, but we use only a single port memory.

- Up to eight instructions could be prefetched and buffered in Lilith, where as in Mini Lilith, only one instruction is prefetched.

- The bus arbitration mechanism is slightly different in Mini Lilith.

- The Arithmetic and Logic Unit is greatly simplified.

Mini Lilith consists of the following modules: Instruction Fetch Unit (IFU), PORT, Memory, Arithmetic and Logic Unit (ALU) and the Central Processing Unit (CPU). These modules are interconnected through an *internal bus*. In many aspects, the architecture of this processor is like that of the simple processor we described before. The memory is smart and uses a Data Acknowledge (*dtack*) handshake signal for asynchronous memory transfers. As in the simple processor, the CPU believes that the units with which it communicates support synchronous data transfers and it is the responsibility of the controller to inhibit the appropriate control signal to the CPU when needed.

Mini Lilith is realized in terms of the following functionally distinct modules: the CPU, PORT, IFU, ALU and MEM. The the submodules and the interconnections among them are indicated in Figure.2. Communication between the submodules takes place primarily through three busses: an *Internal Bus*, which provides a data path shared by the CPU, IFU, PORT and the ALU, an *Address Bus*, on which the PORT and IFU output the address of memory access, and a *data bus*, on which the memory outputs the data requested or the PORT outputs the data to be written into the memory. The controller here is micro programmed, and thus its operation is almost entirely determined by the micro program residing in the micro program memory (*mup_rom* of the CPU). The formal specifications of Mini Lilith and the submodules are given in Appendix D. An informal description follows.

6.1 Abstract Specification of Mini Lilith

The abstract specification of the processor remains almost the same as before. It describes the effect of one more instruction, the indirect jump (JMPIND), where the operand specifies the address at which the jump address is stored. With the abstract machine remaining essentially the same, we just give a different realization specification for the processor.

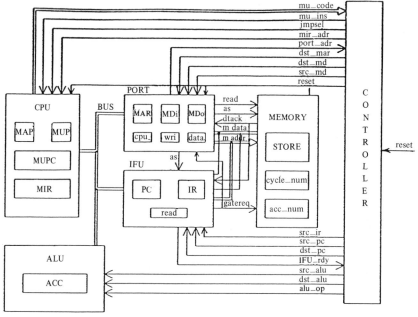

Figure 2

6.2 The CPU

The CPU consists of a micro program counter(mu_PC), an incrementer, a multi-plexor, a micro instruction register(MIR), a map_rom , and micro program rom(mup_rom). The mu_PC contains the address of the current micro instruction. The incrementer takes the contents of mu_PC as its input and outputs $mu\_PC + 1$. The multiplexor selects the source of the address of the next micro instruction from one of: the outputs of the map_rom, the jump field of the micro instruction, and the output of the incrementer. The output of the multiplexor is loaded into the mu_PC, and the contents of the mup_rom at this address are loaded into the MIR. The MIR holds the micro instruction that is currently being executed. The map_rom is a table which maps every macro instruction to its corresponding micro routine address. The address to this ROM is supplied by the internal bus. The micro program resides in the mup_rom.

The CPU takes five inputs, bus, the input from internal bus, mir_adr, which comes from the jump address field of the micro instruction, $jmpsel$ and mu_ins, which specify the operation of the multiplexor and originate from the micro instruction, and $reset$, the external reset input. It produces one output, mu_code, which is the same as the contents of the MIR.

The state of the CPU is modelled as a record of four components, map_rom, mup_rom, mu_PC and mir. Mup_rom and map_rom are read-only memories, whose function was described before. The function of mu_pc and mir is as described earlier.

The Out operation produces the contents of MIR and Next takes the CPU to a new state. If Next is invoked with the *reset* input asserted, the address inputs of the *mup_rom* are forced to zero. This forces the next micro instruction to be taken from address zero, which should contain the first instruction in a sequence that will reset the processor.

6.3   Arithmetic and Logic Unit(ALU)

A simplified ALU, which in the instructions illustrated, simply provides a scratch pad register for storing temporary results, is specified in Appendix D. It takes four inputs, *src_alu*, *alu_op*, *dst_alu* and *bus*. Src_alu, dst_alu and alu_op are obtained by decoding the micro instruction, and bus is the input from the internal bus. It produces one output on the internal bus, which reflects the contents of the scratch pad register.

The state of the ALU is described by a single 16-bit number. The Out operation outputs this state, if the src_alu input is asserted. Otherwise, the output port is in a high impedance state (HiZ). The effect of the Next operation is controlled by the inputs alu_op and dst_alu. When alu_op is neither of SHIFT or ADD, and dst_alu is asserted, the ALU stores the contents of its bus input in its state. In the instructions illustrated, the ALU is used only in this fashion, and hence a more detailed description of the operation of the ALU is skipped here.

6.4   The Instruction Fetch Unit(IFU)

The IFU operates independently of the CPU for most of the time. It prefetches instructions from the memory and stores it in its single word buffer, the Instruction Register(IR), so that the request from the CPU for the next instruction can be serviced without the delay of a memory access. The PC points to the next instruction to be executed. Whenever the IR does not contain this instruction, IFU accesses the memory to fetch it. The IFU has a higher priority over the PORT for accessing the memory, but a memory access started by the PORT is not interrupted by the IFU.

The state of the IFU is characterized by a program counter (PC), an instruction register (IR), which provides a single word buffer for storing the prefetched instruction from the memory, and *read*, which specifies the status of the memory operation initiated by the IFU. If *read* is zero, it means that the IFU is not accessing the memory, and $read = 2$ specifies that the IFU is waiting for the memory to service its request. If $read = 1$, then the IFU is waiting for the PORT to release the memory.

The IFU has eight inputs and four outputs. The inputs *src_ir* and *src_pc* specify that the contents of the IR or PC is to be output on the internal bus. *As* is the input from the PORT that informs the IFU that the PORT is currently accessing the memory. The contents of the internal bus are written into the PC when the Next operation is invoked with the input *dst_pc* asserted. The *dtack* input from the memory informs the IFU that the memory is ready with the data requested. *Reset* is the reset input to the processor, and *m_data* is the data input from the memory. The *ifureq* output of the IFU signals that the IFU is accessing the memory. *IFU_rdy* is asserted when the IFU is ready to service the request for the next instruction from the CPU. IFU outputs the contents of the PC or IR on the internal bus as specified by src_pc or src_ir. The contents of the PC are output on the address bus when the IFU is accessing the memory.

6.5   The Port

The PORT module buffers the movement of data values between the CPU and the memory. It consists of three 16-bit latches which latch the address of memory access, the output data to be written to the memory (in case of a memory write), and the input data read from the memory (in case of a read cycle). The CPU views writing into the PORT or reading from the PORT as equivalent to writing into or reading from the memory. In order to read from the memory, the address of access is to be written into the MAR, and the bus output of the PORT contains the requested data, when port_rdy and src_md are asserted. For writing into the memory, the address of access is to be written into the memory, the data to be written is to be transferred to the MDout register in the following cycle. It is the responsibility of the PORT to assure that the data written into it is ultimately transferred to the memory and that the data read from it corresponds to the actual contents of the memory.

The state of the PORT is modelled as a record of Memory Address Register(MAR), Memory Data In register(MDin), Memory Data Out register(MDout), a state *port_req*, which specifies the status of the memory access initiated by the PORT, and two bits *writing* and *data_read*, which specify the intended direction of data transfer in the current memory cycle, and whether the contents of MDin correspond to the data read from the memory at the address pointed to by MAR.

The inputs *src_md*, *dst_md* and *dst_mar* specify that the contents of MDin are to be output on the internal bus or that the contents of the internal bus are to be stored in the MDout or MAR. The *ifureq* input from the IFU specifies that the IFU is currently accessing the memory. The *dtack* input from memory signals the completion of a memory access. *Reset* is the reset input to the processor. The internal bus is an input or output to the PORT depending on the signals *src_md*, *dst_md* and *dst_mar*. Depending on the intended direction of data transfer, *m_data* serves as an input from the memory or output to the PORT. The output *as* specifies that the PORT is currently accessing the memory. The *r* output specifies the intended direction of data transfer. *Port_rdy* signals that the PORT has finished processing the current memory request. The address of memory access is output on *m_addr*.

6.6 Memory

The memory description remains essentially the same as before. The only difference is that the Out operation takes two additional inputs, the address strobe (*as*), which specifies that the memory is being accessed by the PORT, and *ifureq*, which specifies that the memory is being accessed by the IFU. The minor differences should be obvious from the description of the memory.

6.7 Realization Specification

The behavior of the real module is described in terms of a single operation, Execute. The function of the controller is almost identical in each invocation of Execute. It involves decoding the micro program and delivering the appropriate control signals as specified in the microprogram to the submodules. Every invocation of Execute corresponds to a single micro instruction cycle.

The outputs of the micro program memory are latched into the micro instruction register (MIR), whose fields are decoded by the controller to get information about what is the source of the internal bus, what is its destination, the instructions to the CPU etc.. After decoding, the control inputs are routed to the different modules. The Out operations are invoked on all the modules and all the outputs are determined. These are used to evaluate the next state of the different modules.

The operation of the processor is described here for the Store Program Counter(STPC) instruction. The micro program corresponding to this instruction is given in the appendix. To start with, we assume that the micro instruction labelled STPC is in the MIR at the beginning of execution of this instruction. During the first micro cycle, the word following the instruction, which is pointed to by the PC, is read from the IFU and stored temporarily in the ALU. The CPU automatically goes to the succeeding micro instruction since $mu\_ins = t$. In the next cycle, the data is transferred from the ALU to MAR. In the next cycle, the contents of the PC are transferred to the data out register of the PORT. At this time, the CPU believes that it has written the PC at the address in the MAR. (The integrity constraints on the PORT assure that it eventually gets written into the memory). In the next cycle, the CPU fetches the next instruction and decodes it and branches to the appropriate micro routine.

7 Verification of Mini Lilith

In this section, we outline the proof of correctness of Mini Lilith. Additional lemmas are needed here to effectively isolate the CPU from memory and simplify some of the steps in the actual proof. The formal statement of correctness remains the same as for Simple, but the function F which gives the mapping between the abstract and realization states must be designed more cleverly. Note that the purpose of F is to define the correspondence between the realization state and abstract state before the start of a new instruction. The PC of the abstract and real states differ by one, since the IFU always prefetches the next instruction. The data out latch of the PORT serves as a single word buffer for storing data to be written into the memory, and the controller views writing data into memory as transferring it to the PORT, when the PORT is ready. Hence, the abstract memory is obtained by 'overlaying' the contents of the real memory with the contents of PORT, when buffered data is waiting to be written into the memory. This explains the increased complexity in the mapping function F defined below. Because of the buffering at PORT, data 'written' into the memory in executing an instruction may not be actually written into the memory, even when the execution of the next instruction has started. So, we find it convenient to talk about the 'overlaid' or *compound memory*, $F(R)$.store rather than R.mem in our proofs.

$F(R) =< PC : \text{pc}, STORE : \text{store} >$ WHERE
$\quad PC = (R.\text{ifu.pc} - 1) \bmod 2^{16}$
$\quad STORE[i] = COND\{((R.\text{port.port\_req} == 2) \ AND \ (R.\text{port.writing} == t)$
$\qquad\qquad\qquad AND \ (R.\text{port.mar} == i)) : R.\text{port.mdout}$
$\qquad\qquad ELSE : read(R.\text{mem.store}, i)\}$

Before we start the proof, we need to make the following assumptions.

- We will assume that the code and data memory do not overlap, or, in other words, the code memory can never be overwritten. Since the instructions are prefetched, and since there are no mechanisms for invalidating the contents of the IR if the corresponding memory location is overwritten, the realization specification may behave differently from the abstract specification in the case of programs that modify themselves. Hence, this realization is correct only for programs whose code is re-entrant.

- Another assumption that will be made is that the micro program will not contain any instruction for which more than one of the following conditions hold: $bus\_src = IR$, $bus\_src = MD$, $bus\_dst = MAR$, $bus\_dst = PC$. Some of these combinations, such as $bus\_src = IR$ and $bus\_dst = PC$, $bus\_src = MAR$ and $bus\_dst = MAR$, etc. cannot be implemented with the given modules, since a single port needs to be an output and input port simultaneously. Other combinations, such as $bus\_src = MD$ and $bus\_dst = PC$ can lead to deadlocks between the IFU and the PORT, and hence no liveness property can be proved under these conditions.

With these assumptions, the statement of correctness is restated here for easy reference:

$$\forall R \, \forall n \, \exists n' \; F(E^{n'}(E(R,t),f)) = S^n(S(F(R),t),f)$$

where n and n' are finite integers.

The verification condition can be split into two parts in same way as in the case of the simple processor. The definition of $fetching$ is different as given by:

$$fetching(R) :: [R.\textsf{cpu.mir} = read(R.\textsf{cpu.map\_rom}, read(F(R).\textsf{store}, F(R).\textsf{pc}))]$$

As in the case of Simple, $fetching(R)$ characterizes the realization state at the beginning of an instruction. At this time, the decoded instruction is already in the MIR. The verification condition (1) gets slightly modified, since it is the micro code that initializes the configuration of the processor to correspond to the reset state, and thus reset also involves executing an unspecified number of micro instructions. The two parts of the verification condition are :

$$\forall R \, \exists n \; [(F(E^n(R,t)) = S(F(R),t)) \wedge fetching(E^n(R,t))] \qquad \cdots (1)$$

$$\forall R \, \exists n \; fetching(R) \supset [(F(E^n(R,f)) = S(F(R),f)) \wedge fetching(E^n(R,f))] \qquad \cdots (2)$$

7.1 Towards the Proof of Correctness of Mini Lilith

In addition to the non-deterministic response time of the memory, we have to handle pre-fetching of instructions, buffering of writes at the PORT and the memory sharing protocols of the PORT and IFU. The problem of existential quantification is handled in the same way as in the case of Simple, but we have to come up with additional lemmas to handle the other features of Mini Lilith. In this section, we state the lemmas needed to prove the verification conditions equationally. The proofs of these lemmas are given in Appendix C.

Lemma 2 Safety property of the IFU
When the CPU attempts to read the Instruction Register (IR) of the IFU, if IFUrdy is asserted by IFU, then the data output by the IFU is the same as the contents of the memory pointed to by the Program Counter (PC).

Let $O = Out(R.\text{ifu}, src\_ir : t, src\_pc : f)$ IN
$$[O.\text{ifurdy} = t \supset O.\text{bus} = read(F(R).\text{store}, R.\text{ifu.pc})] \quad \cdots (6)$$

This lemma serves to leave the contents of the IR unspecified in between instructions. It is a statement of correctness of the handshaking protocol for transferring data from the IFU into the CPU.

Lemma 3 Safety property of the PORT
If the CPU does not make a valid attempt to write into the memory, then the compound memory remains unchanged in the succeeding micro cycle.

$$(decode(Out(R.\text{cpu})).\text{bus\_dst} == MD) = f \supset F(R).\text{store} = F(E(R, f)).\text{store} \quad \cdots (8)$$

The purpose of this lemma is to enable us to leave the actual contents of the memory and PORT unspecified, and instead talk about the contents of the compound memory. This is a statement of correctness of the protocol for the transfer of data from the CPU to the PORT and then ultimately into the memory.

Let us define a meta-predicate $WAIT(R, R')$ as:

$(R.\text{cpu} = R'.\text{cpu}) \wedge (R.\text{alu} = R'.\text{alu}) \wedge (F(R).\text{store} = F(R').\text{store}) \wedge$
$(R.\text{ifu.pc} = R'.\text{ifu.pc}) \wedge$
$(R.\text{port.mar} = R'.\text{port.mar}) \wedge (R.\text{port.mdout} = R'.\text{port.mdout})$

$WAIT(R, R')$ holds if R and R' have identical *ALU* and *CPU* components, and the contents of their *PC, MAR, MDout* and the memory are identical. It is clear that

$$WAIT(R_1, R_2) \wedge WAIT(R_2, R_3) \supset WAIT(R_1, R_3).$$

Lemma 4 Liveness property of the PORT and the IFU
A micro instruction may get executed a multiple number of times because the PORT

or the IFU is not ready with the required data. But these units will get ready after a finite time and the result of execution is independent of how many times the instruction was executed. We can consider it as though the micro instruction was executed just once, with all the units ready.

$(decode(Out(R.\text{cpu})).\text{bus\_src} = IR) \supset$
$$\exists n\ WAIT(R, E^n(R, f)) \wedge E^n(R, f).\text{ifu.read} = 0 \qquad \cdots (11)$$

$(decode(Out(R.\text{cpu})).\text{bus\_dst} = MAR) \supset$
$$\exists n\ WAIT(R, E^n(R, f)) \wedge E^n(R, f).\text{port.port\_req} = 0 \qquad \cdots (12)$$

$(decode(Out(R.\text{cpu})).\text{bus\_src} = MD) \wedge R.\text{port.port\_req} = 1 \supset$
$$\exists n\ WAIT(R, E^n(R, f)) \wedge (E^n(R, f).\text{port.port\_req} = 2) \wedge (RT(E^n(R, f)) = 0)$$
$\cdots (13)$

This lemma is the analogue of lemma 1 for this processor. This is the most important of all the lemmas we have stated so far. It lets us abstract away the idle cycles during which nothing of interest gets done, and also lets us give a proof which is dependent only on the fact that the memory responds in a finite time and not on the actual time taken.

Lemma 5 Correctness of IFUrdy and port_rdy protocols
If the CPU tries to transfer data from IFU, or from or to the PORT and the respective module is not ready, as signalled by negating IFUrdy or port_rdy respectively, the state of the CPU, ALU, the compound memory, and the contents of PC, MAR and Data out registers do not change from this micro cycle to the next.

$LET\ s = decode(Out(R.\text{cpu})).\text{bus\_src}, d = decode(Out(R.\text{cpu})).\text{bus\_dst},$
$\quad O = Out(R.\text{ifu}, src\_ir : (s == IR), src\_pc : (s == PC)), R' = E(R, f)\ IN$
$\neg O.\text{ifurdy} \supset WAIT(R, R') \qquad \cdots (9)$

$LET\ s = decode(Out(R.\text{cpu})).\text{bus\_src}, d = decode(Out(R.\text{cpu})).\text{bus\_dst},$
$\quad O' = Out(R.\text{port}, src\_md : (s == MD), dst\_md : (d == MD), dst\_mar : (d == MAR))\ IN$
$\quad \neg O'.\text{port\_rdy} \supset WAIT(R, E(R, f)) \qquad \cdots (10)$

This lemma is useful in simplifying the proof of the liveness lemma. It says that the execution of a micro instruction will be repeated (with all but a few 'unimportant' registers remaining the same) if the data required for completing that instruction is not yet available.

The proof of these lemmas are included in the appendix C. The techniques used in the proof of these lemmas are very much similar to those used in the case of Simple.

7.2 Proof of Correctness

The strategy we use in our proof is identical to what was used in the proof of Simple. The equations used for rewriting are the specifications of the modules, the safety lemmas and the liveness lemma. The liveness lemma is applied whenever the corresponding module is not ready, i.e., the preconditions of (9) or (10) hold; otherwise, the definition of E^n is applied. In addition, we add the following meta-statement which can be proved as a lemma from the definition of F. definition:

$((R.\text{port.writing} == f)OR(R.\text{port.port\_req} == 2)) = t$
$\supset .\text{mem.store} = F(R).\text{store}$

In order to follow the proof, it would be necessary to consult the micro program given in appendix E. The micro program is given in the decoded format for improved readability.

7.2.1 Verification of (1)

$S(F(R), t) = \langle \mathbf{PC} : 0, \mathbf{STORE} : \text{STORE} \rangle$

$F(E^n(E(R, t), f))$

$= F(E^n(\langle \mathbf{CPU} : \langle \text{MAP}, \text{MUP}, \text{MUPC}, RESET\_INS \rangle, \mathbf{IFU} : \langle \text{PC}, \text{IR}, 2 \rangle,$
$\qquad \mathbf{MEM} : \langle \text{STORE}, 0, \text{ACC} \rangle, \mathbf{PORT} : \langle \text{MAR}, \text{MDIN}, \text{MDOUT}, 0, f, f \rangle, \mathbf{ALU} : \text{A} \rangle, f))$

by Premise instantiation and C1, C11, C41, C27, C30, C33

$= F(R_1 : E^{n-1}(\langle \mathbf{CPU} : \langle \text{MAP}, \text{MUP}, NEXT, NEXT\_INS \rangle, \mathbf{IFU} : \langle 0, \_, \_ \rangle,$
$\qquad \mathbf{MEM} : \langle \text{STORE}, 0, \text{ACC} \rangle, \mathbf{PORT} : \langle \text{MAR}, \text{MDIN}, \text{MDOUT}, 0, f, f \rangle, \mathbf{ALU} : \text{A} \rangle, f))$

– from the real specs

$= F(R_2 : E^{n-k}(\langle \mathbf{CPU} : \langle \text{MAP}, \text{MUP}, NEXT, NEXT\_INS \rangle, \mathbf{IFU} : \langle 0, read(\text{STORE}, 0), 0 \rangle,$

$\qquad \mathbf{MEM} : \langle M, \text{N1}, \text{N2} \rangle, \mathbf{PORT} : \langle \text{MAR}, \text{MDIN}, \text{MDOUT}, \_, \_, \_ \rangle, \mathbf{ALU} : \text{A} \rangle, f)),$
$\qquad$ add $F(R_2) = \text{STORE}$ by liveness lemma, and safety of IFU.

$= F(R_3 : \langle \mathbf{CPU} :\langle \mathsf{MAP}, \mathsf{MUP}, read(\mathsf{MAP}, read(\mathsf{STORE}, 0)), JMP \rangle, \mathbf{IFU} :\langle 1, \_, 0 \rangle,$

$\qquad \mathbf{MEM} :\langle \_, \_, \_ \rangle, \mathbf{PORT} :\langle \mathsf{MAR}, \mathsf{MDIN}, \mathsf{MDOUT}, \_, \_, \_ \rangle, \mathbf{ALU} :A \rangle)$, add $F(R_3) =$
STORE

$\qquad$ – by real specs, safety of port and putting $n - k - 1 = 0$

$= \langle \mathbf{PC} :0, \mathbf{MEM} :\mathsf{STORE} \rangle$ – by defn of F, and is same as $S(F(R), t)$

Thus on reset, $fetching(R_3)$ holds, establishing (1).

7.2.2 Verification of (2)

STPC - Store PC

$S(F(R), f) = \langle \mathbf{PC} :inc2(\mathsf{PC}),$

$\qquad \mathbf{MEM} :write(\mathsf{STORE}, read(\mathsf{STORE}, inc(\mathsf{PC})), inc2(\mathsf{PC})) \rangle$

$F(E^n(R : \langle \mathbf{CPU} :\langle \mathsf{MAP}, \mathsf{MUP}, STPC, read(\mathsf{MUP}, STPC) \rangle, \mathbf{IFU} :\langle inc(\mathsf{PC}), \_, \_ \rangle,$

$\qquad \mathbf{MEM} :\langle \_, \_, \_ \rangle, \mathbf{PORT} :\langle \mathsf{MAR}, \mathsf{MDIN}, \mathsf{MDOUT}, \_, \_, \_ \rangle, \mathbf{ALU} :A \rangle, f))$

$\qquad$ add $F(R).\mathsf{store} = \mathsf{STORE}$ – by premise instantiation

$= F(E^{n-k}(R_1 : \langle \mathbf{CPU} :\langle \mathsf{MAP}, \mathsf{MUP}, STPC, read(\mathsf{MUP}, STPC) \rangle, \mathbf{IFU} :\langle inc(\mathsf{PC}), \_, 0 \rangle,$

$\qquad \mathbf{MEM} :\langle \_, \_, \_ \rangle, \mathbf{PORT} :\langle \mathsf{MAR}, \_, \mathsf{MDOUT}, \_, \_, \_ \rangle, \mathbf{ALU} :A \rangle, f)$,

$\qquad$ add $F(R_1).\mathsf{store} = \mathsf{STORE})$ by liveness

$= F(E^{n-k-1}(R_2 : \langle \mathbf{CPU} :\langle \mathsf{MAP}, \mathsf{MUP}, inc(STPC), read(\mathsf{MUP}, inc(STPC)) \rangle,$

$\qquad \mathbf{IFU} :\langle inc2(\mathsf{PC}), \_, \_ \rangle \mathbf{MEM} :\langle \_, \_, \_ \rangle, \mathbf{PORT} :\langle \mathsf{MAR}, \_, \mathsf{MDOUT}, \_, \_, \_ \rangle,$

$\qquad \mathbf{ALU} :read(\mathsf{STORE}, inc(\mathsf{PC})) \rangle, f))$

$\qquad$ add $F(R_2).\mathsf{store} = \mathsf{STORE}$ – by safety of PORT and IFU

$= F(E^{n-k'}(R_3 : \langle \mathbf{CPU} :\langle \mathsf{MAP}, \mathsf{MUP}, inc(STPC), read(\mathsf{MUP}, inc(STPC)) \rangle,$

$\qquad \mathbf{IFU} :\langle inc2(\mathsf{PC}), \_, \_ \rangle, \mathbf{MEM} :\langle \_, \_, \_ \rangle, \mathbf{PORT} :\langle \mathsf{MAR}, \_, \mathsf{MDOUT}, 0, f, \_ \rangle,$

$\qquad \mathbf{ALU} :read(\mathsf{STORE}, inc(\mathsf{PC})) \rangle, f)))$, add $F(R_3).\mathsf{store} = \mathsf{STORE}$ – by liveness

$= F(E^{n-k'-1}(R_4 : \langle \mathbf{CPU} :\langle \mathsf{MAP}, \mathsf{MUP}, inc2(STPC), read(\mathsf{MUP}, inc2((STPC)) \rangle,$

$\qquad \mathbf{IFU} :\langle inc2(\mathsf{PC}), \_, \_ \rangle, \mathbf{MEM} :\langle \_, \_, \_ \rangle, \mathbf{PORT} :\langle read(\mathsf{STORE}, inc(\mathsf{PC})), \_, \mathsf{MDOUT}, 1, f, f \rangle,$

$\qquad \mathbf{ALU} :\_ \rangle, f))$, add $F(R_4).\mathsf{store} = \mathsf{STORE}$ by safety of PORT, real specs

$= F(E^{n-k''-2}(R_5 : \langle \mathbf{CPU} :\langle \mathsf{MAP}, \mathsf{MUP}, inc3(STPC), read(\mathsf{MUP}, inc3(STPC)) \rangle,$

$\qquad \mathbf{IFU} :\langle inc2(\mathsf{PC}), \_, \_ \rangle, \mathbf{MEM} :\langle \mathsf{STORE}, \_, \_ \rangle, \mathbf{PORT} :\langle read(\mathsf{STORE}, inc(\mathsf{PC})), \_,$

$\qquad inc2(\mathsf{PC}), 2, t, f \rangle, \mathbf{ALU} :\_ \rangle, f))$

$\qquad$ – by real specs

$= F(R_6 : \langle \mathbf{CPU} :\langle \mathsf{MAP}, \mathsf{MUP}, MUPC, read(\mathsf{MUP}, MUPC) \rangle, \mathbf{IFU} :\langle inc3(\mathsf{PC}), \_, \_ \rangle,$

$\textbf{MEM} :\langle\_,\_,\_\rangle, \textbf{PORT} :\langle\_,\_,\_,\_,\_,\_\rangle, \textbf{ALU} :\_\rangle)$

with $\mathsf{MUPC} = read(\mathsf{MAP}, read(\mathsf{STORE}, inc2(\mathsf{PC}))$ and $F(R_6).\textsf{store} =$
$write(\mathsf{STORE}, read(\mathsf{STORE}, inc(\mathsf{PC})), inc2(\mathsf{PC}))$

following the last two steps of Reset.

$= \langle\textbf{PC} :inc3(\mathsf{PC}), \textbf{MEM} :write(\mathsf{STORE}, read(\mathsf{STORE}, inc(\mathsf{PC})), inc2(\mathsf{PC}))\rangle$

Thus, (2) holds in this case also.

JMPIND - Indirect Jump

$S(F(R), f) = \langle\textbf{PC} :read(\mathsf{STORE}, read(\mathsf{STORE}, inc(\mathsf{PC}))), \textbf{STORE} :\mathsf{STORE}\rangle$

Following the first few steps in the proof for $STPC$, we have $E^n(R, f)$

$= F(E^{n-k'-1}(R_4 : \langle\textbf{CPU} :\langle\mathsf{MAP}, \mathsf{MUP}, inc2(JMPIND), read(\mathsf{MUP}, inc2(JMPIND))\rangle,$

$\textbf{IFU} :\langle inc2(\mathsf{PC}),\_,\_\rangle, \textbf{MEM} :\langle\_,\_,\_\rangle, \textbf{PORT} :\langle read(\mathsf{STORE},$
$inc(\mathsf{PC})),\_,\mathsf{MDOUT}, 1, f, f\rangle, \textbf{ALU} :\_\rangle, f)),$ add $F(R_4).\textsf{store} = \mathsf{STORE}$

$= F(E^{n-k''}(R_5 : \langle\textbf{CPU} :\langle\mathsf{MAP}, \mathsf{MUP}, inc2(JMPIND), read(\mathsf{MUP}, inc2(JMPIND))\rangle,$

$\textbf{IFU} :\langle inc2(\mathsf{PC}),\_,\_\rangle, \textbf{MEM} :\langle\_, \mathsf{X}, \mathsf{X}\rangle, \textbf{PORT} :\langle read(\mathsf{STORE},$
$inc(\mathsf{PC})),\_,\mathsf{MDOUT}, 2, f, f\rangle, \textbf{ALU} :\_\rangle, f))$
add $F(R_5).\textsf{store} = \mathsf{STORE}$ by liveness.

$= F(E^{n-k''-1}(R_6 : \langle\textbf{CPU} :\langle\mathsf{MAP}, \mathsf{MUP}, inc2(JMPIND), read(\mathsf{MUP}, inc2(JMPIND))\rangle,$

$\textbf{IFU} :\langle inc2(\mathsf{PC}),\_,\_\rangle, \textbf{MEM} :\langle\_,\_,\_\rangle, \textbf{PORT} :\langle read(\mathsf{STORE},$
$inc(\mathsf{PC})),\_,\mathsf{MDOUT}, 0,\_, t\rangle, \textbf{ALU} :\_\rangle, f))$ add $F(R_6).\textsf{store} = \mathsf{STORE}$ by live-

ness.

$= F(E^{n-k''-2}(R_7 : \langle\textbf{CPU} :\langle\mathsf{MAP}, \mathsf{MUP}, inc3(JMPIND), read(\mathsf{MUP}, inc3(JMPIND))\rangle,$

$\textbf{IFU} :\langle read(\mathsf{STORE}, read(\mathsf{STORE}, inc(\mathsf{PC}))),\_,\_\rangle, \textbf{MEM} :\langle\_,\_,\_\rangle,$
$\textbf{PORT} :\langle\_,\_,\_,\_,\_,\_\rangle, \textbf{ALU} :\_\rangle, f))$
, add $F(R_6).\textsf{store} = \mathsf{STORE}$ by safety of PORT and real specs.

$= F(R_8 : \langle\textbf{CPU} :\langle\mathsf{MAP}, \mathsf{MUP}, \mathsf{MUPC}, read(\mathsf{MUP}, \mathsf{MUPC})\rangle,$
$\textbf{IFU} :\langle inc(read(\mathsf{STORE}, read(\mathsf{STORE}, inc(\mathsf{PC})))),\_,\_\rangle, \textbf{MEM} :\langle\_,\_,\_\rangle,$
$\textbf{PORT} :\langle\_,\_,\_,\_,\_,\_\rangle, \textbf{ALU} :\_\rangle)$, add $F(R_7).\textsf{store} = \mathsf{STORE}$ following the
last few steps of reset, with $\mathsf{MUPC} = read(\mathsf{MUP}, read(\mathsf{STORE}, inc2(\mathsf{PC})))$.

$= \langle\textbf{PC} :read(\mathsf{STORE}, read(\mathsf{STORE}, inc(\mathsf{PC}))), \textbf{MEM} :\mathsf{STORE}\rangle$

and thus (2) holds of this instruction also.

8 Conclusions

In this paper we have demonstrated the application of equational techniques for the specification and verification of a microprocessor. A slightly abstracted version of a complex processor, the Lilith Modula II machine, was specified at an abstract and a realization level. Then the design description was verified to be correct with respect to the abstract specification. The reason for choosing the Lilith architecture for the exercise was that it was one of the few large scale designs implemented in hardware which also had a clear and complete design description. We successfully specified the entire design all the way to "off the shelf" components level. We decided to abstract away some of the details of the complete design for the purpose of verification to bring out the essential principles involved more clearly and also because it would have necessitated a proof checker otherwise.

The verification task was carried out by refining the formal statement of correctness into a set of equational assertions. Every equational assertion was verified by simplifying the two sides using an equational interpreter. The equational interpreter and a specification editor were the only two mechanical aids used for the task. We plan to check all the proofs carried out in the paper on the theorem prover ORACLE [12]. ORACLE is a rewriting-based theorem prover which has a functional specification language interface similar to SBL. Some of the strategies used in our proofs (egs., structural induction and case-splitting) are built into ORACLE. The higher order feature supported by ORACLE makes it possible to natuarlly check the other proof strategies, such as iteration induction.

The distinguishing features of our approach in comparison to other microprocessor verification work are the following. The efforts in [2], [1] and [7] use a design in which the memory is dedicated to the CPU and has a fixed response time. Our design permits asynchronous handshake between the memory and other modules in the design. Although the HOL formalism used in [7] is expressive enough to express such an interaction the illustration in [7] does not address this issue. Unlike the approach used by Hunt [6], we handle the asynchronous aspect without an explicit temporal parameter, such as an oracle. We believe our approach is clearer, and will lead to a general method for reasoning about asynchronous interactions within equational framework.

Acknowledgements

The authors wish to acknowledge the contributions made by all the members of the seminar course on hardware verification at Stony Brook in which the present work had its origin. Special thanks are due to David Smith for suggesting the Lilith architecure as an example to try, which turned out to be a good choice, and to N.C. Lee for helping in the specification of parts of the system.

References

[1] Barrow H.G. *VERIFY: A Program for Proving Correctness of Digital Hardware Designs*, Artificial Intelligence 24 (1984), pp. 437-491, Elsevier Science Publishers B.V. (North-Holland).

[2] Gordon M.J. *Proving a Computer Correct* Technical Report No. 42, Computer Laboratory, University of Cambridge, 1983.

[3] Gopalakrishnan G. Smith, D.R., Srivas, M.K. "An Algebraic Approach to the Specification and Realization of VLSI Designs", *Proc. Seventh International Symposium on Computer Hardware Description Languages*, Tokyo August 1985.

[4] Gopalakrishnan G., Srivas, M.K., Smith, D.R., *From Algebraic Specifications to Correct VLSI Circuits* Proc. of IFIP Working Conference on From H.D.L. Descriptions to Guaranteed Correct Circuit Designs Grenoble France September 9-11 1986.

[5] Henderson, P., *Functional Programming: Application and Implementation*, Prentice-Hall, 1980.

[6] Hunt W.A. *The Mechanical Verification of a Microprocessor Design* Proc. of IFIP Working Conference on From H.D.L. Descriptions to Guaranteed Correct Circuit Designs Grenoble France September 9-11 1986.

[7] Joyce J. Birtwistle G. Gordon M., *Proving a Computer Correct in Higher Order Logic*, Research Report No. 85/208/21, University of Calgary, August 1985.

[8] Kapur D. and Sivakumar *Rewrite Rule Laboratory* Proceedings of G.E. Workshop on Rewrite Rule Laboratory, Schnectady, September 1983.

[9] Lee, N. C., "Integration of SBL into a VLSI Design System", PhD thesis, Dept. EE., SUNY, Stony Brook, December 1987.

[10] Lescanne P. *Computer Experiments with the REVE Term Rewriting System Generator*, Proceedings of the *Tenth Annual Symposium on Principles of Programming Languages* Austin, Texas, January 1983.

[11] Manna Z. *Mathematical Theory of Computation* McGraw-Hill Publishing Co., New York, 1974.

[12] Mills C. *Introduction to ORACLE/Caliban* Internal Memo, Odyssey Research Associates Inc. Ithaca, NY 14850.

[13] Proceedings of International Working Conference on *From H.D.L. Descriptions to Guaranteed Correct Circuit Designs* IFIP Grenoble France September 1986.

[14] Srivas, M.K., Smith, D.R., Lee, N.C. "The SBL Users Manual", Tech. Rpt, Dept CS., SUNY, Stony Brook, Dec 1987.

[15] Wirth, N., "The Personal Computer Lilith", Tech. Report, Eidgenossische Technische Hochschule, Institut fur Informatik, Zurich, April 1981.

A Specification of the Simple Processor

```
ABS_MODULE CPU

ABS STATE IS RECORD
                St: {fetch, jmp, store1, store2}
                PC, MAR, MDout: BITVECTOR[16]
                reading, writing: BIT
            END
EXTOP
  Out :: CurSt ==> !addr, !dout, !write  WHERE {
      !addr = CurSt.MAR                                          S
      !dout = CurSt.MDout                                        S
      !write = CurSt.writing }                                   S

  Next :: CurSt, ?din, ?reset ==> newstate WHERE {
      newstate =
        COND{ ?reset: <fetch,0,0,0,t,f>                          S
              ELSE: CASE CurSt.St {
                fetch: <COND {?din = JMPINS: jmp                 S
                              ?din = STPCINS: store1 },inc(PC),inc(PC),MDout,t,f>S
                jmp:    <fetch, ?din, ?din, MDout, t, f>         S
                store1:<store2, inc(PC), ?din, inc(PC), f, t>    S
                store2:<fetch, PC, PC, MDout, t, f> } }  }       S
END CPU


ABS_MODULE MEM(acctime: integer)
ABS_STATE IS RECORD
      store: memory(2**16) of BITVECTOR % memory supports read and write
      cycle_num, acc_num: 0..acctime
    END


EXTOP
  Out :: CurSt, ?write, ?addr ==> !ready, !dout  WHERE {
      !ready = (CurSt.cycle_num == acc_num)                     S
      !dout = COND{(CurSt.cycle_num == acc_num)
                    AND NOT ?write:read(CurSt.store, ?addr)     S
                  ELSE: VECTOR {i: UNK} } }

  Next :: CurSt, ?as, ?write, ?addr, ?din, ?reset ==> newstate WHERE {
```

```
newstate =
    <COND{ (NOT ?reset AND ?write AND (CurSt.cycle_num == CurSt.acc_num)):
                write(CurSt.store, ?addr, ?din)                              S20
            ELSE: CurSt.store },                                             S21
    COND{ ?reset OR (CurSt.cycle_num == CurSt.acc_num): 0                    S22
          ?as AND NOT (CurSt.cycle_num == acc_num) : CurSt.cycle_num+1       S23
          ELSE: CurSt.cycle_num },
    COND{ ?reset OR CurSt.cycle_num == CurSt.acc_num: random(acctime)        S24
          ELSE: CurSt.acc_num}> }                                           S25
          %random returns a random integer between 1 and its argument
END MEM
```

B Proof of Lemma 1

In this section, the proofs of the liveness lemma for Simple and the proof of correctness of the STPC instruction are included.

Lemma 1 Liveness property of the Memory
$$\forall R \, \exists n \, [WAIT(R, E^n(R, f)) \wedge READY(E^n(R, f))]$$

Proof: We will indicate the main steps in proving this lemma. Once we set up the intermediate steps, an equational interpreter could fill in the details.

Proposition 1: $\forall R \, RT(R) \geq 0$

   Basis: $[RT(E(R,t)) \geq 0]$ – follows from:

   $E(R,t)$.mem.cycle_num $= 0$ – from S22

   $E(R,t)$.mem.acc_num $= random(acctime) > 0$ by defn, S24

   Induction Step: $[(RT(R) \geq 0) \supset (RT(E(R,f)) \geq 0)]$.

   Case 1: $RT(R) = 0$: We add the equation R.mem.acc_num $= R$.mem.cycle_num to our system of equations.

   $E(R,f)$.mem.cycle_num $= 0$ – from S22

   $E(R,f)$.mem.acc_num $= random(acctime) > 0$ – by S24 and defn of random

   Thus $RT(E(R,f)) \geq 0$

   Case 2: $RT(R) > 0$: Add R.mem.cycle_num $= R$.mem.acc_num $- j - 1$ with $j \geq 0$

   $E(R,f)$.mem.cycle_num $= R$.mem.acc_num $- j$ – from S23

   $E(R,f)$.mem.acc_num $= R$.mem.acc_num – from S25

   Thus $RT(E(R,f)) = R$.mem.acc_num $- (R$.mem.acc_num $- j) = j \geq 0$

Proposition 2: $[(RT(R) > 0)] \supset [WAIT(R, E(R, f)) \wedge (RT(E(R, f)) = RT(R) - 1)]$:

Add $R.\text{mem.cycle\_num} = R.\text{mem.acc\_num} - 1 - j$, with $j \geq 0$.

$WAIT(R, E(R, f))$: follows from

$R.\text{cpu} = E(R, f).\text{cpu}$ – follows from S18 *(ready = f)*, S5 *(memready = f)*, S6

$R.\text{mem.store} = E(R, f).\text{mem.store}$ – from S20 (not executed)

$R.\text{mem.acc\_num} = E(R, f).\text{mem.acc\_num}$ – from S24 (not executed)

$RT(E(R, f)) = RT(R) - 1$: follows from (3) and from

$E(R, f).\text{mem.cycle\_num} = R.\text{mem.acc\_num} - j$ – from S23

Proposition 3: $[(RT(E(R, f)) = 0) \wedge WAIT(R, E(R, f))] \supset$
$[WAIT(R, E(R, f)) \wedge READY(E(R, f))]$

The antecedent is the same as the consequent of this proposition,

and hence it holds. From these propositions, it follows that $\exists n\ RT(E^n(R, f)) = 0$

Since $[WAIT(R, R_1) \wedge WAIT(R_1, R_2)] \supset WAIT(R, R_2)$ we can conclude

$WAIT(R, E^n(R, f))$. This completes the proof of the lemma.

STPC - Store PC: $read(\text{STORE}, \text{PC}) = STPCINS$

$S(F(R), f) = \langle \mathbf{PC} : inc(inc(\text{PC})),$

$\quad \mathbf{MEM} : write(\text{STORE}, read(\text{STORE}, inc(\text{PC})), inc(inc(\text{PC}))) \rangle$ by S3

$F(E^n(\langle \mathbf{CPU} : \langle FETCH, \text{PC}, \text{PC}, \text{MD}, t, f \rangle,$

$\quad \mathbf{MEM} : \langle \text{STORE}, \text{CN}, \text{AN} \rangle \rangle, f))$ by premise instantiation

$= F(E^{n'}(\langle \mathbf{CPU} : \langle FETCH, \text{PC}, \text{PC}, \text{MD}, t, f \rangle,$

$\quad \mathbf{MEM} : \langle \text{STORE}, \text{AN}, \text{AN} \rangle \rangle, f))$ by liveness lemma

$= F(E^{n'-1}(\langle \mathbf{CPU} : \langle STORE1, inc(\text{PC}), inc(\text{PC}), \text{MD}, t, f \rangle,$

$\mathbf{MEM} : \langle \text{STORE}, 0, \text{Z} \rangle \rangle, f))$ by S4,S9,S10,S11,S5,S18,S19,S6, S14, S7,S21,S22 and S24

$= F(E^{n''}(\langle \mathbf{CPU} : \langle STORE1, inc(\text{PC}), inc(\text{PC}), \text{MD}, t, f \rangle,$

$\quad \mathbf{MEM} : \langle \text{STORE}, \text{Z}, \text{Z} \rangle \rangle, f))$ by liveness lemma

$= F(E^{n''-1}(\langle \mathbf{CPU} : \langle STORE2, inc(inc(\text{PC})), read(\text{STORE}, inc(\text{PC})), inc(inc(\text{PC})), f, t \rangle,$

$\mathbf{MEM} : \langle \text{STORE}, 0, \text{W} \rangle \rangle, f))$ by S4,S9,S10,S11,S5,S18,S19, S6, S16, S7,S21,S22 and S24

$= F(E^{n'''}(\langle \mathbf{CPU} : \langle STORE2, inc(inc(\text{PC})), read(\text{STORE}, inc(\text{PC})), inc(inc(\text{PC})), f, t \rangle,$

$\quad \mathbf{MEM} : \langle \text{STORE}, \text{W}, \text{W} \rangle \rangle, f))$ by liveness lemma

$= F(E^{n'''-1}(\langle \mathbf{CPU} : \langle FETCH, inc(inc(\text{PC})), inc(inc(\text{PC})), inc(inc(\text{PC})), t, f \rangle,$

$$\textbf{MEM} : \langle write(\textsf{STORE}, read(\textsf{STORE}, inc(\textsf{PC})), inc(inc(\textsf{PC}))), 0, \textsf{X} \rangle, f \rangle)$$
$$\text{by S4,S9,S10,S11,S5,S18,S19, S6, S17, S7,S21,S22 and S24}$$
$$= \langle \textbf{PC} : inc(inc(\textsf{PC})), \textbf{MEM} : write(\textsf{STORE}, read(\textsf{STORE}, inc(\textsf{PC})), inc(inc(\textsf{PC})))$$
$$\text{putting } n''' - 1 = 0 \text{ and defn of } F$$

It is clear that the postcondition of (2) holds in this case also.

C   Mini Lilith: Proof of Lemmas

Lemma 2 Safety property of the IFU

Let $O = Out(R.\text{ifu}, src\_ir : t, src\_pc : f)$ IN

$[O.\text{ifurdy} \supset O.\text{bus} = read(F(R).\text{store}, R.\text{ifu.PC})]$ $\qquad\qquad \cdots$ *(6)*

If $IFUrdy$ is asserted when src_ir is asserted, the *bus* has the contents of the memory addressed by the PC. The predicate $SIFU(R)$ is defined by (6).

Proof: The safety property of the IFU is established by induction on the number of times E is invoked on $E(R, t)$. Note that by C5, O.ifurdy is equivalent to R.ifu.read $= 0$, and by C2, the bus output of IFU is the same as the contents of IR when $src\_ir = t$. The lemma should be stated in terms of the state of the IFU rather than its output, but it is intuitively more appealing to state it in this form. We implicitly make use of the fact that $F(R)$.store, R.mem.store and $E(R, f)$.mem.store are all equal as far as the code memory is concerned, since we assumed that the code memory, i.e., the part of the memory accessed by the IFU, is read only.

Basis: $E(R, t)$.ifu.read $= 2$ – from C11

Thus, $SIFU(E(R, t))$ holds vacuously.

Induction Step: $SIFU(R) \supset SIFU(E(R, f))$ is established by:

$\quad$ *Proposition 1:* $(R.\text{ifu.read} = 0) \vee (R.\text{ifu.read} = 1) \vee (R.\text{ifu.read} = 2)$

$\qquad$ Proof is by induction and is straight forward and hence skipped.

$\quad$ *Case 1:* Add R.ifu.read $= 0$:

$\qquad$ *Case 1.1:* Add $src\_ir = f$ and $dst\_pc = f$ at C6

$\qquad\quad$ $R.\text{ifu.pc} = E(R, f).\text{ifu.pc}$ – from C8

$\qquad\quad$ $R.\text{ifu.ir} = E(R, f).\text{ifu.ir}$ – from C10

$\qquad\quad$ $R.\text{ifu.read} = E(R, f).\text{ifu.read}$ – from C16

$\qquad$ Since $R.\text{ifu} = E(R, f).\text{ifu}$, $SIFU(E(R, f))$ holds.

$\qquad$ *Case 1.2:* Add $src\_ir = t$ at C6

$\qquad\quad$ *Case 1.2.1:* Add $as = t$ at C6

$E(R, f)$.ifu.read $= 1$ and hence $SIFU(E(R, f))$ holds vacuously.

Case 1.2.2: Add $as = f$ at C6

$E(R, f)$.ifu.read $= 2$ and hence $SIFU(E(R, f))$ holds vacuously

Case 1.3: Add $dst\_pc = t$ at C6 Proof same as case 1.2

Case 2: Add R.ifu.read $= 1$

Case 2.1: Add $as = t$ at C6

$E(R, t)$.ifu.read $= 1$ and hence $SIFU(E(R, f))$ holds.

Case 2.2: Add $as = f$ at C6

$E(R, t)$.ifu.read $= 2$ and hence $SIFU(E(R, f))$ holds.

Case 3: Add R.ifu.read $= 2$ at C6

Case 3.1: Add R.mem.cycle_num $= R$.mem.acc_num at C36

$E(R, f)$.ifu.ir $= read(R.\text{mem.store}, R.\text{ifu.pc})$

– from C4 ($ifureq = t$), C36, C37, C9

Case 3.2: Add R.mem.cycle_num $= R$.mem.acc_num $- j - 1$ at C36

$E(R, f)$.ifu.read $= 2$ from C36 and C10 and thus $SIFU(E(R, f))$ holds.

In all the cases, it has been shown that $SIFU(E(R, f))$ holds.

Lemma 3 Safety property of the PORT

$decode(Out(R.\text{cpu})).\text{bus\_dst} \neq MD \supset F(R).\text{store} = F(E(R, f)).\text{store}$ $\qquad \cdots$ (8)

If, however, no attempt is made to write into the $MDout$ into $MDout$, the contents of the memory remains unchanged.

Proof:

The proof is by case analysis. It could be easily proved that these cases are exhaustive. Firstly, add $dst\_md = f$.

Case 1: Add R.port.port_req $= 0$:

Case 1.1: Add $dst\_mar = t$:

$E(R, f)$.port.port_req $= 1$ – from C28

Further case splitting is involved here at C38 and C40, but skipped

$F(E(R, f)).\text{store} = E(R, f).\text{mem.store}$ – since $port\_req \neq 2$

$F(R).\text{store} = R.\text{mem.store}$ – since $port\_req \neq 2$

$E(R, f).\text{mem.store} = R.\text{mem.store}$ – from C18, C38 and C40

This completes the proof for case 1.1

Case 1.2: Add $dst\_mar = f$. The proof goes through as in case 1.1.

Case 2: Add R.port.port_req $= 1$

$E(R, f)$.port.writing $= f$ – from C31, and C32

$F(R)$.store $= R$.mem.store – since $port\_req \neq 2$

$F(E(R, f))$.store $= R$.mem.store – since $writing = f$

$E(R, f)$.mem.store $= R$.mem.store – by C18, C38, C40

This completes proof for case 2.

Case 3: Add R.port.port_req $= 2$

    Case 3.1: Add R.port.writing $= f$

        $E(R, f)$.port.writing $= f$ – by C31

        $F(R)$.store $= R$.mem.store – since $writing = f$

        $F(E(R, f))$.store $= E(R, f)$.mem.store – since $writing = f$

        $E(R, f)$.mem.store $= R$.mem.store – by C21, C38

This completes the proof for case 3.1

    Case 3.2: Add R.port.writing $= t$:

        Case 3.2.1: Add R.mem.cycle_num $= R$.mem.acc_num $- j$ with $j > 0$:

            $E(R, f)$.port.mdout $= R$.port.mdout

            $E(R, f)$.port.port_req $= R$.port.port_req – by C36, C27

            $E(R, f)$.port.writing $= R$.port.writing by C31

            $E(R, f)$.port.mar $= R$.port.mar – by C23

            $E(R, f)$.mem.store $= R$.mem.store – by C38, C40

        Hence, $F(R)$.store $= F(E(R, f))$.store

        Case 3.2.2: Add R.mem.cycle_num $= R$.mem.acc_num

            $E(R, f)$.port.port_req $= 0$ – by C36 and C27

            $E(R, f)$.mem.store $= write(R$.mem.store,

                R.port.mar, R.port.mdout) – by C39

    By the definition of F, it is clear that $F(R)$.store $= F(E(R, f))$.store

This establishes the safety property of the PORT.

Lemma 4 $(decode(Out(R.\text{cpu})).\text{bus\_src} = IR) \supset$

    $\exists n\ WAIT(R, E^n(R, f)) \land E^n(R, f).\text{ifu.read} = 0$ $\cdots$ *(11)*

$(decode(Out(R.\text{cpu})).\text{bus\_dst} = MAR) \supset$

    $\exists n\ WAIT(R, E^n(R, f)) \land E^n(R, f).\text{port.port\_req} = 0$ $\cdots$ *(12)*

$(decode(Out(R.\text{cpu})).\text{bus\_src} = MD) \land R.\text{port.port\_req} = 1 \supset$

    $\exists n\ WAIT(R, E^n(R, f)) \land (E^n(R, f).\text{port.port\_req} = 2) \land (RT(E^n(R, f)) = 0)$

$\cdots$ *(13)*

Any micro instruction will be executed repeatedly because of $IFUrdy$ or $port\_rdy$ being negated, and have the last one semantics. Hence, we can ignore the fact that any micro instruction is executed many times in this fashion and take that it is executed once only, and then with both these modules ready.

Proof: By our assumption, $bus\_src = IR, bus\_src = MD$ and $bus\_dst = MAR$ are mutually incompatible, and hence only one of the antecedents in the above lemma could hold at any time. The proof is by case analysis.

Case 1: Add $(decode(Out(R.\text{cpu})).bus\_dst == MAR) = t$
  by our assumption, add $(decode(Out(R.\text{cpu})).bus\_dst == PC) = f$ and
    $(decode(Out(R.\text{cpu})).bus\_dst == MD) = f$ and
    $(decode(Out(R.\text{cpu})).bus\_src == IR) = f$
Case 1.1: Add $R.\text{port.port\_req} = 0$:
  (12) holds with $n = 0$
Case 1.2: Add $R.\text{port.port\_req} = 1$:
  $E(R, f).\text{port.port\_req} = 2$ – from C29
  $WAIT(R, E(R, f))$ holds since $port\_rdy = f$ at C22
  $(decode(Out(E(R, f).\text{cpu})).bus\_dst == MAR) = t$ by $WAIT(E(R, f), R)$
Thus, $E(R, f)$ corresponds to case 1.3, and $WAIT(R, E(R, f))holds$
Case 1.3: Add $R.\text{port.port\_req} = 2$
  Case 1.3.1: Add $R.\text{ifu.read} = 2$
    Proposition 1: $RT(R) \geq 0$
    Proof is identical to that for the simple processor.
  Define a predicate $IFUACC(R)$ as corresponding to this case
  Proposition 2: $[IFUACC(R) \wedge (RT(R) > 0)] \supset$
  $[WAIT(R, E(R, f)) \wedge IFUACC(E(R, f)) \wedge (RT(E(R, f)) = RT(R) - 1)]$
    Add $R.\text{mem.cycle\_num} = R.\text{mem.acc\_num} - 1 - j$, with $j \geq 0$
    $WAIT(E(R, f), R)$ follows from (10) since by C22 $port\_rdy = f$
    $(decode(Out(E(R, f).\text{cpu})).bus\_dst == MAR) = t$ by $WAIT(E(R, f), R)$
      $E(R, f).\text{port.port\_req} = 2$
      $E(R, f).\text{ifu.read} = 2$ by C36 and C16
    Thus $IFUACC(E(R, f))$ holds
      $E(R, f).\text{mem.acc\_num} = R.\text{mem.acc\_num}$ – by C45
      $E(R, f).\text{mem.cycle\_num} = R.\text{mem.acc\_num} - j$ – by C42
    Thus, $RT(E(R, f)) = RT(R) - 1$,
  completing the proof of proposition 2.
  From these two propositions, with $R' = E^m(R, f)$, it follows that
    $IFUACC(R) \supset \exists m \ (RT(R') = 0) \wedge IFUACC(R') \wedge WAIT(R, R')$
  $E(R', f).\text{ifu.read} = 0$ – by C36, C15
  $WAIT(R', E(R', f))$ holds by C22

$(decode(Out(E(R,f).\text{cpu})).\text{bus\_dst} == MAR) = t$ by $WAIT(E(R,f),R)$

$E(R'f)$ corresponds to case 1.3.2

Case 1.3.2: Add $R.\text{ifu.read} = 0$

Define a predicate $PORTACC(R)$ as corresponding to this case

Proposition 3: $[PORTACC(R) \wedge (RT(R) > 0)] \supset$

$[WAIT(R, E(R,f)) \wedge PORTACC(E(R,f)) \wedge (RT(E(R,f)) = RT(R) - 1)]$

Add $R.\text{mem.cycle\_num} = R.\text{mem.acc\_num} - 1 - j$, with $j \geq 0$

$WAIT(E(R,f),R)$ follows from C22 where $port\_rdy = f$

$(decode(Out(E(R,f).\text{cpu})).\text{bus\_dst} == MAR) = t$ by $WAIT(E(R,f),R)$

$E(R,f).\text{port.port\_req} = 2$ – by C36 and C16

$E(R,f).\text{ifu.read} = 0$ since $src\_ir = dst\_pc = f$

Thus $PORTACC(E(R,f))$ holds

$E(R,f).\text{mem.acc\_num} = R.\text{mem.acc\_num}$ – by C45

$E(R,f).\text{mem.cycle\_num} = R.\text{mem.acc\_num} - j$ – by C42

Thus, $RT(E(R,f)) = RT(R) - 1$, completing the proof of proposition 3.

From these propositions 1 and 3, with $R' = E^m(R,f)$, it follows that

$PORTACC(R) \supset \exists m \ (RT(R') = 0) \wedge PORTACC(R') \wedge WAIT(R,R')$

$E(R',f).\text{port.port\_req} = 0$ – by C36, C27

$WAIT(R', E(R',f))$ holds by C22

This completes the proof for case 1.3.2

Case 1.3.3: Add $R.\text{ifu.read} = 1$

The proof is identical to case 1.3.2

This completes the proof for case 1. The proofs for other parts are along the same lines and are skipped to avoid repetition.

Lemma 5 *LET* $s = decode(Out(R.\text{cpu})).\text{bus\_src}, d = decode(Out(R.\text{cpu})).\text{bus\_dst}$,

$O = Out(R.\text{ifu}, src\_ir : (s == IR), src\_pc : (s == PC)), R' = E(R,f)$ *IN*

$\neg O.\text{ifurdy} \supset WAIT(R, R')$ $\qquad \cdots (9)$

LET $s = decode(Out(R.\text{cpu})).\text{bus\_src}, d = decode(Out(R.\text{cpu})).\text{bus\_dst}$,

$O' = Out(R.\text{port}, src\_md : (s == MD), dst\_md : (d == MD), dst\_mar : (d == MAR))$ *IN*

$\neg O'.\text{port\_rdy} \supset WAIT(R, E(R,f))$ $\qquad \cdots (10)$

Proof: The proof is along the same line as the previous two lemmas. The only difference is that, by our assumption on valid micro instructions, we add the equations

$(bus\_dst == MAR) = f$, $(bus\_dst == PC) = f$ and $(bus\_dst == MD) = f$ whenever we add $(bus\_src == IR) = t$.

For the first part of the lemma,

    Case 1: Add $(decode(Out(R.\mathsf{cpu})).\mathsf{bus\_src} == IR) = t$:

        Add $(R.\mathsf{ifu.read} == 0) = f$,

                as this is equivalent to precondition of (9) by C5

        by our assumption, add $(decode(Out(R.\mathsf{cpu})).\mathsf{bus\_dst} == PC) = f$ and

                $(decode(Out(R.\mathsf{cpu})).\mathsf{bus\_dst} == MD) = f$ and

                $(decode(Out(R.\mathsf{cpu})).\mathsf{bus\_dst} == MAR) = f$

        $E(R, f).\mathsf{cpu} = R.\mathsf{cpu}$ – from C5 and C52

        $E(R, f).\mathsf{alu} = R.\mathsf{alu}$ – from C5 and C53

        $F(E(R, f)).\mathsf{store} = F(R).\mathsf{store}$ – by (8) since $dst\_md = f$

        $E(R, f).\mathsf{ifu.pc} = R.\mathsf{ifu.pc}$ – by C8

        $E(R, f).\mathsf{port.mar} = R.\mathsf{port.mar}$ – by C24

        $E(R, f).\mathsf{port.mdout} = R.\mathsf{port.mdout}$ – by C26

    This completes the proof for case 1.

    Case 2: Add $(decode(Out(R.\mathsf{cpu})).\mathsf{bus\_src} == IR) = f$:

        By C5, antecedent of (9) is not satisfied and hence (9) holds vacuously, completing the proof of (9).

The proof is along the same lines for the other part of the lemma, and we skip it here in order to avoid being repetitious.

D   Specification of Mini Lilith

```
ABS_MODULE Lilith

ABS_STATE IS RECORD
                PC: BITVECTOR [16] % program counter
                STORE: memory(2**16) of BITVECTOR [16] % memory is usual memory.
            END                            %
EXTOP
  Step :: CurSt, ?reset ==> newstate WHERE {
  newstate =
    COND{ ?reset: <0, STORE>  % PC becomes zero, STORE unchanged.
         ELSE:
    CASE STORE[PC] {   % only three opcodes:
        JMP: <read(STORE, inc(PC)), STORE>      % 1. jump
```

```
               STPC: <inc(inc(PC)), write(STORE, read(STORE, inc(PC)), inc(inc(PC)))>
               JMPIND:<read(STORE, read(STORE, inc(PC))), STORE> }
END Lilith


ABS_MODULE CPU(mu_adrsize, mu_wdsize, ninstr: INTEGER)


ABS STATE IS RECORD
                 map_rom: VECTOR [0..ninstr-1] OF BITVECTOR[mu_adrsize]
                 muP_rom: VECTOR [0..2**mu_adrsize-1] OF BITVECTOR[mu_wdsize]
                 mu_PC: BITVECTOR[mu_adrsize]
                 mir: BITVECTOR[mu_wdsize]
             END
EXTOP
  Out :: CurSt ==> !mu_code WHERE
      !mu_code = CurSt.mir


  Next :: CurSt, ?bus, ?mir_adr, ?jmpsel, ?reset, ?mu_ins ==> newstate
   WHERE { newstate =
        LETSEQ{
          muP_addr = COND{
             ?jmpsel: CurSt.map_rom[!?bus]
             ?mu_ins: inc(CurSt.mu_PC)
             ELSE: ?mir_adr } IN
          <init_maprom,        % Initial values stored in MAPROM and
           init_muProm,        % muPROM to be specified here
           muP_addr,           % mu_PC
           CurSt.muP_rom[ COND{ ?reset: VECTOR{ i: FALSE} %MIR          C1
                              ELSE: muP_addr} ] > }

END CPU


ABS_MODULE IFU


ABS_STATE IS RECORD
                 PC, IR: BITVECTOR[16]
                 read: 0..2        % read = 0 -> idle, 1 -> waiting for current
             END                   % memory cycle to finish before reaching 2
EXTOP                              % read = 2  -> reading memory
  Out :: CurSt, ?src_pc, ?src_ir ==> !?bus, !m_addr, !ifureq, !IFUrdy WHERE{
    !?bus =
      COND{ ?src_ir: CurSt.IR                                              C2
            ?src_pc: CurSt.PC
            ELSE: VECTOR{ i: HiZ }  }
      !m_addr = COND{ (CurSt.read == 2): CurSt.PC                          C3
                      ELSE: VECTOR{ i: HiZ } }
      !ifureq = (CurSt.read == 2)                                         C4
      !IFUrdy = (NOT ?src_ir) OR (CurSt.read == 0) % if CPU reads IR when  C5
```

```
          }                     % IR is being read from Mem, IFU is not ready
Next :: CurSt, ?as, !?bus, ?dst_pc, ?src_ir, ?dtack,
                               ?m_data, ?reset ==> newstate

   WHERE{ newstate =
        <COND{ ?dst_pc: !?bus                                       C6
               (CurSt.read == 0) AND ?src_ir: inc(CurSt.PC) %PC     C7
                  ELSE: CurSt.PC },                                 C8
           COND{ (CurSt.read==2)AND ?dtack:?m_data % IR             C9
                 ELSE: CurSt.IR },                                  C10
           COND{ ?reset: 2                       % read state       C11
                 ((CurSt.read == 0) AND ?src_ir ) OR ?dst_pc :
                         COND{?as: 1                                C12
                             ELSE: 2 }                              C13
                 (CurSt.read == 1) AND NOT ?as: 2                   C14
                 (CurSt.read==2 AND ?dtack): 0                      C15
                  ELSE: CurSt.read } }                              C16
        }
END IFU


ABS_MODULE PORT

ABS_STATE IS RECORD
                MAR, MDin, MDout: BITVECTOR[16]
                port_req: 0..2
                writing, data_read: BIT
             END
EXTOP
   Out :: CurSt, ?ifureq, ?src_md, ?dst_mar, dst_md
                    ==> !?bus, !m_data, !?m_addr, !r, !?as, !port_rdy WHERE{
     !?bus = COND{ ?src_md: CurSt.MDin
                    ELSE: VECTOR{i: HiZ} }                          C17
     !?as = ((CurSt.port_req == 2) AND  NOT ?ifureq)                C18
     !?m_data = COND{ !?as AND CurSt.writing: CurSt.MDout           C19
                     ELSE: VECTOR{i: HiZ} }
     !m_addr = COND{ !?as: CurSt.MAR                                C20
                    ELSE: VECTOR{i: HiZ} }
     !r = NOT (CurSt.writing)                                       C21
     !port_rdy = NOT (src_md AND NOT (CurSt.port_req == 0 AND CurSt.data_read))
                 AND NOT (dst_mar AND NOT (CurSt.port_req == 0))    C22
          }
   Next :: CurSt, !?bus, !?m_data, ?dst_mar, ?dst_md, ?dtack,
                         ?reset, !?as, ?ifureq ==> newstate
      WHERE { newstate =
         <COND{ CurSt.port_req <> 0: CurSt.MAR % MAR                C23
                 ELSE: COND{ ?dst_mar: !?bus                        C24
                         ELSE: CurSt.MAR } },
```

```
        COND{ ?dtack AND !?as AND NOT CurSt.writing: ?m_data %MDin          C25
               ELSE: CurSt.MDin },

        COND{ ?dst_md AND CurSt.port_req == 1: !?bus  % MDout              C26
               ELSE: CurSt.MDout },

        COND{ ?reset OR (!?as AND ?dtack): 0 % port_req                    C27
               ?dst_mar AND CurSt.port_req == 0: 1                         C28
               CurSt.port_req == 1: 2                                      C29
               ELSE: CurSt.port_req },

        COND{ ?reset : FALSE  % writing                                    C30
               (CurSt.port_req == 2): CurSt.writing                       C31
               (CurSt.port_req == 1) AND ?dst_md: TRUE                    C32
                ELSE: FALSE }

        COND{ ?reset OR (CurSt.port_req == 0 AND dst_mar): FALSE % dataread  C33
               !?as AND ?dtack AND NOT CurSt.writing: TRUE                 C34
               ELSE: CurSt.data_read } >                                   C35
END PORT


ABS_MODULE MEM(acctime: INTEGER)


ABS_STATE IS RECORD
     store: RAM
     cycle_num, acc_num: 0..acctime
   END


EXTOP
  Out :: CurSt, ?as, ?read, ?ifureq, ?m_addr ==> !?m_data, !dtack WHERE {
  !dtack = (CurSt.cycle_num == acc_num) AND (?as OR ?ifureq)             C36
  !m_data = COND{ (?ifureq OR ?read): VECTOR {i: UNK}
                  (CurSt.cycle_num = acc_num): read(CurSt.store, ?m_addr) C37
                  ELSE: VECTOR{i: HiZ} }


  Next :: CurSt, ?read, ?as, ?ifureq, ?m_addr, !?m_data, ?reset ==> newstate
    WHERE { newstate =
        <COND{ ?reset OR ?read OR ?ifureq: CurSt.store % store           C38
               CurSt.cycle_num == acc_num AND ?as:
                    write(CurSt.store, ?m_addr, !?m_data)                 C39
               ELSE: CurSt.store },                                       C40
         COND{ ?reset OR NOT (?as OR ?ifureq): 0 %cycle_num              C41
               (?as OR ?ifureq) AND (CurSt.cycle_num!=acc_num) :
                    CurSt.cycle_num+1                                     C42
               CurSt.cycle_num == acc_num AND NOT (?as OR ?ifureq): 0     C43
               ELSE: CurSt.cycle_num },                                   C44
          COND{ ?reset OR CurSt.cycle_num == CurSt.acc_num: random(acctime) C45
               % random returns a random integer between 1 and its argument
               ELSE: CurSt.acc_num} > }                                   C46
END MEM
```

```
ABS_MODULE ALU

ABS_STATE IS BITVECTOR[16]

EXTOP
  Out:: CurSt, ?src_alu ==> !?bus WHERE{
    !?bus = COND{ ?src_alu: CurSt
                  ELSE: VECTOR{ i: HiZ} } }


  Next:: CurSt, ?alu_op, ?dst_alu, !?bus ==> newstate WHERE {
    newstate = COND{ NOT ?dst_alu: CurSt
                     ELSE: COND { ?alu_op == SHIFT: VECTOR{ i: COND{ i < 8: 0
                                                                     ELSE: !?bus[i-8] } }
                                  ?alu_op ==  ADD:   CurSt + !?bus
                                  ELSE:   !?bus } } }
END ALU

REAL_MODULE Lilith

REAL STATE IS  P = <CPU, IFU, MEM, PORT, ALU>

EXTOP
  Execute:: ?reset ==> newstate  WHERE{
    newstate = LETSEQ{
     code = Out( CPU)     % micro program word in code
     <mu_ins, bus_dst, bus_src, mir_adr, const, alu_op> = decode(code)          C47
            % Since the micro program is given in the decoded format, we need not
            % specify the decode function here
     <ifu_bus, ifu_addr, ifureq, IFUrdy> = Out(IFU, (bus_src==PC), (bus_src==IR)) C48
     <port_bus, port_dat, port_addr, read, as, port_rdy> =
             Out(PORT, ifureq, (bus_src==MD), (bus_dst==MAR), (bus_dst==MD))     C49
     alu_bus = Out(ALU, (bus_src == ALU))
     <m_data, dtack> = Out(MEM, as, read, ifureq, comb(ifu_addr, port_addr))     C50
     data_bus = comb(port_dat, m_data)
     bus = comb(ifu_bus, port_bus, alu_bus, const)          % micro word, msbits=0.
     jmpsel  = (bus_src==IR) AND (bus_dst==MAP)
     NCPU = COND{ ?reset OR (IFUrdy AND port_rdy):
                          Next(CPU, bus, mir_adr, jmpsel, ?reset, mu_ins)        C51
                   ELSE: CPU }                                                    C52
     NALU = COND{ ?reset OR (IFUrdy AND port_rdy):
                                   Next(ALU, alu_op, (bus_dst == ALU), bus)
                   ELSE: ALU }                                                    C53
     NIFU = Next(IFU, as, bus, (bus_dst==PC), (bus_src==IR),
            dtack, data_bus, ?reset)                                             C54
     NMEM = Next(MEM, read, as, ifureq, comb(port_addr, ifu_addr),
```

```
                data_bus, ?reset)                                          C55
      NPORT = Next(PORT, bus, data_bus, (bus_dst==MAR),
                            (bus_dst==MD), dtack, ?reset, as, ifureq)      C56
   IN  <NCPU, NIFU, NMEM, NPORT, NALU> } }
END Lilith
```

E   The Micro Program

```
%
%Addr   mu_ins   mir_adr   bus_src   bus_dst   const   op   comments
%
reset:   f        next     constant    PC        0          % dont care values
not specified
next:    t                   IR        MAP                  % fetch opcode
STPC:    t                   IR        ALU             NOP  % get 16-bit address
         t                   ALU       MAR                  % store PC in that
         t                   PC        MD                   % location
         t                   IR        MAP                  % fetch opcode
JMPIND:  t                   IR        ALU             NOP  % get 16-bit address
         t                   ALU       MAR                  % jump to the addr
         t                   MD        PC                   % specified at this location
         t                   IR        MAP                  % fetch opcode
```

Contents

OBJ as a Theorem Prover
with Applications to Hardware Verification*

Joseph A. Goguen[†]

SRI International, Menlo Park CA 94025, and

Center for the Study of Language and Information,

Stanford University 94305

Abstract: This paper states, justifies, and illustrates some new techniques for proving theorems with a functional programming language, like OBJ, that has a rigorous algebraic semantics. These techniques avoid the complexities of both higher order logic and Knuth-Bendix completion, and instead support a style of user interaction in which: (1) OBJ's powerful reduction engine does the routine work automatically, in such a way that partially successful proofs return information that often suggests what to try next; and (2) OBJ's flexible facilities for hierarchical and generic modules describe complex proof strategies in a style that is familiar from experience with programming languages. New results in this paper include: a simple extension of first order equational logic that allows universal quantification over functions instead of just over ground elements; a technique for eliminating both first and second order universal quantifiers; a completeness theorem; a technique for transforming conditional equations to unconditional equations; some very useful structural induction principles; and some techniques for reasoning about parameterized modules. These tools for reasoning about (first order) functions are powerful enough to justify algorithms that automatically generate OBJ code for verifying some rather broad classes of (second order) assertions. The paper features some hardware verification examples, but also includes some inductive proofs for the natural numbers and an OBJ-style parameterized module verification. Parameterized module verification supports a flexible style of proof reuse, and this example also illustrates the rigorous treatment of exceptions in OBJ. The hardware verification examples illustrate what seems a very promising approach to both combinational and sequential circuits, using standard gates and/or bidirectional components (such as MOS transistors), possibly with "don't care" conditions. Order sorted algebra plays an important role in many of the examples.

1   Introduction

This paper provides some general techniques for reducing proofs of sentences in equational logic, possibly having (universal) quantification over functions, to ground term reductions that can be performed by an equational logic based programming

*Supported by Office of Naval Research Contracts N00014-85-C-0417 and N00014-86-C-0450, NSF Grant CCR-8707155, and a gift from the System Development Foundation.

†Address from September 1988 onward: Oxford University Computing Laboratory, Programming Research Group, 8-11 Keble Road, Oxford OX1 3QD, England (UK).

language like OBJ [12,17,26]. A **proof score** for a sentence s is a program P, such that if P executes correctly, then s is proved; this paper uses OBJ programs, consisting of some declarations (for modules with sorts, subsorts, functions, and equations) and some reductions, such that if all the reductions evaluate to true, then s is true. This uses OBJ in two distinct ways: as a reduction engine, to do the calculations; and as a language for expressing the structure and content of proofs. OBJ's associative and/or commutative rewriting capability is especially useful for the first purpose, and OBJ's program structuring facilities, including module inheritance (possibly multiple) and parameterization, are especially useful for the second. The basic structure of a proof is summarized in the hierarchical (import/export) acyclic graph structure of its proof score. In general, basic abstract data types are defined near the top of this graph, followed by general hypotheses (if any), while the reduction commands occur at leaf nodes.

Equational logic is attractive as the foundation for a theorem prover because of the simplicity and elegance of equational reasoning, which is just the substitution of equals for equals. Moreover, a good deal of relevant theory has already been developed, including the extensive literature on abstract data types. For example, any computable function over any computable data structure can be defined in equational logic [1], and *order sorted* equational logic [16,22,23] also encompasses the *partial* computable functions[1]. Of course, not everything of interest can be expressed using just equational logic, but much more can be expressed than you might think. In particular, many typical results about functional programs and most of the usual hardware verification examples fall within this framework.

Eqlog [21] would be even better than OBJ as a framework for equational theorem proving, since it can compute values for existentially quantified variables; but unfortunately, we do not yet have an implementation of Eqlog. The examples in this paper use OBJ3 [26], the most current implementation of OBJ. Although OBJ3 is based on order sorted equational logic, the theoretical apparatus of this paper is developed in unsorted equational logic for expository simplicity; all results extend without essential difficulty to the order sorted case.

The appendices to this paper contain examples that illustrate the theoretical results. Appendix B contains hardware verifications, for both combinational and sequential circuits. These examples illustrate some general proof techniques, which suffice to verify (and sometimes to refute) equational properties of a broad range of circuits, including bidirectional circuits such as MOS transistors. They also allow "don't care" conditions. A connection with unification is pointed out, and the existence of multi-stable circuits (e.g., memory units, such as flip-flops) is related to the existence of more than one most general unifier. Appendix C includes some inductive proofs for the natural numbers, and Appendix D verifies an OBJ-style parameterized module.

1.1 Principles

The following principles may help to explain the research program behind this paper:

1. The only satisfactory way to validate a theorem prover for a classical logic is to give precise notions of *model* and *satisfaction*, and then demonstrate the *soundness* of all proposed rules of inference with respect to these essentially

---

[1]This is an as yet unpublished result of Dr. José Meseguer.

semantic notions[2]. In particular, it is not acceptable to define some "expressive" (i.e., complex) syntax, write some code that is driven by it, and then call the result a "theorem prover" if it usually prints TRUE when you want it to. Similarly, it is not a good idea to "give a semantics" after the fact, since it will probably be too complex to be useful. Also, it is dangerous to try to combine several logics, unless precise and reasonably simple notions of model and satisfaction are known for the combination. In summary,

Semantics should come first.

2. Because we are primarily concerned with *truth*, i.e., with properties of models, i.e., with *semantics*, we regard *proof as a necessary nuisance* that is needed to effectively establish true facts. In particular, we prefer to justify proof scores with semantic arguments, rather than with syntactic arguments, because usually they are easier and more intuitive. In the same vein, it is often enough to know that a set of rules of deduction *exists* for some logic having the important semantic property of *completeness*; then we can ignore the syntactic details in justifying further proof techniques.

3. Most computer science applications concern properties of a *standard* model for some set of sentences, rather than properties of *all* its models. We formalize the notion of "standard model" with *initiality*, and we show that the systematic use of initiality can give some very simple proofs, both at the "object level" (properties of objects) and at the "meta level" (justifying proof scores). As noted above, equational logic is powerful enough to characterize any standard model of interest.

4. For any given application, the simplest possible logic should be used. Since most hardware verifications only require universally quantified equations, we should try to use just equational logic. Perhaps surprisingly, a few simple tricks make first order reduction adequate for most applications.

Although these principles seem fairly obvious (at least for classical logic), many of the largest (i.e., best-funded) theorem proving projects violate one or more of them. On the other hand, there are many excellent projects that uphold high standards of rigor, including the work of Milner and others on LCF [29,39], the Boyer-Moore theorem prover [3], work of Stickel on the Prolog Technology Theorem Prover [44], work of Gordon on HOL [27], and work of Constanble and others on NuPRL [33]; there is also some very interesting recent work on ELF [30] and IPE [40] from the University of Edinburgh.

Two additional principles relate to the user interface:

5. User interaction should facilitate the proof process, and in particular, partially successful proofs should provide information that helps the user determine what to do next; for example, the output might suggest a new lemma that would further advance the proof.

---

[2]I am indebted to Ian Mason for pointing out that this remark could be seen as critical of constructivist approaches like those of Martin-Löf and others who identify the syntax and semantics of logical systems. In fact, I am very sympathetic to such positions, but note that soundness problems reappear in this context as *consistency* problems.

6. Semi-automatic theorem proving is preferable to either fully automatic or fully manual theorem proving, both of which can be very painful to use, the former because the user often has to trick built-in heuristics into doing what he wants, and the latter because of the extremely tedious level of detail involved.

It also seems interesting and appropriate to apply all these principles to logic-based programming languages (e.g., as in [21]) and to database query languages.

1.2 OBJ

OBJ is a declarative[3] programming and specification language whose development began at UCLA [18,24], and then continued at SRI International [12,17,26] and several other sites around the world [13,8,42,10]. Our latest implementation of OBJ is OBJ3, which has mathematical semantics given by order sorted equational logic, with a powerful type system featuring subtypes and overloading that supports precise error definition and handling [23]. In addition, OBJ has user-definable abstract data types with user-definable mixfix syntax, and a uniquely powerful parameterized module facility that includes "views" and "module expressions" [17]. The specific features of Version 1.0 of OBJ3 are described in some detail in [26].

OBJ is a wide-spectrum language that integrates coding, specification and design into a single framework. In particular, an OBJ text P may be of two kinds:

1. an **object**, whose intended interpretation is a *standard model* of P; and

2. a **theory**, whose intended interpretation is the *variety* of *all* models of P.

The second case, which is sometimes called "loose" semantics, generally appears in an auxiliary role, because most applications involve particular data structures and particular functions over them. (But curiously, the second case has been much more extensively studied in the theorem proving literature.) A basic principle is that *standard models are initial models*. Here we exploit this principle for equational logic, but it applies much more generally, for example, to standard models of Horn clause logic as used in (pure) Prolog, and of Horn clause logic with equality as used in Eqlog [21].

I wish to emphasize that it is only because the semantics of OBJ *is* the semantics of equational logic that OBJ can be used directly as a theorem prover for equational logic. It is *not* true that any other functional programming language will do just as well; most of them have an operational semantics based on higher order rewriting, but do not have a declarative, logical semantics. It is also important that the OBJ module facility has a rigorous semantics. Originally based on colimits of data theories, as in the Clear specification language [6], this paper simplifies that semantics using the notion of free extension.

Of course, OBJ was not originally designed as a theorem prover, but rather as a programming and specification language. As a result, some things have to be done by hand that would be done automatically in a fully developed verification environment. However, it would not be very difficult to extend OBJ with a proof score construction facility that would make it comparable to familiar theorem provers. In particular, such a system could guarantee the validity of any proof it constructs, and could also support incremental proof management.

---

[3]This means that its statements assert properties that solutions should have; i.e., they describe the problem.

1.3   The Meaning of Verification

Practical interest in verification technology is particularly acute for so-called "critical systems," which have the property that incorrect operation may result in loss of human life, compromise of national security, massive loss of property, etc. Typical examples are heart pacemakers, automobile brake controllers, encryption systems, and electronic fund transfer systems. In this context, it is especially important to understand the limitations of verification, both those limitations that are inherent in the nature of verification, and those that are due to the current state of the art. Unfortunately, there is a tendency to play down, or even cover up, these limitations, due to the lure of fame and fortune. When verification is just an academic exercise, this does little harm; but there is cause for serious concern when manufacturers make advertising claims about the reliability of a critical system (or component thereof) based on its having been verified. As Dr. Avra Cohn [9] says,

> The use of the word 'verified' must *under no circumstances* be allowed to confer a false sense of security.

This is because it may lead to taking severe risks unnecessarily and unintentionaly. Indeed, one might well argue that to knowingly make false or misleading claims about the reliability of a critical system should be a criminal offense. It is certainly a grave moral offense.

We should realize that nothing in the real world can have the certainty of mathematical truth. Although we might like to think that '$8 \times 7 = 56$' is always and incontrovertibly true in some mathematical heaven, an actual human being can sometimes misremember the multiplication table, and an actual machine can sometimes drop a bit, break a connector, or burn out a power supply. The most that can truthfully be asserted of a "verified chip" is that certain kinds of design errors have been made far less likely. Of course, this is not trivial, and is very much worth pursuing, but there remains a long chain of assumptions that must be satisfied in order that an actual physical instance of a chip whose logic has been verified will operate as intended *in situ*, including the following: the chip must be correctly fabricated; it must be correctly installed; it must be fed correct data and given correct power; the electronic circuits that realize its logic must be correctly designed, and used only within their design limits; the analog circuits that support communication must operate correctly; there must not be excessive electromagnetic radiation around the chip; etc., etc. In addition, there will often be assumptions involving human factors, for example, that the user does not override warning signals, and does not incorrectly interpert the output.

However, I wish to concentrate on certain logical issues that are involved. A major point, emphasized by Cohn [9], is that verification is a relationship between two models, that is, between two mathematical abstractions, one of the chip, and the other of the designers' intentions. However, such models can never capture either the totality of any particular chip, or even of the designers intentions for that chip, because these are necessarily complex abstractions, and errors can be made in constructing them. In particular, since the languages in which designs and specifications are usually expressed today are relatively informal, there must be a translation into some formal language, and errors may be introduced either by the designers or the verifiers misunderstanding these informal languages. In addition, the theorem prover must correctly implement some logical system, and the formalism for representing the chip in logic must be correct with respect to that system. Unless a theorem prover is rigorously based on a precise and well-understood logical system, there is little basis for confidence in its "proofs." For example, it is seductive but dangerous to "combine" several different logical systems, since the combination may fail to have any obvious notions of model and satisfaction, even though the components do have them. Even a theorem prover with a sound logical basis is likely to have some bugs in its code, because it is after all a very complex real system.

Moreover, it is always possible that the assumptions about how formal sentences represent physical devices are flawed in some subtle ways, for example, relating to signal strength, and it is usually possible to use a theorem prover incorrectly, for example, to give it erroneous input, or to interpret its output incorrectly. Finally, we must note that the current state of the art is not adequate to support the verification of really large or complex systems, although recent advances have been both rapid and significant, and the future looks promising.

2 Signatures, Algebras and Terms

This section is an introduction to universal (or general) algebra. For expositional simplicity, it is restricted to the unsorted case. More detail, including the generalization to many sorted and to order sorted algebra, can be found in many places, for example, [5,36,32].

Let ω denote the set $\{0, 1, 2, ...\}$ of all natural numbers. Then a **signature** Σ is an indexed family $\{\Sigma_n \mid n \in \omega\}$ of sets. $f \in \Sigma_n$ is called a **function symbol** of **arity** n; in particular, $f \in \Sigma_0$ is a **constant symbol**. Define $\Sigma' \subseteq \Sigma$ to mean that $\Sigma'_n \subseteq \Sigma_n$ for each $n \in \omega$. A useful construction on signatures is their **union**, defined by

$$(\Sigma \cup \Sigma')_n = \Sigma_n \cup \Sigma'_n.$$

A common special case involves first viewing a set X as a signature consisting only of constants, i.e., $X_n = \emptyset$ for $n > 0$. Then we may write

$$\Sigma(X) = \Sigma \cup X.$$

A Σ-**algebra** A consists of a set $|A|$ called the **carrier** of A (usually denoted just A for simplicity), and an **interpretation** of Σ in A, which is a family of functions $i_n : \Sigma_n \to [A^n \to A]$ for each $n \in \omega$, interpreting the function symbols in Σ_n as actual functions on A. Note that A^0 is some (arbitrary) one point set, e.g., $\{*\}$, and that for each $c \in \Sigma_0$ we identify $i_0(c)$ with $i_0(c)(*)$, a point in A. Usually we write just f for $i_n(f)$ in A.

Given Σ-algebras A and B, a Σ-**homomorphism** $h : A \to B$ is a function $h : |A| \to |B|$ such that

$$h(f(a_1, ..., a_n)) = f(h(a_1), ..., h(a_n))$$

for each $f \in \Sigma_n$ and in particular, such that $h(c) = c$ for $c \in \Sigma_0$. A Σ-homomorphism $h : A \to B$ is a Σ-**isomorphism** iff there is another Σ-homomorphism $g : B \to A$ such that $g(h(a)) = a$ for all $a \in A$ and $h(g(b)) = b$ for all $b \in B$, iff h is bijective, i.e., both injective and surjective. In this case let us write $A \cong B$.

Define the set T_Σ of all Σ-terms to be the smallest set of *strings* over the alphabet $\Sigma \cup \{(,)\}$ (where $($ and $)$ are special symbols disjoint from Σ) such that

- $\Sigma_0 \subseteq T_\Sigma$ and

- given $t_1, \ldots, t_n \in T_\Sigma$ and $\sigma \in \Sigma$, then $\sigma(t_1 \ldots t_n) \in T_\Sigma$.

We can now view T_Σ as a Σ-algebra as follows:

- For $\sigma \in \Sigma_0$, let $(T_\Sigma)_\sigma$ be the string σ.

- For $\sigma \in \Sigma_n$, let $(T_\Sigma)_\sigma$ be the function sending $t_1, \ldots, t_n$ to the string $\sigma(t_1 \ldots t_n)$.

Thus, $\sigma(t_1, \ldots, t_n) = \sigma(t_1 \ldots t_n)$, and from here on we prefer to use the first notation. The key property of T_Σ is its *initiality*:

Theorem 1 For any Σ-algebra A, there is a unique Σ-homomorphism $T_\Sigma \to A$. $\square$

3 Equations, Presentations and Satisfaction

This section generalizes ordinary equational logic, which only permits quantification over constants, to an **extended case** which permits *quantification over functions*. Although this is a kind of second order quantification, it should be seen as taking first order equational logic toward its limit, rather than as an incursion into the second order realm; what is essential is that the terms themselves are first order. We will see that this extension can be very useful. However, the mathematics is an easy extension of the ordinary case; indeed, it is hard to see why it has not been thought of before. The result is a powerful first order calculus for reasoning about (first order) functions that is, in my view, much more satisfactory than trying to use the λ-calculus or some other overly expressive (and thus harder to use) tool.

Definition 2 A Σ-equation is a signature Φ of **variable symbols** (disjoint from Σ), plus two $(\Sigma \cup \Phi)$-terms; let us write such an equation abstractly in the form
$$(\forall \Phi)\ t = t'$$
and concretely in the form
$$(\forall x, y, f, g)\ t = t'$$
when $\Phi = \{x, y, f, g\}$ and the ranks of x, y, f, g can (presumably) be inferred from their uses in t and in t'. $\square$

An example of this kind of equation arises in a denotational style semantics for expressions, where one would normally have to write equations
$$(\forall e, e')\ [[e + e']](\rho) = [[e]](\rho) + [[e']](\rho)$$
$$(\forall e, e')\ [[e - e']](\rho) = [[e]](\rho) - [[e']](\rho)$$
$$(\forall e, e')\ [[e \times e']](\rho) = [[e]](\rho) \times [[e']](\rho)$$
....
instead of the following much simpler equation[4] which quantifies over the binary function symbol $*$,
$$(\forall e, e', *)\ [[e * e']](\rho) = [[e]](\rho) * [[e']](\rho).$$

In the ordinary case, only Φ_0 is nonempty, and Φ can be identified with a set X of (ordinary) variables. Ordinary equations are written abstractly in the form
$$(\forall X)\ t = t'$$
where t, t' are $\Sigma(X)$-terms, and concretely in the form
$$(\forall x, y, z)\ t = t'$$
when for example $X = \{x, y, z\}$.

A **presentation** is a pair $\langle \Sigma, E \rangle$, consisting of a signature Σ and a set E of Σ-equations; a Σ-presentation is a presentation whose signature is Σ, and a **ordinary**

---

[4]This equation actually has a somewhat different meaning from the finite set of equations given above, since it asserts the homomorphic property for *any possible* function $*$, including for example functions defined by terms; however, we can get the other meaning by using a conditional equation; see later for the definition of extended conditional equations.

presentation in a presentation such that E contains only ordinary equations. An OBJ program P defines a presentation $\langle \Sigma, E \rangle$ where Σ and E are both finite and (at present) E is ordinary; the details of how P yields $\langle \Sigma, E \rangle$, which involve theories, views, colimits, etc., need not concern us here, and from now on we simply identify P with its presentation $\langle \Sigma, E \rangle$ and ignore the concrete syntax of OBJ3.

Given a Σ-algebra A and an interpretation $f: \Phi \to A$ in A of the variable symbols in Φ, there is a unique extension to a $(\Sigma \cup \Phi)$-homomorphism $\overline{f}: T_{\Sigma \cup \Phi} \to A$, by the initiality of $T_{\Sigma \cup \Phi}$ and by using f to extend the interpretation function i of A from Σ to $\Sigma \cup \Phi$. Of course, a Σ-term t with variables in Φ is just an element of $T_{\Sigma \cup \Phi}$. Then

Definition 3 a Σ-algebra A **satisfies** Σ-equation $(\forall \Phi) \ t = t'$ iff for any interpretation $f: \Phi \to A$ we have that $\overline{f}(t) = \overline{f}(t')$ in A. In this case we write
$$A \models_\Sigma (\forall \Phi) \ t = t'$$
sometimes omitting the subscript Σ for simplicity. A Σ-algebra A **satisfies** a set E of Σ-equations iff it satisfies each $e \in E$, and in this case we write
$$A \models_\Sigma E$$
and say A is a P-algebra where $P = \langle \Sigma, E \rangle$. □

Proposition 4 Given Σ-algebras A and B, and given a Σ-equation e,
$$A \cong B \text{ implies } (A \models e \text{ iff } B \models e).$$

Proof: Let $h: A \to B$ be an isomorphism, let e be $(\forall \Phi) \ t = t'$, and let $a: \Phi \to A$ be an interpretation for Φ in A. Then
$$\overline{a}(t) = \overline{a}(t') \text{ iff } h(\overline{a}(t)) = h(\overline{a}(t'))$$
and any interpretation $b: \Phi \to B$ is of the form $h \circ \overline{a}$ for some $a: \Phi \to A$. □

Proposition 5 If $(\forall \Phi) \ t = t'$ is a Σ-equation and if $\Phi \subseteq \Psi$, then $(\forall \Psi) \ t = t'$ is also a Σ-equation, and for every Σ-algebra A,
$$A \models_\Sigma (\forall \Phi) \ t = t' \text{ iff } A \models_\Sigma (\forall \Psi) \ t = t'.$$
□

This result justifies adding extra constants and function symbols in a proof score under certain conditions. The proof makes a good exercise in the definitions involved.

4   Initial Truth

Given a presentation $P = \langle \Sigma, E \rangle$, then a P-algebra T is an **initial** P-algebra iff there is a *unique* Σ-homomorphism from T to any other P-algebra A. The next result is a nice example of initiality at work.

Proposition 6 Given a presentation $P = \langle \Sigma, E \rangle$, any two initial P-algebras A and B are Σ-isomorphic.

Proof: Since A and B are both initial, there are Σ-homomorphisms $f: A \to B$ and $g: B \to A$. Therefore, we have Σ-homomorphisms $f \circ g: A \to A$ and $g \circ f: B \to B$, where "$\circ$" denotes composition, and since the identity on A is a Σ-homomorphism, we necessarily have $g \circ g = id_A$. Similarly, $f \circ g = id_B$. □

The following result is well known (e.g., see [36]):

Theorem 7 If P is ordinary, then there is an initial P-algebra, denoted T_P. $\square$

(We only need to know that T_P exists, but it can be constructed as the quotient algebra of T_Σ by the Σ-congruence generated by E, which is the least congruence that contains the relation constructed by making all possible ground substitutions for all variable symbols.)

The *Abstract Data Type* (ADT) defined by P is the class of all initial P-algebras [25,36]. By Propositions 4 and 6, exactly the same equations are true of any initial P-algebra as any other, which is fortunate, because usually these are exactly the properties in which we are most interested. We therefore *define* **initial satisfaction** as follows:

(0) $P \models\!\!\!\models e$ iff $T_P \models e$.

Both of these are *semantic* properties. Since anything that is true of all models is certainly true of initial models, we have

(1) $P \models e$ implies $P \models\!\!\!\models e$.

However, the converse does *not* hold. For example, let Σ contain a constant 0, a unary function symbol s, and a binary function symbol $+$, which we will write with infix notation for convenience; let E contain the equations

$(\forall n)\ 0 + n = n$

$(\forall m, n)\ s(m) + n = s(m + n)$.

Then the commutative equation

$(\forall m, n)\ m + n = n + m$

holds in T_P but does not hold in *every* P-algebra. For example, it does not hold in the Σ-algebra A with carrier all strings of a's and b's, with $0 \in \Sigma$ denoting empty string in A, with $m + n$ denoting the concatenation of the string n after the string m, and with s sending a string m to the string $a + m$.

The following is another important property of initial algebras, also with a very simple proof:

Proposition 8 If $P = \langle \Sigma, E \rangle$ is a ordinary presentation, then an initial P-algebra has no proper sub-Σ-algebras.

Proof: If A is an initial P-algebra and $B \subseteq A$ is a sub-Σ-algebra, then B is a P-algebra. Then by initiality, the inclusion homomorphism is an isomorphism. $\square$

We will later see that this result is the foundation for proofs by induction.

5 Deduction

Necessary and sufficient syntactic criteria for $\models$ are well known for ordinary equational logic (where there is only quantification over constants). Let us write

$E \vdash_\Sigma e$

or equivalently when $P = \langle \Sigma, E \rangle$, also

$P \vdash e$

to mean that we can prove the equation e from E using some such proof system. The detailed rules of deduction need not concern us here, all we need is that they exist and that they are *sound* and *complete*, i.e., that

(2) $E \models_\Sigma (\forall X)\ t = t'$ iff $E \vdash_\Sigma (\forall X)\ t = t'$

for any signature Σ and set E of ordinary Σ-equations ([2] gives the classical rules for the unsorted case, [20] and [36] develop the many sorted case, and [22] develops the order sorted case). By (1), we therefore have

(3) $E \vdash_\Sigma e$ implies $E \models_\Sigma e$.

Unfortunately, it is in general *impossible* to give effective necessary and sufficient purely syntactic criteria for $\models_\Sigma$ [35]. However, Section 8 gives some useful sufficient criteria.

6   Variables as Constants

In the technical development of equational logic, variables are defined to be unconstrained constants. We can exploit this to justify proof scores that prove equations with variables by regarding the variables as constants and using only ground term deduction. The results in this section are new in the context of equational logic, although an analog of Theorem 9 is familiar in first order logic as a "Theorem of Constants."

Theorem 9 Given disjoint signatures Σ and Φ, given a set E of Σ-equations, and given $t, t' \in T_{\Sigma \cup \Phi}$ we have

(4) $E \models_\Sigma (\forall \Phi)\ t = t'$ iff $E \models_{\Sigma \cup \Phi} (\forall \emptyset)\ t = t'$.

Proof: Each condition is equivalent to the condition

$\overline{f}(t) = \overline{f}(t')$ for every $(\Sigma \cup \Phi)$-algebra A satisfying E and any every $f : \Phi \to A$, where $\overline{f} : T_{\Sigma \cup \Phi} \to A$ is the unique homomorphism. $\square$

It is pleasing that the proof is so simple. This is because it is based entirely on the *semantics* of satisfaction, rather than on any particular choice of rules of deduction, and exploits the power of *initiality*.

Of course, using (2) we also get

(4') $E \models_\Sigma (\forall \Phi)\ t = t'$ iff $E \vdash_{\Sigma \cup \Phi} (\forall \emptyset)\ t = t'$.

A further generalization of (4) involves another signature Ψ which is disjoint from Σ and Φ,

(5) $E \models_\Sigma (\forall \Phi \cup \Psi)\ t = t'$ iff $E \models_{\Sigma \cup \Phi} (\forall \Psi)\ t = t'$.

The special case of (4) with just first order quantification is: Given a signature Σ, a set X disjoint from Σ, a set E of Σ-equations, and $t, t' \in T_{\Sigma(X)}$, then

(6) $E \models_\Sigma (\forall X)\ t = t'$ iff $E \models_{\Sigma(X)} (\forall \emptyset)\ t = t'$.

We can regard Theorem 9 as justifying a new rule of deduction. For this purpose, let $\vdash^1$ denote the syntactic derivability relation defined by some complete set of rules $\mathcal{R}^1$ for ordinary equations, and let $\vdash^2$ be defined by $\mathcal{R}^1$ plus the rule

(R) $E \vdash^1_{\Sigma \cup \Phi} (\forall \emptyset)\ t = t'$ implies $E \vdash^2 (\forall \Phi)\ t = t'$.

Then, we have the following completeness result:

Theorem 10 If $\mathcal{R}^1$ is some set of rules of deduction for ordinary equational logic, defining a relation $\vdash^1$ that is complete in the sense of (2), then $\mathcal{R}^2 = \mathcal{R}^1 \cup \{(R)\}$ is complete for extended equations, in the sense that it defines a relation $\vdash^2$ such that

(7) $E \models_\Sigma (\forall \Phi)\ t = t'$ iff $E \vdash^2_\Sigma (\forall \Phi)\ t = t'$

for any signature Σ and set E of ordinary Σ-equations. $\square$

I am not presently sure whether a complete set of rules of deduction can be given for extended equational logic as such. However, there is little need for such a result, because one does not need (or want) extended equations to define abstract data types, and one can eliminate them in other cases.

7 Reduction

OBJ does *reduction*, that is, *left-to-right* deduction, by treating the equations in P as *rewrite rules* (see [32] for a nice survey of rewrite rule theory). Given a presentation $P = \langle \Sigma, E \rangle$ and two (ground) Σ-terms t, t', let us write[5]

$$P \vdash t \Rightarrow t'$$

to indicate that t can be obtained from t' in exactly one step using some rewrite rule in P. Similarly, let us write

$$P \vdash t \overset{*}{\Rightarrow} t'$$

to indicate that P rewrites t to t' in any number of steps (including zero) using rules in P. These are purely syntactic notions. Then,

(8) $P \vdash t \overset{*}{\Rightarrow} t'$ implies $P \vdash t = t'$

so that by (2),

(8') $P \vdash t \overset{*}{\Rightarrow} t'$ implies $P \models t = t'$.

P is said to be **Church-Rosser** as a set of rewrite rules iff $P \vdash t \overset{*}{\Rightarrow} t_1$ and $P \vdash t \overset{*}{\Rightarrow} t_2$ imply there is some t' such that $P \vdash t_1 \overset{*}{\Rightarrow} t'$ and $P \vdash t_2 \overset{*}{\Rightarrow} t'$. Also, P is said to be **terminating** iff there is no infinite sequence of terms, t_i for $i \in \omega$, such that $P \vdash t_i \Rightarrow t_{i+1}$. Finally, P is called **canonical** iff it is both Church-Rosser and terminating.

If P is a canonical presentation, then every Σ-term t has a **canonical** (also called **reduced**) form $[t]$, written $[t]_P$ for added clarity, a term to which t can be rewritten and to which no further rewrite rules apply; moreover, such canonical forms are *unique*. Then for t, t' both Σ-terms, we certainly have

(9) $[t]_P = [t']_P$ implies $P \vdash (\forall \emptyset)\ t = t'$.

((12) below shows that this can be strengthened to "iff.")

OBJ has a built-in Boolean-valued function == on ground terms that decides, under certain mild assumptions[6], whether or not $P \vdash (\forall \emptyset)\ t = t'$, by comparing the reduced forms of t and t' for syntactic identity. To avoid worrying about these assumptions here, we regard == as a relation on terms t, t' that holds iff the reduced forms of t, t' are syntactically identical; that is, we *define*

(10) $P \vdash t$ == t' iff $[t]_P = [t']_P$.

The following basic result was first proved in [15]:

Theorem 11 If P is canonical, then the set of all reduced ground terms is an initial P-algebra. □

The next result contains some assertions that are important for justifying certain proof scores; the first two results say that to prove a certain semantic property, it is necessary and sufficient that certain reductions give the same result.

Corollary 12 Given a canonical presentation $P = \langle \Sigma, E \rangle$ and ground Σ-terms t, t', we have

(11) $P \models (\forall \emptyset)\ t = t'$ iff $P \vdash t$ == t'.

(12) $P \models (\forall \emptyset)\ t = t'$ iff $P \vdash t$ == t'.

(13) $P \models (\forall \emptyset)\ t = t'$ iff $P \models (\forall \emptyset)\ t = t'$.

Proof: (11) is immediate from Theorem 11.

---

[5] This should be regarded as a 3-ary relation.

[6] The assumptions are that P is canonical and that the Boolean sort is protected [17]; however, the latter cannot be stated satisfactorily in unsorted logic, because it involves viewing == as a Bool-valued function on each defined sort.

To prove (12), we note by (10) that (9) gives one half, and it remains only to prove the contrapositive of (9), i.e., that

$$[t]_P \neq [t']_P \text{ implies } P \not\models (\forall \emptyset) \ t = t'.$$

But by (11), we have

$$[t]_P \neq [t']_P \text{ implies } P \not\models (\forall \emptyset) \ t = t'$$

which by (1) implies

$$P \not\models (\forall \emptyset) \ t = t'$$

as desired.

(13) now follows from (11) and (12). $\square$

Result (13) gives a very convenient way to check ground equations in initial algebras. The classical application of term rewriting to theorem proving goes back to the seminal paper of Knuth and Bendix [34]: a canonical term rewriting system P decides the validity of ground equations in the theory defined by P, i.e., (12). Of course, this immediately extends to equations with variables, by (4). For example, Knuth and Bendix give a decision procedure for equations in the theory of groups. Similarly, (11) gives a decision procedure for ground term equations under $P \models_{\cong}$. These decision procedures apply to objects and theories in OBJ. We will see below that the restriction to ground equations is a significant limitation for validity in the initial model. However, the discussion in Section 8 shows how to transcend this limitation in many important cases, by using induction.

From (12) and (4) we get

(14) $E \models_\Sigma (\forall \Phi) \ t = t'$ iff $E \vdash_{\Sigma \cup \Phi} t == t'$

which can be very useful in practice. An interesting corollary of (13) and (4) is

(15) $E \models_\Sigma (\forall \Phi) \ t = t'$ iff $E \models_{\Sigma \cup \Phi} (\forall \emptyset) \ t = t'$.

However, it is *not* true that

($\times$) $E \models_\Sigma (\forall \Phi) \ t = t'$ iff $E \models_\Sigma (\forall \Phi) \ t = t'$

as can be seen by considering, for example, the commutative law when $\langle \Sigma, E \rangle$ is the theory of groups (the initial group, which has just one element, is commutative, but not every group is commutative). Thus, it is also *not* true in general that

($\times$) $E \models_\Sigma (\forall \Phi) \ t = t'$ iff $E \vdash_{\Sigma \cup \Phi} t == t'$.

However, it *is* true for any Φ-sentence s that

(16) $E \models_\Sigma (\forall \Phi)s$ if $E \models_{\Sigma \cup \Phi} (\forall \emptyset)s$,

because $E \models_{\Sigma \cup \Phi} (\forall \emptyset)s$ is equivalent to $E \models_\Sigma (\forall \Phi)s$, which implies $T_{\Sigma, E} \models_\Sigma (\forall \Phi)s$, which is equivalent to the first condition.

An important variant of term rewriting called **rewriting modulo** involves rewriting *equivalence classes* of terms rather than individual terms. For example, in rewriting modulo associativity of $+$, terms that differ only in the association of subterms that are the arguments of $+$'s are considered equivalent, e.g., $(a + (b + a)) + c$ and $a + (b + (a + c))$. Rewriting modulo associativity allows rewrites that would be impossible if associativity were regarded as another rewrite rule; however, as far as semantics (i.e., models and satisfaction) is concerned, associativity is just another equation for $+$. (See [32] for extended discussion of these issues.)

It makes sense for a rewrite system to be **canonical modulo** some equations, and all the expected results carry over. OBJ allows rewriting modulo associativity, modulo commutativity, and modulo both associativity and commutativity. These three cases are indicated by providing the keywords assoc and/or comm as "attributes" after the operator's syntax declaration. The third case is often called **AC rewriting**. Appendix A gives a decision procedure for the predicate calculus using AC rewriting, based on ideas of Hsiang [31]. AC rewriting can be very helpful in verification, because it automatically handles the grouping and ordering of (certain) terms.

A limitation of the current OBJ3 implementation is that it provides no way to check whether a presentation is canonical. Although the Knuth-Bendix algorithm can often check the Church-Rosser property, and can sometimes even convert a given presentation to an equivalent one that has this property [34,32], OBJ would need order sorted Knuth-Bendix and unification algorithms, and these are not yet available. A complication in this area is the need to check termination, for which no general method exists. However, we have found that in practice, users nearly always write initial model presentations that are easily seen to be canonical, because they just define primitive recursive functions over free constructors (possibly modulo AC). Of course, termination remains an interesting area for research, even if nothing fancy is needed for most problems that arise in practice. Although substantial difficulties can arise with the termination of presentations for varieties, in practice such presentations are both simpler and rarer than initial model presentations.

It is not necessary for the presentations used in OBJ proof scores to be canonical. If a presentation is not canonical, then some some valid formulae may not reduce to true; but because *reduction is always sound*, we certainly do have a proof whenever the the given reductions do evaluate to true. For example, adding the equation of Lemma 3 in Section C.3 to the original presentation destroys the Church-Rosser property; but this makes no difference to the validity of the proof that uses it. On the other hand, it is necessary for the presentation to be canonical if we want a *decision procedure*, for which failure of an expression to reduce to true implies that the theorem really is false.

8 Induction

In general, pure equational deduction is inadequate for proving hard properties of standard models, and it is necessary to use induction or other techniques. Fortunately, there are some very useful "structural induction" principles for initial models which allow proving that something holds for all arguments by proving that it holds for each constructor whenever it holds for the arguments to that constructor [4]. The results in this section follow [36], and justify the use of induction in proof scores.

Given a signature Σ and a Σ-algebra A, a **signature of constructors** for A is a subsignature $\Pi \subseteq \Sigma$ such that the unique Π-homomorphism $T_\Pi \to A$ is surjective. If a Σ-algebra A has a signature of constructors, then it has a **minimal** signature of constructors, i.e., a subsignature $\Pi \subseteq \Sigma$ such that no proper subsignature $\Pi' \subset \Pi$ has $T_{\Pi'} \to A$ surjective; there may well be more than one minimal signature. Π is a **signature of free constructors** if the Π-homomorphism $T_\Pi \to A$ is a Π-isomorphism.

Given a presentation $P = \langle \Sigma, E \rangle$, a **signature of constructors** for P is a signature of constructors for $A = T_P$. Every presentation has a signature of constructors, and also at least one minimal signature of constructors. Although presentations need not in general have a signature of free constructors, the presentations that arise in practice often have a unique minimal signature of free constructors. Other things being equal, a minimal signature of constructors will require less effort in proofs. Our main result about induction is the following:

Theorem 13 [Structural Induction I] Given a presentation $P = \langle \Sigma, E \rangle$ and a signature Π of constructors for P, let $V \subseteq T_P$. Then $V = T_P$ if:

0. $c \in \Pi_0$ implies $[c] \in V$.
1. $f \in \Pi_n$ for $n > 0$ and $[t_i] \in V$ for $i = 1, ..., n$ imply $[f(t_1, ..., t_n)] \in V$.

Proof: Since T_P has no proper subalgebras (Proposition 8) and $V \subseteq T_P$, we need only show that V is closed under Π; but that is exactly what conditions 0. and 1. say. $\square$

This simple proof is possible because we have taken initial algebra semantics as our starting point. The usual formulation of induction now follows easily:

Corollary 14 [**Structural Induction II**] Given a presentation $P = \langle \Sigma, E \rangle$ and a signature Π of constructors for P, let $Q(x)$ be a $(\Sigma \cup \{x\})$-sentence. Then $E \models_\Sigma (\forall x) Q(x)$ if:

0. $c \in \Pi_0$ implies $E \models_\Sigma Q(c)$.
1. $f \in \Pi_n$ for $n > 0$ and $t_i \in T_\Sigma$ for $i = 1, ..., n$ imply $E \models_\Sigma Q(t_1) \wedge ... \wedge Q(t_n) \Rightarrow Q(f(t_1, ..., t_n))$.

Proof: This follows from Theorem 13 by letting $V = \{x \mid E \models_\Sigma Q(x)\}$. $\square$

As with Theorem 10, we can regard Corollary 14 as providing semantic justification for a proof rule. To this end, writing
$$E \models_\Sigma e$$
to indicate that e can be proved from E using this rule plus one of the usual rule sets for $\vdash_\Sigma$, the new proof rule can be written

(I) If $E \models_\Sigma (\forall \emptyset) \, t(c) = t'(c)$ for each $c \in \Pi_0$, and if $E \models_\Sigma (\forall \emptyset) \, t(t_i) = t'(t_i)$ for $1 \leq i \leq n$ and $f \in \Pi_n$ for $n > 0$ imply $E \models_\Sigma (\forall \emptyset) \, t(f(t_1, ..., t_n)) = t'(f(t_1, ..., t_n))$, then $E \models (\forall x) \, t = t'$.

Corollary 14 (plus soundness of $\vdash$) shows that $\models_\Sigma$ is *sound*, i.e., that
$$(17) \quad E \models_\Sigma e \text{ implies } E \models_\Sigma e,$$
but of course $\models_\Sigma$ cannot be complete, i.e., in general the converse of (17) does not hold [35].

Corollary 14 justifies the use of simple induction in proof scores, and is used in many examples in the appendices.

Experience indicates that it is generally easier to prove things with structural induction, as justified by the above results, than with so-called inductionless induction [37,15,36] using Knuth-Bendix completion; Garland and Guttag [14] report a similar experience, and of course, induction is quite basic to the Boyer-Moore theorem prover [3]. In particular, simple structural induction arguments do not require showing the termination of new rule sets, and do not produce an uncontrollable explosion of strange new rules that often seem to gradually become less and less relevant.

Similarly, it is usually much easier to exploit the close connection between rewrite rules, initiality, and induction than to try to remain within a "loose" semantics framework by axiomatizing a standard model with explicit reachability and induction schemata, since the first of these requires existential quantification (e.g., Skolem functions) and the second requires second order quantifiers. Note that these two properties correspond to the "no junk" and "no confusion" conditions that together are equivalent to initiality; see [5,36]. In fact, induction is simply not valid for pure loose semantics, and inductive proof techniques necessarily have semantics in standard (initial) models, whether or not this is explicitly acknowledged by (or even known to) the designers and users of a given verification system.

9 Conditional Equations

The definition and result in this section justify proof scores that prove a conditional equation by proving a certain associated unconditional ground equation.

Definition 15 A conditional (extended) Σ-equation consists of a signature Φ disjoint from Σ, a set E of pairs of $(\Sigma \cup \Phi)$-terms, and a pair t, t' for $(\Sigma \cup \Phi)$-terms; we will use the notation
$$(\forall \Phi)(E \Rightarrow t = t').$$
Now given a set E' of (extended) Σ-equations, we define
$$E' \models_\Sigma (\forall \Phi)(E \Rightarrow t = t')$$
to mean that, given any Σ-algebra A that satisfies E' and given any interpretation $f \colon \Phi \to A$, if $\overline{f}(t_1) = \overline{f}(t_2)$ for each $\langle t_1, t_2 \rangle \in E$, then $\overline{f}(t) = \overline{f}(t')$ in A. $\square$

The following result gives us a technique for proving conditional equations with ground term deduction, and hence with ground term reduction.

Theorem 16 Given a conditional extended Σ-equation $(\forall \Phi)(E \Rightarrow t = t')$ and a set E' of (extended) Σ-equations, then
$$(18)\quad E' \models_\Sigma (\forall \Phi)(E \Rightarrow t = t') \text{ iff } (E \cup E') \models_{\Sigma \cup \Phi} (\forall \emptyset) t = t'.$$
Proof: Each condition is equivalent to the condition

$$\overline{f}(t) = \overline{f}(t') \text{ for every } (\Sigma \cup \Phi)\text{-algebra } A \text{ that satisfies } (\forall \emptyset) e \text{ for each } e \in (E \cup E'),$$

where $\overline{f} \colon T_{\Sigma \cup \Phi} \to A$ is the unique homomorphism. $\square$

Once again, initiality gives a very simple proof.

10 Reducible Sentences

We can generalize and systematize the results of the preceding sections by defining some classes of sentences for which OBJ proof scores can be generated automatically. Although this class of sentences captures the power of the proof techniques developed above, it is not the most general possible; for example, the next section introduces some entirely new considerations for parameterized modules, and still other issues are raised by negation. By necessity, the development in this section contains many recursive definitions; fortunately, they are all fairly simple, and so are the proofs, all of which have been omitted.

Definition 17 For a fixed signature Σ, the sets P_Φ^Σ of **positive universal Σ-sentences over** a signature Φ of **variable symbols**, are defined by the following recursive rules:

1. Given $t, t' \in T_{\Sigma \cup \Phi}$ then $(t = t')$ is in P_Φ^Σ.

2. Given $s \in P_\Phi^\Sigma$ and $s' \in P_\Psi^\Sigma$ then $(s \wedge s')$ is in $P_{\Phi \cup \Psi}^\Sigma$.

3. Given $s \in P_\Phi^\Sigma$ and $s' \in P_\Psi^\Sigma$ then $(s \Rightarrow s')$ is in $P_{\Phi \cup \Psi}^\Sigma$.

4. Given signatures Φ and Ψ, with Ψ disjoint from $\Sigma \cup \Phi$, and given $s \in P_{\Phi \cup \Psi}^\Sigma$ then $(\forall \Psi) s$ is in P_Φ^Σ.

Given a positive universal Σ-sentence s, let $\mathcal{V}(s)$ be the least signature Φ such that $s \in P_\Phi^\Sigma$. Also, let us call the sentences in $P^\Sigma = P_\emptyset^\Sigma$ **closed**, and call all others **open**. A sentence of the form $(t = t')$ for $t, t' \in T_\Sigma$ is called a **ground Σ-equation**. □

Our next goal is to define the semantics of closed sentences; but first, we need to define the application of an assignment to a sentence.

Definition 18 For any disjoint signatures Σ and Φ, and any assignment $a \colon \Phi \to A$ to a Σ-algebra A, let $a^* \colon T_{\Sigma \cup \Phi} \to A$ be the unique $(\Sigma \cup \Phi)$-homomorphism. Then a^* extends to the sets P^Σ by the following rules:

1. Given $s = (t = t') \in P_\Phi^\Sigma$ let $a^*(s) = (a^*(t) = a^*(t'))$.

2. Given $s = (s' \wedge s'') \in P_\Phi^\Sigma$ let $a^*(s) = (a^*(s') \wedge a^*(s''))$.

3. Given $s = (s' \Rightarrow s'') \in P_\Phi^\Sigma$ let $a^*(s) = (a^*(s') \Rightarrow a^*(s''))$.

4. Given $s = ((\forall \Psi) s') \in P_\Phi^\Sigma$ with Ψ disjoint from $\Sigma \cup \Phi$, let $a^*(s) = (\forall \Psi)(a^*(s'))$.

□

Definition 19 For any signature Σ and Σ-algebra A, the **satisfaction** relation, $A \models_\Sigma s$, is defined for $s \in P^\Sigma$, by the following rules:

1. Given $t, t' \in T_\Sigma$ then $A \models_\Sigma (t = t')$ iff $a(t) = a(t')$ where $a \colon T_\Sigma \to A$ is the unique Σ-homomorphism.

2. Given $s, s' \in P^\Sigma$ then $A \models_\Sigma (s \wedge s')$ iff $(A \models_\Sigma s$ and $A \models_\Sigma s')$.

3. Given $s, s' \in P^\Sigma$ then $A \models_\Sigma (s \Rightarrow s')$ iff $(A \models_\Sigma s$ implies $A \models_\Sigma s')$.

4. Given $s \in P_\Psi^\Sigma$ with Ψ disjoint from Σ, then $A \models_\Sigma (\forall \Psi) s$ iff $A \models_{\Sigma \cup \Psi} a^*(s)$ for all assignments $a \colon \Psi \to A$.

Given two closed Σ-sentences s and s', let us define
$$s \models_\Sigma s' \text{ iff for all } \Sigma\text{-algebras } A \ (A \models_\Sigma s \text{ implies } A \models_\Sigma s'),$$
and
$$s \models_\Sigma s' \text{ iff } T_{\Sigma, \{s\}} \models_\Sigma s',$$
provided, of course, that s does have an initial model $T_{\Sigma, \{s\}}$. This happens, for example, if s is a conjunction of closed conditional equations. □

We can now show

Proposition 20 For any Σ-algebra A,

1. If $(t = t')$ is a ground Σ-equation, then
$$A \models_\Sigma (t = t') \text{ iff } A \models_\Sigma (\forall \emptyset)(t = t').$$

2. If $s_1, s_2, s_3 \in P^\Sigma$, then
$$A \models_\Sigma (s_1 \wedge (s_2 \wedge s_3)) \text{ iff } A \models_\Sigma ((s_1 \wedge s_2) \wedge s_3).$$
This implies that we can use associative syntax for $\wedge$.

3. If Φ and Ψ are disjoint from each other and from Σ, and if $s \in P^\Sigma$, then
$$A \models_\Sigma (\forall\Phi)(\forall\Psi)s \text{ iff } A \models_\Sigma (\forall\Phi \cup \Psi)s.$$

$\square$

Clearly, we could have handled disjunction, negation and existential quantifiers in the same style. However, we are only interested in two limited classes of sentence that can be verified by reduction. Before defining these classes, we introduce the class C^Σ of conjunctions of ground Σ-equations.

Definition 21 For any signature Σ, define C^Σ by the following rules:

1. Given $t, t' \in T_{\Sigma \cup \Psi}$ then $(\forall\Psi)(t = t')$ is in C^Σ.

2. Given $s, s' \in C^\Sigma$ then $(s \wedge s')$ is in C^Σ.

Note that $C^\Sigma \subseteq P^\Sigma$. $\square$

Definition 22 The **reducible Σ-sentences**, $\mathcal{R}_\Phi^\Sigma$, are defined by the following rules:

1. Given $t, t' \in T_{\Sigma \cup \Phi}$ then $(t = t')$ is in $\mathcal{R}_\Phi^\Sigma$.

2. Given $s \in \mathcal{R}_\Phi^\Sigma$ and $s' \in \mathcal{R}_\Psi^\Sigma$ then $(s \wedge s')$ is in $\mathcal{R}_{\Phi \cup \Psi}^\Sigma$.

3. Given $s \in C^{\Sigma \cup \Phi}$ and $s' \in \mathcal{R}_\Phi^\Sigma$ then $(s \Rightarrow s')$ is in $\mathcal{R}_\Phi^\Sigma$.

4. Given signatures Φ and Ψ, with Ψ disjoint from $\Sigma \cup \Phi$, and given $s \in \mathcal{R}_{\Phi \cup \Psi}^\Sigma$ then $(\forall\Psi)s$ is in $\mathcal{R}_\Phi^\Sigma$.

Now let $\mathcal{R}^\Sigma = \mathcal{R}_\emptyset^\Sigma$. Then $\mathcal{R}_\Phi^\Sigma \subseteq P_\Phi^\Sigma$ and $C^\Sigma \subseteq \mathcal{R}^\Sigma \subseteq P^\Sigma$. $\square$

Definition 23 A **loose atomic assertion** has the form $s \models_\Sigma s'$ for $s, s' \in P^\Sigma$; call s its **antecedent** and s' its **consequent**. Then the set $\mathcal{L}$ of **loose assertions** is defined by the following rules:

1. Every loose atomic assertion is in $\mathcal{L}$.

2. Given $a, a' \in \mathcal{L}$ then $(a \wedge a')$ is in $\mathcal{L}$.

3. Given $a, a' \in \mathcal{L}$ then $(a \Rightarrow a')$ is in $\mathcal{L}$.

(Notice that each loose assertion is either true or false.) $\square$

In reasoning about proof scores, it is convenient to avoid the actual details of OBJ, and instead use the following, more abstract notion.

Definition 24 A loose assertion is **grounded** iff each of its atomic assertions has a C sentence as its antecedent, and a single ground equation as its consequent. $\square$

We call the following results **semantic transformation rules**, since we use them to transform one loose assertion into another; they can even be expressed as OBJ rewrite rules to translate loose assertions into grounded assertions; a second step would then transform them into OBJ programs. The purpose of these rules is similar to that of LCF tactics [29], but they do not involve higher order functions. The proofs are easy generalizations of results already shown.

Proposition 25 For any $s, s_1, s_2 \in P^\Sigma$ and $r \in P_\Phi^\Sigma$, with Φ disjoint from Σ,

T1. $s \models_\Sigma (s_1 \wedge s_2)$ iff $(s \models_\Sigma s_1) \wedge (s \models_\Sigma s_2)$.

T2. $s \models_\Sigma (s_1 \Rightarrow s_2)$ iff $(s \wedge s_1) \models_\Sigma s_2$.

T3. $s \models_\Sigma (\forall \Phi) r$ iff $s \models_{\Sigma \cup \Phi} r$.

T4. If $a \colon \Phi \to \Psi$ is an isomorphism of signatures Φ and Ψ that are disjoint from Σ, then $s \models_\Sigma (\forall \Phi) r$ iff $s \models_\Sigma (\forall \Psi) a^*(r)$.

$\square$

Notice that these transformations always produce equivalent sentences. The main result is the following:

Theorem 26 Any loose atomic assertion whose antecedent is in C and whose consequent is in $\mathcal{R}$ transforms into an equivalent grounded assertion. $\square$

It follows that if each antecedent sentence in the resulting grounded assertion is canonical as a term rewriting system, then the original assertion is true iff the transformed assertion evaluates to **true**. This gives a decision procedure for a fairly broad class of second order quantified assertions. Moreover, this decision procedure can be implemented by further translating the assertion into OBJ code whose objects build up the antecedents of assertions, and whose reductions evaluate their consequents. We now extend the above to encompass satisfaction in standard models.

Definition 27 An **atomic assertion** is either a loose atomic assertion $s \models_\Sigma s'$ or else an **initial atomic assertion** of the form $s \cong_\Sigma s'$ for $s, s' \in P^\Sigma$; call s its **antecedent** and s' its **consequent**. Then the set S of **assertions** is defined by the following rules:

1. Every atomic assertion is in S.

2. Given $a, a' \in S$ then $(a \wedge a')$ is in S.

3. Given $a, a' \in S$ then $(a \Rightarrow a')$ is in S.

An atomic assertion is **grounded** iff its antecedent is in C and its consequent is a single ground equation. Note that $\mathcal{L} \subseteq S$, and that each assertion must be either true or false. $\square$

Proposition 28 For any $s, s' \in P^\Sigma$,

T5. $s \cong_\Sigma s'$ if $s \models_\Sigma s'$.

$\square$

Using this transformation in conjuction with those of Proposition 25 gives a verification method applicable to all atomic assertions; however, because the transformed assertion is not necessarily equivalent, we do not get a decision procedure.

Theorem 29 Any atomic assertion whose antecedent is in C and whose consequent is in $\mathcal{R}$ transforms into a grounded loose assertion. $\square$

It follows that if all the antecedent sentences in the transformed assertion are canonical as term rewriting systems, then the original assertion is true if the transformed assertion evaluates to **true**. It cannot be expected that the transformed assertion always succeeds when the original assertion is actually true; in general, inductive arguments may be needed. However, we have found in practice that this procedure works for the assertions that arise in hardware verification.

11 Verifying Parameterized Modules

The results in this section justify techniques for verifying the correctness of OBJ-style parameterized modules. This is important because it extends reusability from code to proofs. In particular, once a property of a parameterized module has been verified (in the sense made precise below), we know that it is true of any instantiation of the module, which saves us from having to prove it separately for each instantiation. We first generalize (by relativizing it to a parameter presentation) the development that culminates in the soundness proof for the induction rule in Section 8.

Definition 30 A **parameterized presentation** g is a pair $\langle R, P \rangle$ of presentations, where if $R = \langle \Sigma^\circ, E^\circ \rangle$ and $P = \langle \Sigma, E \rangle$ then $\Sigma^\circ \subseteq \Sigma$ and $E^\circ \subseteq E$; we will write $g \colon R \subseteq P$, and call R the **requirement** or **parameter** presentation and P the **full** or **body** presentation. $\square$

Intuitively, a parameterized presentation freely extends any given R-algebra A to a P-algebra. The best examples of this use many sorted algebra, or even order sorted algebra; for example, R might specify a set with a binary Boolean-valued function satisfying the axioms for a partial order, and then P might give a sorting algorithm for lists over such sets. Let us now make precise the notion of "free" that is involved.

Definition 31 Given a parameterized presentation $g \colon R \subseteq P$ and an R-algebra A, let us say that a g-**extension** of A is a Σ°-homomorphism $e \colon A \to B$ where B is a P-algebra. Given another g-extension $e' \colon A \to B'$ of A, a **morphism** of g-extensions of A is a Σ-homomorphism $h \colon B \to B'$ such that $h \circ e = e'$. If h is an inclusion, then we say that B is a **sub-g-extension** of B'. Also, $e \colon A \to B$ is a **free** g-extension of A iff there is a *unique* morphism from it to any other g-extension of A. $\square$

This is an *initiality* property[7], and as you might expect, we have

Proposition 32 Given a parameterized presentation $g : R \to P$, any two free g-extensions of a given R-algebra are Σ-isomorphic. Moreover, if B is a free g-extension of A, then B has no proper sub-g-extensions of A.

Proof: We omit the first assertion, which is proved just as Proposition 6. The proof of the second assertion relativizes the proof of Proposition 8. Let $e \colon A \to B$ be a free g-extension of A, let $i \colon B' \subseteq B$ be the inclusion, and assume that e factors as $i \circ e'$ where $e' \colon A \to B'$. Then i is a morphism of g-extensions, and since e is initial, there is a morphism $f \colon e \to e'$ such that $i \circ h = id_e$. This implies that i is surjective, and therefore an isomorphism. $\square$

The following result about free extensions can be proved using a term algebra construction[8], much like Theorem 7:

Theorem 33 Given a parameterized presentation $g \colon R \to P$, any every R-algebra A has a free g-extension; let us denote a choice of one such by $T_g(A)$. $\square$

---

[7]It is initiality in the comma category $(A \downarrow U_g)$, where U_g is the forgetful functor from P-algebras to R-algebras.

[8]In fact, T_g is a functor from the R-algebras to the P-algebras, left adjoint to U_g.

Definition 34 Given a parameterized presentation $g\colon R \to P$ and a Σ-sentence s, we define
$$g \models_\Sigma s \text{ iff } (A \models_{\Sigma^\circ} R \text{ implies } T_g(A) \models_\Sigma s).$$
Then, a **signature of constructors** for g is a subsignature $\Pi \subseteq \Sigma$ disjoint from Σ° such that the unique $(\Sigma^\circ \cup \Pi)$-homomorphism $h_A\colon T_{\Sigma^\circ \cup \Pi} \to T_g(A)$ is surjective for any R-algebra A. Moreover, Π is a **free signature of constructors** for g iff h_A is always an isomorphism, and Π is a **minimal signature of constructors** for g iff no proper subsignature is a signature of constructors for g. $\square$

Theorem 35 [Structural Induction III] Given a parameterized presentation $g\colon R \to P$, an R-algebra A, and a signature Π of constructors for g, let $V \subseteq T_g(A)$. Then $V = T_g(A)$ if:

  0. $c \in \Pi_0$ implies $[c] \in V$.

  1. $f \in \Pi_n$ for $n > 0$ and $[t_i] \in V$ for $i = 1, ..., n$ imply $[f(t_1, ..., t_n)] \in V$.

Proof: Since $T_g(A)$ has no proper sub-g-extensions (by Proposition 32) and since $V \subseteq T_g(A)$, we need only show that V is closed under Π; but that is exactly what conditions 0. and 1. say. $\square$

Corollary 36 [Structural Induction IV] Given a parameterized presentation $g\colon R \to P$ and a signature Π of constructors for g, let $Q(x)$ be a $T_{\Sigma^\circ \cup \{x\}}$-sentence. Then $g \models_\Sigma (\forall x) Q(x)$ if:

  0. $c \in \Pi_0$ implies $g \models_\Sigma Q(c)$.

  1. $f \in \Pi_n$ for $n > 0$ and $t_i \in T_\Sigma$ for $i = 1, ..., n$ imply $g \models_\Sigma Q(t_1) \wedge ... \wedge Q(t_n) \Rightarrow Q(f(t_1, ..., t_n))$.

Proof: For any R-algebra A, we must show that $T_g(A) \models_\Sigma (\forall x) Q(x)$. Choosing one such A, let $V = \{x \in T_g(A) \mid g \models_\Sigma Q(x)\}$ and then apply Theorem 35. $\square$

The OBJ-style parameterized module[9] verified in Appendix D uses the above in justifying its proof score. This semantics for parameterized modules is a simplification of the semantics based on colimits in the category of theories that has been given for the Clear specification language [6]. It seems possible that [14] discusses something similar to Theorem 36, but it is difficult to be certain, because no explicit model theoretic criteria for correctness are given; this means, of course, that it is impossible to prove soundness of the proof rules given in [14].

Many other results also extend to parameterized modules. We first generalize result T1 of Proposition 25.

Proposition 37 Given a parameterized presentation $g\colon R \to P$ and Σ-sentences s_1 and s_2, then

  T6. $g \models_\Sigma (s_1 \wedge s_2)$ iff $(g \models_\Sigma s_1) \wedge (g \models_\Sigma s_2)$.

---

[9]OBJ parameterized modules actually assume that g is **protecting** in the sense that each free extension $A \to U_g(T_g(A))$ is an isomorphism. However, this assumption is not needed for the results of this paper.

Proof: We have that
$$(A \models R) \Rightarrow T_g(A) \models s_1 \wedge s_2$$
iff (using T1 of Proposition 25)
$$(A \models R) \Rightarrow (T_g(A) \models s_1) \wedge (T_g(A) \models s_2)$$
iff
$$(A \models R \Rightarrow T_g(A) \models s_1) \wedge (A \models R \Rightarrow T_g(A) \models s_2).$$
□

The next result is rather powerful, and perhaps even surprising.

Theorem 38 Given a parameterized presentation $g \colon R \to P$ and $\Sigma(X)$-sentences s_1 and s_2,

T7. $g \models_\Sigma (\forall X)(s_1 \Rightarrow s_2)$ if $E \cup \{s_1\} \models_{\Sigma(X)} s_2$

where E is the set of equations in P.

Proof: Assume that $A \models_{\Sigma_\circ} R$ and let $f : X \to T_g(A)$ be an assignment such that $T_g(A) \models_{\Sigma(X)} s_1$. Then $T_g(A)$ can be seens as a P'-algebra, where $P' = \langle \Sigma(X), E \cup \{s_1\} \rangle$, and therefore satisfies s_2 by hypothesis. □

It seems worthwhile to explicitly bring out the following two special cases of this general result:

Corollary 39 Given a parameterized presentation $g \colon R \to P$, $\Sigma(X)$-equation e, and a set E' of $\Sigma(X)$-equations, then

T8. $g \models_\Sigma (\forall X)e$ if $E \models_{\Sigma(X)} e$

T9. $g \models_\Sigma (\forall X)(E' \Rightarrow e)$ if $(E \cup E') \models_{\Sigma(X)} e$

where E is the set of equations in P. □

Parameterized module verification techniques are not part of the usual toolkit for hardware verification. However, these techniques could be quite useful for certain problems. For example, one might verify n-bit ripple carry adders, subtractors, etc., for arbitrary n, and then apply these results to some specific device having a specific value of n, such as $n = 12$. Reusing these proofs should be much easier than redoing them. Perhaps the Sobel edge detection chip discussed in [38] would be a good candidate for this approach.

12   Discussion

The body of this paper has stated and justified some techniques for using a powerful rewrite engine like OBJ as a theorem prover, including a new extension of first order equational logic to second order universal quantification; we see this as an extension of first order equational logic toward its natural limit, rather than an incursion into the second order realm (order sorted algebra extends first order equational logic in an orthogonal direction). The paper proves some results about this logic, including a completeness theorem, and suggests using these results to justify **proof scores**, which are OBJ programs such that if all their reductions evaluate to true, then some theorem is verified. In particular, we show that proof scores can be automatically generated for two fairly broad classes of sentence. The justifications of proof techniques have been greatly eased by exploiting semantics, and especially initiality, instead of relying on purely syntactic arguments. We also suggest that the verification of parameterized modules as a novel and very general technique for resuing proofs.

The appendices to this paper give some sample proofs using these techniques. Many of these examples use order sorted algebra, and rewriting modulo associativity and commutativity, the details of which are not explicitly developed in this paper. The proof scores have been constructed by the user (i.e., by the author). Although only the OBJ code is given, rather than a transcript of its execution, all these examples have been run in OBJ3 exactly as shown, and all the results have been exactly what they should be. I wish to note that it took remarkably little user time to do these examples, just an hour or two in each case but one[10]; moreover, most of this time was spent tracking down minor syntactic errors and then rerunning the code[11]. It is fortunate that the class of assertions for which proof scores can be generated automatically includes the assertions of greatest interest in hardware verification, as illustrated in Appendix B. Another promising application area is verifying (first order) functional programs, as illustrated by the verification of the map function in Appendix D.

The following summarizes some observations on theorem proving that arise from my experience with various versions of OBJ and some other theorem proving systems:

1. **Strong typing is very useful**. Without strong typing, it is necessary that theorems about the natural numbers (for example) somehow exclude integers, lists, stacks, etc. Also, the correctness of combined decision procedures becomes a serious worry for untyped logic, whereas it is not an issue when different procedures apply to different sorts. Of course, a typed language is more expressive and convenient, and perhaps most important, theorem proving can be more efficient, because typing can eliminate much useless search.

2. **A simple way to describe proof strategies is desirable.** Completely automatic theorem provers can be very painful to use. Because such provers slavishly follow their own heuristics, users must somehow trick them into finding proofs when the built-in heuristics are not adequate. (But automatic theorem proving remains an interesting research problem.) On the other hand, complex meta-languages for describing proof strategies can be difficult to use, and fully manual proof checkers are also very painful, because of the enormous amount of detail that users must manage. OBJ3 program structure, using module hierarchies and generic module, seem much easier to use than these alternatives.

3. **Incremental construction of proof environments can save time.** Otherwise, much work may have to be repeated whenever an error is encountered.

4. **Useful feedback** from failed proof attempts can make a tremendous difference in the practical use of a theorem prover. It is very frustrating to

---

[10]The ripple carry adder example took me two or three days (i.e, 16-24 hours), spread over some weeks, most of which was caused by uncovering and fixing some new bugs in OBJ3's associative/commutative rewriting algorithms.

[11]However, in some cases it took *much* longer to justify the proof score; indeed, trying to justify proof scores is exactly what led to the theoretical results given in this paper.

get unproved, a timeout, or a core dump, as feedback. The feedback from Knuth-Bendix completion procedures is hardly more helpful, often consisting of enormously many new rules, most of which are silly; in fact, it is not only hard to understand the output from completion proceduress, but such procedures are also very hard to guide, may not terminate, and can generated unorientable rules.

5. **Rigorous logic-based semantics is essential.** Otherwise, it is difficult to justify that what is being done is theorem proving at all. This applies to facilities for structuring theories and proofs (e.g., parameterized modules), as well as to the basic sentences, and Section 11 gives a precise semantics for OBJ-style parameterized modules.

The use of OBJ suggested in this paper is a novel style of interaction with a theorem prover, that seems to be an interesting compromise among various current approaches. Three main levels are involved:

1. The most concrete level is the *routine calculation* that is done by OBJ's powerful reduction engine completely automatically, when a proof score is executed.

2. The second level is the *construction of proof scores*, either automatically by some algorithm already known to be correct for some class of assertions, or else by the user. A proof score has the explicit structure of an OBJ3 program, which is (relatively) easy to write, to read, and to modify. Moreover, when a reduction fails to yield true, it is often easy to get further information that can suggest what to try next, for example, proving a new lemma.

3. The most abstract level concerns the *correctness of general methods for constructing valid proof scores*. For example, Corollary 14 states the correctness of proofs by induction. Such proofs are done in ordinary mathematics, and thus fall under the usual "social processes" [11] of discussion, criticism, refinement and generalization, by the community of those who are interested in such things. This seems highly appropriate, because such meta-level proofs, although they need not be lengthy or detailed, can be rather sophisticated (e.g., using initial algebra semantics) and thus difficult to verify mechanically. Moreover, this is exactly the aspect of verification that can benefit most from discussion, and least from being buried inside of code; it is also the aspect most likely to be useful to other theorem proving projects.

My preliminary conclusion is that this style of interaction may require substantially less user time than other styles; however, more experience is needed, especially with larger examples.

It would be useful to construct a verification interface for OBJ to generate proof scores that use only techniques already shown correct. For example, a score for an inductive proof could be automatically generated from knowing the constructors, what to prove, and what variable to do induction over. Such a system would guarantee the validity of any proof that it constructs, and could also support incremental proof management. For example, it would be convenient to be able to retrofit a new lemma back into a partially successful reduction whose result suggested using that lemma. Section 10 justifies a way to compile proof scores for certain classes of sentences, which include many important problems in hardware verification.

This paper views OBJ3 not just as an implementation of equational logic, but rather as a powerful term rewriting engine that can do the "grunt work" for a number of different logical systems, including second order universal equational logic, embedded in a very expressive language for structuring proofs. Second order universal equational logic seems an especially good choice for hardware description, simulation and verification, because in general, both the description and the specification consist of universally quantified equations (possibly with second order variables). It would be interesting to consider using constructive type theory as a framework for proving the correctness of OBJ proof scores; in fact, it should not be too difficult to define such a type theory within OBJ itself. A program to generate proof scores from expressions in a suitable sublanguage of the type theory could also be written in OBJ (after all, it is a function), and then everything could be done in OBJ!

Of course, I do not claim that OBJ is the ideal all-purpose theorem prover, because there are many things that it cannot handle; however, I do recommend it for consideration in the surprisingly many cases where equational reasoning is sufficient, and in particular, for hardware verification. Eqlog [21] would be even better than OBJ for some applications, since it can compute values for existentially quantified variables; but unfortunately, we do not yet have an implementation of Eqlog. It should also be noted that OBJ3 is currently supported only by a prototype interpreter, and greater speed would certainly be desirable for large applications; in particular, some form of compilation would be very desirable.

A Decision Procedure for Propositional Calculus

The examples in these appendices assume some familiarity with OBJ3; for details, see [17,13,12] and especially [26], which gives the current syntax. All reductions evaluate to true unless otherwise indicated.

This first appendix gives a decision procedure for the propositional calculus, based on ideas of Hsiang [31]. Some examples using it directly are given, and it is also used later for hardware verification. This is a good example of code reuse, since the original was written (by David Plaisted) years before we thought of using it for hardware verification. In reading this code, note that the prec attribute of an operator declares its precedence, with a lower number indicating tighter binding, and that *** and ***> indicate comments; the latter prints its content during execution.

```
obj PROPC is sort Prop .
  protecting TRUTH + QID .
  subsorts Id Bool < Prop .

  op _and_ : Prop Prop -> Prop [assoc comm prec 2] .
  op _xor_ : Prop Prop -> Prop [assoc comm prec 3] .
  vars p q r : Prop .
  eq p and false = false .
  eq p and true = p .
  eq p and p = p .
  eq p xor false = p .
  eq p xor p = false .
```

```
   eq p and (q xor r) = (p and q) xor (p and r)  .

   op _or_  : Prop Prop -> Prop [assoc comm prec 7]  .
   op not_  : Prop -> Prop [prec 1]  .
   op _implies_  : Prop Prop -> Prop [prec 9]  .
   op _iff_  : Prop Prop -> Prop [assoc prec 11]  .
   eq p or q = (p and q) xor p xor q  .
   eq not p = p xor true  .
   eq p implies q = (p and q) xor p xor true  .
   eq p iff q = p xor q xor true  .
endo
```

The following are some tests of this decision procedure. An expression is a tautology iff it reduces to true; otherwise the reduced form gives simplified necessary and sufficient conditions for the original expression to be true.

```
reduce 'a or not 'a  .
reduce 'a or 'b iff 'b or 'a  .
reduce 'a implies ('b implies 'a).
reduce 'a implies 'b iff not 'b implies not 'a  .
reduce not('a or 'b) iff not 'a and not 'b  .
reduce 'c or 'c and 'd iff 'c  .
reduce 'a iff 'a iff 'a iff 'a  .
reduce 'p implies 'q and 'r iff ('p implies 'q) and ('p implies 'r).

reduce 'a iff not 'b  .                 ***> should be: 'a xor 'b
reduce 'a and 'b xor 'c xor 'b and 'a  . ***> should be: 'c
reduce 'a implies ('b implies not 'a).
   ***> should be: true xor ('b and 'a)
reduce 'a and not 'a  .                 ***> should be: false
reduce 'a iff 'a iff 'a  .              ***> should be: 'a
```

B   Hardware Verification

Hardware verification seems an especially promising application for equational logic, because circuits are generally designed to exhibit some definite causal behavior, and it is natural to describe such behavior by functions. Furthermore, both the circuit description and its purported behavior can generally be given just by equations. Moreover, equational logic is (relatively) simple and well-understood, has reasonably efficient algorithms for many problems, supports bidirectional circuits through the bidirectionality of equality, and easily extends to second order universal quantification, as shown in the body of this paper.

This appendix verifies five simple circuits: first, a 1-bit full adder implemented with AND, OR and XOR (exclusive or) gates; second, a CMOS XOR circuit; third, a 1-bit CMOS memory cell; fourth, a series connection of two unit-delay inverters; and fifth, an n-bit ripple carry adder. The first three examples illustrate a general approach to verifying combinational circuits, and also demonstrate that bidirectionality and "don't care" conditions are handled naturally. The fourth example

Figure 1: Power and Ground

extends the method to sequential circuits, and the fifth example makes good use of OBJ's abstract data type facilities. The research reported here is still at an exploratory stage, and no very large or very subtle examples have yet been attempted. I thank Mike Gordon for his clear expositions of hardware verification with higher order logic, from which the present exposition has taken several examples and much inspiration. [41] is an early paper applying OBJ to hardware specification and testing.

B.1 Combinational Circuits

The wires in circuits are modelled by variables in this paper. Combinational circuits do not require modelling time dependency, so the voltage levels on wires can be modelled by constants. The simple circuit models in this paper have just two voltage levels, called "power" and "ground," represented by the truth values true and false respectively, and diagramed as shown in Figure 1. In a circuit diagram, any collection of lines that are directly connected share the same voltage level, and thus are equal to the same variable. Although this ignores issues of load (i.e., current flow), timing, and capacitance, it seems likely that some of these issues could be handled by using a more liberal set of values on wires, for example, in the style of Glenn Winskel [45]. It also seems likely that these enlarged value sets could also be handled with rewrite rules.

This subsection treats three simple combinational circuits. The first is a 1-bit full adder, the second is a CMOS implementation of an XOR gate, and the third is a 1-bit CMOS memory cell. The first illustrates the verification of unidirectional circuits, while the second and third illustrate the verification of bidirectional circuits. The third example has the interesting property of bistability.

B.1.1 Unidirectional Circuits

The simplest combinational circuits have what we shall call **triangular form**, in which there is given a set of input variables, say $i_1, ..., i_n$, and a *sequencing* of the non-input variables, say $p_1, ..., p_m$, such that each p_k is a Boolean function of the inputs and of the p_j with $j < k$,

$$p_k = f_k(i_1, ..., i_n, p_1, ..., p_{k-1}).$$

(Of course, this equation is implicitly universally quantified over all the variables that occur in it.) Some of the p_k are output variables, and the others are internal "test point" variables. By induction, they can all be expressed as functions of the input variables only. Notice that the whole *system* of equations,

$$p_1 = f_1(i_1, ..., i_n)$$
....
$$p_k = f_k(i_1, ..., i_n, p_1, ..., p_{k-1})$$
....

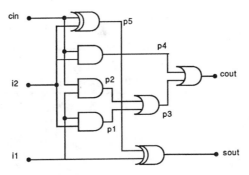

Figure 2: 1-bit Full Adder

$$p_m = f_m(i_1, ..., i_n, p_1, ..., p_{m-1})$$

is canonical, i.e., is Church-Rosser and terminating, when regarded as a set of rewrite rules, by virtue of its form. Therefore, an equation $t = t'$ can be verified for the circuit simply by checking whether or not its two sides reduce to the same canonical form. It is convenient to do this in OBJ by evaluating the expression

    t == t'

using the decision procedure for the propositional calculus given in Appendix A to simplify Boolean expressions. Then the equation $t = t'$ is true of a circuit described by the above system of equations iff the expression t == t' reduces to true using the equations as rewrite rules, and is false iff it reduces to false.

B.1.2 1-bit Full Adder

Figure 2 shows a circuit diagram for a 1-bit full adder, producing sum and carry output bits from three inputs, one of which is a previous carry bit. The following system of equations is exactly equivalent to this diagram; it is in triangular form, with input variables i1, i2, and cin.

    p1 = i1 and i2
    p2 = i1 and cin
    p3 = p1 or p2
    p4 = cin and i2
    p5 = cin xor i2
    cout = p3 or p4
    sout = i1 xor p5

If we want to verify the following two properties of this circuit,

    cout = (i1 and i2) or (i1 and cin) or (i2 and cin)
    sout = (i1 and i2 and cin) or (i1 and not i2 and not cin) or
           (not i1 and i2 and not cin) or (not i1 and not i2 and cin)

then the following is an appropriate OBJ proof score:

```
obj FADD is extending PROPC .
  ops i1 i2 cin p1 p2 p3 p4 p5 cout sout : -> Prop .
```

Figure 3: MOS Transistors

```
   eq p1 = i1 and i2 .
   eq p2 = i1 and cin .
   eq p3 = p1 or p2 .
   eq p4 = cin and i2 .
   eq p5 = cin xor i2 .
   eq cout = p3 or p4 .
   eq sout = i1 xor p5 .
endo

*** correctness of cout
reduce cout iff (i1 and i2) or (i1 and cin) or (i2 and cin) .

*** correctness of sout
reduce sout iff (i1 and i2 and cin) or
                (i1 and not i2 and not cin) or
                (not i1 and i2 and not cin) or
                (not i1 and not i2 and cin) .
```

This score has been run and it works. No manual application of inference rules is needed; OBJ does all the work. In fact, the OBJ proof score could be produced *entirely automatically* from just the circuit diagram and the equations to be proved, and it should be clear that the same holds for any circuit in triangular form. By contrast, [7] gives a six step one and a half page proof outline for just the sout formula.

B.1.3   XOR Gate

It seems not to be obvious that a function-based formalism can effectively handle bidirectional circuits, such as MOS transistors. However, equational logic is actually based upon equality, which is a bidirectional relation; moroever, we have not only simple equalities, but also conditional equalites available for modelling circuits.

This subsection shows that the circuit of Figure 4 realizes the exclusive or (xor) function using six transistors. i1 and i2 are input wires, while p1, p2 and p3 are internal wires, and o is an output. Examination of Figure 4 shows that, despite a much larger number of lines, there are only six connected components (since devices disconnect lines), and hence there are only six wires, exactly the ones named above. Recall that the two voltage levels, "power" and "ground," are represented by true and false respectively, and are diagrammed as in Figure 1.

Besides wires and the two voltage levels, this circuit has two kinds of transistor, the n-transistor and the p-transistor, as shown in Figure 3. These are modelled by the conditional equations

    a = b if g = true

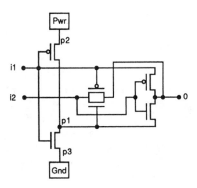

Figure 4: XOR Circuit

---

and
```
    a = b if g = false
```
respectively. Following OBJ notation, from here on we will write these equations
in the forms
```
    a = b if g
```
and
```
    a = b if not g
```
respectively[12]. To model the entire circuit, we just write one such equation for each
transistor, plus two equations for the two fixed voltages, obtaining the following
system:

```
p1 = p2 if not i1
p1 = p3 if i1
o = i1 if not i2
o = p1 if i2
o = i2 if not i1
o = i2 if p1
p2 = true
p3 = false
```

Since this circuit is supposed to implement xor, we want to prove that it satisfies
the equation
```
    o = i1 xor i2 .
```
The situation is as follows: we are given a system of (conditional) equations
describing the circuit; we are given an expression describing the output variable in
terms of the input variables; and we want to prove that this expression is part of
a solution to the given system. To verify this, we need to know the behavior of all
non-input variables. In this example, the only interesting one is
```
    p1 = not i1
```
since p2 and p3 are already given as true and false, respectively. To check that
these equations solve the given system, we just substitute them into the system

---

[12][28] points out that the transistor model used there, which is the same as that used here (although
the representations are quite different), is not adequate for guaranteeing that a circuit is useful in
practice, since it fails to model load (i.e., current flow). However, the *logical* aspect of circuits *is*
adequately modelled, and logical correctness is certainly necessary for practical utility.

and then simplify. This is analoguous to solving a system of linear equations. Such a system can be **underdetermined**, in the sense that there are more variables (degrees of freedom) than there are (independent) equations, so that a most general solution consists of some expressions in some parameter variables (these correspond to our input variables); to check that some such expressions really are a solution, we just substitute them into each equation and then check if its two sides are equal.

This is actually more than an analogy, because systems of linear equations over the field Z_2 describe circuits that do not involve conjunctions of variables. Although the method for verifying that some system of parameterized expressions is a solution by substituting into the given equations does not depend on linearity, standard methods for finding the solution to a system of equations do depend on linearity for their completeness; nevertheless, Gaussian elimination (for example) can often be used to solve the non-linear systems that result from logic circuits.

The problem of verifying a proposed solution is inherently simpler than the problem of finding a solution, and can be done by letting OBJ *reduce* each side of each equation, using the proposed solution as rewrite rules. An interesting complication is that for conditional equations, we must assume the condition is true when doing the reductions. The following is an OBJ3 proof script for the XOR circuit of Figure 4, again using the decision procedure for the propositional calculus from Appendix A to simplify Boolean expressions:

```
obj BEH is extending PROPC .
  ops i1 i2 p1 p2 p3 o : -> Prop .
  eq p1 = not i1 .
  eq o = i1 xor i2 .
  eq p2 = true .
  eq p3 = false .
endo

*** p1 = p2 if not i1 .
obj BEH1 is using BEH .
  eq i1 = false .
endo
reduce p1 == p2 .

*** p1 = p3 if i1 .
obj BEH2 is using BEH .
  eq i1 = true .
endo
reduce p1 == p3 .

*** o = i1 if not i2 .
obj BEH3 is using BEH .
  eq i2 = false .
endo
reduce o == i1 .

*** o = p1 if i2 .
obj BEH4 is using BEH .
  eq i2 = true .
```

```
endo
reduce o == p1 .

*** o = i2 if not i1 .
obj BEH5 is using BEH .
  eq i1 = false .
endo
reduce o == i2 .

*** o = i2 if p1 .
obj BEH6 is using BEH .
  eq i1 = false .
endo
reduce o == i2 .
```

Let us now consider in more detail how to determine when the condition of a conditional equation is true: first, substitute the functional specifications into the condition; then compute its disjunctive normal form; next, let each disjunct define a *case*, in which the input variables that occur positively are true, while those that occur negatively are false; and finally, check that in each such case, the two sides evaluate to the same reduced form, using the given values for the variables that belong to that case. For example, the condition of the sixth equation is p1; substituting the functional specification of p1 yields not i1, which is already in disjunctive normal form, with only one disjunct and hence only one case, in which i1 = false; the object BEH6 sets up this assumption before evaluating the equation. Clearly, more complex conditions can arise, but they can be handled the same way.

We can summarize this method of proof as follows: first, regard each variable as a new constant; then, given functional specifications for each non-input wire[13] as a Boolean function of the input variables, regard them as rewrite rules (where each lefthand side is a single variable); and finally, verify that each equation in the circuit description circuit is satisfied, by reducing its lefthand and righthand sides, under each disjunctive case where its condition is true.

I wish to emphasize that this algorithm does not require any human intervention; the OBJ proof score can be generated automatically from a circuit diagram and its functional spec. Moreover, it is a *decision procedure* for satisfaction of specifications by circuits, since a circuit fails to satisfy a specification iff some reduction of some equation evaluates to false. Also, exactly the same method works for combinational circuits built from any (combinational) components, whether bidirectional or not. Moreover, this approach may be more efficient, at least in some cases, than the more usual approaches.

A circuit is **consistent** (with respect to a given set of inputs) iff for each choice of input values, there is at least one possible value for each non-input variable that satisfies its system of equations. A consistent circuit is **underdetermined** (with respect to a given set of inputs) iff for some choice of input values, there is more than one possible value for some non-input variable, satisfying its system of equations. For example, our transistor models are consistent and underdetermined (choosing

---

[13]This is not a demanding requirement, because a circuit designer should certainly know what all his wires are supposed to do; indeed, the redundancy involved in checking the mutual consistency of the functional specifications for all non-input wires seems highly desirable in itself.

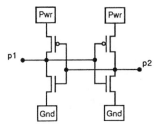

Figure 5: 1-bit Cell

---

say a and g as inputs). Also, a circuit consisting of an inverter with its output connected to its input is inconsistent, since its equation

    p = not p

has no solutions.

It is interesting to notice that this method can treat incomplete behavioral specifications and incomplete circuit designs. An incomplete specification is a set of non-trivially conditional equations, of the form

$$p = t(i_1, ..., i_n) \text{ if } t'(i_1, ..., i_n)$$

where p is a non-input variable, where $i_1, ..., i_n$ are the input variables, and where t and t' are Boolean functions. The cases where $t'(i_1, ..., i_n)$ is not true are the "don't care" conditions.

B.1.4   Cell

Figure 5 shows a CMOS circuit for a simple 1-bit memory cell. This circuit is **bistable**, in the sense that its set of equations has *two solutions*; in particular, it is underdetermined. The equation system for this circuit is

    p1 = true if not p2
    p1 = false if p2
    p2 = true if not p1
    p2 = false if p1

which is equivalent to

    p1 = not p2
    p2 = not p1

as well as to

    p1 = not p2 .

This circuit has no designated input or output wires; indeed, the wires p1 and p2 can be used bidirectionally, for either reading or writing. Since solutions are Boolean expressions in the input variables, and since there are only two such expressions that depend on *no* variables, namely true and false, these are the only possible solutions for this circuit, and it is clear that they do indeed satisfy the equations.

For underdetermined systems, it is usually more convenient to introduce auxiliary parameter variables. In this case, we can use a single auxiliary parameter q to write the solution as

```
p1 = not q
p2 = q
```

This situation may remind some readers of unification, and indeed, it *is* a special case of unification understood in a sufficiently broad sense, for example, as in [19]: solutions are unifiers that satisfy the additional restriction that all the input variables (if there are any) must appear among the unification variables. Of course, it is possible that an unwise choice of input variables will produce inconsistency. For example, any system that contains the equation

```
i1 = not i1
```

is unsolvable, if both i1 and i2 are input variables. To avoid this *impasse,* one of them could be considered internal, or else we could consider the above equation as a constraint on solutions.

Now here is an OBJ3 proof score for the above circuit. What makes it just a little more subtle than the previous examples is that the cases when the condition of a conditional equations are true must be expressed in terms of the parameter variables; in this case, there is just one, namely q.

```
obj BEH is extending PROPC .
  ops p1 p2 q : -> Prop .
  eq p1 = not q .
  eq p2 = q .
endo

*** p1 = true if not p2 .
obj BEH1 is using BEH .
  eq q = false .
endo
reduce p1 == true .

*** p1 = false if p2 .
obj BEH2 is using BEH .
  eq q = true .
endo
reduce p1 == false .

*** p2 = true if not p1 .
obj BEH3 is using BEH .
  eq q = true .
endo
reduce p2 == true .

*** p2 = false if p1 .
obj BEH4 is using BEH .
  eq q = false .
endo
reduce p2 == false .
```

Figure 6: Series Connected Inverters

B.2 Sequential Circuits

Sequential circuits have behaviors that vary with time, and thus require modelling wires with time-varying values, e.g., as streams of Boolean values. We can still apply the previous method to obtain a system of equations from a circuit diagram, but the variables that model wires now take values that are functions from Time (which is a copy of the natural numbers) to truth values, i.e., to Prop.

Any verification of a combinational circuit "lifts" to a verification of the same circuit viewed as (potentially) sequential. All we have to do is replace each wire variable, whether an input i_k or a non-input p_k, by a function of type Time -> Prop, evaluated at an arbitrary time t, e.g., by $f_k(t)$. Then it is easy to see that exactly the same proof works for the lifted system.

B.2.1 Series Connected Inverters

This subsection proves that the series connection of two NOT gates (i.e., inverters), each with one unit delay, has the same effect as a two unit delay; see Figure 6. The system of equations involved here is

$$f_1(t+1) = \overline{f_0}(t)$$
$$f_2(t+1) = \overline{f_1}(t)$$

with (implicit) universal quantification over f_0, f_1, f_2 and t, with the f_i have type Time -> Prop, where $\overline{f_i}$ indicates the negation of f_i, and where f_0 is the input variable, f_2 is the output variable, and f_1 is an internal variable. The behavior is supposed to be

$$f_2(t+2) = f_0(t)$$

and it is clear that the solution equation for f_1 is given by the first equation of the system,

$$f_1(t+1) = \overline{f_0}(t).$$

An OBJ proof score for proving that these expressions indeed solve the given system is given below. In the object TIME, note that 0 is the initial time, and s is the successor function. The object 2NOT is the composite of the parameterized object NOT with itself; the requirement theory of NOT is WIRE, which allows any input stream. This is a nice illustratation of how to use OBJ-style parameterized modules and module expressions for hardware description.

The assertion that is verified in the OBJ3 proof score below has the form

$$E \models_\Sigma (\forall \Phi) r$$

where Σ is the union of the signatures of the objects PROPC and TIME, E is the union of their equations, Φ is the signature containing three functions f_0, f_1, f_2 from Time to Prop, and r is of the form $(s_1 \wedge s_2) \Rightarrow s_3$, where

$$s_1 = (\forall t) f_1(s\ t) = \text{not } f_0(t)$$
$$s_2 = (\forall t) f_2(s\ t) = \text{not } f_1(t)$$
$$s_3 = (\forall t) f_2(s\ s\ t) = f_0(t).$$

To prove the assertion, by the transformation rules T5 and T3 of Section 10, it suffices to prove

$$E \models_{\Sigma \cup \Phi} (\forall \emptyset) r$$

which by rule T2 is equivalent to

$$E \wedge s_1 \wedge s_2 \models_{\Sigma \cup \Phi} s_3$$

which by rule T3 can be verified with the reduction

$$E \wedge s_1 \wedge s_2 \models_{\Sigma \cup \Phi \cup \{t\}} f_2(s \ s \ t) == f_0(t)$$

which is exactly what the proof score below does.

It is also worth noting that the form of the sentence proved here is typical of a very large class of hardware verification problems, namely

$$E \models_{\Sigma} (\forall \Phi)(C \Rightarrow e)$$

where E defines the abstract data types of the problem, where C is a conjunction of equations defining the circuit, where e is an equation to be proved, and where Φ may involve second order quantification.

```
obj TIME is sort Time .
  op 0 : -> Time .
  op s_ : Time -> Time .
endo

th WIRE is protecting TIME + PROPC .
  op f1 : Time -> Prop .
endth
obj NOT[W :: WIRE] is
  op f2 : Time -> Prop .
  var T : Time .
  eq f2(0) = false .
  eq f2(s T) = not f1(T) .
endo

obj F is extending TIME + PROPC .
  op t : -> Time .
  op f0 : Time -> Prop .
endo

make 2NOT is NOT[NOT[F]*(op f2 to f1)] endm
reduce f2(s s t) iff f0(t) .
```

B.3 n-bit Ripple Carry Adder

This section verifies a ripple carry adder of arbitrary width, i.e., shows that it really adds, making good use of OBJ's abstract data type capabilities. Of course the ADT of natural numbers with addition is needed, and the use of multiplication and exponentiation should not be surprising; but it is interesting that the integers with subtraction are really very convenient for this example. n-bit wide busses are represented by lists of Booleans of length n; this data type is defined using order sorted algebra so that inductive proofs have the 1-bit case for their base, and (the equivalent of) cons for their induction step. The theorem to be proved is that

$$(\forall w_1 w_2) \ |w_1| = |w_2| \Rightarrow \#sout^*(w_1, w_2) + 2^{|w_1|} * \#cout(w_1, w_2) = \#w_1 + \#w_2$$

where w_1 and w_2 range over lists of Booleans, where $|w|$ and $\#w$ respectively denote the length, and the number denoted by, a Boolean list w, where $sout^*(w_1, w_2)$ represents the content of the output bus, and where $cout(w_1, w_2)$ is the carry bit. What this formula says in words is that, given two input buses of the same width, the number on the output bus plus the carry (as highest bit) equals the sum of the numbers on the input buses. The proofs of the lemmas are by straightforward case analysis[14] or induction, and are omitted here. The proof of the main result is by induction on the width of the input busses, starting from width 1.

```
obj INT is sorts Int Nat .
  subsort Nat < Int .
  ops 0 1 2 : -> Nat .
  op s_ : Nat -> Nat [prec 1] .
  ops (s_)(p_) : Int -> Int [prec 1] .
  op (_+_) : Nat Nat -> Nat [assoc comm prec 3] .
  op (_+_) : Int Int -> Int [assoc comm prec 3] .
  op (_*_) : Nat Nat -> Nat [assoc comm prec 2] .
  op (_*_) : Int Int -> Int [assoc comm prec 2] .
  op (_-_) : Int Int -> Int [prec 4] .
  op -_ : Int -> Int [prec 1] .
  vars I J K : Int .
  eq 1 = s 0 .
  eq 2 = s 1 .
  eq s p I = I .
  eq p s I = I .
  eq I + 0 = I .
  eq I + s J = s(I + J) .
  eq I + p J = p(I + J) .
  eq I * 0 = 0 .
  eq I * s J = I * J + I .
  eq I * p J = I * J - I .
  eq I * (J + K) = I * J + I * K .
  eq - 0 = 0 .
  eq - - I = I .
  eq - s I = p - I .
  eq - p I = s - I .
  eq I - J = I + - J .
  eq I + - I = 0 .
  eq -(I + J) = - I - J .
  eq I * - J = -(I * J) .
  op 2**_ : Nat -> Nat [prec 1] .
  var N : Nat .
  eq 2** 0 = 1 .
  eq 2** s N = 2** N * 2 .
endo

obj BUS is sort Bus .
  extending PROPC + INT .
```

---

[14]The validity of case analysis as a proof technique follows from the validity of induction, since a case analysis proof is just an inductive proof that doesn't actually use the inductive hypothesis.

```
  subsort Prop < Bus .
  op __ : Prop Bus -> Bus .
  op |_| : Bus -> Nat .
  var B : Prop .
  var W : Bus .
  eq | B | = 1 .
  eq | B W | = s | W | .
  op #_ : Bus -> Int [prec 1] .   *** is really -> Nat
  eq # false = 0 .
  eq # true = 1 .
  eq #(B W) = 2** | W | * # B + # W .
endo


***> full adder
obj FADD is extending PROPC .
  ops cout sout : Prop Prop Prop -> Prop .
  vars I1 I2 C : Prop .
  eq sout(I1,I2,C) =  I1 xor I2 xor C .
  eq cout(I1,I2,C) = I1 and I2 xor I1 and C xor I2 and C .
endo


***> n-bit ripple carry adder
obj NADD is protecting FADD + BUS .
  ops cout sout : Bus Bus -> Prop .
  op sout* : Bus Bus -> Bus .
  vars B1 B2 : Prop .
  vars W1 W2 : Bus .
  eq cout(B1,B2) = cout(B1,B2,false) .
  eq sout(B1,B2) = sout(B1,B2,false) .
  eq cout(B1 W1,B2 W2) = cout(B1,B2,cout(W1,W2)) .
  eq sout(B1 W1,B2 W2) = sout(B1,B2,cout(W1,W2)) .
  eq sout*(B1,B2) = sout(B1,B2) .
  eq sout*(B1 W1,B2 W2) = sout(B1 W1,B2 W2) sout*(W1,W2) .
endo


obj LEMMAS is protecting NADD .
  vars B1 B2 : Prop .
  eq #(B1 and B2) = # B1 * # B2 .
  eq #(B1 xor B2) = # B1 + # B2 - #(B1 and B2)* 2 .
      *** would write up if # : Bus -> Nat .
  vars W1 W2 : Bus .
  ceq | sout*(W1,W2)| = | W1 | if | W1 | == | W2 | .
endo


***> base case
obj VARS is extending LEMMAS .
  ops b1 b2 : -> Prop .
endo
reduce # sout*(b1,b2)+ # cout(b1,b2)* 2 == # b1 + # b2 .
```

```
***> induction step
obj INDHYP is extending LEMMAS .
  ops b1 b2 : -> Prop .
  ops w1 w2 : -> Bus .
  op n : -> Nat .
  eq | w1 | = n .
  eq | w2 | = n .
  eq # sout*(w1,w2) + 2** n * # cout(w1,w2) = # w1 + # w2 .
endo

reduce #(b1 w1)+ #(b2 w2) ==
    # sout*(b1 w1,b2 w2) + 2** | b1 w1 | * # cout(b1 w1,b2 w2) .
```

In reducing the above expression to true, OBJ3 does 174 rewrites, many of which are associative-commutative (there are also many more rewrites that fail). This takes about thirteen minutes on a Sun 3/75 with 8 Mbytes of memory, but could be much faster with more sophisticated associative-commutative rewriting. One would certainly prefer to have this calculation done mechanically, rather than do it oneself by hand. I suspect that this proof is about two orders of magnitude easier in OBJ than in a fully manual proof system.

B.4   Discussion

It is interesting to compare our method for representing circuits with the popular method which represents components using higher order relations, and represents connections using existential quantification as in the usual definition of the composition of relations [27,28,7]; see also [43]. By contrast, the representation suggested here uses no relations other than equality, and represents interconnection by equality of wires. For sequential circuits, both methods represent wires as variables that range over functions, and both methods use second order quantifiers. However, the results of this paper surprisingly show that existential quantification and higher order relations can be avoided in favor of our simple extension of ordinary equational logic by universal quantification over functions.

It has been argued [7] that the higher order logic approach has many benefits, including: (1) natural definitions of data types (using Peano style induction principles); (2) the possibility of leaving certain values undefined (such as the initial output of a delay); and (3) dealing with bidirectional devices. However, these benefits can be realized more simply using only extended equational logic: (1) It is well known that initial algebra semantics supports elegant data type definitions, and this paper justifies some structural induction principles for such definitions. (2) It is also easy to leave values undefined; for example, if we omit the first equation of the NOT object, then the reduced form of the initial value of an inverter with output g is just g(0); conditional equations can also be used. (3) Although it is advantageous to exploit causality (in the form of an input/output distinction) whenever possible, we are not limited to that case, because equality is bidirectional. Moreover, because equational logic is simpler than higher order logic, in general, its proofs are also simpler. Of course, our method also has its limits; in particular, it cannot verify existentially quantified sentences. Fortunately, the problems of greatest interest in hardware verification do not require such features.

C   Inductive Proofs

The proof scores in this appendix use the inductive proof techniques developed in Section 8. The first two examples import the following code for the natural numbers with addition:

```
obj NAT is sort Nat .
  op 0 : -> Nat .
  op s_ : Nat -> Nat [prec 1] .
  op _+_ : Nat Nat -> Nat [prec 3] .
  vars M N : Nat .
  eq M + 0 = M .
  eq M + s N = s(M + N) .
endo
```

C.1   Associativity of Addition

This subsection proves that addition of natural numbers is associative.

```
obj VARS is extending NAT .
  ops l m n : -> Nat .
endo
***> base case, n=0: 1+(m+0)=(1+m)+0
reduce 1 + (m + 0) == (1 + m) + 0 .
***> induction step
obj STEP is using VARS .
  eq 1 + (m + n) = (1 + m) + n .
endo
reduce 1 + (m + s n) == (1 + m) + s n .
*** thus we can assert
obj ASSOC is protecting NAT .
  vars L M N : Nat .
  eq L + (M + N) = (L + M) + N .
endo
```

C.2   Commutativity of Addition

This subsection proves that addition of natural numbers is commutative.

```
obj VARS is extending NAT .
  ops m n : -> Nat .
endo

***> first lemma0: 0 + n = n, by induction on n
***> base for lemma0, n=0
reduce 0 + 0 == 0 .
***> induction step
obj STEP0 is using VARS .
```

```
   eq 0 + n = n .
endo
reduce 0 + s n == s n .
*** thus we can assert
obj LEMMAO is protecting NAT .
  vars N : Nat .
  eq 0 + N = N .
endo
```

```
***> show lemma1: s m + n = s(m + n), again by induction on n
***> base for lemma1, n=0
reduce in VARS : s m + 0 == s(m + 0) .
***> induction step
obj STEP1 is using VARS .
  eq s m + n = s(m + n) .
endo
reduce s m + s n == s(m + s n) .
*** thus we can assert
obj LEMMA1 is protecting NAT .
  vars M N : Nat .
  eq s M + N = s(M + N).
endo
```

```
***> show m + n = n + m, again by induction on n
***> base case, n=0
reduce in VARS + LEMMAO : m + 0 == 0 + m .
***> induction step
obj STEP is using VARS + LEMMA1 .
  eq m + n = n + m .
endo
reduce m + s n == s n + m .
```

Of course, we must not assert commutativity as a rewrite rule, or we would get non-terminating behavior. The above two proofs show that we are entitled to use associative-commutative rewriting for +, and we do so in the next subsection.

It is interesting to contrast the above proofs with corresponding proofs due to Paulson in Cambridge LCF [39]. The LCF proofs are much more complex, in part because LCF allows partial functions, and then must prove them total, whereas functions are automatically total (on their domain) in equational logic.

C.3   Formula for $1 + \ldots + n$

We now give a standard inductive proof over the natural numbers, the formula for the sum of the first n positive natural numbers,

$$1 + 2 + \ldots + n = n(n+1)/2.$$

Here we take advantage of the two results proven above by giving + the attributes assoc and comm; the score, as given, will not work if either (or both) of these attributes are deleted. This application of associative/commutative rewriting saves the user from having to worry about the ordering and grouping of subterms within + expressions.

```
obj NAT is sort Nat .
  op 0 : -> Nat .
  op s_ : Nat -> Nat [prec 1] .
  op _+_ : Nat Nat -> Nat [assoc comm prec 3] .
  vars M N : Nat .
  eq M + 0 = M .
  eq M + s N = s(M + N).
  op _*_ : Nat Nat -> Nat [prec 2] .
  eq M * 0 = 0 .
  eq M * s N = M * N + M .
endo
obj VARS is protecting NAT .
  ops m n : -> Nat .
endo

***> first show two lemmas, 0*n=0 and sm*n=m*n+n
***> base for first lemma
reduce 0 * 0 == 0 .
***> induction step for first lemma
obj HYP1 is using VARS .
  eq 0 * n = 0 .
endo
reduce 0 * s n == 0 .
*** thus we can assert
obj LEMMA1 is protecting NAT .
  vars N : Nat .
  eq 0 * N = 0 .
endo

***> base for second lemma
reduce in VARS + LEMMA1 : s n * 0 == n * 0 + 0 .
***> induction step for second lemma
obj HYP2 is using VARS .
  eq s m * n = m * n + n .
endo
reduce s m * s n == (m * s n)+ s n .
*** so we can assert
obj LEMMA2 is protecting NAT .
  vars M N : Nat .
  eq s M * N = M * N + N .
endo

obj SUM is protecting NAT .
  op sum : Nat -> Nat .
  var N : Nat .
  eq sum(0) = 0 .
  eq sum(s N) = s N + sum(N) .
endo
***> show sum(n)+sum(n)=n*sn
```

```
***> base case
reduce in SUM + LEMMA1 : sum(0) + sum(0) == 0 * s 0 .
***> induction step
obj HYP is using SUM + VARS .
  eq sum(n) + sum(n) = n * s n .
endo
reduce in HYP + LEMMA2 : sum(s n) + sum(s n) == s n * s s n .
```

C.4 Fermat's Little Theorem for $p = 3$

The so-called "little Fermat theorem" says that
$$x^p \equiv x \pmod{p}$$
for any prime p, i.e., that the remainder of x^p by p equals the remainder of x by p.
The following OBJ3 proof score for the case $p = 3$ assumes we have already shown
that multiplication is associative and commutative. This is a nice example of an
inductive proof where there are non-trivial relations among the constructors[15].

```
obj NAT is sort Nat .
  op 0 : -> Nat .
  op s_ : Nat -> Nat [prec 1] .
  op _+_ : Nat Nat -> Nat [assoc comm prec 3] .
  vars L M N : Nat .
  eq M + 0 = M .
  eq M + s N = s(M + N) .
  op _*_ : Nat Nat -> Nat [assoc comm prec 2] .
  eq M * 0 = 0 .
  eq M * s N = M * N + M .
  eq L * (M + N) = L * M + L * N .
  eq M + M + M = 0 .
endo
***> base case, x = 0
reduce 0 * 0 * 0 == 0 .
***> induction step
obj VARS is extending NAT .
  op x : -> Nat .
  eq x * x * x = x .
endo
reduce s x * s x * s x == s x .
```

D Parameterized Module Verification

This section verifies an OBJ-style parameterized module MAP which defines a typical (higher-order) functional programming construct, namely map, which applies a
function f to all elements of a list L. This involves a parameter presentation FN
with one sort Elt and one operation f: Elt -> Elt, FN is contained in another

---

[15]I thank Dr. Immanuel Kounalis for doubting that OBJ3 could handle non-trivial relations on
constructors, and then presenting the challenge to prove this result.

presentation MAP having (say) signature Σ; let g denote the inclusion. Then what we want to show is that
$$g \models_\Sigma (\forall N, L)(N \leq |L|) \Rightarrow nth(N, map(L)) = f(nth(N, L))$$
where L has sort List and N has sort Nat, where $nth(N, L)$ is the N-th element of list L, and where to simplify notation, we write $N \leq |L|$ instead of
$$N \leq length(L) = true \ .$$

The justification of the proof score for this example is something of a technical *tour de force*, involving many of the proof techniques given in this paper. The first step is to apply the relativized structural induction technique of Theorem 36, using the variable L. Thus, we seek to prove
$$g \models_\Sigma (\forall L)Q(L),$$
where
$$Q(L) = (\forall N)(N \leq |L| \Rightarrow q(N, L))$$
where $q(N, L)$ is the equation
$$nth(N, map(L)) = f(nth(N, L))) \ .$$

The constructors for sort List are *nil* and *cons*, so the base case for the induction is
$$g \models_\Sigma Q(nil)$$
i.e.,
$$g \models_\Sigma (\forall N)(N \leq |nil|) \Rightarrow q(N, nil)$$
which is vacuously true because $|nil| = 0$ while N must be positive.

Because it may seem preferable to start an induction with a non-vacuous case, the proof score given below actually checks the case of lists of length one, i.e., it proves
$$g \models_\Sigma (\forall E) \ Q(cons(E, nil))$$
where E has sort Elt, i.e.,
$$g \models_\Sigma (\forall E, N)(N \leq 1 \Rightarrow q(N, cons(E, nil)))$$
i.e.,
$$g \models_\Sigma (\forall E) \ nth(1, map(cons(E, nil))) = f(nth(1, cons(E, nil)))$$
since necessarily $N = 1$. For this, it is sufficient by T8 to check that
$$\mathcal{E} \models_{\Sigma(e)} nth(1, map(cons(e, nil))) = f(nth(1, cons(e, nil)))$$
where $\mathcal{E}$ is the set of equations of P. This can be done by reduction.

The induction step is more interesting. We must show that
$$g \models_\Sigma (\forall E, L) \ Q(L) \Rightarrow Q(cons(E, L))$$
for which by T7, it is sufficient to show that
$$\mathcal{E} \cup \{(\forall N)N \leq |l| \Rightarrow q(N, l)\} \models_{\Sigma(e,l)} (\forall N)(N \leq |cons(e, l)| \Rightarrow q(N, cons(e, l))).$$
At this point, the inadequacy of our unsorted notation becomes apparent. Although E and L have been made "loose," N remains "tight," i.e., it retains initial algebra semantics, so that we can still do induction over it. Actually, a case analysis for N of sort Pos is sufficient here: N must be either 1 or else $s \, P$ for P of sort Pos. For the first case, by T2 we can show that
$$\mathcal{E} \cup \{(\forall N)N \leq |l| \Rightarrow q(N, l)\} \cup \{1 \leq |cons(e, l)|\} \models_{\Sigma(e,l)} q(1, cons(e, l))$$
i.e., that
$$\mathcal{E} \cup \{(\forall N)N \leq |l| \Rightarrow q(N, l)\} \models_{\Sigma(e,l)} nth(1, map(cons(e, l))) = f(nth(1, cons(e, l)))$$
since $|cons(e, l)| \geq 1$ anyway.

For the second case, using T3 and T2, we can show that
$$\mathcal{E} \cup \{(\forall N)(N \leq |l| \Rightarrow q(N, l))\} \cup \{s \, n \leq |cons(e, l)|\} \models_{\Sigma(e,l,n)} q(s \, n, cons(e, l))$$
i.e., since $|cons(e, l)| = s \, |l|$, that
$$\mathcal{E} \cup \{(\forall N)(N \leq |l| \Rightarrow q(N, l)), (n \leq |l|)\} \models_{\Sigma(e,l,n)}$$
$$nth(s \, n, map(cons(e, l))) = f(nth(s \, n, cons(e, l)))$$

which can be checked by reduction. This is where we might be glad to let OBJ do the work.

An interesting feature of this example is its treatment of exceptions. Actually, this one example illustrates two different approaches to exceptions. The first, in the parameterized object LIST, defines the functions car and cdr only on a subsort of lists, namely the non-empty lists. Then, for example, car(nil) gives a run-time type error, which is handled with retracts (for which, see [12,22]). On the other hand, if we try to take the n-th element of a list with length less than n, we get the exception message toolong, by the third equation of the object NTH. This too will be encapsulated in a retract. More ambitiously, we could have declared an "error supersort" ErrElt of Elt, given nth value sort ErrElt, and used a *sort constraint* [22] to declare that nth(N,L) actually has sort Elt whenever $N \leq$ length(L); unfortunately sort constraints are not yet implemented in OBJ3.

It is also interesting to notice the parameterized module instantiations in this example, which are LIST[E], NTH[F], and MAP[SETUPFN]; in each case, a default view is used [17]. This example uses a version of the natural numbers that is somewhat more sophisticated than the one used before, since it has a subsort Pos of Nat for the non-zero natural numbers.

```
obj NAT is sorts Nat Pos .
  subsort Pos < Nat .
  op 0 : -> Nat .
  op 1 : -> Pos .
  op s_ : Nat -> Pos .
  eq 1 = s 0 .
  op _<=_ : Nat Nat -> Bool .
  vars M N : Nat .
  eq M <= M = true .
  eq s M <= 0 = false .
  eq 0 <= M = true .
  eq s M <= s N = M <= N .
endo

obj LIST[E :: TRIV] is sorts List NeList .
  subsort NeList < List .
  op nil : -> List .
  op cons : Elt List -> NeList .
  op car_ : NeList -> Elt .
  op cdr_: NeList -> List .
  var E : Elt .
  var L : List .
  eq car cons(E,L) = E .
  eq cdr cons(E,L) = L .
endo

obj NTH[E :: TRIV] is protecting LIST[E] + NAT .
  sort ErrElt .
  subsort Elt < ErrElt .
  op length_ : List -> Nat .
  op nth : Pos List -> Elt .
```

```
  op toolong : -> ErrElt .
  var E : Elt .
  var L : List .
  eq length nil = 0 .
  eq length cons(E,L) = s length L .
  var P : Pos .
  eq nth(P,nil) = toolong .
  eq nth(s 0,cons(E,L)) = E .
  eq nth(s P,cons(E,L)) = nth(P,L) .
endo

th FN is sort Elt .
  op f : Elt -> Elt .
endth
obj MAP[F :: FN] is protecting NTH[F] .
  op map_ : List -> List .
  var E : Elt .
  var L : List .
  eq map nil = nil .
  eq map cons(E,L) = cons(f(E),map L) .
endo

obj SETUPFN is sort Elt .
  op f : Elt -> Elt .
endo
obj SETUP is extending MAP[SETUPFN] .
  op e : -> Elt .
  op l : -> List .
  op p : -> Pos .
  var P : Pos .
  ceq nth(P,map l) = f(nth(P,l)) if P <= length l .
  eq p <= length l = true .
endo

***> base case
reduce nth(1,map cons(e,nil)) == f(nth(1,cons(e,nil))) .

***> induction step, by case analysis on P: Pos
reduce nth(1,map cons(e,l)) == f(nth(1,cons(e,l))) .
reduce nth(s p,map cons(e,l)) == f(nth(s p,cons(e,l))) .

***> thus we can assert:
obj MAP-FACT[F :: FN] is protecting MAP[F] .
  var L : List .
  var P : Pos .
  ceq nth(P, map L) = f(nth(P,L)) if P <= length L .
endo
```

Acknowledgements

I wish to thank Dr. José Meseguer for his comments on a draft of this paper, which prevented several bugs from getting into print, and Ms. Victoria Stavridou for providing the impetus that got OBJ started on hardware verification (see [43]) and for her many helpful comments. Mr. Timothy Winkler deserves great credit for his preseverence in enhancing the OBJ3 implementation, particularly its module expressions and associative/commutative rewriting.

References

[1] Jan Bergstra and John Tucker. Characterization of computable data types by means of a finite equational specification method. In *Automata, Languages and Programming, Seventh Colloquium*, pages 76–90, Springer-Verlag, 1980. Lecture Notes in Computer Science, Volume 81.

[2] Garrett Birkhoff. On the structure of abstract algebras. *Proceedings of the Cambridge Philosophical Society*, 31:433–454, 1935.

[3] Robert Boyer and J Moore. *A Computational Logic*. Academic Press, 1980.

[4] Rod Burstall. Proving properties of programs by structural induction. *Computer Journal*, 12(1):41–48, 1969.

[5] Rod Burstall and Joseph Goguen. Algebras, theories and freeness: an introduction for computer scientists. In Manfred Wirsing and Gunther Schmidt, editors, *Theoretical Foundations of Programming Methodology*, pages 329–350, Reidel, 1982. Proceedings, 1981 Marktoberdorf NATO Summer School, NATO Advanced Study Institute Series, Volume C91.

[6] Rod Burstall and Joseph Goguen. The semantics of Clear, a specification language. In Dines Bjorner, editor, *Proceedings of the 1979 Copenhagen Winter School on Abstract Software Specification*, pages 292–332, Springer-Verlag, 1980. Lecture Notes in Computer Science, Volume 86.

[7] Albert Camilleri, Michael Gordon, and Tom Melham. *Hardware Verification Using Higher-Order Logic*. Technical Report 91, University of Cambridge, Computer Laboratory, June 1986.

[8] Carlo Cavenathi, Marco De Zanet, and Giancarlo Mauri. MC-OBJ: a C interpreter for OBJ. *Note di Software*, 36/37:16–26, October 1988. In Italian.

[9] Avra Cohn. Correctness properties of the Viper block model: the second level. In P. A. Subramanyan, editor, *Proceedings, Second Banff Workshop on Hardware Verification*, Springer-Verlag, to appear 1988.

[10] Derek Coleman, Robin Gallimore, and Victoria Stavridou. The design of a rewrite rule interpreter from algebraic specifications. *IEE Software Engineering Journal*, July:95–104, 1987.

[11] R. A. DeMillo, Richard Lipton, and Alan Perlis. Social processes and proofs of theorems and programs. In *Proceedings, Fourth Symposium on Principles of Programming Languages*, pages 206–214, Association for Computing Machinery, 1977.

[12] Kokichi Futatsugi, Joseph Goguen, Jean-Pierre Jouannaud, and José Meseguer. Principles of OBJ2. In Brian Reid, editor, *Proceedings of 12th ACM Symposium on Principles of Programming Languages*, pages 52–66, Association for Computing Machinery, 1985.

[13] Kokichi Futatsugi, Joseph Goguen, José Meseguer, and Koji Okada. Parameterized programming in OBJ2. In Robert Balzer, editor, *Proceedings, Ninth International Conference on Software Engineering*, pages 51–60, IEEE Computer Society Press, March 1987.

[14] Stephen Garland and John Guttag. Inductive methods for reasoning about abstract data types. In *Proceedings, Fifteenth Symposium on Principles of Programming Languages*, pages 219–229, ACM, January 1988.

[15] Joseph Goguen. How to prove algebraic inductive hypotheses without induction: with applications to the correctness of data type representations. In Wolfgang Bibel and Robert Kowalski, editors, *Proceedings, Fifth Conference on Automated Deduction*, pages 356–373, Springer-Verlag, 1980. Lecture Notes in Computer Science, Volume 87.

[16] Joseph Goguen. *Order Sorted Algebra*. Technical Report 14, UCLA Computer Science Department, 1978. Semantics and Theory of Computation Series.

[17] Joseph Goguen. Parameterized programming. *Transactions on Software Engineering*, SE-10(5):528–543, September 1984.

[18] Joseph Goguen. Some design principles and theory for OBJ-0, a language for expressing and executing algebraic specifications of programs. In Edward Blum, Manfred Paul, and Satsoru Takasu, editors, *Proceedings, Mathematical Studies of Information Processing*, pages 425–473, Springer-Verlag, 1979. Lecture Notes in Computer Science, Volume 75; Proceedings of a Workshop held August 1978.

[19] Joseph Goguen. What is unification? — a categorical view of substitution, equation and solution. In Maurice Nivat and Hassan Aït-Kaci, editors, *Resolution of Equations in Algebraic Structures*, Academic Press, to appear, 1988.

[20] Joseph Goguen and José Meseguer. Completeness of many-sorted equational logic. *Houston Journal of Mathematics*, 11(3):307–334, 1985. Preliminary versions have appeared in: *SIGPLAN Notices*, July 1981, Volume 16, Number 7, pages 24-37; SRI Computer Science Lab Technical Report CSL-135, May 1982; and Report CSLI-84-15, Center for the Study of Language and Information, Stanford University, September 1984.

[21] Joseph Goguen and José Meseguer. Eqlog: equality, types, and generic modules for logic programming. In Douglas DeGroot and Gary Lindstrom, editors, *Logic Programming: Functions, Relations and Equations*, pages 295–363, Prentice-Hall, 1986. An earlier version appears in *Journal of Logic Programming*, Volume 1, Number 2, pages 179-210, September 1984.

[22] Joseph Goguen and José Meseguer. *Order-Sorted Algebra I: Partial and Overloaded Operations, Errors and Inheritance*. Technical Report to appear, SRI International, Computer Science Lab, 1988. Given as lecture at Seminar on Types, Carnegie-Mellon University, June 1983.

[23] Joseph Goguen and José Meseguer. Order-sorted algebra solves the constructor selector, multiple representation and coercion problems. In *Proceedings, Second Symposium on Logic in Computer Science*, pages 18–29, IEEE Computer Society Press, 1987. Also Technical Report CSLI-87-92, Center for the Study of Language and Information, Stanford University, March 1987.

[24] Joseph Goguen and Joseph Tardo. An introduction to OBJ: a language for writing and testing software specifications. In Marvin Zelkowitz, editor, *Specification of Reliable Software*, pages 170–189, IEEE Press, 1979. Reprinted in *Software Specification Techniques*, Nehan Gehani and Andrew McGettrick, Eds., Addison-Wesley, 1985, pages 391-420.

[25] Joseph Goguen, James Thatcher, and Eric Wagner. *An Initial Algebra Approach to the Specification, Correctness and Implementation of Abstract Data Types*. Technical Report RC 6487, IBM T. J. Watson Research Center, October 1976. Appears in *Current Trends in Programming Methodology, IV*, Raymond Yeh, Ed., Prentice-Hall, 1978, pages 80-149.

[26] Joseph Goguen and Timothy Winkler. *Introducing OBJ3*. Technical Report, SRI International, Computer Science Lab, to appear 1988.

[27] Michael Gordon. *HOL: A Machine Oriented Formulation of Higher-Order Logic*. Technical Report 85, University of Cambridge, Computer Laboratory, July 1985.

[28] Michael Gordon. Why higher-order logic is a good formalism for specifying and verifying hardware. In George Milne and P. A. Subrahmanyam, editors, *Formal Aspects of VLSI Design*, North-Holland, 1986.

[29] Michael Gordon, Robin Milner, and Christopher Wadsworth. *Edinburgh LCF*. Springer-Verlag, 1979. Lecture Notes in Computer Science, Volume 78.

[30] Robert Harper, Furio Honsell, and Gordon Plotkin. A framework for defining logics. In *Proceedings, Second Symposium on Logic in Computer Science*, pages 194–204, IEEE Computer Society Press, 1987.

[31] Jieh Hsiang. *Refutational Theorem Proving using Term Rewriting Systems*. PhD thesis, Univeristy of Illinois at Champaign-Urbana, 1981.

[32] Gérard Huet and Derek Oppen. Equations and rewrite rules: a survey. In Ron Book, editor, *Formal Language Theory: Perspectives and Open Problems*, Academic Press, 1980.

[33] Robert Constable *et al. Implementing Mathematics with the NuPRL Proof Development System*. Prentice-Hall, 1986.

[34] Donald Knuth and Peter Bendix. Simple word problems in universal algebra. In J. Leech, editor, *Computational Problems in Abstract Algebra*, Pergamon Press, 1970.

[35] David MacQueen and Donald Sannella. Completeness of proof systems for equational specifications. *IEEE Transactions on Software Engineering*, SE-11(5):454–461, May 1985.

[36] José Meseguer and Joseph Goguen. Initiality, induction and computability. In Maurice Nivat and John Reynolds, editors, *Algebraic Methods in Semantics*, pages 459–541, Cambridge University Press, 1985.

[37] David Musser. On proving inductive properties of abstract data types. In *Proceedings, 7th Symposium on Principles of Programming Languages*, Association for Computing Machinery, 1980.

[38] Paliath Narendran and Jonathan Stillman. Formal verification of the Sobel image processing chip. In P. A. Subramanyan, editor, *Proceedings, Second Banff Workshop on Hardware Verification*, Springer-Verlag, to appear 1988.

[39] Lawrence Paulson. *Logic and Computation: Interactive Proof with Cambridge LCF*. Cambridge University Press, 1987. Cambridge Tracts in Theoretical Computer Science, Volume 2.

[40] Brian Ritchie and Paul Taylor. The interactive proof editor: an experiment in interactive theorem proving. In P. A. Subramanyan, editor, *Proceedings, Second Banff Workshop on Hardware Verification*, Springer-Verlag, to appear 1988.

[41] A.B.C. Sampaio and Kamran Parsaye. The formal specification and testing of expanded hardware building blocks. In *Proceedings, ACM Computer Science Conference*, Association for Computing Machinery, 1981.

[42] S. Sridhar. An implementation of OBJ2: an object-oriented language for abstract program specification. In K. V. Nori, editor, *Proceedings, Sixth Conference on Foundations of Software Technology and Theoretical Computer Science*, pages 81–95, Springer-Verlag, 1986. Lecture Notes in Computer Science, Volume 241.

[43] Victoria Stavridou. *Specifying in OBJ, Verifying in REVE, and Some Ideas about Time*. Technical Report Draft, Department of Computer Science, University of Manchester, 1987.

[44] Mark Stickel. A Prolog technology theorem prover. In *First International Symposium on Logic Programming*, Association for Computing Machinery, February 1984.

[45] Glynn Winskel. Relating two models of hardware. In *Proceedings, Second Summer Conference on Category Theory and Computer Science*, pages 98–113, Laboratory for Computer Science, University of Edinburgh, 1987.

Formal Verification in m-EVES

Bill Pase and Mark Saaltink

Trusted Systems Group
I. P. Sharp Associates Limited
265 Carling Avenue, Suite 600
Ottawa, Ontario K1S 2E1
CANADA[*]

July 1988

Abstract

The Trusted Systems Group at I.P. Sharp Associates Limited has developed a prototype system, called m-EVES, for formally verifying software. Programs are written in a new language, called m-Verdi, and proved with the help of a theorem prover called m-NEVER. The logic for proving programs has been proved sound.

This paper describes the logic and the theorem prover, and presents two transcripts of the application of the system to simple problems.

Keywords: Automated deduction, Formal specification, Formal verification, Hardware verification, Logic of programs, m-EVES, m-NEVER, m-Verdi, Program correctness, Program specification, Program verification systems, Theorem proving.

1 Introduction

m-EVES (an Environment for Verifying and Evaluating Software) is a prototype formal verification system developed by the Trusted Systems Group of I.P. Sharp Associates Limited.[1]

The development of m-EVES has followed three principal streams:

- the design of a new specification and implementation language (with the supporting mathematics), called m-Verdi,

- the implementation of a new theorem prover, called m-NEVER, and

- the development of compilers for the m-Verdi language.

In this paper, we present a brief description of the language, logic, and theorem prover. Two examples are used to illustrate the capabilities of the system. More detailed descriptions of m-EVES and its components can be found in the references.

---

[*]IPSA Conference Paper CP-88-5402-35

Telephone: (613) 236–9942.   Arpanet: bill@ipsa.arpa, mark@ipsa.arpa

[1]The development of m-EVES was funded by the Canadian Department of National Defence and by SPAWAR, of the United States Navy.

2 m-Verdi Language and Logic

m-Verdi is a notation for expressing programs and their specifications. The design goals of m-Verdi included:

- It must admit a sound logic for proving properties of programs.

- The language should be fairly low-level (to support systems programming).

- The design should consolidate existing technology.

Roughly speaking, m-Verdi is the merger of First-Order Predicate Calculus and Pascal, with additions to overcome some shortcomings of that combination:

- A facility for definitions extends Predicate Calculus.

- A mechanism for packaging and information hiding supplements both Predicate Calculus and Pascal.

- An *environment* allows a systems program access to the underlying machine.

Programs are specified and proved in the Floyd-Hoare style, using a pre/post specification of desired behaviour. m-Verdi provides a strong logic; termination must be proved. In addition, proofs of executable procedures must take into account the limitations of arithmetic on finite machines. (It is also possible to ignore these limitations when proving non-executable procedures. As their name implies, these procedures can not be run, nor can they influence a running program in any way. However, proving these procedures, ignoring arithmetic limitations, can be a useful first step in a program proof.) Storage exhaustion is not addressed by the logic.

A denotational semantics defines the meaning of m-Verdi. The m-Verdi logic has been proved sound relative to this semantics.

m-Verdi is completely described in [Cra 87a]. A compiler for m-Verdi has been written and runs on VAX computers under the VMS[2] or UNIX[3] operating systems [Mei 87a] [Mei 87b]. The formal semantics, logic, and soundness proof appear in [Saa 87].

2.1 Expressions

The m-Verdi language and logic are modelled on Predicate Calculus, and can be characterized as Many-Sorted First-Order Predicate Calculus with the addition of imperative features and a definition mechanism. As in Predicate Calculus, an m-Verdi expression can be a constant, a variable, a quantification, or the application of a function to argument expressions. In addition, m-Verdi has equality tests (infix = and <>) and conditional expressions.

Each term has a unique type. Types are simply unanalysed names; two types are the same if the names are identical. Each function has a *signature* that indicates the required types of its arguments and the type of its result. For example, function **not** takes a single Boolean and returns a Boolean; function **plus** takes two integers and returns an integer; function **int'le** takes two integers and returns a Boolean.

m-Verdi does not follow Predicate Calculus in distinguishing terms and formulas. Instead, there are only terms, with terms of type Bool serving as formulas.

The relationship between Predicate Calculus and m-Verdi is illustrated in this table:

---

[2]VAX and VMS are trademarks of Digital Equipment Corporation.
[3]UNIX is a trademark of AT&T Bell Laboratories.

Predicate Calculus	m-Verdi
$f(x, y) = z$	`f(x, y) = z`
$x + 1 < y$	`int'lt (plus(x, 1), y)`
$P(x) \vee Q(y)$	`or (P(x), Q(y))`
$\forall x\, P(x, y)$	`all x: P(x, y)`
—	`if x = y then 1 else 0 end if`
—	`R (all x: P(x, y))`

Many of the concepts of Predicate Calculus are used in m-Verdi, for example, vocabulary (sometimes called language), axioms, theories, and models. The model theory for expressions is quite conventional. In particular, all functions are total.

2.2   Declarations

m-Verdi extends Predicate Calculus with a definitional capability, allowing the addition of new types, variables, constants, functions, and axioms. Such additions define an *extension* of a theory. Obviously, the arbitrary addition of axioms would lead to unsoundness, so there are constraints on the types of extensions that can be made. In m-Verdi, extensions are specified by *declarations*. A declaration is *acceptable* when every model of the original theory can be expanded to a model of the extended theory. (That is, the extension is model-conservative.) This guarantees two important properties:

- Names and terms in the original theory can retain their meanings.

- Meanings can be found for the new names, so that the new axioms are satisfied.

To demonstrate that a declaration is acceptable, a proof may be required.

Every theory must be an extension of the *initial theory*, which describes the predefined names of m-Verdi. The types, constants, functions, and axioms of this theory are therefore available for use in all m-Verdi programs and proofs. The initial theory defines m-Verdi's basic types (`Int`, `Bool`, `Char`, and `Ordinal`) along with basic functions acting on those types.

Since declarations extend theories, different theories are in force at different times. To keep track of the different theories, the m-Verdi logic tags each sentence with a description of the theory in effect. These theory descriptions are just declarations that extend the initial theory to the desired theory. Thus the m-Verdi sentence $D \mid e$ expresses that e is true in the extension $\text{Th}(D)$ of the initial theory by the declarations in D. The normal Predicate Calculus rules of inference are expressed in this system in the obvious way. For example, *modus ponens* is expressed as

$$\frac{D \mid p \qquad D \mid \texttt{implies}(p, q)}{D \mid q}$$

Instead of having "non-logical" rules of inference specific to each theory, there is a rule for axioms:

$$\frac{}{D \mid e} \qquad \text{if } e \in \text{Axioms}(\text{Th}(D))$$

There are also sentences of the form $D \mid d$, expressing that d is an acceptable declaration in theory $\text{Th}(D)$, and there are rules of inference for accepting declarations. For example, an axiom declaration adds a single axiom to a theory. It is acceptable only if the axiom is already valid in the theory. (Thus, an axiom declaration expresses a lemma: once proved and added to the theory, it can be drawn out and used in subsequent proofs.) We therefore have the rule:

$$\frac{D \mid e}{D \mid \texttt{axiom } e}$$

For most elementary declarations d, there is a declaration rule of the form

$$\frac{D' \mid e}{D \mid d},$$

where e is an expression we call the *proof obligation*. (D' is usually the same as D.)
  Successive declarations specify successive extensions:

$$\frac{D \mid d \qquad D; d \mid d'}{D \mid d; d'}$$

Observe how the second declaration must be accepted in a richer theory $(\mathrm{Th}(D; d))$ than the first $(\mathrm{Th}(D))$.

  A variable declaration **var** **x**: T adds an infinite family of variable names (**x**, **x'**0, **x'**1, ...), all of type T, to a vocabulary. This declaration is always acceptable if it is well-formed. (In this case, **x** must not already be in the vocabulary and T must be the name of a type.) The new variables can subsequently appear in quantifications or in parameter lists in other declarations. Each such use binds the variable; *there is no global state in m-Verdi.*

  A constant declaration **const** **c**: T := **e** adds a new constant symbol **c** of type T to a vocabulary, and adds the axiom **c** = **e** which expresses the value of the constant. This declaration is always acceptable if it is well-formed.

  Type declarations add many symbols to a vocabulary. In addition to the type symbol itself, there are usually primitive functions that act on the new type. For example, the record declaration

```
type T = record
   field a: Int
   field b: Bool
end record
```

adds type symbol T, functions **T'a** (of type T → Int) and **T'b** (of type T → Bool) for accessing the fields of record values, and so on; we write **T'a(x)** where in Pascal one would write **x.a**. Similarly, if T is declared to be an array type, function **T'extract** is defined and we write **T'extract(x, i)** where in Pascal one would write **x[i]**. m-Verdi's type constructors are enumerations, records, arrays, and restriction types.

  Function declarations add new function symbols and an axiom about the new function. The body of the function declaration is an expression. As with constant definitions, the added axiom is derived directly from the declaration. However, for recursive function declarations, this extension is not guaranteed to be acceptable. For example, the declarations

```
var x: int
function f (x): int =
  begin
    plus (1, f(x))
  end f
```

would add the axiom `all x: f (x) = plus (1, f(x))`. Clearly, this extension is not acceptable (if the original theory is consistent). Following Boyer and Moore [BM 79], we use *measures* to show that recursions in function definitions are well-founded, and thus that the extension is acceptable. Unlike Boyer and Moore, we require the measure to be given explicitly in the function declaration. A measure is an expression of type ordinal over the parameters of the function. To accept a recursive function declaration, the measure must be shown to decrease in every recursive call. (There are examples later in this paper.) In practice, the measure expression is usually the application of function `ordinal'val` to an integer expression (`ordinal'val` maps integers into the natural numbers, mapping negative numbers to zero).

Package declarations allow for information hiding and thus abstraction. A package declaration consists of two parts: a *header* and a *body*. Each part consists of a sequence of declarations. The extension to the theory is determined by the header. The body is used to show that the extension is acceptable, by defining an interpretation of the extended theory. (Abusing terminology, we sometimes refer to the body as the model of the package.) In the package header, *stub declarations* can be used. These declarations add new symbols but no new axioms. The desired axioms can then be added explicitly by axiom declarations in the header.

To accept a package, two phases of proof are required:

- The body must be accepted.

- Every axiom added by the header must be shown to hold in the theory defined by the model.

Note that the header itself does not need to be accepted. These requirements can be stated more formally as

$$\frac{D \mid d' \quad D; d' \mid a_1 \quad \cdots \quad D; d' \mid a_n}{D \mid \textbf{package } d \textbf{ model } d' \textbf{ end package}}$$

where $\{a_1, \ldots a_n\} = \text{Axioms}(\text{Th}(D; d)) \setminus \text{Axioms}(\text{Th}(D))$.

This package mechanism allows quite strict control over how much information is "exported" from a package, as in this trivial example:

```
package
  const c: Int
  axiom int'ge (c, 3)
model
  const c: Int = 5
end package
```

This package specifies an extension of a theory that adds constant `c`, along with the fact that `c` is at least 3. The model plays no role in this extension, but allows us to accept the declaration by providing a particular `c` that satisfies the axiom.

2.3   Imperative Features

There is an additional class of symbols not yet mentioned: procedures. Vocabularies can contain procedure names; theories can contain procedure axioms (which are pre/post specifications of individual procedures). Models associate meanings with procedures.

Procedure declarations define new procedures. Procedure bodies are composed of *commands*. The m-Verdi commands are assignment, procedure call, abort, conditional commands, loop with exit, return, annotation, and block (possibly with local variable bindings).

The meaning of a command body is defined using a denotational semantics. The procedure declaration rule uses a predicate transformer similar to the weakest precondition function. (The antecedent of the procedure declaration rule therefore looks like a "verification condition" for the procedure body.) Internal annotations of the procedure body are used in generating the precondition. (In particular, loops must be furnished with invariants and measures. The loop measure is used to show termination of the loop.)

Several features of m-Verdi make it rather unconventional as a programming language:

- There are no global variables.

- Functions are total; there is never a run-time failure in the evaluation of an expression.

- Function declarations have expression bodies.

Each of these design choices reflects compromises that must be made in order for m-Verdi to serve the dual role of a programming and specification language.

2.4   Environments

In most programming languages, programs achieve their effects through the use of standard predefined procedures. A similar approach is adopted in m-Verdi: programs use special procedures to have effects. These procedures are not, however, standard or predefined. Instead, they are specified in an *environment* declaration at the start of the m-Verdi program. We refer to these procedures as *interface procedures*. Each interface procedure's specification describes how calling that procedure will cause the machine's state to change.

The implementation of the interface procedures lies beyond the boundaries of what can be accomplished using m-Verdi. The interface procedures must be implemented by means other than by m-Verdi programs. They may, for example, be implemented in assembly language and then linked together with an m-Verdi program.

An environment is similar to a package specification. The same sorts of declarations are allowed in both, and their meanings are similar — a promise to provide values for the declared names that will satisfy the axioms. In contrast to a package, however, the fulfillment of that promise is not expressed in m-Verdi and can not be proved in the m-Verdi logic. The environment is therefore an expression of the axiomatic basis on which the proof of a program rests.

Since environments specify assumptions, the environment of any proved program must be scrutinized with care. If the axioms are inconsistent, the proof of the program establishes nothing, since the program can never be furnished with a model of the environment.

3   The Prover m-Never

The theorem proving system grew out of our desire to have a state-of-the-art prover as part of our verification system. Currently, m-NEVER [PK 87] consists of the following major components: a simplifier, a rewriter, an invoker, a reducer, user commands, and support for input/output and database management. For the most part, the design is not radical, and we have minimized the research needed by adopting ideas which have been published in the field of Automated Deduction. Our design has primarily been influenced by four theorem proving systems with which we have had some experience. These four systems are:

- the Bledsoe-Bruell prover [BB 73],

- the Stanford Pascal Verifier [Luc 79],

- the Boyer-Moore theorem prover [BM 79], and

- the Affirm theorem prover [Tho 81].

Each of these systems has had a distinct influence on our work. From the Stanford system we developed most of our ideas for the deduction techniques we use for handling equalities and integers [Nel 81]. The Affirm and Boyer-Moore systems are the inspiration for our general strategies for the traversal of formulas and the automatic capabilities. Finally, much of our philosophy with respect to the incorporation of both manual and automatic capabilities was derived from the early work of Bledsoe and Bruell.

3.1 Design Philosophy

m-NEVER is neither a fully automatic nor a fully manual theorem prover. Although m-NEVER provides powerful deductive techniques for the automatic proof of theorems, it also includes simple user steps which permit its use as a system more akin to a proof checker than a theorem prover. This idea of combining the manual and automatic capabilities is most closely related to Bledsoe's and Bruell's work. It represents a tacit admission that, for some proofs, it may be essential for the user to resort to hand steps, since the automatic capabilities of m-NEVER may be inadequate.

Combining the manual and automatic functions within a single system creates the possibility of a synergy between abilities of the system (fast and accurate) and the user (the necessary insight). By allowing the user to back up and selectively override specific decisions the prover may make, we hope to enhance this synergy between man and machine.

3.2 User Interaction

User interaction with m-NEVER consists of adding declarations to the database and performing proof steps upon the resulting proof obligations. As declarations are added, events which represent the declarations are entered into the database. The result is a new current theory which is the previous theory extended to include the new declarations. All the proof obligations arising from the new declarations must be proved.

3.2.1 Prover Database

The prover database contains a representation of a current theory, heuristic information associated with the theory, and records of complete and partial proofs. The entire database can be saved and later restored.

The current theory is represented as events in the database. Each axiom or declared name (i.e., an element of the vocabulary) is represented by a sequence of events. The current theory may be extended using declarations and contracted using undoing commands. In addition, events containing heuristic information may be disabled, which prevents the prover from automatically applying them. A disabled event can later be re-enabled.

3.2.2 Proof Steps

To prove a formula, the formula must first be made current. Subsequently, proof steps are used to transform the current formula to a new formula. Some of the proof steps use the prover's automatic capabilities and may apply complex transformations. Other proof steps, considered to be manual steps, apply simpler transformations. Finally, there are macro proof steps, each of which may perform a combination of the above proof steps.

The proof steps are not an implementation of the primitive rules of inference defined by the logic. Instead, they are derived from those inference rules.

3.3 Declaration of Axioms

Axioms can have heuristic information that allows m-NEVER to apply them automatically. There are three types of axioms: facts, rewrite rules, and forward rules.

3.3.1 Facts

Facts are axioms which can be assumed during a proof of some other conjecture with the use command. They differ from rewrite and forward rules in that facts are not used automatically.

Example of a fact:

```
axiom MAXINT_VALUE () =
  begin INT'GT (MAXINT, 1000)
  end MAXINT_VALUE;
```

3.3.2 Rewrite Rules

Rewrite rules allow the user to extend the prover's automatic capabilities. A rewrite rule specifies the replacement of an expression, called the left side, by another expression, called the right side. Furthermore, this replacement may be restricted to occur only under certain conditions.

The application of a rewrite rule to a subexpression of the formula being rewritten proceeds as follows. First, an attempt is made to match the left side of the rule being considered to the subexpression. If the match succeeds and the rule is conditional, then an attempt is made to prove the condition. If there is no condition, or the proof of the condition succeeds, then the subexpression is replaced by the right side of the rule. All variables in the right side are replaced by the expressions determined by the pattern match and the proof of the condition.

Occasionally, it is useful to have rewrite rules in which the right side is a permutation of the left side. Such permutative rules are allowed. When such a rule is applied, the replacement occurs only if the result is alphabetically smaller than the subexpression being rewritten.

Unconditional rewrite rules have formulas of the following form: `LEFTSIDE =`
`RIGHTSIDE`. Two abbreviations are allowed. First, if the right side is `TRUE` then the
formula may be given as `LEFTSIDE`. Second, if the right side is `FALSE` then the formula
may be given as `NOT (LEFTSIDE)`. Neither of these abbreviations is allowed if the
left side is an equality. The expression `X <> Y` is handled the same as `NOT (X =`
`Y)`. Conditional rules have formulas of the following form: `IMPLIES (CONDITION,`
`LEFTSIDE = RIGHTSIDE)`.

Example of a rewrite rule and its components:

```
axiom NTH_IN_INT (IO_INT, SO_INT) =
  rule
  begin IMPLIES (AND (INT'LT (0, IO_INT),
                      INT'LE (IO_INT, LENGTH_INT (SO_INT))),
                 IN_INT (NTH_INT (IO_INT, SO_INT), SO_INT))
  end NTH_IN_INT;
```

```
Condition:
AND (INT'LT (0, IO_INT),
     INT'LE (IO_INT, LENGTH_INT (SO_INT)))
```

```
Left side: IN_INT (NTH_INT (IO_INT, SO_INT), SO_INT)
```

```
Right side: TRUE
```

3.3.3   Forward Rules

Forward rules also allow the user to extend the prover's automatic capabilities. A
forward rule specifies a formula to be assumed whenever the rule is triggered. Furthermore, this assumption may be restricted to occur only under certain conditions.

The application of a forward rule to a subexpression of the formula being rewritten
proceeds as follows. First, an attempt is made to match a trigger of the rule being
considered to the subexpression. If a match succeeds and the rule is conditional,
then an attempt is made to prove the condition. If there is no condition, or the proof
of the condition succeeds, then the assumption is performed. All variables in the
formula are replaced by the expressions determined by the pattern match and the
proof of the condition.

A forward rule is an axiom of any form. However, if it has the form of an
implication, it is a conditional forward rule.

Example of a forward rule and its components:

```
axiom TACK_HEAD_TAIL_INT (SO_INT) =
  frule
  triggers (HEAD_INT (SO_INT), TAIL_INT (SO_INT))
  begin IMPLIES (SO_INT <> EMPTY_INT,
                 TACK_INT (HEAD_INT (SO_INT), TAIL_INT (SO_INT))
                 = SO_INT)
  end TACK_HEAD_TAIL_INT;
```

```
Triggers: HEAD_INT (SO_INT), TAIL_INT (SO_INT)
```

```
Condition: SO_INT <> EMPTY_INT
```

Assumption:
TACK_INT (HEAD_INT (SO_INT), TAIL_INT (SO_INT)) = SO_INT

3.4 Declaration of Functions

Functions permit the extension of the logical system used by the prover by introducing new symbols which may then be used within an expression. Functions may be introduced as stubs, or may be provided with complete definitions which may be recursive.

In the simpler case, the function declaration provides the argument and return types for the function, without any definition. The new function may be left uninterpreted or alternatively axioms may be given which provide a complete or partial description of the function.

In the more complex case, the function declaration not only provides the argument and return types but also includes an expression, called the body, which is the definition of the function. Optionally, a precondition can also be supplied as a part of the declaration. When the prover encounters an occurrence of a function with a definition in an expression and its precondition can be proved, the function will be expanded into its defining expression.

Function definitions may be recursive, though mutual recursion is not allowed. For recursive functions, an ordinal measure is obligatory. This measure is used by the system for the generation of the proof obligation for the recursive function. Should the prover encounter a recursively defined function during the traversal of an expression, it will make a heuristic decision as to whether or not to invoke the function. Alternatively, the user may direct the prover to perform a specific invocation either to perform the proof by hand or to perform some invocation which the prover heuristics have rejected.

Example of a function and its components:

```
function INDEX_INT (EO_INT, SO_INT): INT =
  pre IN_INT (EO_INT, SO_INT)
  measure ORDINAL'VAL (LENGTH_INT (SO_INT))
  begin if EO_INT = HEAD_INT (SO_INT)
          then 1
          else PLUS (1,
                     INDEX_INT (EO_INT,
                                TAIL_INT (SO_INT))) end if
  end INDEX_INT;

Name: INDEX_INT
Parameters: EO_INT, SO_INT
Return Type: INT
Precondition: IN_INT (EO_INT, SO_INT)
Ordinal Measure: ORDINAL'VAL (LENGTH_INT (SO_INT))
Body:
if EO_INT = HEAD_INT (SO_INT)
  then 1
  else PLUS (1, INDEX_INT (EO_INT, TAIL_INT (SO_INT))) end if
```

3.5 Application of Axioms

3.5.1 Simplification

The process of simplification attempts to reduce an expression to one which the system considers to be simpler. In general, the resulting simplified expression is smaller than the original. Propositional tautologies are always detected. In addition, the simplification process reasons about equalities, integers, and quantifiers. Simplification is performed by the `simplify` commands.

3.5.2 Rewriting

The rewriting process applies both rewrite and forward rules. While traversing a formula being rewritten, an attempt is made to match patterns of rewrite rules and triggers of forward rules against subexpressions of the formula. If a pattern or trigger matches a subexpression, an attempt is made to apply the associated rewrite or forward rule.

Rewriting is performed by the `rewrite` commands.

3.5.3 Pattern Matching

Pattern matching is the process whereby the prover selects rules for application. The selected rules may be either rewrite rules or forward rules. For a rewrite rule, the pattern is the left side of the rule. For a forward rule, a list of patterns may be supplied; these patterns are known as the trigger expressions for the forward rule.

Pattern matching compares the pattern to the current subexpression being traversed by the prover. If there exists a set of substitutions for the variables occurring within the pattern which make the pattern equal to the subexpression, then the pattern is said to have matched the subexpression with those substitutions.

3.6 Application of Functions

3.6.1 Invocation

Automatic invocation is the process of expanding function definitions. While traversing a formula being reduced, an attempt is made to expand calls to functions with definitions, recursively, using the functions' definitions.

Function definitions may also be explicitly invoked by the user using the `invoke` command.

3.6.2 Reduction

Reduction is the main strategy used by the automatic theorem prover. It consists of a single traversal of the expression to be proved, which results in an equivalent new expression which is considered to be simpler than the original.

Prior to the traversal, the expression to be reduced is first converted into the prover's internal representation. The most significant conversion is the replacement of all Boolean connectives by their equivalent "if" expressions.

After the expression has been converted to the internal representation, it is traversed in an innermost-leftmost order. At each subexpression, any of the following may occur: the subexpression may be replaced by a simpler but equivalent expression; it may trigger any number of forward rules; it may be rewritten by an applicable rewrite rule; or, if it is an application of a function with a definition, the function may be invoked. If any of these occur, the resulting expression is recursively reduced until no further reductions take place.

Following the traversal the resulting expression is converted from the internal form to an output form that is more familiar to the user.

The following commands are based on reduction: `simplify`, `rewrite`, and `reduce`.

3.7 Induction

The induction technique performed by the theorem prover is that of Boyer-Moore. Normally, induction schemes are heuristically chosen based on calls to recursive functions within the current formula. However, the user may direct the prover using the `induct` command with an explicit term on which to induct. The following is an example of a user-directed induction.

First, introduce a recursive function definition for exponentiation.

```
function EXP (I, K): INT =
  pre INT'GE (K, 0)
  measure ORDINAL'VAL (K)
  begin if K = 0
          then 1
          else TIMES (I, EXP (I, MINUS (K, 1))) end if
  end EXP;
```

The recursive call to `EXP` is governed by the precondition, `INT'GE (K, 0)`, and the "if" test negated, `K <> 0`.

Next, we begin a proof of a conjecture involving `EXP`.

```
try
 IMPLIES (AND (INT'GT (I, 0),
               INT'GE (J, 0)),
          INT'GT (EXP (I, J), 0));
Beginning proof of ...
IMPLIES (AND (INT'GT (I, 0),
              INT'GE (J, 0)),
         INT'GT (EXP (I, J), 0))
```

We now perform an explicit induction step, giving the term on which to induct.

```
induct on EXP (I, J);
Inducting using the following scheme ...
AND (IMPLIES (NOT (AND (J <> 0,
                        INT'GE (J, 0))),
              *P* (I, J)),
     IMPLIES (AND (J <> 0,
                   INT'GE (J, 0),
                   *P* (I, MINUS (J, 1))),
              *P* (I, J)))

produces ...
AND (IMPLIES (NOT (AND (J <> 0,
                        INT'GE (J, 0))),
              IMPLIES (AND (INT'GT (I, 0),
```

```
                    INT'GE (J, 0)),
                 INT'GT (EXP (I, J), 0)))),
   IMPLIES (AND (J <> 0,
              INT'GE (J, 0),
              IMPLIES (AND (INT'GT (I, 0),
                           INT'GE (MINUS (J, 1), 0)),
                     INT'GT (EXP (I, MINUS (J, 1)), 0)))),
           IMPLIES (AND (INT'GT (I, 0),
                        INT'GE (J, 0)),
                  INT'GT (EXP (I, J), 0))))
```

The above induction scheme consists of a base case and an induction case, and follows directly from the definition of EXP. The induction case follows from the recursive call in the definition. Thus the conjecture, with J replaced by MINUS (J, 1), is assumed together with the condition governing the recursion. In the base case, the governing condition is the negation of the condition of the induction case.

Note that for the above example the heuristic would have chosen the same induction scheme, since there is only one call to a recursive function in the conjecture, and that call is exactly the term specified.

Commands that may perform an induction step include induct and prove by induction.

4 The System m-EVES

m-EVES currently runs on a Symbolics Lisp Machine. Interaction with m-EVES may occur through either an editor interface or a command processor, using a language called ECL (an EVES Command Language). This language is essentially an extension of m-Verdi to include m-EVES commands (which, for the most part, are commands to invoke portions of the theorem prover).

The prover commands used in our examples are described in the Appendices. In the examples that follow, the text presented in typewriter font arises from interaction with m-EVES. Input commands are terminated by semicolons. We have slightly edited the interaction to take into account formatting restrictions for this paper.

5 A Simple Program Verification

In our first example, we will write an algorithm which calculates x^y, for $y \geq 0$, where x and y are integers. We use the fact that when y is even, we have $x^y = (x^2)^{y/2}$. In this example, we are ignoring issues arising from running the program on a finite architecture; the procedure we will verify is not executable.

The example begins with m-EVES in its initial state. We start by introducing some variables for later use:

```
!var X, Y, RESULT: INT;
X, Y, RESULT
```

The above declaration introduces the variables X, Y and RESULT. These variable symbols are now in the vocabulary and are available for use in quantifications or parameter lists. (Recall that these are *not* global variables.)

Each m-EVES command starts with the prompt "!" and is terminated by a semicolon. m-EVES responds in a manner appropriate to the command invoked. For example, for a declaration, m-EVES will respond by printing the declared symbols.

We now introduce functions and lemmas that are used to annotate and prove the algorithm.

```
!function EXP (X, Y): INT =
  pre INT'GE (Y, 0)
  measure ORDINAL'VAL (Y)
  begin if Y = 0 then 1 else TIMES (X, EXP (X, MINUS (Y, 1))) end if
  end EXP;
EXP
```

The function EXP formally describes the concept of exponentiation. EXP is defined recursively. The pre-condition restricts the definition of EXP to those instances where the exponent is non-negative.

Since the definition of EXP is recursive, it has a proof obligation:

```
Beginning proof of EXP ...
IMPLIES (AND (Y <> 0,
              INT'GE (Y, 0)),
         ORDINAL'LT (ORDINAL'VAL (MINUS (Y, 1)), ORDINAL'VAL (Y)))
!reduce;

Which simplifies
when rewriting with ORDINAL'LT_6 to ...
TRUE
```

This proof uses a rewrite rule called ORDINAL'LT_6, which is defined in the initial theory. The rewrite rule replaces the ordinal relation with an integer relation. The integer relation is reduced to TRUE using integer and propositional simplification.

Before advancing to the statement and proof of the fundamental property of exponentiation, we need to introduce two simple lemmas that will be needed in the later proofs. These lemmas express properties of integer division that are non-trivial consequences of its definition. (While m-NEVER has built-in knowledge of linear arithmetic, its initial knowledge of multiplication and division is minimal. In fact, all that is known about integer division is its recursive definition.)

The following axiom is described as being a FRULE (forward rule) so that it can be applied automatically in later proofs.

```
!axiom DIV_NONNEGATIVE (X, Y) =
  FRULE
  triggers (DIV (X, Y))
  begin IMPLIES (AND (INT'GE (X, 0),
                      INT'GT (Y, 0)),
                 INT'GE (DIV (X, Y), 0))
  end DIV_NONNEGATIVE;
DIV_NONNEGATIVE
```

The proof of the axiom proceeds by induction. The **prove by induction** command initially tries to prove the result without using induction. When that fails, it applies an induction scheme. The chosen induction scheme is described by the proposition containing *P*. After the scheme is applied, the resulting formula is reduced.

```
Beginning proof of DIV_NONNEGATIVE ...
IMPLIES (AND (INT'GE (X, 0),
             INT'GT (Y, 0)),
        INT'GE (DIV (X, Y), 0))
!prove by induction;

Which simplifies to ...
IMPLIES (INT'GE (X, 0),
        OR (INT'LE (Y, 0),
            INT'GE (DIV (X, Y), 0)))

Returning to :
Beginning proof of DIV_NONNEGATIVE ...
IMPLIES (AND (INT'GE (X, 0),
             INT'GT (Y, 0)),
        INT'GE (DIV (X, Y), 0))

Inducting using the following scheme ...
AND (IMPLIES (NOT (AND (NOT (INT'LT (X, Y)),
                       INT'GE (X, 0),
                       INT'GT (Y, 0))),
             *P* (X, Y)),
     IMPLIES (AND (NOT (INT'LT (X, Y)),
                  INT'GE (X, 0),
                  INT'GT (Y, 0),
                  *P* (MINUS (X, Y), Y)),
             *P* (X, Y)))
produces ...
AND (IMPLIES (NOT (AND (NOT (INT'LT (X, Y)),
                       INT'GE (X, 0),
                       INT'GT (Y, 0))),
             IMPLIES (AND (INT'GE (X, 0),
                          INT'GT (Y, 0)),
                     INT'GE (DIV (X, Y), 0))),
     IMPLIES (AND (NOT (INT'LT (X, Y)),
                  INT'GE (X, 0),
                  INT'GT (Y, 0),
                  IMPLIES (AND (INT'GE (MINUS (X, Y), 0),
                               INT'GT (Y, 0)),
                          INT'GE (DIV (MINUS (X, Y), Y), 0))),
             IMPLIES (AND (INT'GE (X, 0),
                          INT'GT (Y, 0)),
                     INT'GE (DIV (X, Y), 0))))

Which simplifies
with invocation of DIV to ...
TRUE
```

Here, we introduce another property of DIV. This fact will be needed to prove the termination of the loop in the exponentiation procedure.

```
!axiom DIV_SMALLER (X, Y) =
```

```
        FRULE
        triggers (DIV (X, Y))
        begin IMPLIES (AND (INT'GT (X, 0),
                            INT'GT (Y, 1)),
                       INT'LT (DIV (X, Y), X))
        end DIV_SMALLER;
DIV_SMALLER

Beginning proof of DIV_SMALLER ...
IMPLIES (AND (INT'GT (X, 0),
             INT'GT (Y, 1)),
         INT'LT (DIV (X, Y), X))
!prove by induction;
```

We omit that portion of the transcript describing the induction since the induction scheme is the same as that chosen above.

```
...

Which simplifies
with invocation of DIV
with the assumptions DIV_NONNEGATIVE to ...
TRUE
```

The next axiom formalizes the crucial property about exponentiation, namely, $x^y = (x^2)^{y/2}$. This axiom is described as a RULE (rewrite rule) rather than a FRULE.

```
!axiom EXP_EVEN (X, Y) =
    RULE
    begin IMPLIES (AND (INT'GE (Y, 0),
                        MOD (Y, 2) = 0),
                   EXP (TIMES (X, X), DIV (Y, 2)) = EXP (X, Y))
    end EXP_EVEN;
EXP_EVEN
```

As with the previous axioms, the axiom EXP_EVEN can be proved by induction.

```
Beginning proof of EXP_EVEN ...
IMPLIES (AND (INT'GE (Y, 0),
             MOD (Y, 2) = 0),
         EXP (TIMES (X, X), DIV (Y, 2)) = EXP (X, Y))
!prove by induction;

Inducting using the following scheme ...
AND (IMPLIES (NOT (AND (NOT (INT'LT (Y, 2)),
                        INT'GE (Y, 0),
                        INT'GT (2, 0))),
              *P* (X, Y)),
     IMPLIES (AND (NOT (INT'LT (Y, 2)),
                   INT'GE (Y, 0),
                   INT'GT (2, 0),
```

```
                        *P* (X, MINUS (Y, 2))),
                *P* (X, Y)))
produces ...
AND (IMPLIES (NOT (AND (NOT (INT'LT (Y, 2)),
                        INT'GE (Y, 0),
                        INT'GT (2, 0))),
              IMPLIES (AND (INT'GE (Y, 0),
                           MOD (Y, 2) = 0),
                       EXP (TIMES (X, X), DIV (Y, 2)) = EXP (X, Y))),
     IMPLIES (AND (NOT (INT'LT (Y, 2)),
                   INT'GE (Y, 0),
                   INT'GT (2, 0),
                   IMPLIES (AND (INT'GE (MINUS (Y, 2), 0),
                                MOD (MINUS (Y, 2), 2) = 0),
                            EXP (TIMES (X, X), DIV (MINUS (Y, 2), 2))
                            = EXP (X, MINUS (Y, 2)))),
              IMPLIES (AND (INT'GE (Y, 0),
                           MOD (Y, 2) = 0),
                       EXP (TIMES (X, X), DIV (Y, 2)) = EXP (X, Y))))
```

```
Which simplifies
with invocation of EXP, DIV, MOD
with the assumptions DIV_SMALLER, DIV_NONNEGATIVE to ...
IMPLIES (AND (INT'GE (Y, 2),
             MOD (MINUS (Y, 2), 2) = 0,
             EXP (TIMES (X, X), DIV (MINUS (Y, 2), 2))
             = EXP (X, MINUS (Y, 2))),
         TIMES (X, TIMES (X, EXP (TIMES (X, X),
                                   DIV (MINUS (Y, 2), 2)))))
         = EXP (X, Y))
```

To complete the proof, we need to invoke the function EXP (X, Y) twice. Unfortunately, the prover heuristics were unable to perform this task automatically. We invoke it once manually. Then the reduce command can complete the proof.

```
!invoke EXP (X, Y);
```

```
Invoking EXP (X, Y) gives ...
IMPLIES (AND (INT'GE (Y, 2),
             MOD (MINUS (Y, 2), 2) = 0,
             EXP (TIMES (X, X), DIV (MINUS (Y, 2), 2))
             = EXP (X, MINUS (Y, 2))),
         TIMES (X, TIMES (X, EXP (TIMES (X, X),
                                   DIV (MINUS (Y, 2), 2)))))
         = if INT'GE (Y, 0)
              then if Y = 0
                      then 1
                      else TIMES (X, EXP (X, MINUS (Y, 1))) end if
              else EXP (X, Y) end if)
!reduce;
```

```
Which simplifies
```

with invocation of EXP
with the assumptions DIV_NONNEGATIVE to ...
TRUE

Now that we have described the concept of exponentiation and proved properties about exponentiation and division, we are ready to describe the algorithm for determining x^y. The algorithm is stated in the form of a procedure declaration.

```
!procedure FAST_EXPONENTIATION (pvar X, pvar Y, pvar RESULT) =
  initial (X'0 = X, Y'0 = Y)
  pre  INT'GE (Y, 0)
  post RESULT = EXP (X'0, Y'0)
  begin
    RESULT := 1
    loop
      invariant AND (INT'GE (Y, 0),
                     TIMES (RESULT, EXP (X, Y)) = EXP (X'0, Y'0))
      measure ORDINAL'VAL (Y)
        exit when Y = 0
        if MOD (Y, 2) = 0
          then X := TIMES (X, X)
               Y := DIV (Y, 2)
          else RESULT := TIMES (X, RESULT)
               Y := MINUS (Y, 1)
          end if
      end loop
  end FAST_EXPONENTIATION;
FAST_EXPONENTIATION
```

The variable symbols X, Y and RESULT are *bound* in the formal parameter list. The keyword **pvar** indicates that the values associated with these variables may be modified as the algorithm is executed.

The initial clause uses variable symbols X'0 and Y'0 to capture the values of the variables X and Y, respectively, on entry to the procedure.

The application of pre, post and invariant annotations are conventional. It is intended that the pre-condition and post-condition will be satisfied on entry and exit of each call to the procedure, respectively. The invariant is to be satisfied each time the loop body is executed.

The measure expression will be used to prove that the loop will terminate. For each iteration through the loop body, the value of the measure expression must decrease.

The proof obligation for a procedure consists of proving that when the pre-condition is satisfied the procedure will terminate and satisfy the post-condition.

For the FAST_EXPONENTIATION procedure the proof obligation is:

```
Beginning proof of FAST_EXPONENTIATION ...
IMPLIES (
  AND (INT'GE (Y, 0),
       X'0 = X,
       Y'0 = Y),
  AND (INT'GE (Y, 0),
       TIMES (1, EXP (X, Y)) = EXP (X'0, Y'0),
```

```
all Y'1, RESULT, X'1, X'2, Y'2:
  IMPLIES (
    AND (INT'GE (Y'1, 0),
         TIMES (RESULT, EXP (X'1, Y'1)) = EXP (X'2, Y'2)),
      if      Y'1 = 0
        then RESULT = EXP (X'2, Y'2)
      elseif MOD (Y'1, 2) = 0
        then AND (INT'GE (DIV (Y'1, 2), 0),
                  TIMES (RESULT,
                         EXP (TIMES (X'1, X'1),
                              DIV (Y'1, 2)))
                  = EXP (X'2, Y'2),
                  ORDINAL'LT (ORDINAL'VAL (DIV (Y'1, 2)),
                              ORDINAL'VAL (Y'1)))
      else    AND (INT'GE (MINUS (Y'1, 1), 0),
                   TIMES (X'1,
                          RESULT,
                          EXP (X'1, MINUS (Y'1, 1)))
                   = EXP (X'2, Y'2),
                   ORDINAL'LT (ORDINAL'VAL (MINUS (Y'1, 1)),
                               ORDINAL'VAL (Y'1))) end if)))
```

The structure of this proposition mirrors the structure of the procedure body. This formula expresses the correctness and termination of the procedure bodies.

The axioms we have defined about EXP and DIV are sufficient for the theorem prover to reduce the proposition to TRUE automatically.

```
!reduce;
```

```
Which simplifies
with invocation of EXP
when rewriting with ORDINAL'LT_6, EXP_EVEN
with the assumptions DIV_NONNEGATIVE, DIV_SMALLER to ...
TRUE
```

6   A Hardware Specification Verification

In this second example, we mimic Gordon's verification of a ripple-carry adder built from CMOS components [Gor 86]. The same method of representing hardware components by predicates is used as in Gordon's proof. As there, we ignore delays and model values on lines with truth values. We will not further explain the proof method here, but will focus instead on the details of the proof in m-EVES.

We begin with m-EVES in its initial state, and start the development by declaring some variables that will be needed later.

```
!var IN, OUT, IN0, IN1, OUT0, OUT1, CTL, P, Q, R,
     P1, P2, P3, P4, P5, P6, P7, P8, P9, P10: BOOL;
IN, OUT, IN0, IN1, OUT0, OUT1, CTL, P, Q, R,
P1, P2, P3, P4, P5, P6, P7, P8, P9, P10

!var CIN, COUT: BOOL;
```

```
CIN, COUT

!var I, J: INT;
I, J
```

Next, we define predicates (Boolean-valued functions) describing the two types of CMOS transistors. Since these definitions are not recursive, their proof obligations are trivially true (and the system suppresses them).

```
!function PTRAN_REL (CTL, IN, OUT): BOOL =
  begin IMPLIES (NOT (CTL),
                 OUT = IN)
  end PTRAN_REL;
PTRAN_REL

!function NTRAN_REL (CTL, IN, OUT): BOOL =
  begin IMPLIES (CTL, OUT = IN)
  end NTRAN_REL;
NTRAN_REL
```

A full adder has three input lines (here IN0, IN1 and CIN) and two output lines (here OUT and COUT). Each line represents a binary digit according to the coding scheme:

```
!function BIT_VAL (P): INT =
  begin if P then 1 else 0 end if
  end BIT_VAL;
BIT_VAL
```

It is desired that the inputs and outputs satisfy the relation:

```
PLUS (TIMES (2, BIT_VAL (COUT)),
      BIT_VAL (OUT))
= PLUS (BIT_VAL (CIN),
        PLUS (BIT_VAL (IN0), BIT_VAL (IN1))))
```

Figure 1 shows the full adder implementation we will verify. This circuit is specified by the following formula:

```
!function FULL_ADDER_REL (IN0, IN1, CIN, OUT, COUT): BOOL =
  begin some P1, P2, P3, P4, P5, P6, P7, P8, P9, P10:
          AND (PTRAN_REL (IN0, TRUE, P7),
               PTRAN_REL (IN1, TRUE, P7),
               PTRAN_REL (CIN, P7, P1),
               NTRAN_REL (CIN, P1, P9),
               NTRAN_REL (IN1, P9, FALSE),
               NTRAN_REL (IN0, P9, FALSE),
               PTRAN_REL (IN0, TRUE, P8),
               PTRAN_REL (IN1, P8, P1),
               NTRAN_REL (IN1, P1, P10),
               NTRAN_REL (IN0, P10, FALSE),
               PTRAN_REL (P1, TRUE, COUT),
               NTRAN_REL (P1, COUT, FALSE),
```

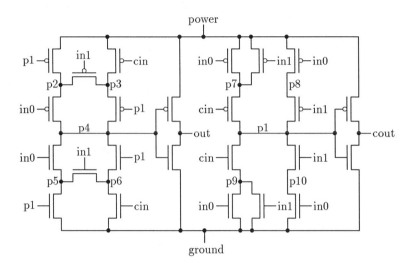

Figure 1: The Full Adder Circuit

```
                PTRAN_REL (P1, TRUE, P2),
                PTRAN_REL (INO, P2, P4),
                PTRAN_REL (IN1, P2, P3),
                PTRAN_REL (CIN, TRUE, P3),
                PTRAN_REL (P1, P3, P4),
                NTRAN_REL (P1, P4, P6),
                NTRAN_REL (CIN, P6, FALSE),
                NTRAN_REL (IN1, P6, P5),
                NTRAN_REL (INO, P4, P5),
                NTRAN_REL (P1, P5, FALSE),
                PTRAN_REL (P4, TRUE, OUT),
                NTRAN_REL (P4, OUT, FALSE))
        end FULL_ADDER_REL;
FULL_ADDER_REL
```

In this formula we have used the names indicated on the diagram to label the connections. The existential quantifier in this formula is used to hide the internal connections. Note that the power lines are represented by **true** and the ground lines by **false**.

The correctness of this circuit is established by the proof of the following theorem:

```
!axiom FULL_ADDER_SPEC (INO, IN1, CIN, OUT, COUT) =
   rule
   begin FULL_ADDER_REL (INO, IN1, CIN, OUT, COUT)
          = (PLUS (TIMES (2, BIT_VAL (COUT)), BIT_VAL (OUT))
             = PLUS (BIT_VAL (CIN), PLUS (BIT_VAL (INO),
                                          BIT_VAL (IN1))))
    end FULL_ADDER_SPEC;
FULL_ADDER_SPEC
```

```
Beginning proof of FULL_ADDER_SPEC ...
FULL_ADDER_REL (IN0, IN1, CIN, OUT, COUT)
 = (PLUS (TIMES (2, BIT_VAL (COUT)), BIT_VAL (OUT))
     = PLUS (BIT_VAL (CIN),
             PLUS (BIT_VAL (IN0), BIT_VAL (IN1))))
!reduce;

Which simplifies
with invocation of FULL_ADDER_REL,
                   NTRAN_REL, PTRAN_REL, BIT_VAL to ...
TRUE
```

This theorem has been expressed as a rewrite rule to facilitate the proofs of circuits containing full adders (as will be seen later).

Next, we will combine a number of full adders to make a 16-bit ripple-carry adder. It will be convenient to combine the 16 wires carrying a 16-bit value into a conceptual unit. We use an array:

```
!type BUS = array 0 .. 15 of BOOL;
BUS

Beginning proof of BUS ...
INT'LE (0, 15)
!reduce;

Which simplifies to ...
TRUE
```

To accept an array declaration, it is necessary to show that the low bound (of indexes) is not greater than the high bound. This proof is trivial.

This array declaration has added type BUS, function BUS'EXTRACT of type BUS × INT → BOOL (for indexing into the array), and other functions not used in this example.

We will need several variables of type BUS:

```
!var B, B0, B1, B2, B3, BIN, BOUT, CB: BUS;
B, B0, B1, B2, B3, BIN, BOUT, CB
```

A bus of 16 lines represents a number in base 2, using each successive bit value as a digit. A bus value B will represent the number $B[0]+2 \cdot B[1]+4 \cdot B[2]+\cdots+2^{15} \cdot B[15]$. This sum could be written explicitly, but we will, instead, define a formula for the n-term sum of this series, and then specialize that to the 16 terms we use here. To this end, we first need to define the function $i \mapsto 2^i$. The recursive definition is easily proved acceptable:

```
!function E2 (I): INT =
  pre INT'GE (I, 0)
  measure ORDINAL'VAL (I)
  begin if I = 0 then 1 else TIMES (2, E2 (MINUS (I, 1))) end if
  end E2;
E2
```

```
Beginning proof of E2 ...
IMPLIES (AND (I <> 0,
             INT'GE (I, 0)),
        ORDINAL'LT (ORDINAL'VAL (MINUS (I, 1)), ORDINAL'VAL (I)))
!reduce;

Which simplifies
when rewriting with ORDINAL'LT_6 to ...
TRUE
```

Next, we define the value represented by the low $i + 1$ bits of the bus. Again, the recursive definition is trivial to justify:

```
!function BUS_PARTIAL_VALUE (B, I): INT =
  pre INT'GE (I, 0)
  measure ORDINAL'VAL (I)
  begin if I = 0
          then BIT_VAL (BUS'EXTRACT (B, 0))
          else PLUS (TIMES (E2 (I), BIT_VAL (BUS'EXTRACT (B, I))),
                     BUS_PARTIAL_VALUE (B, MINUS (I, 1))) end if
  end BUS_PARTIAL_VALUE;
BUS_PARTIAL_VALUE
```

```
Beginning proof of BUS_PARTIAL_VALUE ...
IMPLIES (AND (I <> 0,
             INT'GE (I, 0)),
        ORDINAL'LT (ORDINAL'VAL (MINUS (I, 1)), ORDINAL'VAL (I)))
!reduce;

Which simplifies
when rewriting with ORDINAL'LT_6 to ...
TRUE
```

Finally, we define the value represented by a 16-bit bus:

```
function BUS_VALUE (B): INT =
  begin BUS_PARTIAL_VALUE (B, 15)
  end BUS_VALUE;
BUS_VALUE
```

The 16-bit adder is formed by joining 16 full adders, connecting the carry-out of the ith adder to the carry-in of the $i + 1$st. We use a fourth bus, CB, to connect the carry signals (see Figure 2).

Again, the formula for this circuit could be written explicitly, but we instead use an inductive definition of the low $i + 1$ bits of the circuit. (So, for example, the part of the circuit to the right of the ellipsis in Figure 2 is described by ADDER_PART_REL (CIN, B0, B1, BOUT, CB, 1).)

```
!function ADDER_PART_REL (CIN, B0, B1, BOUT, CB, I): BOOL =
  pre INT'GE (I, 0)
  measure ORDINAL'VAL (I)
```

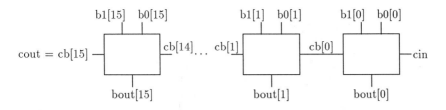

Figure 2: The 16-Bit Ripple-Carry Adder

```
begin if I = 0
       then FULL_ADDER_REL (BUS'EXTRACT (B0, 0),
                            BUS'EXTRACT (B1, 0),
                            CIN,
                            BUS'EXTRACT (BOUT, 0),
                            BUS'EXTRACT (CB, 0))
       else AND (ADDER_PART_REL (CIN, B0, B1, BOUT,
                                 CB, MINUS (I, 1)),
                 FULL_ADDER_REL (BUS'EXTRACT (B0, I),
                                 BUS'EXTRACT (B1, I),
                                 BUS'EXTRACT (CB, MINUS (I, 1)),
                                 BUS'EXTRACT (BOUT, I),
                                 BUS'EXTRACT (CB, I))) end if
   end ADDER_PART_REL;
ADDER_PART_REL

Beginning proof of ADDER_PART_REL ...
IMPLIES (AND (I <> 0,
              INT'GE (I, 0)),
         ORDINAL'LT (ORDINAL'VAL (MINUS (I, 1)),
                     ORDINAL'VAL (I)))
!reduce;

Which simplifies
when rewriting with ORDINAL'LT_6 to ...
TRUE
```

Now the 16-bit adder can be defined as a 16-bit adder part with CB[15] connected to COUT, and the entire carry bus hidden:

```
!function ADD16_REL (CIN, B0, B1, BOUT, COUT): BOOL =
  begin some CB:
        AND (ADDER_PART_REL (CIN, B0, B1, BOUT, CB, 15),
             COUT = BUS'EXTRACT (CB, 15))
   end ADD16_REL;
ADD16_REL
```

To verify the adder, we first prove that each i-bit adder part performs an i-bit add. This is another proof by induction. Some of the intermediate formulas have been deleted from the transcript:

```
!axiom ADDER_PART_SPEC (CIN, B0, B1, BOUT, CB, I) =
  begin IMPLIES (AND (INT'LE (0, I),
                      INT'LE (I, 15),
                      ADDER_PART_REL (CIN, B0, B1, BOUT, CB, I)),
                 PLUS (BUS_PARTIAL_VALUE (BOUT, I),
                       TIMES (E2 (PLUS (I, 1)),
                       BIT_VAL (BUS'EXTRACT (CB, I))))
                 = PLUS (PLUS (BUS_PARTIAL_VALUE (B0, I),
                               BUS_PARTIAL_VALUE (B1, I)),
                         BIT_VAL (CIN)))
  end ADDER_PART_SPEC;
ADDER_PART_SPEC

Beginning proof of ADDER_PART_SPEC ...
IMPLIES (AND (INT'LE (0, I),
             INT'LE (I, 15),
             ADDER_PART_REL (CIN, B0, B1, BOUT, CB, I)),
         PLUS (BUS_PARTIAL_VALUE (BOUT, I),
               TIMES (E2 (PLUS (I, 1)),
               BIT_VAL (BUS'EXTRACT (CB, I))))
         = PLUS (PLUS (BUS_PARTIAL_VALUE (B0, I),
                       BUS_PARTIAL_VALUE (B1, I)),
                 BIT_VAL (CIN)))
!prove by induction;

Which simplifies
with invocation of BIT_VAL, E2 to ...
...

Returning to :
Beginning proof of ADDER_PART_SPEC ...
IMPLIES (AND (INT'LE (0, I),
             INT'LE (I, 15),
             ADDER_PART_REL (CIN, B0, B1, BOUT, CB, I)),
         PLUS (BUS_PARTIAL_VALUE (BOUT, I),
               TIMES (E2 (PLUS (I, 1)),
                      BIT_VAL (BUS'EXTRACT (CB, I))))
         = PLUS (PLUS (BUS_PARTIAL_VALUE (B0, I),
                       BUS_PARTIAL_VALUE (B1, I)),
                 BIT_VAL (CIN)))

Inducting using the following scheme ...
AND (IMPLIES (NOT (AND (I <> 0,
                        INT'GE (I, 0))),
              *P* (B0, B1, BOUT, CB, CIN, I)),
     IMPLIES (AND (I <> 0,
                   INT'GE (I, 0),
                   *P* (B0, B1, BOUT, CB, CIN, MINUS (I, 1))),
              *P* (B0, B1, BOUT, CB, CIN, I)))

produces ...
```

...

```
Which simplifies
with invocation of E2, BUS_PARTIAL_VALUE,
                    ADDER_PART_REL, BIT_VAL
when rewriting with FULL_ADDER_SPEC to ...
TRUE
```

We can now state and prove the correctness of the adder circuit:

```
!axiom ADD16_SPEC (CIN, B0, B1, BOUT, COUT) =
  begin IMPLIES (ADD16_REL (CIN, B0, B1, BOUT, COUT),
                PLUS (BUS_VALUE (BOUT),
                      TIMES (E2 (16), BIT_VAL (COUT)))
              = PLUS (PLUS (BUS_VALUE (B0),
                           BUS_VALUE (B1)),
                      BIT_VAL (CIN)))
    end ADD16_SPEC;
ADD16_SPEC
```

```
Beginning proof of ADD16_SPEC ...
IMPLIES (ADD16_REL (CIN, B0, B1, BOUT, COUT),
         PLUS (BUS_VALUE (BOUT), TIMES (E2 (16), BIT_VAL (COUT)))
       = PLUS (PLUS (BUS_VALUE (B0), BUS_VALUE (B1)),
               BIT_VAL (CIN)))
```

This is a simple consequence of the preceding lemma. Before we can apply that lemma, we first must perform some hand steps:

```
!invoke ADD16_REL;
```

```
Invoking ADD16_REL gives ...
IMPLIES (some CB:
            AND (ADDER_PART_REL (CIN, B0, B1, BOUT, CB, 15),
                 COUT = BUS'EXTRACT (CB, 15)),
         PLUS (BUS_VALUE (BOUT), TIMES (E2 (16), BIT_VAL (COUT)))
       = PLUS (PLUS (BUS_VALUE (B0), BUS_VALUE (B1)),
                    BIT_VAL (CIN)))
```

```
!prenex;
```

```
Prenexes to ...
all CB:
  IMPLIES (AND (ADDER_PART_REL (CIN, B0, B1, BOUT, CB, 15),
               COUT = BUS'EXTRACT (CB, 15)),
          PLUS (BUS_VALUE (BOUT), TIMES (E2 (16), BIT_VAL (COUT)))
        = PLUS (PLUS (BUS_VALUE (B0), BUS_VALUE (B1)),
                     BIT_VAL (CIN)))
```

```
!open;
```

```
Opening the formula results in ...
IMPLIES (AND (ADDER_PART_REL (CIN, B0, B1, BOUT, CB, 15),
```

```
        COUT = BUS'EXTRACT (CB, 15)),
    PLUS (BUS_VALUE (BOUT), TIMES (E2 (16), BIT_VAL (COUT)))
    = PLUS (PLUS (BUS_VALUE (B0), BUS_VALUE (B1)),
                    BIT_VAL (CIN)))
```

Now the adder-part lemma is used. We instantiate some of the variables (the prover is unable to find all the correct instantiations automatically). The previous steps of the proof have made it possible to mention CB in the instantiation list.

```
!use ADDER_PART_SPEC B0 = B0, B1 = B1, BOUT = BOUT, CB = CB, I = 15;
```

```
Assuming ADDER_PART_SPEC with the instantiations: B0 = B0
B1 = B1
BOUT = BOUT
CB = CB
I = 15 generates ...
IMPLIES (AND (all CIN'0:
                IMPLIES (AND (INT'LE (0, 15),
                             INT'LE (15, 15),
                             ADDER_PART_REL (CIN'0, B0, B1,
                                             BOUT, CB, 15)),
                        PLUS (BUS_PARTIAL_VALUE (BOUT, 15),
                             TIMES (E2 (PLUS (15, 1)),
                                     BIT_VAL (BUS'EXTRACT (CB, 15))))
                        = PLUS (PLUS (BUS_PARTIAL_VALUE (B0, 15),
                                     BUS_PARTIAL_VALUE (B1, 15)),
                                BIT_VAL (CIN'0))),
            ADDER_PART_REL (CIN, B0, B1, BOUT, CB, 15),
            COUT = BUS'EXTRACT (CB, 15)),
        PLUS (BUS_VALUE (BOUT), TIMES (E2 (16), BIT_VAL (COUT)))
        = PLUS (PLUS (BUS_VALUE (B0), BUS_VALUE (B1)),
                        BIT_VAL (CIN)))
```

The proof is almost complete. We need only invoke the definition of BUS_VALUE to have a trivial proposition:

```
!invoke BUS_VALUE;
```

```
Invoking BUS_VALUE gives ...
IMPLIES (AND (all CIN'0:
                IMPLIES (AND (INT'LE (0, 15),
                             INT'LE (15, 15),
                             ADDER_PART_REL (CIN'0, B0, B1,
                                             BOUT, CB, 15)),
                        PLUS (BUS_PARTIAL_VALUE (BOUT, 15),
                             TIMES (E2 (PLUS (15, 1)),
                                     BIT_VAL (BUS'EXTRACT (CB, 15))))
                        = PLUS (PLUS (BUS_PARTIAL_VALUE (B0, 15),
                                     BUS_PARTIAL_VALUE (B1, 15)),
                                BIT_VAL (CIN'0))),
            ADDER_PART_REL (CIN, B0, B1, BOUT, CB, 15),
            COUT = BUS'EXTRACT (CB, 15)),
```

```
    PLUS (BUS_PARTIAL_VALUE (BOUT, 15),
          TIMES (E2 (16), BIT_VAL (COUT)))
    = PLUS (PLUS (BUS_PARTIAL_VALUE (B0, 15),
                  BUS_PARTIAL_VALUE (B1, 15)),
            BIT_VAL (CIN)))
!simplify;
```

```
Which simplifies
with the instantiation CIN'0 = CIN to ...
TRUE
```

This final proof illustrates the use of hand steps to complete a complex proof that can not be proved fully automatically.

7 Conclusions

To date, the system has been used to prove, amongst others, the following examples:

- micro-Flow Modulator: A Flow Modulator is a program which filters and transforms (usually, this means sanitizes) messages. Four variations of a simple Flow Modulator have been proved.

- Low Water Mark: The Low Water Mark system has been described in [Che 81]. We have proved the non-interference property for this system and have also verified its implementation. Non-interference generally captures the concept that a process, which is running at some fixed security classification, can not be affected by processes with higher classifications.

- Network Security: Specification of the security of a communication network using propositions generated by the Gypsy Verification Environment [Goo 85].

- Sorting Programs: We have proved the verification conditions of a selection sort program and have proved a recursive function implementation of a linear insertion sort program. We have also proved both a recursive function implementation of quicksort, and a complete (non-recursive) quicksort program using a bounded stack.

- The specification of a secure terminal as written by Dick Kemmerer using an equational approach [Kem 86]. (Our effort uncovered two errors in Kemmerer's specification.)

- The specification of a library transaction system as written by Dick Kemmerer [Kem 85]. (This effort revealed that Kemmerer's original specification was incomplete.)

- Saddleback Search: Verification conditions for the saddleback search program as taken from the Gypsy Verification Environment.

- The equational style specification of the symbol table problem described by Guttag, *et al.* [GHM 78].

- Various examples including a program that determines the minimum value of an array, an implementation of Euclid's GCD and a solution to David Gries' tsquare problem [Gri 82].

- Axiomatic descriptions of theories of finite sets and finite sequences, along with model theoretic proofs of the consistency of these descriptions.

The m-EVES project has resulted in the following achievements:

- The design and implementation of a new specification and implementation language, m-Verdi, which has a complete, formal mathematical basis and supporting logic.

- The design of a sound logical system which is sufficiently powerful to handle linguistic features commonly used in the formal verification process.

- The development of a new theorem prover, m-NEVER, which incorporates significant functionality drawn from the theorem proving literature.

- The development of production quality compilers for m-Verdi.

The development of m-EVES was completed in November, 1987. Future work (resulting in the development of EVES) will be directed at increasing the expressibility of m-Verdi (resulting in Verdi) and the logic; the writing of a compiler for Verdi; the continuing evolution of the theorem prover (resulting in NEVER); the continuing evolution of the interface; the porting of the system to other hardware bases; and the continued application of the system to various examples.

References

[BB 73]   W. Bledsoe and P. Bruell. A Man-Machine Theorem Proving System. In *Proceedings 3rd IJCAI*, Stanford University, 1973.

[BM 79]   Bob Boyer and J Strother Moore. *A Computational Logic*. Academic Press, 1979.

[Che 81]   M. Cheheyl, *et al.* Verifying Security. *Computing Surveys* 13(4), September 1981.

[Cra 87a]   Dan Craigen. *A Description of m-Verdi*. TR-87-5420-02, I.P. Sharp Associates Limited, November 1987.

[Cra 87b]   Dan Craigen and Mark Saaltink. *An m-Verdi User's Guide*. TR-87-5420-12, I.P. Sharp Associates Limited, November 1987.

[Goo 85]   Don Good. *Proving Network Security*. Internal Note #125-B, Institute for Computing Science, University of Texas at Austin, 1985.

[Gor 86]   Mike Gordon. Why Higher-Order Logic is a Good Formalism for Specifying and Verifying Hardware. In G.J. Milne and P.A. Subrahmanyam (Eds.), *Formal Aspects of VLSI Design*, North-Holland, 1986.

[Gri 82]   David Gries. A Note on a Standard Strategy for Developing Loop Invariants and Loops. In *Science of Computer Programming* 2(3), December 1982.

[GHM 78]   John Guttag, *et al.* Abstract Data Types and Software Validation. *CACM* 21(12), December 1978.

[Kem 85] Dick Kemmerer. Testing Formal Specifications to Detect Design Errors. *IEEE Transactions on Software Engineering* 21(1), January 1985.

[Kem 86] Dick Kemmerer, *et al. Verification Assessment Study Final Report.* Technical Report C3-CR01-86, National Computer Security Center, Maryland.

[Luc 79] D.C. Luckham, *et al. Stanford Pascal Verifier User Manual.* STAN-CS-79-731, Stanford University Computer Science Department, March 1979.

[Mei 87a] Irwin Meisels. *VAX/VMS m-Verdi Compiler User's Manual.* TR-87-5420-10, I.P. Sharp Associates Limited, October 1987.

[Mei 87b] Irwin Meisels. *VAX/Unix m-Verdi Compiler User's Manual.* TR-87-5420-11, I.P. Sharp Associates Limited, October 1987.

[Nel 81] Greg Nelson. *Techniques for Program Verification.* CSL-81-10, Xerox PARC, June 1981.

[PK 87] Bill Pase and Sentot Kromodimoeljo. *NEVER: An Interactive Theorem Prover.* CP-87-5402-20, I.P. Sharp Associates Limited, January 1987.

[Saa 87] Mark Saaltink. *The Mathematics of m-Verdi.* FR-87-5420-03, I.P. Sharp Associates Limited, November 1987.

[Tho 81] D. Thompson, *et al. AFFIRM Reference Manual.* USC Information Sciences Institute, Marina Del Rey, California, 1981.

A   Appendix – Declarations

A.1   Var

`[prog] var EVENT-NAME-OR-LIST : TYPE ;`

Declares the symbol(s) `EVENT-NAME-OR-LIST` to be variable(s) of the given `TYPE`. Variables may be used as parameters for facts, rewrite, and forward rules, functions, and quantifications. They may also be used as program variables in procedures.

A.2   Const

`[prog] const EVENT-NAME : TYPE [= VALUE] ;`

Declares the symbol `EVENT-NAME` to be a constant of the given `TYPE`. If `VALUE` is supplied then the constant is an abbreviation for that `VALUE` wherever it is used. This fact is stored in the database in the form of a rewrite rule that is automatically applied whenever the symbol is encountered during rewriting. (For a constant named `C1`, the rewrite rule is named `C1'VALUE`.)

A.3   Type

`[prog] type EVENT-NAME [= BODY] ;`

Defines `EVENT-NAME` to be a valid type symbol to the prover. The new type symbol can then be used in future declarations. The following are built-in types: `INT` and `BOOL`. (The axioms for these types are not accessible to the user.)

The optional `BODY` must be an enumerated, record, array, or restriction type definition.

A.4 Fact

```
axiom EVENT-NAME ( [ARGLIST] ) =
    [PRAGMA-DEFINITION]
    begin FORMULA end EVENT-NAME ;
```

Introduces a new fact to the theorem prover database. The ARGLIST is a list of variables which occur free within the formula; these variables are considered to be universally quantified. Although the prover does not force the user to prove a fact, it does remember whether or not it has been proved. The EVENT-NAME is displayed, provided the addition of the fact succeeded.

A.5 Rule

```
axiom EVENT-NAME ( [ARGLIST] ) =
    [PRAGMA-DEFINITION]
    rule
    begin FORMULA end EVENT-NAME ;
```

Adds EVENT-NAME as a rewrite rule to the prover database. If any errors are found in the rule then the event is not added. The ARGLIST consists of variables which are universally quantified over the formula. Rewrite rules may be applied automatically by the REWRITE commands.

A.6 Frule

```
axiom EVENT-NAME ( [ARGLIST] ) =
    [PRAGMA-DEFINITION]
    frule
    triggers ( TRIGGERS )
    begin FORMULA end EVENT-NAME ;
```

Adds EVENT-NAME as a forward rule to the prover database. The ARGLIST consists of variables which are universally quantified over the FORMULA. The TRIGGERS are used as triggering patterns for the forward rule. Forward rules are applied automatically by the REWRITE commands.

A.7 Function

```
[prog] function EVENT-NAME ( [ARGLIST] ): RETURN-TYPE
    [=
    [PRAGMA-DEFINITION]
    [[pre PRE] [measure MEASURE] begin BODY end EVENT-NAME]] ;
```

Adds a function symbol with or without a definition to the theorem prover. For a function without a definition, the user must supply the EVENT-NAME, ARGLIST and RETURN-TYPE. The signature of the function is a mapping from the argument types to the return type. To introduce a function with a definition, the user need only add the BODY. The MEASURE is an expression of type ORDINAL used for recursive functions. If the PRE is omitted, it is equivalent to specifying the pre as TRUE. Functions with definitions may be invoked by the INVOKE and REDUCE commands.

A.8 Procedure

```
[prog] procedure EVENT-NAME ( [ARGS] ) =
    [PRAGMA-DEFINITION]
    [initial INIT-CLAUSE]
    [pre PRE]
    [post POST]
    begin BODY end EVENT-NAME ;

  ARGS: MODE VARIABLE-NAME [, MODE VARIABLE-NAME]*
```

Introduces a procedure to the theorem prover database. If no errors are detected, the procedure is added to the database.

B   Appendix – Proof Steps

B.1   Equality Substitute

```
equality substitute [EXPRESSION] ;
```

Substitutes, for EXPRESSION, its equal in appropriate contexts of the current formula. The expression must appear as the left or right side of an equality within the current formula. The resulting formula is equivalent to the original. A user may invoke equality substitute without supplying a value for EXPRESSION, in which case a heuristic is used to substitute equalities automatically.

B.2   Induct

```
induct [on TERM] ;
```

Tries to apply an induction scheme to the current formula. The Boyer-Moore induction technique is used. Normally, the induction scheme is heuristically chosen based on calls to recursive functions within the current formula. However, the user may direct the prover by explicitly specifying the TERM on which to induct.

B.3   Instantiate

```
instantiate INSTANTIATIONS ;
```

Performs the given INSTANTIATIONS on the current formula. To allow the instantiations to occur, the scopes of quantifiers in the formula may be extended or contracted. Logical equivalence is maintained by keeping the uninstantiated subexpression(s) as extra conjunct(s) or disjunct(s). The requested instantiations may be disallowed, in which case the command has no effect.

B.4   Invoke

```
invoke FUNCTION-OR-EXPRESSION ;
```

Opens the function definition wherever it occurs within the formula. If the object being invoked is a function with a precondition then the invocation is a conditional form on the precondition, the expanded form, and the original form. `invoke` works for functions which have been disabled. `invoke` may be applied to an expression rather than to a function, in which case it works like a selective invoke in that occurrences of the expression in the formula are replaced by the expanded version.

B.5   Open

```
open ;
```

If the current formula is contained within a series of universal quantifiers, the universal quantifiers are removed and the quantified variables become free variables. This may reduce the proof effort; however, the automatic part of the prover will not search for refutations.

B.6   Prenex

```
prenex ;
```

Places the current formula into prenex form as far as possible. If the result of this command is a completely prenexed formula with only universal quantifiers, the `open` command may be used to make the formula quantifier free.

B.7   Prove by Induction

```
prove by induction ;
```

Tries to prove the current formula with automatic induction. A non-inductive proof is first attempted. If this fails, the theorem prover will try to apply an induction scheme to the formula. The theorem prover may discard the non-inductive proof prior to the induction.

B.8   Rearrange

```
rearrange ;
```

Tries to rearrange the current formula. The rearrangement consists of reordering the arguments of the conjunctions and disjunctions of the current formula. In general, equalities and simple expressions are placed before more complex arguments such as quantified expressions.

Rearranging can be useful because the prover is sensitive to the ordering within formulas. As a result, the prover is often more likely to obtain a proof, and may do so more efficiently, if a formula has been rearranged using the `rearrange` command.

B.9   Reduce

```
reduce ;
```

Reduces the current formula. Reduction consists of simplification, rewriting, and invocation. Reduction forms the kernel of the automatic prover.

B.10 Rewrite

`rewrite ;`

Rewrites and simplifies the current formula. Conditional rewrite rules may be applied, provided their condition can be proved using only simplification and rewriting. This command also applies any forward rules which are triggered and whose condition is provable.

B.11 Simplify

`simplify ;`

Simplifies the current formula. This may substitute equalities as well as trying to instantiate variables in order to find a proof.

B.12 Split

`split PREDICATE ;`

Performs a case split on the current formula with the supplied `PREDICATE`. This results in a new formula of the form `if PREDICATE then FORMULA else FORMULA end if`, provided there are no references to the quantified variables of the formula within the predicate. If there are, `split` performs a case split on the largest subformulas within the scope of the referenced quantified variables. In effect, a case split causes the prover to work on the current formula under two cases: in the first, the predicate explicitly is assumed; and, in the second, the predicate explicitly is denied.

`Split` may also be used to place a specific hypothesis before a subexpression. This proof step may be required because of the sensitivity of the prover towards the ordering of subexpressions within the formula being reduced.

B.13 Trivial Simplify

`trivial simplify ;`

Performs a simple tautology check and propositional simplification.

B.14 Try

`try [EVENT-OR-FORMULA] ;`

Starts the proof of `EVENT-OR-FORMULA` by making it current. If `EVENT-OR-FORMULA` is the name of some event, the event's proof obligation is made current. Should any error occur, `try` will not have any effect. If some proof is already in progress, its partial proof will be saved so it may be continued later. If an event is tried which has previously been tried, its partial proof is resumed from the point it was left. If no `EVENT-OR-FORMULA` is supplied to `try`, the user is given a menu of unproved events.

B.15   Use

`use EVENT-NAME [INSTANTIATIONS] ;`

Adds the axiom associated with `EVENT-NAME` to the current formula as an assumption. This results in a new formula of the form `IMPLIES (ASSUMPTION, FORMULA)` where the assumption is the axiom instantiated with `INSTANTIATIONS`. Assume is most often used for assuming a `fact`, but it can also be used for assuming the axiom associated with a function, rule or frule.

The Interactive Proof Editor
An Experiment in Interactive Theorem Proving

Brian Ritchie and Paul Taylor

1   Introduction

The past thirty years have seen a diversity of strategies applied to the problem of using computers as an effective vehicle for the performance of mathematical proofs. As the development of the discipline has reflected many trends, it might be useful to note some of the more important:

- The activity of working through examples has led to an increasing appreciation of the profundity of the basic problem. Often proofs presented by mathematicians appear simpler than is actually the case, possibly because absolutely formal proofs are seldom attempted.

- Today there is a deeper theoretical understanding of the activity of proof itself; computer based theorem proving should perhaps be seen as applied proof theory. An outstanding example of this is the Curry-Howard propositions as types isomorphism, which underlies systems such as AutoMath [3], Nuprl [13], Constructions [6] and Edinburgh Logical Framework [9].

- The capabilities of the equipment used by today's researchers exceeds by several orders that which was available in the sixties: compare the teletype (or batch mode) of the sixties with today's workstations with bitmap screens and mouse.

The wide range of approaches have resulted in many different styles of system, each of which have their adherents and each have supplied important paradigms with influences far beyond their immediate setting. One of todays tasks is to build proof systems integrating many of the important ideas so as to make the activity of proof accessible to a wider community.

Computer based proof assistants have a long history at Edinburgh (for instance the LCF [8] system), and recent years have seen the active development of interactive proof editors . This paper describes one of the initial results of this work, the Edinburgh IPE, a system built to explore the application of attribute grammars to proof (following the experience of Reps and Alpern with the Synthesiser Generator [15, 16]) and the possibility of performing "proof by clicking". It was implemented by Brian Ritchie, Tatsuya Hagino, John Cartmell and Claire Jones with the supervision of Rod Burstall and funding from SERC. Experiments with the IPE have highlighted the advantages of workstation technology for exploring proofs interactively, but also indicate a variety of problems which have to be tackled in order to provide a powerful yet natural proof environment.

2 Some Proofs with the IPE

The IPE runs on a SUN3 workstation under X-windows. We will give a few simple examples to indicate how "proof by clicking" has been realized within the IPE. As "proof by clicking" obviously makes extensive use of the SUN's mouse, throughout this paper we shall use **LM**, **MM** and **RM** to abbreviate references to the various mouse buttons (Left, Middle and Right). Upon start up the user is presented with an empty proof window (buffer) as in Figure 1.

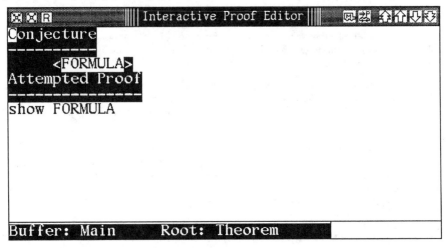

Figure 1

Notice that the display is broken into two areas, one for the conjectured formula and the other for the proof. Initially, as we have yet to conjecture anything, the conjecture is represented by a dummy displayed in angle brackets (`<FORMULA>`). Clicking **MM** at this location starts a simple editor which can be used to enter the formula you wish to prove.

For instance, type in the formula `A&B&C->(B|C)&A`: the character "`|`" is used to input the Or-operator, it is displayed with the conventional symbol. Upon exiting the editor the formula becomes the goal of the proof. Hence the first line of the proof is

    `show A&B&C->(B∨C)&A`

The proof is developed by clicking **MM** on this first goal formula. The IPE will generate a subgoal from the goal by applying an appropriate rule (here Implies Introduction), which it selects by the top connective (an implication) of the formula. The second line and new goal of the proof will now be

    `show A&B&C entails (B∨C)&A`

note that the subformula `A&B&C` has become an premise which hopefully will entail `(B∨C)&A` as a conclusion. Clicking **MM** at the premise `A&B&C` will cause IPE to generate a new goal, and further clicks will produce the complete proof shown in Figure 2.

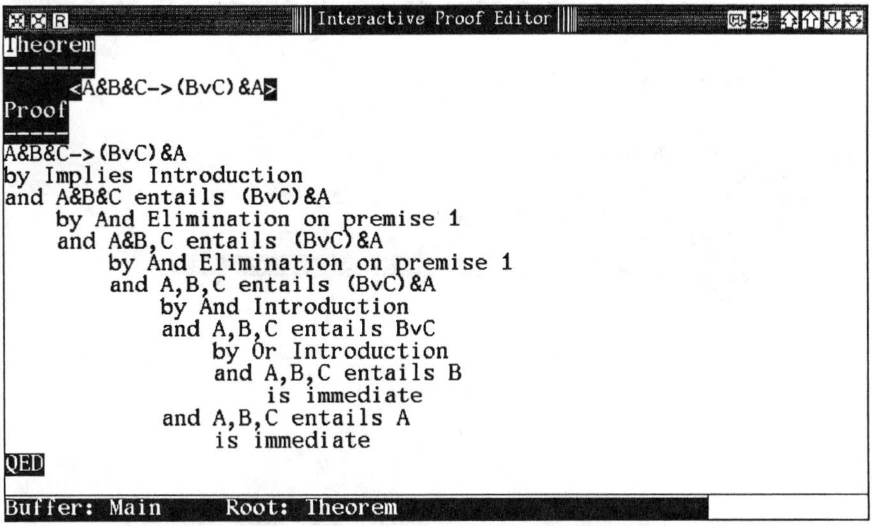

Figure 2

You should now see the style of proof which is implemented in the IPE. At any stage in the proof we have a number of outstanding goals to satisfy. A goal is a sequent, a collection of formulae which have been assumed (the premises) and a formula which hopefully they entail (the conclusion). Clicking **MM** at the conclusion or a premise will refine the goal with the appropriate introduction or elimination rule (Figure 6 gives a complete table of the built in rules).

It is possible to go back to the conjecture and edit it (notice it is displayed within angle brackets indicating editable data), for example change it to **A&B&C -> B&(C|A)**; the attribute evaluation in IPE will cause the proof to be reapplied with this new goal, much of it will succeed so only a few more clicks will be needed to complete the proof of the new conjecture. Similarly it is possible to explore alternative paths through the proof, by clicking upon the formulae in a different sequence: for instance we could first expand the operators in the conclusion.

2.1 Proofs, Quantifiers and Entering Terms

To achieve proofs of formulae involving quantifiers it is necessary for the user to supply some data to the IPE. Consider for instance

$$\forall x A(x) \& \exists a (A(a) \to B(a)) \to \exists z B(z)$$

which is input to IPE via the sequence

`!x A(x) & ?a(A(a)->B(a)) -> ?z B(z)`

By clicking in sequence on **->**, **&** and $\exists$ the proof tree shown in Figure 3 will be obtained.

Figure 3

You should notice that the last rule, an Exist Elimination, has introduced a new free variable a; this should be thought of as a name for the object whose existence was asserted by the existential premise. The next step must exploit the fact that with any term x we may assume A(x), in particular x may be the free variable a. Clicking on the premise ∀x A(x) invokes All Elimination extending the proof tree as follows:

```
use All Elimination with <TERM_1> on premise 1
and show A(TERM_1),A(a)->B(a) entails ∃zB(z)
```

TERM_1 is enclosed by angle brackets, indicating data which may be edited; so by clicking MM at this point to start up the editor, and by typing in a, we make the appropriate instantiation.

Two further clicks and entering one more term will give the complete proof shown in Figure 4.

2.2 Using Facts, Rewriting and Proof by Induction

The proofs described in the previous sections have all been of tautologies in intuitionistic predicate logic - the logic built into the IPE. However it is possible to extend IPE with theories by supplying a collection of UNIX files. Usually the IPE will be set up with a collection of predefined theories amongst which should be List, a theory of lists.

We now describe how you would prove a new fact about lists with the IPE. Type in as a conjecture for IPE the following formula.

```
length(x@y) = length(x) + length(y)
```

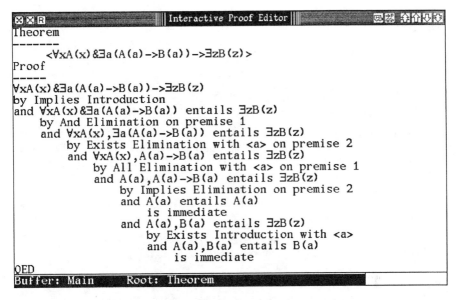

Figure 4

The formula expresses the fact that when you append a list to the end of another list the length of the combined list is the sum of the lengths of the component lists (here @ represents the appending operation). If you now attempt to prove this conjecture by clicking on the goal you will see IPE does nothing, apart from printing a rather obscure error message ("Immediate - Not so"); and indeed we can make no progress with this proof until the facts about lists have been loaded.

Loading a theory is one of the actions accessed via the menu. Pressing **RM** reveals the top level menu of a walk-through menu, this has a submenu of **Theory** operations, one of which can be used to load the theory of lists. Choosing this extends the IPE with the operations, axioms and rewrite rules significant for the theory of lists. Clicking **MM** upon the goal no longer results in the error message: in fact, as we are trying to prove an equality, IPE breaks our equality into two, a standard tactic for tackling proofs of general equalities. Here however this tactic is irrelevant; what is needed is a proof by induction over the structure of lists. Hence it is necessary to apply one of the facts about lists to our goal.

Again the facts of a theory are also accessed via the menu. To exploit a fact it is first necessary to indicate where we wish to apply the fact (in this case to the original goal). A formula can be selected for some action by clicking upon it with **LM**, selections becoming highlighted in the display. Choosing the action **Choose from All Facts** from the **Facts** submenu, allows the goal to be seen as an instance of the schema for list induction. To achieve a proof by induction it is necessary to select a variable in the goal as the induction variable — thus you will see an editable term <InductionVariable>. Using the editor this can be replaced with x.

This leaves two goal formulae to prove:
the base case,

```
length(nil@y)=(length(nil)+length(y))
```

and the step case

```
∀L((length(L@y)=length(L)+length(y)->
    ∀a length(a:L@y)=length(a:L)+length(y))
```

The first of these is an equality, which follows directly by simplifying its left hand term using the definitions of the operations length and @. This action can be achieved by selecting the operation "Rewrite" from from the main menu for this goal. Having disposed of the first goal; the second can be tackled in a similar fashion once the logical constants have been stripped away. This can be done by repeated clicks with **MM** or more immediately by switching on autoprove (by clicking **LM** on the icon depicting a racing car). The outstanding goal now becomes

```
show length(L@y)=length(L)+length(y) entails
     length(a:L@y)=length(a:L)+length(y)
```

which becomes

```
show length(L@y)=length(L)+length(y) entails
     S(length(L@y))=S(length(L)+length(y))
```

upon rewriting the conclusion. We are almost finished; notice that the equality of the conclusion is a consequence of the premise if we use the premise to simplify it. This we do by clicking upon this premise with **MM**. Figure 5 shows the complete proof.

```
⊗⊠⊟                        ‖‖‖ Interactive Proof Editor ‖‖‖              ⊡⊠ ⊠⊞⊠⊞⊠
Theorem
        ◀length(x@y)=(length(x)+length(y))▆
Proof
────
length(x@y)=(length(x)+length(y))
by axiom <ListInduction> on conclusion
    with length(x@y)=(length(x)+length(y)) for phi(InductionVariable)
    and  <x> for InductionVariable
and length(nil@y)=(length(nil)+length(y))
    by rewriting
    and length(y)=length(y)
        is immediate
and ∀L(length(L@y)=(length(L)+length(y))->∀a length(a:L@y)=(length(a:L)+length(y)))
    by All Introduction with <L>
    and length(L@y)=(length(L)+length(y))->∀a length(a:L@y)=(length(a:L)+length(y))
        by Implies Introduction
        and length(L@y)=(length(L)+length(y))
            entails ∀a length(a:L@y)=(length(a:L)+length(y))
            by All Introduction with <a>
            and length(L@y)=(length(L)+length(y))
                entails length(a:L@y)=(length(a:L)+length(y))
                by rewriting
                and length(L@y)=(length(L)+length(y))
                    entails S(length(L@y))=S(length(L)+length(y))
                    by axiom <Subst> on premise 1
                        with length(L)+length(y) for y
                        and  length(L@y) for x
                    and S(length(L)+length(y))=S(length(L)+length(y))
                        is immediate
QED

Buffer: Main      Root: Theorem
```

Figure 5

3 The Features of the IPE

These three proofs above illustrate the main characteristics of IPE, we will now discuss the features of the IPE more systematically. This account is mainly intended to provide a user's viewpoint of the system, so the details of the internals of the implementation will only be touched upon briefly. An extended account of these may be found in [17].

3.1 The Logic, Proofs and the Proof Display

Proofs in the IPE are generated in a top down fashion with goals being refined into subgoals. The logic is intuitionistic first order logic defined as a natural deduction system, each logical connective being defined by a pair of refinement tactics implementing an elimination rule and an introduction rule. Negation is defined by $\neg A$ is **A -> false**, and the negation elimination rule. Goals in the proof are represented by sequents, which can be thought of as conforming to the SML type definition

```
type sequent = {premises:formula list,conclusion:formula}
```

The introduction rules refine the formula in the conclusion whilst the elimination rules refine a formula from the premises. Figure 6 tabulates the rules implemented in IPE.

Structural Rules

Immediate

$$\frac{}{\Gamma, A \; entails \; A}$$

Weakening

$$\frac{\Gamma, A \; entails \; C}{\Gamma \; entails \; C}$$

Duplication

$$\frac{\Gamma, A \; entails \; C}{\Gamma, A, A \; entails \; C}$$

Rules For Logical Connectives

Connective	*Elimination*	*Introduction*
And &	$\dfrac{\Gamma, A\&B \; entails \; C}{\Gamma, A, B \; entails \; C}$	$\dfrac{\Gamma \; entails \; A\&B}{\Gamma \; entails \; A \quad \Gamma \; entails \; B}$
Or $\vee$ (|)	$\dfrac{\Gamma, A \vee B \; entails \; C}{\Gamma, A \; entails \; C \quad \Gamma, B \; entails \; C}$	$\dfrac{\Gamma \; entails \; A \vee B}{\Gamma \; entails \; A \quad or \quad \Gamma \; entails \; B}$
Implies ->	$\dfrac{\Gamma, A \rightarrow B \; entails \; C}{\Gamma \; entails \; A \quad \Gamma, B \; entails \; C}$	$\dfrac{\Gamma \; entails \; A \rightarrow B}{\Gamma, A \; entails \; B}$
Not $\neg$ ($\tilde{\ }$)	$\dfrac{\Gamma, \neg A \; entails \; C}{\Gamma \; entails \; A}$	$\dfrac{\Gamma \; entails \; \neg A}{\Gamma, A \; entails \; contradiction}$
Forall $\forall$ (!)	$\dfrac{\Gamma, \forall x A(x) \; entails \; C}{\Gamma, A(t) \; entails \; C}$	$\dfrac{\Gamma \; entails \; \forall x A(x)}{\Gamma \; entails \; A(x')}$
Exists $\exists$ (?)	$\dfrac{\Gamma, \exists x A(x) \; entails \; C}{\Gamma, A(x') \; entails \; C}$	$\dfrac{\Gamma \; entails \; \exists x A(x)}{\Gamma \; entails \; A(t)}$

For the quantifier rules:-

x' indicates a new free variable name introduced by IPE

t indicates an arbitrary term which the user must input

Figure 6

Notice the rules are presented with the goal above the bar and the subgoals below the bar, which is consistent with the way proofs evolve in a refinement editor, but opposite to the usual convention for presenting deduction rules. The rules are a fairly direct translation of those found in Gentzen [7], and are complete for intuitionistic logic. The Duplicate rule is necessary to ensure completeness of the logic, because in IPE the application of an elimination rule to a premise removes that premise from the subgoal sequents. This makes the search that the immediate tactic has to perform more efficient and, more importantly, prevents the AutoProving from looping; but removing the duplication rule would render certain tautologies unprovable ($\neg\neg(A \vee \neg A)$ for instance).

As noted in the previous section proofs in IPE are extended by clicking the mouse upon a premise or conclusion. Internally this results in the construction of an attributed tree for an attribute grammar. An attribute grammar is a context free grammar where each syntactic category has an associated set of attributes, and with each production is associated an equation which defines how the attributes can be calculated. Attributes are classified as being either inherited (defined prior to the elaboration of a production of the grammar), or synthesised (computed from the attributes of the subtrees). Attribute grammars were originated by Knuth [10] to define programming language semantics.

In proof editors there is a correspondence between the proof rules for the logic and the productions of an attribute grammar, moreover the structure of the proof rule ultimately defines the equation which computes the attribute. The advantage of this technique in the implementation of proof editors is that the effect of modifying parameters can be propagated through a proof, enabling the user to see interactively the effect of changing the conjecture, modifying the instantiation of a variable or using an alternative axiom or lemma. Attribute grammars form a reasonable vehicle for implementing interactive proofs as algorithms exist to incrementally recalculate attributes (see [14]). Thus if a new branch is grafted into the proof tree only the minimum computation necessary to restore the consistency of the attributes will occur.

Internally the IPE grammar makes use of many attributes, and it is unnecessary for the user to have knowledge of any of the attributes. However it is possibly easier to understand operation of IPE if you are aware of certain of the attributes of the proof grammar:

- **Goal** is an inherited attribute of a proof - the goal sequent to which it is being applied.

- **Proven** is a synthesised attribute for a proof - a boolean which is true when each branch of the proof is resolved.

- **Appropriate** is a synthesised attribute for a proof - a boolean which is true when the expansion operation is consistent with the goal.

- **Self** is an inherited attribute for Formulae, Variables or Terms - essentially the string which defines the value of these items and which can be edited by the user.

The IPE display gives a complete presentation of the status of the current proof tree. The sequent associated with each node is shown, plus the rule used to expand it to its subgoals. The rules embodied in IPE will result in a binary and/or tree. Whilst an Or-node is unresolved both of its branches will be displayed, so that the alternative branch can be expanded if desired, but upon the successful conclusion of an Or-branch the other branch will disappear. Editable inputs to the proof (instantiations of terms and formulae, and the names of lemmas) are displayed within angle brackets and can be changed by invoking the text editor, by clicking **MM** at that point. For certain operations such as saving and restoring proofs or copying into and applying proof buffers, it is necessary that a node in the proof tree be selected. This can be done by clicking **LM** at the word "show" or the proof expansion rule associated with that node.

Figure 7

Figure 7 presents the display for an attempted proof of $\neg(A\&B) \rightarrow \neg A \vee \neg B$ in the context of the theory Classical. Note that an OR-branch has been completely explored whilst its companion is still unexpanded. From the highlighting we can see the attempted proof of the OR-branch is the currently selected point of the proof tree.

Of course this form of display will rapidly fill the screen. It is necessary to scroll the screen over such a display, and focus in and out to selected nodes of the proof tree. This capability is provided by the icons to right and left of the titlebar.

3.2 AutoMove and AutoProve

In its basic mode of operation the path taken by IPE through a proof is the sole responsibility of the user, however as one becomes more experienced with the system this mode of operation quickly becomes tedious. The **Automove** and **Autoprove** modes (which can be switched on and off by the use of icons on the titlebar) provide a facility for the system to assist in the elaboration of the proof. The Automove is purely a housekeeping device; it highlights the next unresolved branch in a depth first traversal of the proof tree. Whilst simple, this is very useful, in the context of the IPE's proof display, for directing the user's attention to nodes requiring work.

The Autoprove mode is more sophisticated. In keeping with the IPE's philosophy of leaving control of the proof solely in the hands of the user it is not an automatic theorem proving strategy. Rather the intention is that it should carry out sequences of proof expansions which are inevitable in some sense (i.e. no real choice is involved). As a corollary to this property Autoprove should not take the proof down a blind alley (i.e. a path which fails but which has a successful alternate path), for the choice of which point to backtrack to might present difficulties. A simple tactic, which has these properties, is to repeatedly expand a sequent whilst it is not immediately valid and there is just one valid expansion. However consideration of a proof of A&B->B&A shows that this *BoringTactic* is too simple. After one step we have

    show A&B entails B&A

and further progress is blocked because of the tactic's inability to choose between using *AndElimination* and *AndIntroduction*. Autoprove should continue to elaborate the proof by those expansions where the order of application doesn't matter, because the rules commute in some sense. To implement this for IPE required the introduction of tactics and tacticals as in the LCF system, see [8, 17]. (This tactical subsystem is invisible and unavailable to the user.)

The autoprove tactic is then defined as

AlternateTac(*BoringTactic*,AlternateTac(*AutoIntros,AutoElim*))

where

fun AlternateTac (t1,t2) = REPEAT(ORELSE(THEN(t1,t2),t2))

REPEAT, THEN and ORELSE are essentially the standard tacticals defined in LCF [8], so AlternateTac will alternately swop between trying tactic t1 and tactic t2 until both fail. *AutoIntros* tries to expand the goal by an introduction rule (but not *OrIntro*), if it is not an immediate consequence. Similarly *AutoElim* will try to expand by an elimination rule (but only for And and Exist).

The resulting tactic is remarkably successful, often what appear to be quite subtle tautologies can be proved with just one or two clicks with Autoprove mode engaged. Also, routine sequences which strip away logical constants are well adapted to the use of Autoprove. However the tactic is not completely foolproof, as it may for example expand the wrong branch of an OR-node. As Autoprove rapidly changes the look of the display it can be quite confusing for beginners to use.

3.3   Proof Buffers and Reapplying Proofs

As noted previously the incremental re-evaluation of attributes makes it feasible to graft arbitrary chunks of tree into the proof tree. Thus there is no need to simply develop a proof with single basic steps; fragments of proofs should be reusable to prove an analogous subgoal of a proof. To implement this facility IPE has been provided with the concept of a buffer. Buffers can contain arbitrary fragments of an attributed proof tree - i.e. a "Theorem" - a formula plus an associated proof, or a "Proof" an actual fragment of a proof tree, or a "Formula" a simple formula. Operations with buffers require consistency between the type of the currently selected node of the display and the type of the buffer. Of course a buffer in IPE can be completely independent of any other buffer, in the sense that the user can spawn a new Theorem buffer. Thus in the course of a proof you can switch your attention to another buffer for the proof of some relevant lemma. The important operations with buffers are

Change To Buffer Displays an alternative buffer, either pre-existing or new.

Copy Copy the selected subtree of the display into a buffer.

Apply Graft the contents of a buffer at the currently selected node of the display.

The IPE also has what is in effect an implicit buffer, invisible to the user, which retains the subtree excised by the last operation that overwrote the proof tree. The operation **Yank** applies this buffer at the currently selected node of the proof. This gives a one step undo facility, which can be quite useful as "proof by clicking" also makes unintended changes easy. Yanking itself will not affect the yank buffer.

As attributed proof trees tend to become rather large internal data structures, the IPE also has the notion of a proof skeleton. A skeleton is a sequence which records sufficient information regarding the proof steps to recreate the proof in the same context. Like buffers, skeletons reflect the type of node of the attributed tree selected at their creation. Skeletons can be written to and read from files, and so provide an ability to retain a proof state between sessions, or to maintain a kind of tactic library. However the skeletons probably should preserve more information about the context to provide a worthwhile reprove facility.

Additionally there is a facility to print out a conventional bottom up presentation of the proof, to provide documentation for completed proofs.

3.4   The Theory Database, Using Facts and Rewriting

Of course most proofs occur relative to the hypotheses of some theory, and IPE has the capacity for working with a theory database. Theories are not completely integrated into the system, partly because of certain shortcomings of the interface between UNIX and the version of ML used to implement IPE. An IPE theory is a UNIX directory containing a file, the ".environment" file, recording names of constants and operations specific to the theory and the names of subtheories upon which it depends, and further files each of which define an axiom or lemma of the theory. Axioms are created externally of the system, whilst lemmas are created by the system but both have the same textual form. A typical Axiom file is shown below:

```
axiom ListInduction is
    phi(nil) & (!L (phi(L) -> (!a phi(a:L)))) -> phi(InductionVariable)
generic terms InductionVariable
generic formulae phi(L)
```

Notice that certain formulae and terms are classified as generic, this implies that their value can be set by a substitution, which is either input interactively by the user or discovered via a matching algorithm. This enables axiom schemas to be defined and universally quantified expressions to be presented as schemas. Once a theory (and recursively all its subtheories) has been loaded into the system its facts are available for use in proofs. Facts may be used to resolve some goal sequent: the conclusion or a premise has to be selected from the display of the sequent by clicking **LM**. Selecting the conclusion lets the fact be used in a backwards direction, whilst selecting a premise lets the fact be used in a forwards direction. The IPE expands a fact by use of Implies and And eliminations to transform it into a form which is usually most convenient for the progress of the proof. A variety of mechanisms have been supplied to aid the user in selecting the most appropriate fact.

- **Choose Fact by Name** Here the user must select a fact from a list of names of all the available facts. All the generic terms and formulae have to be filled in interactively.

- **Choose Matching Fact** Here the system will present a list of all the matches between the facts and the goal. The user can then select which of these should be used to extend the proof.

- **Choose from all Facts** Here each fact is interactively presented in turn to the the user, showing the form it takes after the best partial match between it and the formula selected from the sequent. Again the user can choose which fact should be used in the proof.

- **Match Chosen Fact** This is similar to choosing a fact by name, but now a list of facts is passed on to the matcher which can supply the substitution for the generic variables.

The order in which facts are searched reflects the order in which the theories have been loaded into the system. Each of these operations make use of a chooser window, such as shown in Figure 8

Quite frequently proofs involve equalities. To prove these at the lowest levels of the logic is quite tedious so the IPE has a number of adaptations for this case.

- If an equality in a goal doesn't directly match one of the facts of the current theory, the usual tactic is to use transitivity to decompose the goal into two equalities. This is the default action initiated by clicking upon a conclusion.

- A premise which is an equality is often used to simplify the conclusion. Hence clicking upon an equality premise will substitute any occurrence in the conclusion of the premise's left hand term by its right hand term.

- Often equalities in theories of data have a computational intent, they describe how terms should be simplified to a normal form. To take advantage of this observation the IPE has been extended with a very simple rewriting facility. Theorems which are equalities can be added to a list of rewrite rules. The menu action **Rewrite** will apply a simple leftmost outermost rewriting function, with all the loaded rules, to simplify the terms of an equality. It is up to the user to take care that the set of rewrite rules has nice termination properties.

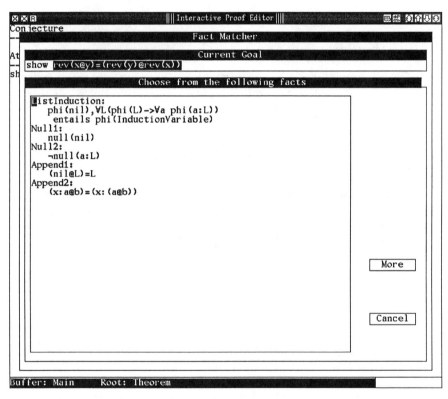

Figure 8: A Chooser

4 Some Reflections on Interactive Proof Development

4.1 Using IPE

The IPE was a first attempt in applying workstation and attribute grammar technology in the construction of interfaces to proof assistants. It has shown that "proof by clicking" provides a very immediate entry into the activity of proof. An understanding of the possibilities and mechanics of proof, can be developed incrementally through a graded sequence of examples. The ease with which proofs are constructed, and the lack of penalty involved in jumping about the search space, allows students to develop an understanding of the logical operations via experiment. Also presenting the more sophisticated operations of IPE via structured menus allows students to unfold as much or as little of the system as their experience requires. Compare this with the tactical style of proof embodied in system such as LCF, where the activity of proof is guarded by an intimidating array of operations with obscure technical titles, presupposing some understanding of logic, and the use of a new programming language ML.

Although IPE clearly has advantages as a teaching tool, more work is required to prove its worth in a more general setting. The most complex case studies attempted have been of a simple parser after Cohn [1] and a simple compiler for expressions [5]. This is in the main a reflection of the difficulties in formulating problems in the IPE's logic which lacks notions of set or type. However it is possible to see how IPE style proof editors would affect the proof process.

IPE lends itself to an exploratory style of proof. Thus one will often attempt a proof without a clear understanding of which lemmas will be necessary for its successful completion. However a lemma, or a new axiom itself, or a change to the conjecture will frequently suggest itself in the course of the proof. The mutability of proof objects and the facility to stack proofs embodied in the IPE, shows itself to be particularly useful in such situations. This is in marked contrast to the style of proof adopted and recommended by practitioners of LCF and Boyer-Moore theorem provers (see [12, 4]) where a plan of the key lemmas and concepts is required in advance of the actual proof. Thus a proof will frequently reveal an error or inconsistency or incompleteness in the formulation of a definition. This is perhaps a true reflection of the significance of the role of proof in formal semantics: it is the chief experimental tool available to reassure formalists of the validity of their formalisations. See for instance Avra Cohn's discussion of the verification of the Viper specification [2], or in a wider setting Lakatos's analysis of role of proof in mathematics [11]. It is tempting to conjecture that formal methods will only play a prominent role in information technology when accessible systems which support realistic proofs become generally available.

4.2 The Structure of Proofs

The IPE development identified various classes of features of general application in a proof development environment. It is interesting to speculate how these can be made more powerful and yet more consistent with one another. Indeed it would be a worthwhile exercise to make a more formal analysis of the proof process in the light of the IPE experience. The collection of attributes associated with a proof tree in IPE is essentially very simple: with the goal and input terms as inherited attributes and the status (proved or unresolved) of the goal a synthesised attribute. There are other attributes concerned with how the proof is displayed on the screen but of no direct relevance to the proof. However certain enhancements suggest that a more sophisticated collection of attributes is required.

At present the instantiation of terms must be supplied by the user of IPE ; except in the case of application of a fact where matching can supply the instantiation of the generic terms of a ForAll Elimination. This means that the IPE can provide no help for the user in satisfying existential goals - or in specialising universals in the most general setting. Hence Autoprove will never completely resolve a conjecture involving quantifiers, and so for example user input is necessary to prove something as simple as

$$\forall y\ A(y)\ \rightarrow\ \forall x\ A(x)$$

as the system does not test equivalence of formulae up to α-conversion. An obvious solution to this would be to extend the Immediacy test with the facility to unify a premise with a conclusion. This would imply the necessity for a further synthesised attribute of a proof recording the substitutions discovered by unification. However such an enhancement requires care, because the substitution discovered in a subtree might be relevant to the proof recorded in a parallel subtree: so the attribute must propagate both up and down the proof tree. There is a danger of defining a circular attribute, especially if the user is still to be allowed to interactively input instantiations for terms. There is further discussion of this problem in [17].

Elimination rules are applied at a particular premise of a goal sequent, so in IPE each elimination rule is defined not solely by its connective, but also in terms of the absolute position of the premise upon which it acts. This causes no problems whilst developing a proof, but if we wish to reapply a proof in a slightly different context it causes obstructions. Hence we see in the case of elimination rules the need for another inherited attribute identifying the position of the relevant premise, so the user can edit the proof to conform to the new situation. Alternatively the position of the premise might be a synthesised attribute whose value is discovered by investigating the structure of each of the premises. Of course the real significance in a hypothesis is in its information content - conveyed by its structure. In informal practise we often recognise this by attaching names to hypotheses and axioms in the course of a proof, would this be useful in this situation for the IPE? More precisely it is how the subformulae of a premise are exploited subsequently in a proof which is the real significance of a premise; it might be possible to synthesis a minimal pattern encoding this data, which the IPE could use when reapplying a proof.

The simple model of proof editing implemented in IPE assumes that proofs grow in a tree like manner. Actual proofs may be much more complex, for instance an extra step may have to be inserted into a proof, perhaps because the conjecture is edited to include an extra hypothesis. The effects of even such a simple change might propagate throughout the proof (the positions of all the hypotheses may be shifted one place). Frequently one will elaborate a sequence of steps in slightly the wrong order, and then discover a proof step is blocked because it should have occurred earlier. Although buffers and yanking provide a capability for dealing with this, sometimes a need is felt for higher level proof operations.

Further questions arise when one considers how theory structures should be integrated into the system. Should there be attributes expressing the dependency of a theorem upon a theory? Should the maintenance of the consistency of the theory database be via the attribute system?

4.3 The Interface

For IPE the display is a vital component of the system, since the path through the proof is selected by clicking upon formulae. In order to backtrack, some alternative formula must be selected from a previous state of the proof. This feature has the unfortunate consequence that all of the information relevant to the proof has to be present within the display. As a result the screen rapidly becomes packed with redundant data. Additionally a proof is often dependent upon facts contained within a particular theory, and there is no convenient method for displaying this larger context concurrently with the local state of the proof. Thus devising an effective display for an interactive proof editor is an interesting challenge. At any point in the proof both local and global data are of interest: to choose the next step it is necessary to see all the local data in the context; whilst to backtrack in the search space it is necessary to know how the proof developed, and at what stage the premises were introduced into the context (i.e. what was the subgoal at that time). The display might also convey other information of interest, such as whether a premise has been exploited in a proof.

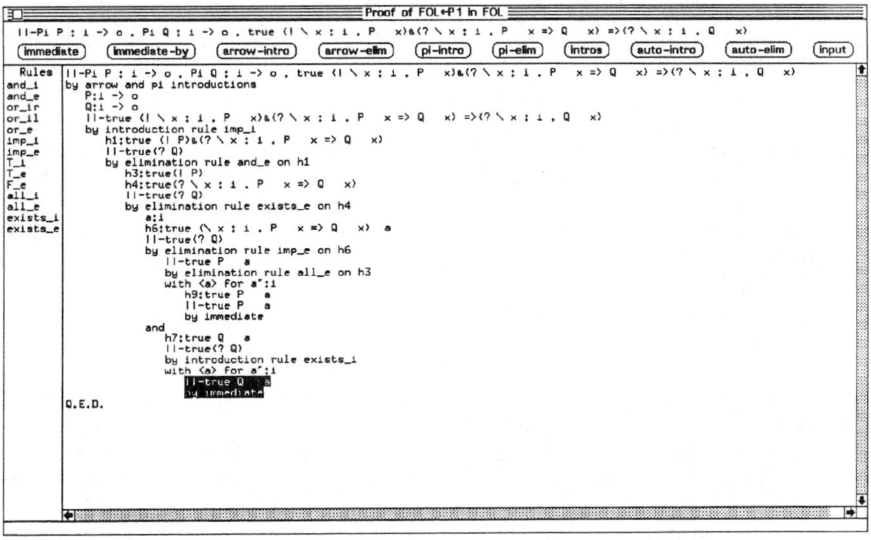

Figure 9

Figure 9 shows an alternative style of display (devised by Lin Huimin and Tatsuya Hagino for use with a prototype), which is somewhat more economic than the IPE's in presenting the proof data. However yet more efficient layouts which make better use of the capabilities of the technology can be imagined e.g. different shadings or colour, multiple displays. We haven't considered the issue of pretty printing formulae in IPE, however how a formula is displayed can often be significant in deciding the next step of a proof, especially as formulae become large and complex. Similarly it is possible to imagine alternative schemes for associating the operations with the mouse buttons: the IPE could be driven using a mouse with just a single button and a variety of menus. These interface aspects may appear to have little theoretical relevance, but can significantly affect the way users understand and exploit the system.

4.4   The Role of Tactics

It has already been noted that IPE was built to explore an alternative to tactic driven proof systems, but we also pointed out that there is an underlying tactic level to the IPE. The questions naturally arise

> "Should this tactic level be made available to users of the IPE? "
> "If it is, in what way should it be made accessible? "

Initially the immediacy of proof by clicking, especially with the Autoprove feature, leads one to question the need for certain classes of syntactic tactics and hence possibly all tactics. Still the Autoprove and the Rewrite tactics, despite their simplicity, have proved their worth in the system. But perhaps they are useful because they have been used only in simple settings: sometimes one might need to switch off part of the Autoprove, and there are situations where one would like to supply interactively the rules Rewrite uses. Again it might be useful to be able to supply a default tactic for a non-logical operator (as is already the case with equality) or, in general, define a default tactic to be used in the context of a particular theory. All these considerations suggest that the answer to the first question should be "Yes", but the creation of tactics should be guarded. Users shouldn't have the ability to usurp the lowest level of operation of the system and hence compromise the consistency of the logic. Again it should be possible to backtrack over a tactic, or interrupt a tactic and still maintain the consistency of the proof and all its attributes.

Of course even if the problems of integrating tactics into the system were solved, a mechanism also has to provided to create new tactics. The simplest solution might be to make ML and the tactic level directly accessible to the users; however that seems somehow contrary to the spirit of an IPE style of system, and would almost certainly interfere with the attribute consistency of the system. An alternative might be to provide the ability to create buffers where fragments of proof trees (or proof tree skeletons) may be constructed by elaborating a null goal, by selecting from the basic operations embodied in the IPE. In addition it might be possible to provide iterators, conditionals and facilities to preset some of the proof inputs to make it look more like a language of tacticals. The standard attribute evaluation mechanism would still have to be invoked to apply these buffers.

The facility the IPE already has for copying fragments of proofs into buffers and saving proof skeletons provides an ability to create tactics. These tactics are in some sense "proofs by example". Considering how this works in the case of an inductive proof, might point out the attractiveness of such a scheme for creating tactics and also some of its shortcomings. A proof by induction tactic will have the form :-

> Break into cases by inductive definition of the data
> Simplify the base case
> With Each Step Case
>   Clear the logical operators
>   Simplify the conclusion of the goal
>   Simplify the conclusion by the induction premises

In the course of an inductive proof this will be created interactively, possibly using Autoprove, but certainly using Rewrite; saved as proof skeleton and reapplied in a different setting by reading in the proof skeleton. However what information should the proof skeleton record? Presently the skeleton records the trace generated by Autoprove, not a reference to the use of it. This is fine for Autoprove, but for Rewrite the skeleton must record the use of the tactic not its trace. This behaviour is vital, as the details of simplifications of terms will differ for every proof. Thus if a proof by example is to be reusable, it must have within itself some flexibility, generated by the appeal to general tactics. List induction is similar in form to Natural Number induction, so one would hope that by just editing the name of the axiom we could reuse this proof by induction tactic, however this will not work because the skeleton records the trace of Autoprove, not its use. Thus even this simplistic method of creating tactics requires a richer set of built in tactics and more flexible parameterisation of proofs to be worthwhile.

5 Conclusion

IPE revealed the challenge involved in the construction of truly interactive yet powerful proof development environments. The problems that arise combine both theoretical and practical issues, whilst the systems themselves are truly multilayered: a logical kernel, controlled by a proof management system, with which one interacts via a window based interface. Appropriate solutions to these can only be found by experiments with prototypes, supported by the development of the underlying theory. Whilst further development of IPE itself has ceased, work on its successors is well under way at LFCS. Future editors will support refinement proofs, based upon use of typed lambda calculus to represent logics as in Constructions and Edinburgh LF.

References

[1] Cohn, A.J. and Milner, R., On Using Edinburgh LCF to Prove the Correctness of a Parsing Algorithm, University of Edinburgh Computer Science Department Internal Report CSR-113-82.

[2] Cohn, A.J., A Proof of Correctness of the Viper Microprocessor, Proceedings of the Calgary Hardware Verification Workshop, Calgary 1987.

[3] de Bruijn, N.G., A Survey of the Project Automath, To H.B. Curry: Essays in Combinatory Logic, Lambda Calculus and Formalism, Academic Press 1980.

[4] The User's Manual for A Computational Logic, Boyer, R.S. and Moore, J.S., University of Texas at Austin 1987.

[5] Burstall, R.M., Taylor, P. and Jones, C., Interactive Proof Editing with the Edinburgh IPE, course organised by the Laboratory for Foundations of Computer Science, University of Edinburgh 1987.

[6] Coquand, T. and Huet, G., Constructions:a higher-order proof system for mechanising mathematics, EUROCAL'85: European Conference on Computer Algebra, 151-184, Springer-Verlag 1985.

[7] Gentzen, G.,Investigations into Logical Deduction, in: The Collected Works of Gerhart Gentzen, edited by M. E. Szarbo, North Holland 1969.

[8] Gordon, M., Milner, R. and Wadsworth, C. Edinburgh LCF: A Mechanised Logic of Computation, Springer-Verlag 1978.

[9] Harper, R., Honsell, F. and Plotkin, G., A Framework for Defining Logics. IEEE Symposium on Logic in Computer Science, Ithaca 1987.

[10] Knuth, D.E., The Semantics of Context-Free Languages, Math. Syst. Theory 2, 1968.

[11] Lakatos I., Proofs and Refutations, Cambridge University Press 1976.

[12] Paulson L.C., Interactive Theorem Proving with Cambridge LCF, Technical Report 80, University of Cambridge Computer Laboratory 1985.

[13] The PRL Group (Constable et. al.) Implementing Mathematics with the Nuprl Proof Development System, Prentice Hall, New Jersey 1986.

[14] Reps, T., Generating Language Based Environments, Ph.D. Thesis, Cornell University 1982, and MIT Press 1984.

[15] Reps, T. and Alpern, B., Interactive Proof Checking, ACM Symposium on Principles of Programming Languages, Salt Lake City 1984.

[16] Reps, T. and Teitelbaum T., The Synthesiser Generator. Proceeding ACM SIGPLAN Symposium on Software Development Environments, Pittsburgh, 1984

[17] Ritchie B., The Design and Implementation of an Interactive Proof Editor, Ph.D. Thesis , Edinburgh University 1987.

An Overview of the Edinburgh Logical Framework

Arnon Avron, Furio Honsell and Ian A. Mason
Laboratory for Foundations of Computer Science*

August 29, 1988

1 Introduction

In recent years there has been a growing interest in using computers as an aid for correctly manipulating logical systems, as well as using formal systems for correctly designing computers. However, implementing a proof environment for a specific logical system is both complex and time-consuming, this—together with the proliferation of logics—suggests that a uniform and reliable alternative is desirable. One such alternative is the Edinburgh Logical Framework (LF), currently under development at the LFCS. The LF is a logic-independent tool which, given a specification for a logical system, synthesizes a proof editor and checker for that system. Its specification language is based on a general theory of logics, which enables one to capture uniformities and idiosyncrasies of a large class of logics without sacrificing generality for tractability. Peculiarities (such as side conditions on rule application, variable occurrence or formula formation) are expressed at the level of the specification.

The paper [7] describes the basic features of the LF, while the paper [1] provides a broader illustration of its applicability and discusses to what extent it is successful. This paper serves as an introduction to the LF and summarizes the main points made in [1]. It is organized as follows. In section 2 we provide an outline of the LF specification language. This is done in somewhat more detail than is necessary on first reading. In section 3 we give a simple example of a specification. In section 4 we discuss the general LF paradigm for specifying a logical system. The subsequent sections illustrate this paradigm. Section 5 deals with modal logics, section 6 deals with various lambda calculii and in section 7 we discuss program logics.

2 The LF Specification Language

The LF specification language is a weak constructive type theory, more specifically a Π-typed λ-calculus, closely related to AUT-PI and AUT-QE [4], to Martin Löf's early type theories and to Meyer and Reinhold's λ^π [10]. It is a calculus for establishing the correctness and equivalence (i.e. definitional equality) of certain constructions. These constructions involve four kinds of objects: functions; types, i.e. assertions about functions; type valued functions, i.e. predicates of the assertion language; and kinds,

---

*Computer Science Department, Edinburgh University, Edinburgh, EH9 3JZ.

i.e. assertions about typed valued functions. The only type (or kind) constructor is the dependent product, Π. The abstract syntax is given by the following:

Kinds	K	$::=$	Type $\mid \Pi_{x:A}.\,K$
Type Families	A	$::=$	$c \mid \Pi_{x:A}.\,B \mid \lambda x : A.B \mid AM$
Objects	M	$::=$	$c \mid x \mid \lambda x : A.M \mid MN$

We let M and N range over expressions for objects, A and B for types and families of types, K for kinds, x and y over variables, and c over constants. Types are used to classify objects, while kinds are used to classify types and type families. The kind Type classifies the basic types. The other kinds classify type families, i.e. functions from objects to types. Any function definable in the system has a type as domain, while its range can either be a type, if it is an object, or a kind, if it is a family of types. The LF type theory is therefore predicative.

The theory we shall deal with is a formal system for deriving assertions of one of the following shapes:

$$\begin{array}{ll} \Gamma \vdash_\Sigma K & K \text{ is a kind} \\ \Gamma \vdash_\Sigma A : K & A \text{ has kind } K \\ \Gamma \vdash_\Sigma M : A & M \text{ has type } A \end{array}$$

where the syntax for Signatures and contexts is specified by the following grammar:

| Signatures | Σ | $::=$ | $\langle\rangle \mid \Sigma, c : K \mid \Sigma, c : A$ |
| Contexts | Γ | $::=$ | $\langle\rangle \mid \Gamma, x : A$ |

We write $A \to B$ for $\Pi_{x:A}.\,B$ when x does not occur free in B. The inference rules of the LF type theory are listed below. They are grouped according to which of the three forms of assertions they concern, α-conversion is assumed throughout.

1. **Valid Kinds**

(1) $$\frac{}{\vdash \text{Type}}$$

(2) $$\frac{\Gamma \vdash_\Sigma A : \text{Type} \quad \Gamma, x : A \vdash_\Sigma K}{\Gamma \vdash_\Sigma \Pi_{x:A}.\,K}$$

2. **Valid Elements of a Kind**

(3) $$\frac{\Gamma \vdash_\Sigma \text{Type} \quad c : K \in \Sigma}{\Gamma \vdash_\Sigma c : K}$$

(4) $$\frac{\vdash_\Sigma K \quad c \notin \text{Dom}(\Sigma)}{\vdash_{\Sigma,c:K} \text{Type}}$$

(5) $$\frac{\Gamma \vdash_\Sigma A : \text{Type} \quad \Gamma, x : A \vdash_\Sigma B : \text{Type}}{\Gamma \vdash_\Sigma \Pi_{x:A}.\,B : \text{Type}}$$

(6) $$\frac{\Gamma \vdash_\Sigma A : \text{Type} \quad \Gamma, x : A \vdash_\Sigma B : K}{\Gamma \vdash_\Sigma \lambda x : A.B : \Pi_{x:A}.\,K}$$

(7)
$$\frac{\Gamma \vdash_\Sigma B : \prod_{x:A} . K \quad \Gamma \vdash_\Sigma N : A}{\Gamma \vdash_\Sigma BN : [N/x]K}$$

(8)
$$\frac{\Gamma \vdash_\Sigma A : K \quad \Gamma \vdash_\Sigma K' \quad K =_{\beta\eta} K'}{\Gamma \vdash_\Sigma A : K'}$$

3. Valid Elements of a Type

(9)
$$\frac{\vdash_\Sigma A : \text{Type} \quad c \notin \text{Dom}(\Sigma)}{\vdash_{\Sigma, c:A} \text{Type}}$$

(10)
$$\frac{\Gamma \vdash_\Sigma A : \text{Type} \quad x \notin \text{Dom}(\Gamma)}{\Gamma, x : A \vdash_\Sigma \text{Type}}$$

(11)
$$\frac{\Gamma \vdash_\Sigma \text{Type} \quad M : A \in \Sigma \cup \Gamma}{\Gamma \vdash_\Sigma M : A}$$

(12)
$$\frac{\Gamma \vdash_\Sigma A : \text{Type} \quad \Gamma, x : A \vdash_\Sigma M : B}{\Gamma \vdash_\Sigma \lambda x : A.M : \prod_{x:A} . B}$$

(13)
$$\frac{\Gamma \vdash_\Sigma M : \prod_{x:A} . B \quad \Gamma \vdash_\Sigma N : A}{\Gamma \vdash_\Sigma MN : [N/x]B}$$

(14)
$$\frac{\Gamma \vdash_\Sigma M : A \quad \Gamma \vdash_\Sigma A' : \text{Type} \quad A =_{\beta\eta} A'}{\Gamma \vdash_\Sigma M : A'}$$

A term is said to be *well–typed in a signature and context* if it can be shown to either be a kind, have a kind, or have a type in that signature and context. A term is *well–typed* if it is well-typed in some signature and context. The notion of $\beta\eta$–contraction, written $\to_{\beta\eta}$, can be defined both at the level of objects and at the level of types and families of types in the obvious way. Rules (8) and (14) make use of a relation $=_{\beta\eta}$ between terms which is defined as follows: $M =_{\beta\eta} N$ iff $M \to^*_{\beta\eta} P$ and $N \to^*_{\beta\eta} P$ for some term P. The following theorem from [7] summarizes the basic theoretical facts about LF (here α ranges over the basic assertions of the type theory):

Theorem 1

1. **Thinning:** *thinning is an admissible rule: if* $\Gamma \vdash_\Sigma \alpha$ *and* $\Gamma, \Gamma' \vdash_{\Sigma, \Sigma'} \text{Type}$, *then* $\Gamma, \Gamma' \vdash_{\Sigma, \Sigma'} \alpha$.

2. **Transitivity:** *transitivity is an admissible rule: if* $\Gamma \vdash_\Sigma M : A$ *and* $\Gamma, x : A, \Delta \vdash_\Sigma \alpha$, *then* $\Gamma, [M/x]\Delta \vdash_\Sigma [M/x]\alpha$.

3. **Uniqueness of types and kinds:** *if* $\Gamma \vdash_\Sigma M : A$ *and* $\Gamma \vdash_\Sigma M : A'$, *then* $A =_{\beta\eta} A'$, *and similarly for kinds.*

4. **Subject reduction:** *if* $\Gamma \vdash_\Sigma M : A$ *and* $M \to^*_{\beta\eta} M'$, *then* $\Gamma \vdash_\Sigma M' : A$, *and similarly for types.*

5. **Confluence:** *all well–typed terms are Church–Rosser, while in general this is false.*

6. **Strong Normalization:** *all well–typed terms are strongly normalizing.*

7. **Decidability:** *each of the three relations defined by the inference system of the LF is decidable, as is the property of being well–typed.*

8. **Predicativity:** *if $\Gamma \vdash_\Sigma M : A$ then the type free λ-term obtained by erasing all type information from M can be typed in the Curry type assignment system.*

3   A Simple Specification of a Logic

Encoding the classical propositional calculus (with the connectives $\neg$ and $\supset$) provides a simple example of how, in general, a logic is encoded. The set of well formed formulas, o, is represented as a basic type. This is achieved by declaring it to be of kind Type. The connectives then correspond to unary and binary functions on o. This is summarized in the following.

- Syntactic Category of Formulas

$$o \; : \; \text{Type}$$

- Operations

$$\neg \; : \; o \to o$$
$$\supset \; : \; o \to o \to o$$

Propositional atoms (or variables) are then simply constants (or LF variables) of type o. The judgement (or predicate encoding the assertion) that a formula is provable is encoded as a family of types.

- Judgement

$$T \; : \; o \to \text{Type}$$

The set of proofs of a formula Φ is identified, in this encoding, with the basic type $T(\Phi)$. Consequently showing that a formula Φ is provable reduces to producing a term of type $T(\Phi)$. Similarly checking whether or not an object M is a proof of Φ reduces to checking whether or not the object M has type $T(\Phi)$. This style of encoding is refered to as *the judgements as types principle* in the paper [7].

There are several systems for constructing proofs in the propositional calculus. Two well known examples, which we shall treat in this section, are Hilbert style and natural deduction style systems. Encoding a Hilbert style system involves introducing a constant for each axiom schema together with a constant corresponding to modus ponens. The following is one such system.

- Hilbert Style Axioms and Rules

$$
\begin{array}{lll}
A_1 & : & \prod_{\Phi_1, \Phi_2 : o} \quad T(\Phi_1 \supset (\Phi_2 \supset \Phi_1)) \\
A_2 & : & \prod_{\Phi_1, \Phi_2, \Phi_3 : o} \quad T(\Phi_1 \supset (\Phi_2 \supset \Phi_3) \supset (\Phi_1 \supset \Phi_2) \supset (\Phi_1 \supset \Phi_3)) \\
A_3 & : & \prod_{\Phi_1, \Phi_2 : o} \quad T((\neg \Phi_1 \supset \neg \Phi_2) \supset (\Phi_2 \supset \Phi_1)) \\
MP & : & \prod_{\Phi_1, \Phi_2 : o} \quad T(\Phi_1 \supset \Phi_2) \to T(\Phi_1) \to T(\Phi_2)
\end{array}
$$

Encoding a natural deduction style system, on the other hand, involves introducing constants corresponding to each introduction and elimination rule. In this case the following declarations suffice.

- Natural Deduction Style Rules

$$\supset E \quad : \quad \Pi_{\Phi_1,\Phi_2:o} \quad T(\Phi_1 \supset \Phi_2) \rightarrow T(\Phi_1) \rightarrow T(\Phi_2)$$
$$\supset I \quad : \quad \Pi_{\Phi,\Psi:o} \quad (T(\Phi) \rightarrow T(\Psi)) \rightarrow T(\Phi \supset \Psi)$$
$$\neg I \quad : \quad \Pi_{\Phi,\Psi:o} \quad (T(\Phi) \rightarrow T(\Psi)) \rightarrow (T(\Phi) \rightarrow T(\neg\Psi)) \rightarrow T(\neg\Phi)$$
$$\neg E \quad : \quad \Pi_{\Phi:o} \quad T(\neg\neg\Phi) \rightarrow T(\Phi)$$

In both versions schemata are encoded as functions mapping formulas to the instatiated axioms and rules. Instantiation of a schema is thus simply application. For example, if $\Phi : o$ is a formula, then $\supset I(\Phi)(\Phi)$ inhabits the type $(T(\Phi) \rightarrow T(\Phi)) \rightarrow T(\Phi \supset \Phi)$. Hence to show that $\Phi \supset \Phi$ is provable, it suffices to show that $T(\Phi \supset \Phi)$ is inhabited, which reduces to producing an element of type $T(\Phi) \rightarrow T(\Phi)$. The identity function $\lambda x : T(\Phi).x$ satisfies this requirement. Thus

$$\supset I(\Phi)(\Phi)(\lambda x : T(\Phi).x) : T(\Phi \supset \Phi).$$

The $\supset I$ rule also provides an example of how assumption discharge is handled in the LF. To prove $\Phi \supset \Psi$ it suffices to assume that Φ is true, by introducing a variable $x : T(\Phi)$, and producing a proof of Ψ, i.e. an object $B(x) : T(\Psi)$. The initial assumption that Φ is true is discharged by forming $\lambda x : T(\Phi).B(x)$ and supplying it to $\supset I$. Thus both assumption discharge and the schematic nature of rules and axioms is handled by lambda abstraction.

One of the positive features of the LF is that proofs of theorems, derivable rules, axioms and basic rules are treated uniformly, they are all simply LF objects of a particular type. So for example, in the natural deduction style system we could introduce a new connective encoding conjunction together with constants encoding its introduction and elimination rules as follows:

- Conjunction

$$\wedge \quad : \quad o \rightarrow o \rightarrow o$$

- Introduction and Elimination Rules for Conjuction

$$\wedge I \quad : \quad \Pi_{\Phi,\Psi:o} \quad T(\Phi) \rightarrow T(\Psi) \rightarrow T(\Phi \wedge \Psi)$$
$$\wedge E_l \quad : \quad \Pi_{\Phi,\Psi:o} \quad T(\Phi \wedge \Psi) \rightarrow T(\Phi)$$
$$\wedge E_r \quad : \quad \Pi_{\Phi,\Psi:o} \quad T(\Phi \wedge \Psi) \rightarrow T(\Psi)$$

A more frugal, but equivalent approach, would be to let $\wedge$ denote the LF term

$$\lambda x : o.\lambda y : o.\neg(x \supset \neg y).$$

Thus $\wedge$ is of type $o \rightarrow o \rightarrow o$. One could then construct LF terms, out of the constants governing $\neg$ and $\supset$, which inhabited the same type as the rules for conjunction above. For example, suppose that $x : T(\Phi)$ and $y : T(\Psi)$. Then letting

$$\Delta_\Psi \quad = \lambda z : T(\Phi \supset \neg\Psi).y$$
$$\Delta_{\neg\Psi} \quad = \lambda z : T(\Phi \supset \neg\Psi).\supset E(\Phi)(\neg\Psi)(z)(x)$$

we have that

$$\Delta_\Psi \quad : T(\Phi \supset \neg \Psi) \to T(\Psi)$$
$$\Delta_{\neg \Psi} \quad : T(\Phi \supset \neg \Psi) \to T(\neg \Psi).$$

Applying $\neg$I yields

$$\Delta \ = \neg I(\Phi \supset \neg \Psi)(\Psi)(\Delta_\Psi)(\Delta_{\neg \Psi}),$$

which is a term of type $T(\neg(\Phi \supset \neg \Psi))$, which by definition is the same as $\Delta : T(\Phi \wedge \Psi)$. Discharging the assumptions regarding the truth of Φ and Ψ gives

$$\lambda x : T(\Phi).\lambda y : T(\Psi).\Delta \ : \ T(\Phi) \to T(\Psi) \to T(\Phi \wedge \Psi),$$

which is the same type as the constant $\wedge$I above.

Thus derived rules and basic rules are treated as equal in the LF. The case is somewhat different for admissible rules however. Both the encoded natural deduction system and the encoded Hilbert system are equivalent in the sense that

$$\Phi_1, \ldots, \Phi_n \vdash \Psi$$

in the classical propositional calculus iff there exists an LF term Δ such that, assuming $p_1 : o, \ldots, p_m : o$ (the $p_1, \ldots, p_m$ being the atomic variables occurring in $\Phi_1, \ldots \Phi_n$ and Ψ), we can show that

$$\Delta : T(\Phi_1) \to \cdots \to T(\Phi_n) \to T(\Psi)$$

in either of the above LF encodings (we use the same symbol for denoting a formula and the corresponding term in LF).

This does not mean they have that same *higher order rules*. For example in the Hilbert system the type

$$\Pi_{\Phi, \Psi:o}(T(\Phi) \to T(\Psi)) \to T(\Phi \supset \Psi)$$

corresponds to the non-trivial direction of the deduction theorem. The theorem is proved by induction on the complexity of proofs. Since no such proof procedure is available in the LF presentation, we cannot produce a term which inhabits this type. In the natural deduction system the type is inhabited by the basic rule $\supset$I.

We can extend either of the above specifications to encompass full predicate logic. In this presentation we will be content with extending the natural deduction style system. We begin by adding a type, i, encoding the set of individuals (over which the quantifiers will range) together with the equality predicate on them.

- Additional Syntactic Category

$$i \ : \ \text{Type}$$

- Additional Operation

$$= \ : \ i \to i \to o$$

The rules governing equality are then easily stated.

- Equality Rules

$$E_0 \; : \; \Pi_{x:i} \quad T(x = x)$$
$$E_1 \; : \; \Pi_{\substack{x,y:i \\ t:i\to i}} \quad T(x = y) \to T(t(x) = t(y))$$
$$E_2 \; : \; \Pi_{\substack{x,y:i \\ \Phi:i\to o}} \quad T(x = y) \to T(\Phi(x)) \to T(\Phi(y))$$

Note that the LF encoding of substitution is carried out by lambda abstraction and application. Thus the LF encoding of the rule

$$\frac{x = y}{\Phi[x/z] = \Phi[y/z]}$$

is schematic not in $\Phi : o$ but rather in $\Psi = \lambda z.\Phi$ of type $i \to o$. Thus $\Phi[x/z]$ is simply encoded as $\Psi(x)$. Binding operators are handled similarly. We provide three examples of how one handles binding operators in the LF. Apart from the quantifiers $\exists$ and $\forall$ we include ϵ, a version of Hilbert's choice operator. The only rule concerning the choice operator is the introduction rule:

$$\frac{\exists x \Phi(x)}{\Phi(\epsilon x \Phi(x))}$$

The binding operators are encoded as follows.

- Binding Operators

$$\epsilon \; : \; (i \to o) \to i$$
$$\exists \; : \; (i \to o) \to o$$
$$\forall \; : \; (i \to o) \to o$$

If x is a variable of type i, then $x = x$ is a term of type o. We can bind x by λ-abstraction obtaining an object of type $i \to o$, $\lambda x : i.x = x$. The binding operators applied to this give

1. $\epsilon(\lambda x : i.x = x)$, which represents the first–order term $\epsilon x.x = x$.

2. $\exists(\lambda x : i.x = x)$, which represents the first–order formula $\exists x.x = x$.

3. $\forall(\lambda x : i.x = x)$, which represents the first–order formula $\forall x.x = x$.

Representing binding operators as constructors of higher order type allows for α-conversion and substitution to be taken care of by the LF, rather than axiomatized by the encoding. Another consequence of this method of encoding is that we can identify the variables of the object logic with those of the LF (of type i). It also allows for a smooth representation of instantiating a quantified formula (and generalizing and instantiated one) as the constants encoding the introduction and elimination rules below indicate.

- Introduction and Elimination Rules

$$\epsilon I \; : \; \Pi_{\Phi:i\to o} \quad T(\exists(\Phi)) \to T(\Phi(\epsilon(\Phi)))$$
$$\exists E \; : \; \Pi_{\substack{\Phi:i\to o \\ \Psi:o}} \quad T(\exists(\Phi)) \to (\Pi_{x:i}T(\Phi(x)) \to T(\Psi)) \to T(\Psi)$$
$$\exists I \; : \; \Pi_{\substack{\Phi:i\to o \\ t:i}} \quad T(\Phi(t)) \to T(\exists(\Phi))$$
$$\forall E \; : \; \Pi_{\substack{\Phi:i\to o \\ t:i}} \quad T(\forall(\Phi)) \to \Phi(t)$$
$$\forall I \; : \; \Pi_{\Phi:i\to o} \quad (\Pi_{t:i}T(\Phi(t))) \to T(\forall(\Phi))$$

Note that side conditions on variable occurrence are handled by the scoping conventions of the underlying lambda calculus. For example, the introduction rule for $\forall$ is usually stated as

$$\frac{\Phi(x)}{\forall x \Phi(x)}$$

with the side condition that the variable x does not occur free in any assumption that Φ depends on. The side condition is encoded in the LF by requiring a parametric proof, which instantiated at any variable $x : i$ yields a proof of $\Phi(x)$. This slight rewording of the rule is easily seen to be equivalent to the classical formulation.

An example of a proof in this system is

$$\Delta \; : \; \Pi_{\Phi : i \to o} \; T(\forall(\Phi)) \to T(\exists(\Phi))$$

where Δ is the following term

$$\lambda \Phi : i \to o \; . \; \lambda \rho : T(\forall(\Phi)) \; . \; \exists I(\Phi)(\epsilon(\Phi))(\forall E(\Phi)(\epsilon(\Phi))(\rho))$$

We should point out that if we removed the choice operator from the signature this example would no longer be provable, unless one explicitly added a constant of type i. The adequacy of this representation is expressed in the following theorem.

Theorem 2 *Letting*

$$\Gamma = \{x_1 : i, \ldots x_n : i, X_1 : i \to o, \ldots X_m : i \to o\}$$

the following hold:

1. *$\Gamma \vdash M : i$ iff $\Phi_\Gamma(M)$ is a well formed term of first order logic with a choice operator whose only free individual variables are among $x_1, \ldots, x_n$ and whose unary relations are among the $X_1, \ldots, X_m$.*

2. *$\Gamma \vdash M : o$ iff $\Phi_\Gamma(M)$ is a well formed formula of first order logic with a choice operator whose only free individual variables are among $x_1, \ldots, x_n$ and whose unary relations are among the $X_1, \ldots, X_m$.*

3. *$(\exists M)(\Gamma \cup \{y_1 : \mathrm{True}(\Psi_1), \ldots, y_k : \mathrm{True}(\Psi_k)\} \vdash M : \mathrm{True}(\Psi))$*

$$\text{iff}$$

$$\Phi_\Gamma(\Psi_1), \ldots, \Phi_\Gamma(\Psi_k) \vdash \Phi_\Gamma(\Psi).$$

where Φ_Γ is a bijective function

$$\Phi_\Gamma : \Xi_\Gamma(i) \cup \Xi_\Gamma(o) \to \epsilon 1^o$$

to be defined shortly. $\epsilon 1^o$ denotes the collection of terms and formulas of first order logic with a choice operator whose only free individual variables are among $x_1, \ldots, x_n$ and whose unary relations are among the $X_1, \ldots, X_m$. $\Xi_\Gamma(\tau)$ is the set of long $\beta\eta$ normal forms of type τ in the context Γ. Finally

$$\Phi_\Gamma(M) = \begin{cases} x & \text{if } M \equiv x \\ \Phi_\Gamma(M') = \Phi_\Gamma(N) & \text{if } M \equiv (M' = N) \\ \epsilon(\Phi_{\Gamma \cup \{x : i\}}(P[x])) & \text{if } M \equiv \epsilon(\lambda x : i.P[x]) \\ \neg \Phi_\Gamma(M') & \text{if } M \equiv \neg M' \\ \Phi_\Gamma(M') \supset \Phi_\Gamma(N) & \text{if } M \equiv M' \supset N \\ \forall x.\Phi_{\Gamma \cup \{x : i\}}(M'[x]) & \text{if } M \equiv \forall(\lambda x : i.M'[x]) \\ \exists x.\Phi_{\Gamma \cup \{x : i\}}(M'[x]) & \text{if } M \equiv \exists(\lambda x : i.M'[x]) \\ X(\Phi_\Gamma(M')) & \text{if } M \equiv X(M'). \end{cases}$$

Throughout this paper we will identify terms up to α-equivalence, and assume that in notations such as $\forall x.\Phi_{\Gamma \cup \{x::i\}}(M'[x])$ we have $x \notin \text{dom}(\Gamma)$.

4   The LF Paradigm for Specifying a Logical System

In this section we outline the LF paradigm for specifying a logic and elaborate on some of the points made in the previous section. A typed λ-calculus is used as the specification language for formal systems because syntax and rules are typically presented schematically. Moreover rules are usually treated as functional objects, mapping proofs of premisses to proofs of conclusions and proofs of lemmas (or conjectures) to complete proofs. Proof checking reduces to checking the correctness of instantiations of schemas and the application of rules (both basic and derived) to premisses or proofs thereof.

Consequently the LF, whenever possible, reduces: all forms of dependency and parameterization involved in defining and using a formal system to λ-abstractions; all forms of schematic instantiation to λ-application; all forms of substitution to β-reduction. These reductions are carried out in such a way that the correctness of any of the above activities can be enforced through type matching and checking.

In the LF language a logical system is specified by a finite list of typed constants, called a *signature*. The syntax (for a given logical system) is encoded, into the signature, by introducing a type for each syntactic category and a constant of appropriate functional type for each expression constructor. Object language variables and schematic variables are then modelled by LF variables of the appropriate type. A schematic expression, of a given syntactic category, in certain schematic variables is expressed as the λ-abstraction of that expression with respect to those variables. Finally binding operators are modelled as expression constructors with arguments of functional type.

The LF paradigm for specifying and handling rules and proofs is centred on the notion of *judgement*. This notion was introduced by Martin-Löf [8] and corresponds to the notion of *assertion* of a formal system. However the LF does not commit itself to the intuitionistic viewpoint and extends the meaning of this notion. That part of the signature encoding the rules of a logical system is a list of declarations of judgement types of the appropriate kind (corresponding to the assertions of the system) and of constants of the appropriate higher order judgement type (corresponding to the rules and axioms of the system). Rule schemas are modelled by means of λ-abstractions. One of the major benefits of this approach is that proofs of theorems and of derived rules are treated on the same logical level.

An LF type encodes an *open concept*, i.e. no induction principle over the type is available. This implies that the notion of proof actually encoded is not merely the notion of *proof of logical theorem* in a fixed system. Using a judgement J, we encode a *consequence relation* definable in the formal system under consideration. A term of type $J(\Phi) \rightarrow J(\Psi)$ encodes a proof that J holds of Ψ follows from the assumption that J holds of Φ. It does not just encode a function which transfers a proof that Φ is a logical theorem to a proof that Ψ is also a logical theorem. The system may even lack logical theorems altogether. A proof of a hypothetical judgement therefore corresponds to either a rule of derivation or a derivable rule of the system, not simply a rule of proof or an admissible rule.

The consequence relations which are directly encoded by the LF's judgements are ordinary single-conclusioned consequence relations [2]. A proof of a sequent $\Phi_1, \ldots, \Phi_n \vdash \Psi$ is encoded by a term of type $J(\Phi_1) \rightarrow J(\Phi_2) \rightarrow \cdots J(\Phi_n) \rightarrow J(\Psi)$, where J is a judgement that is induced by $\vdash$. Note, however, that the type structure of the LF makes it possible to formulate and prove also higher-order logical facts about an internalized consequence relation or even logical facts relating two or more such relations, e.g Modal logics.

A number of classical side conditions on rules concerning binding operators and connectives can be handled unproblematically. In other cases additional judgements need to be introduced. We remark that the implicit identification the LF utilizes between object language variables and schematic variables (over terms of the language) will often suggest similarities between the LF translation of a system and a denotational model of that system (see for example the internalizations of the various lambda calculii). It is an *LF thesis* that well-behaved natural-deduction formalisms are those that can be directly encoded, and also that given a formal representation of a consequence relation the notion of a derivable rule (of arbitrary order) is *defined* by the non-emptiness of the type encoding its specification in the corresponding signature.

An LF specification of a formal system is satisfactory only if *adequate*, i.e. if for each syntactic and proof theoretic category of the system there is a *compositional* surjection from the ($\beta - \eta$ equivalence classes of an) LF type, corresponding to that category, onto the category itself.

5 Modal Logics

Standard presentations of modal logics are problematic even in the simplest case, Hilbert systems. In these systems one usually has, apart from axiom schemes, two rules of inference: Modus Ponens and necessitation (from Φ infer $\Box\Phi$). The latter, however, is taken to be only a *rule of proof*: its application to a premise is permitted only if the premise does not depend on any assumptions (i.e. is a theorem). It is not the case, therefore, that $\Box\Phi$ *follows* from Φ in such systems. Thus it would not be sound to encode the necessitation rule by simply introducing a constant, Nec, of type $\prod_{\Phi:o} T(\Phi) \rightarrow T(\Box\Phi)$ (where T corresponds to the intended consequence relation).

The solution to this problem illustrates the power gained by simultaneously employing different judgements. In the case of S4 we introduce *two* judgements: True and Valid. The first corresponds to the intended consequence relation, $\vdash_t$, in which necessitation is only a rule of proof. The second one corresponds to the consequence relation, $\vdash_v$, obtained by taking both rules as pure rules of derivation. The complete signature is:

- Syntactic Categories

    o : Type

- Operations

    $\neg$: $o \rightarrow o$
    $\supset$: $o \rightarrow o \rightarrow o$
    $\Box$: $o \rightarrow o$

- Judgements

$$\begin{array}{lll}
\text{True} & : & o \to \text{Type} \\
\text{Valid} & : & o \to \text{Type}
\end{array}$$

- Axioms and Rules

$$\begin{array}{llll}
\text{C} & : & \prod_{\Phi:o} & \text{Valid}(\Phi) \to \text{True}(\Phi) \\
\text{A}_1 & : & \prod_{\Phi_1,\Phi_2:o} & \text{Valid}(\Phi_1 \supset (\Phi_2 \supset \Phi_1)) \\
\text{A}_2 & : & \prod_{\Phi_1,\Phi_2,\Phi_3:o} & \text{Valid}(\Phi_1 \supset (\Phi_2 \supset \Phi_3) \supset (\Phi_1 \supset \Phi_2) \supset (\Phi_1 \supset \Phi_3)) \\
\text{A}_3 & : & \prod_{\Phi_1,\Phi_2:o} & \text{Valid}((\neg\Phi_1 \supset \neg\Phi_2) \supset (\Phi_2 \supset \Phi_1)) \\
\text{A}_4 & : & \prod_{\Phi:o} & \text{Valid}(\Box\Phi \supset \Phi) \\
\text{A}_5 & : & \prod_{\Phi_1,\Phi_2:o} & \text{Valid}(\Box(\Phi_1 \supset \Phi_2) \supset (\Box\Phi_1 \supset \Box\Phi_2)) \\
\text{A}_6 & : & \prod_{\Phi:o} & \text{Valid}(\Box\Phi \supset \Box\Box\Phi) \\
\text{MP}_T & : & \prod_{\Phi_1,\Phi_2:o} & \text{True}(\Phi_1 \supset \Phi_2) \to \text{True}(\Phi_1) \to \text{True}(\Phi_2) \\
\text{MP}_V & : & \prod_{\Phi_1,\Phi_2:o} & \text{Valid}(\Phi_1 \supset \Phi_2) \to \text{Valid}(\Phi_1) \to \text{Valid}(\Phi_2) \\
\text{Nec} & : & \prod_{\Phi:o} & \text{Valid}(\Phi) \to \text{Valid}(\Box\Phi)
\end{array}$$

An example of a proof is $\Delta : \prod_{\Phi:o} \text{True}(\Box(\Box\Phi \supset \Phi) \supset \Box\Phi) \to \text{True}(\Phi)$, where Δ is the following:

$$\lambda\Phi : o.\lambda x : \text{True}(\Box(\Box\Phi \supset \Phi) \supset \Box\Phi).\text{MP}_T(\Box\Phi)(\Phi)(\text{C}(\Box\Phi \supset \Phi)(\text{A}_4(\Phi)))$$
$$(\text{MP}_T(\Box(\Box\Phi \supset \Phi))(\Box\Phi)(x)(\text{C}(\Box(\Box\Phi \supset \Phi))(\text{Nec}(\Box\Phi \supset \Phi)(\text{A}_4(\Phi)))))$$

$\Box(\Box\Phi \supset \Phi) \supset \Box\Phi$ is the characteristic axiom of GL— the famous modal system for provability in Peano arithmetic. The above is a proof that in S4 any Φ-instance of this formula actually entails Φ.

Theorem 3 *In the Hilbert-type system which is obtained by adding $\Phi_1, \ldots, \Phi_n$ as axioms to S4,*

$$\Psi_1, \ldots, \Psi_m \vdash \vartheta$$

if and only if there exists a term Δ of type

$$(\dagger) \quad \text{Valid}(\Phi_1) \to \cdots \to \text{Valid}(\Phi_n) \to \text{True}(\Psi_1) \to \cdots \text{True}(\Psi_m) \to \text{True}(\vartheta)$$

in the context $p_1 : o, \ldots, p_m : o$ and the above signature. Where $p_1, \ldots, p_m$ are the atomic variables occuring in $\Phi_1, \ldots \Phi_n, \Psi_1, \ldots \Psi_m$ and Ψ.

The above method for handling rules of proof is not specific to modal logic. In the above case the consequence relations have natural semantic interpretations in terms of Kripke models: $\bar{\Phi} \vdash_v \Psi$ iff Ψ is *valid* in any frame in which all the $\bar{\Phi}$ are valid (i.e. true in all worlds); $\bar{\Phi} \vdash_t \Psi$ iff Ψ is *true* in every world in which all the $\bar{\Phi}$ are true. Moreover, in the LF we can express and prove logical facts concerning *both* internalized consequence relations. For example a term of type $\dagger$ encodes a proof that ϑ is true in any world in which the $\bar{\Psi}$ are all true, provided this world belongs to a frame in which the $\bar{\Phi}$ are all valid.

We turn now to the natural deduction formulation of S4, presented by Prawitz in [12]. It is obtained from the usual natural deduction formulation of classical propositional calculus by the addition of the following two rules:

$$(\Box\ \text{Intro}) \quad \frac{\Phi}{\Box\Phi} \qquad\qquad \frac{\Box\Phi}{\Phi} \quad (\Box\ \text{Elim})$$

The first rule has a side condition on its application. Prawitz gives several possible versions of this side condition. In the first one, for example, all assumptions on which Φ depends should be modal (i.e. the main connective is $\square$). In all versions the side condition makes this rule *impure*. This impurity is of the *second degree* [2]. Thus we lack the coherence, which the LF paradigm expects, between the formulation of the rules of a system and the consequence relation represented by it. In the Hilbert style presentation we can ignore the intended consequence relation of truth and still encode all proofs of theorems using only one judgement, validity. This is not possible here since the introduction rule for implication is not sound for validity.

A compact solution to this problem is to encode proofs of theorems using *two* judgements and model implication introduction in a more elaborate way. The two judgements are Taut and Valid , both of type $o \rightarrow$ Type. Taut encodes the usual consequence relation of classical propositional logic. Valid encodes the consequence relation of validity. The constants of the specification then fall into three groups: those corresponding to pure tautological inferences; those corresponding to the modal rules; and those which relate the two sorts of inferences. The resulting signature has the same syntactic categories and operations as the previous example. We omit two groups of rules. The first group simply states that Taut behaves like truth in the usual natural deduction presentation of classical propositional calculus. The second group states that Valid(Φ) is equivalent to Valid($\square\Phi$). The crucial rules are:

$$
\begin{array}{lcl}
\text{C} & : & \prod_{\Phi:o} \quad \text{Taut}(\Phi) \rightarrow \text{Valid}(\Phi) \\
\text{R} & : & \prod_{\Phi_1,\Phi_2:o} \ (\text{Taut}(\Phi_1) \rightarrow \text{Valid}(\Phi_2)) \rightarrow (\text{Valid}(\Phi_1) \rightarrow \text{Valid}(\Phi_2)) \\
\supset\text{I}_V & : & \prod_{\Phi_1,\Phi_2:o} \ (\text{Valid}(\square\Phi_1) \rightarrow \text{Valid}(\Phi_2)) \rightarrow \text{Valid}(\square\Phi_1 \supset \Phi_2)
\end{array}
$$

6 Theories of Functions

We now discuss the main issues which arise in encoding functional calculii, such as λ-calculus, call-by-value-λ-calculus, λ-I-calculus and linear λ-calculus. While all these systems are of interest from the point of view of functional programming, the latter two are interesting also from a purely logical point of view. Systems such as relevance and linear logic have consequence relations with weaker structural rules than those implicit in the LF type theory, at least when the constructor $\rightarrow$ is used to encode the $\vdash$. For example, in the case of relevance logic the implication introduction is sound only for λ_I-abstraction. Therefore if we do not introduce in the LF new primitive abstraction operators, then we essentially have to implement this calculus prior to encoding the logic.

We begin by discussing the case of the classical λ-calculus. To this end we define a basic LF type, o, encoding the set of λ-terms together with a judgement, $M = N$, intended to encode the assertion that the term M is $\alpha - \beta$-equal to the term N. In order to encode the β-reduction rule it is convenient to encode the λ-constructor as $\Lambda : (o \rightarrow o) \rightarrow o$. In doing so we take care of, at the level of the metalanguage, the operation of capture avoiding substitution which is normally used in formulating the β-rule. Finally we introduce the constant App : $o \rightarrow o \rightarrow o$ encoding application. All this is summarized in the following.

- Syntactic Category

  $o \ : \ $ Type

- Operations

$$\Lambda \quad : \quad (o \to o) \to o$$
$$\mathrm{App} \quad : \quad o \to o \to o$$

- Judgements

$$= \ : \quad o \to o \to \mathrm{Type}$$

Encoding the β and congruence rules is now routine. We also encode the ξ rule which is classically formulated as

$$\frac{M = N}{\lambda x.M = \lambda x.N}$$

in the following.

- Axioms and Rules

$$
\begin{array}{llll}
E_0 & : & \Pi_{x:o} & x = x \\
E_1 & : & \Pi_{x,y:o} & x = y \to y = x \\
E_2 & : & \Pi_{x,y,z:o} & x = y \to y = z \to x = z \\
E_3 & : & \Pi_{x,y,x',y':o} & x = y \to x' = y' \to \mathrm{App}(x,x') = \mathrm{App}(y,y') \\
\beta & : & \Pi_{\substack{x:o\to o \\ y:o}} & \mathrm{App}(\Lambda(x),y) = xy \\
\xi & : & \Pi_{x,y:o\to o} & (\Pi_{z:o}xz = yz) \to \Lambda(x) = \Lambda(y)
\end{array}
$$

Notice that there is no counterpart to α-conversion in the above signature. The fact that we have encoded the classical λ-calculus is the expressed by the following theorem.

Theorem 4 *The following hold:*

1. $x_1 : o, \dots, x_n : o \vdash_{\Sigma_\Lambda} M : o$ iff $\Phi_\Gamma(M) \in \Lambda$

2. $(\exists P)(x_1 : o, \dots, x_n : o \vdash_{\Sigma_\Lambda} P : M = N)$ iff $\vdash_\lambda \Phi_\Gamma(M) = \Phi_\Gamma(N)$

where $M \in \Xi_\Gamma(o)$, $\Xi_\Gamma(o)$ is the set of normal forms of type o in the context Γ,

$$\Gamma = x_1 : o, \dots, x_n : o$$

and

$$\Phi_\Gamma : \Xi_\Gamma(o) \longrightarrow \Lambda[x_1, \dots, x_n]$$

is a bijective function defined as follows

$$\Phi_\Gamma(M) = \begin{cases} x & \textit{if } M = x \\ \Phi_\Gamma(M')\Phi_\Gamma(N) & \textit{if } M = \mathrm{App}(M', N) \\ \lambda x.\Phi_{\Gamma,x:o}(M'[x]) & \textit{if } M = \Lambda(\lambda x.M'[x]) \end{cases}$$

It is interesting to consider the possibility of extending $\vdash_\lambda$ from a unary to a binary consequence relation, and hence extend the above theorem to proofs from assumptions. However, the consequence relation that is encoded by considering the inhabitability of types like

$$M_1 = N_1 \to \dots \to M_n = N_n \to M = N$$

is not exactly as one would hope. The way the ξ rule has been encoded is responsible for the discrepancy. For example, in the classical λ-calculus one can show that

$$x(\Delta\Delta) = x(\Delta\Delta\Delta) \vdash_\lambda \lambda x.x(\Delta\Delta) = \lambda x.x(\Delta\Delta\Delta)$$

where Δ is the term $\lambda z.zz$. In contrast to this there is no way of showing in the above signature that

$$\Phi_{\{x:o\}}(x(\Delta\Delta)) = \Phi_{\{x:o\}}(x(\Delta\Delta\Delta)) \to \Phi_\emptyset(\lambda x.x(\Delta\Delta)) = \Phi_\emptyset(\lambda x.x(\Delta\Delta\Delta)).$$

However we can show that

$$\prod_{x:o}(\Phi_{\{x:o\}}(x(\Delta\Delta)) = \Phi_{\{x:o\}}(x(\Delta\Delta\Delta))) \to \Phi_\emptyset(\lambda x.x(\Delta\Delta)) = \Phi_\emptyset(\lambda x.x(\Delta\Delta\Delta)).$$

The underlying reason for this discrepancy is that traditionally free variables in assumptions are implicitly taken as universally quantified. The consequence relation defined by such a convention is often referred to as the consequence relation of validity. On the contrary in the LF encoding we are forced to indicate explicitly if our assumptions are universally quantified, as in the case for the ξ-rule. We are in fact encoding the consequence relation of truth. If we were to encode the consequence relation of validity, problems similar to those encountered in Hoare's logic (see section 7) would arise. To summarize, if universal quantification in assumptions is not made explicit by means of Π, then the ξ-rule can only be utilized as a rule of proof. This situation corresponds to the encoding of the theory of λ-algebras. It is important to note that the consequence relations of truth and validity either for λ-algebras or for the λ-calculus all coincide if only closed assumptions are considered. In the remainder of this section we shall only discuss the truth-consequence relations.

The call-by-value λ-calculus (λ_v-calculus) [11] differs from the traditional λ-calculus in the formulation of the β-reduction rule:

$$(\lambda x \quad . \quad M)N = M[x := N]$$

provided that N is a value, i.e. either a variable or an abstraction.

The immediate problem in encoding the call-by-value λ-calculus is expressing the syntactic notion of being a variable. The solution is inspired by a denotational model for the calculus [5] where the functions are strict and the variables range only over $D - \{\perp\}$. The syntax is modelled using two syntactic categories, v for values and o for expressions, together with a mapping $! : v \to o$. This illustrates the general technique for handling subcategories in LF. The only bindable type is v, with binding operator $\Lambda_v : (v \to o) \to v$. The full signature is:

- Syntactic Categories

$$o \; : \; \text{Type}$$
$$v \; : \; \text{Type}$$

- Operations

$$! \quad : \; v \to o$$
$$\Lambda_v \quad : \; (v \to o) \to v$$
$$\text{App} \; : \; o \to o \to o$$

- Judgement

$$= \; : \; o \to o \to \text{Type}$$

- Axioms and Rules

$$
\begin{array}{lll}
E_0 & : & \Pi_{x:o} & x = x \\
E_1 & : & \Pi_{x,y:o} & x = y \to y = x \\
E_2 & : & \Pi_{x,y,z:o} & x = y \to y = z \to x = z \\
E_3 & : & \Pi_{x,y,x',y':o} & x = y \to x' = y' \to \text{App}(x,x') = \text{App}(y,y') \\
\beta_v & : & \Pi_{\substack{x:v\to o \\ y:v}} & \text{App}(\Lambda_v(x)!,y!) = xy \\
\xi_v & : & \Pi_{x,y:v\to o} & (\Pi_{z:v}xz = yz) \to \Lambda_v(x)! = \Lambda_v(y)! \\
\eta_v & : & \Pi_{x:v} & \Lambda_v(\lambda y : v.\text{App}(x!,y!)) = x!
\end{array}
$$

- Example of a Proof

$$\beta_v(!) \; : \; \Pi_{x:v}\text{App}(\Lambda_v(!)!,x!) = x!$$

The example of a proof included above demonstrates that

$$\Lambda_v(!)!$$

behaves like the identity (with respect to App) over values. It is worth noting that in this setting the correct version of the η-rule suggests itself more naturally than in the original presentation.

The set, Λ_I, of terms of the λ_I-calculus is defined as in the classical calculus, except for the abstraction clause: if $M \in \Lambda_I$ and $x \in M$, then $\lambda x.M \in \Lambda_I$.

The problem of encoding the λ_I-calculus is enforcing the binding constructor λ_I to be defined only on *relevant* schemes. Two solutions can be given, again inspired by denotational models of the calculus. The first follows quite closely the model presented in [5]. A new constant $\perp: o$ is introduced together with rules governing its behaviour. The predicate *being a relevant function* is encoded as:

$$\text{Rel}_1 \equiv \lambda x : o \to o.x(\perp) = \perp \, .$$

The λ_I-constructor is $\Lambda_I : \prod_{x:o\to o} \text{Rel}_1(x) \to o$. Constants appearing in the rule, other than the ones mentioned explicitly, are as in the previous signature. It is interesting to notice the role of judgements in the definition of the syntax.

The second approach is a generalization of the previous one. No $\perp$ constant is needed. The idea is to axiomatize the predicate $x \in M$ by introducing a new judgement $\in$ and appropriate rules. The predicate *being a relevant function* is encoded as:

$$\text{Rel}_2 \equiv \lambda x : o \to o. \prod_{z,y:o} z \in y \to z \in xy$$

The λ_I constructor is encoded as follows: $\Lambda_I : \prod_{x:o\to o} .\text{Rel}_2(x) \to o$.

The set Λ_L of terms of the linear-λ-calculus is inductively defined as follows:

- $x \in \Lambda_L^*$.

- If $M \in \Lambda_L^*$ and x occurs free in M exactly once then $\lambda x.M \in \Lambda_L^*$.

- If $M, N \in \Lambda_L^*$ then $MN \in \Lambda_L^*$.

Encoding the linear λ-calculus exploits the notion of a function being linear. A function $f : X \to Y$, where X and Y are upper semi-lattices with a least element, is linear iff its strict and distributive. This solution is based on an idea of Gordon Plotkin. The predicate *being a distributive function* is encoded as

$$L = \lambda x : o \to o. \prod_{z,w:o} x(z \vee w) = x(z) \vee x(w),$$

the λ-linear constructor is then encoded by $\Lambda_L : \prod_{x:o \to o} x(\bot) = \bot \to L(x) \to o$. Of course the full signature includes enough rules to axiomatize the notion of upper semi-lattice with least element.

7 Program Logics

Program logics such as Hoare's logic and dynamic logic exhibit an unusual overloading of variables. In both these logics variables play two roles, behaving in some instances as *logical variables* ranging over the data domain, and in other instances as assignable *identifiers* or *locations*. A typical example, from dynamic logic, is

$$\forall x > 0[\text{while}(x > 0, x := x - 1)]x = 0.$$

It not only illustrates the dual nature of variables but also the difficulties in defining the notion of a free and bound variable. The occurrence of x in the while test is, in a sense, bound by both the quantifier and the assignment. Nevertheless even in the somewhat simpler case of Hoare's logic for a simple assignment language (whose only control primitives are assignment and sequencing), problems arise. The assignment axiom for this system is $\{p[t/x]\}x := t\{p\}$. As usual, $p[t/x]$ stands for the result of substituting t for the free occurrences of x in p.

There are, at least, two complications one must deal with in encoding this logic. Firstly we must distinguish between the variables of the first order logic and the variables of the programming language. We cannot model := as an object of type $i \to i \to w$ since this would allow expressions like $0 := 1$. A new type l, corresponding to locations, is introduced together with a function $! : l \to o$, called bang, which takes a location to its contents. Secondly, note that := is a binding operator. In the assignment axiom free occurrences of x in p are bound by the assignment operator $x := t$. This is not true of those occurrences in t either in $p[t/x]$ or in the assignment. One could even claim that it is an example of a binding operator which does not α-convert. α-conversion does not appear to be in the spirit of Hoare's logic, since one wants to reason about the identifier x not some α-conversion of it. This has the consequence that simply modelling the assignment axiom by

$$\text{Ass} : \prod_{\substack{x:l,t:i \\ p:i \to o}} \vdash_h \{p(t)\}x := t\{p(x!)\}$$

would be incorrect, e.g. $\text{Ass}(y)(1)(\lambda u.\neg(y! = u)) : \quad \vdash_h \{\neg(y! = 1)\}y := 1\{\neg(y! = y!)\}$.

The problem, intuitively, is that $\{p(t)\}x := t\{p(x!)\}$ can be false because the assignment $x := t$ can alter the meaning of the predicate $\lambda z : i.p(z)$. One solution to this problem is to incorporate syntactic notions explicitly into the theory. We do this by adding three new judgements, $\natural_l, \natural_i$ and $\natural_o$, concerning non-interference along the lines of [13], $\natural_x$ is of type $l \to (x \to \text{TYPE})$. The intuitive meaning of the judgements can be explained, using infix notation, as follows: $x \natural_l y$ is interpreted as meaning that x and y denote distinct identifiers or locations. $x \natural_i t$ is interpreted as meaning that no assignment to the location denoted by x effects the value of the term denoted by t. This of course is equivalent to saying that the location or identifier denoted by x does not occur in the term denoted by t. $x\sharp_o e$ is interpreted as meaning that no assignment to the location denoted by x effects the value or meaning of the formula denoted by e. Again this is equivalent to saying that the location or identifier denoted by x does not occur freely in the formula denoted by e (note that it cannot occur bound). The corrected version of the assignment axiom may be written as follows.

$$\text{Ass} : \Pi_{\substack{x:l,t:i \\ \Phi:i\to o}} x\sharp_o \forall \Phi \to (\vdash_h \{\Phi(t)\}x := t\{\Phi(x!)\})$$

This solution, see [9], takes the notion of a free variable as primitive, another solution is to encode *substituting a term for all free occurrences of a banged location in terms and formulas*. This would involve introducing two new operations (rather than the two judgements $\sharp_i$ and $\sharp_o$) sub_i and sub_o, where sub_x is of type $i \to l \to x \to x$, and $sub_x(t, y, z)$ represents the result of substituting the term t for all free occurrences of $y!$ in z. To axiomatize these operations, in particular the base case, one must still retain the judgement $\sharp_l$, and so in some sense the two solutions are dual. There is little reason, on the face of it, to choose one over the other. We should point out, however, that to correctly formalize more complex versions of Hoare's logic, for example one in which recursive procedure calls were allowed, it would be necessary to incorporate the notion on non-interference anyway. Thus in the long run the first solution seems most suited to Hoare's logic. On the other hand in dynamic logic the substitution approach may be more natural, since there is no clear notion of free and bound variables in that logic.

Presentations of these systems in the literature are not uniform and often important syntactic decisions are not entirely motivated. Such systems may even benefit from the analysis required to encode them.

Another approach is not to reason about Hoare triples directly but rather deal primarily with functions from state to triples. Explicitly we deal with objects obtained from triples by abstracting the program locations. Thus we must restrict our attention to assertions concerning programs built up from a fixed finite number of such locations. In the case we present here this number is two, the judgement $\vdash$ is therefore o f type $(l \to l \to h) \to$ Type, and the sequencing and assignment axioms are:

$$
\begin{aligned}
\text{Ass}_1 \quad &: \quad \Pi_{\substack{t:l\to l\to i \\ \Phi:i\to i\to o}} \vdash \lambda x : l \;.\; \lambda y : l \;.\; \{\Phi(t(x,y), y!)\}x := t(x,y)\{\Phi(x!, y!)\} \\
\text{Ass}_2 \quad &: \quad \Pi_{\substack{t:l\to l\to i \\ \Phi:i\to i\to o}} \vdash \lambda y : l \;.\; \lambda x : l \;.\; \{\Phi(t(x,y), y!)\}x := t(x,y)\{\Phi(x!, y!)\} \\
\text{Seq} \quad &: \quad \Pi_{\substack{\Phi_0,\Phi_1,\Phi_2:l\to l\to o \\ w_1,w_2:l\to l\to w}} \\
& \qquad (\vdash \lambda x : l \;.\; \lambda y : l.\{\Phi_0(x,y)\}w_1(x,y)\{\Phi_1(x,y)\}) \to \\
& \qquad (\vdash \lambda x : l \;.\; \lambda y : l.\{\Phi_1(x,y)\}w_2(x,y)\{\Phi_2(x,y)\}) \to \\
& \qquad (\vdash \lambda x : l \;.\; \lambda y.\;: l.\{\Phi_0(x,y)\}w_1(x,y); w_2(x,y)\{\Phi_2(x,y)\})
\end{aligned}
$$

The question "Which solution is best?" is rather a philosophical one, and the reply depends somewhat on the aims of the answerer. We only point out that the syntactic judgements in the first solution are axiomatizable in such a way as to ensure that if they can be proved, then such a proof is unique. In other words the search space for these subsystems is linear, and so extremely suitable for automation, perhaps behind the naive users back.

References

[1] Arnon Avron, Furio Honsell and Ian A. Mason. *Using Typed Lambda Calculus to Implement Formal Systems on a Machine.* Technical Report, Laboratory for the Foundations of Computer Science, Edinburgh University, 1987. ECS-LFCS-87-31.

[2] Arnon Avron. *Simple Consequence Relations.* Technical Report, Laboratory for the Foundations of Computer Science, Edinburgh University, 1987. ECS-LFCS-87-30.

[3] J. Barwise and S. Feferman, editors. *Model–Theoretic Logics.* Perspectives in Mathematical Logic, Springer-Verlag, 1985.

[4] Nicolas G. de Bruijn. *A survey of the project AUTOMATH.* In J. P. Seldin and J. R. Hindley, editors, To H. B. Curry: Essays in Combinatory Logic, Lambda Calculus, and Formalism, pages 589–606, Academic Press, 1980.

[5] Mariangiola Dezani, Furio Honsell and Simonetta Ronchi della Rocca. *Models for Theories of Functions Strictly Depending on all their Arguments.* Journal of Symbolic Logic 51:3, 1986. Abstract.

[6] Jean-Yves Girard. *Linear Logic.* Theoretical Computer Science. volume 50, 1987, pp 1-102.

[7] Robert Harper, Furio Honsell, Gordon Plotkin. *A Framework for Defining Logics.* Proceedings of the Second Annual Conference on Logic in Computer Science, Cornell, 1987.

[8] Per Martin-Löf. *On the Meanings of the Logical Constants and the Justifications of the Logical Laws.* Technical Report 2, Scuola di Specializzazione in Logica Matematica, Dipartimento di Matematica, Università di Siena, 1985.

[9] Ian A. Mason. *Hoare's Logic in the LF.* Technical Report, Laboratory for the Foundations of Computer Science, Edinburgh University, 1987. ECS-LFCS-87-32.

[10] Albert Meyer and Mark Reinhold. *'Type' is not a type: preliminary report.* In Proceedings of the 13th ACM Symposium on the Principles of Programming Languages, 1986.

[11] Gordon Plotkin. *Call–by–name, Call–by–value and the λ–calculus.* Theoretical Computer Science, 1:125–159, 1975.

[12] Dag Prawitz. *Natural Deduction: A Proof-Theoretical Study.* Almquist & Wiksell, Stockholm, 1965.

[13] John Reynolds. *Syntactic Control of Interference.* Conference Record of the Fifth Annual Symposium on Principles of Programming Languages, Tucson, 1978.

Automating Recursive Type Definitions in Higher Order Logic

Thomas F. Melham

University of Cambridge Computer Laboratory
New Museums Site, Pembroke Street
Cambridge, CB2 3QG, England.

Abstract: *The expressive power of higher order logic makes it possible to define a wide variety of types within the logic and to prove theorems that state the properties of these types concisely and abstractly. This paper contains a tutorial introduction to the logical basis for such type definitions. Examples are given of the formal definitions in logic of several simple types. A method is then described for systematically defining any instance of a certain class of commonly-used recursive types. The automation of this method in HOL, an interactive system for generating proofs in higher order logic, is also discussed.*

Introduction

Recursive structures, such as lists and trees, are widely used by computer scientists in formal reasoning about the properties of both hardware and software systems. The aim of this paper is to show how recursive structures of this kind can be defined in *higher order logic*, the logical formalism used by the HOL interactive proof-generating system [7].

Higher order logic is a typed logic; each variable in the logic has an associated logical *type* which specifies the kind of values it ranges over. Sets which contain recursive structures such as lists and trees can be represented in higher order logic by extending the syntax of types in the logic with new type expressions that denote these sets. In the version of higher order logic supported by the HOL system, this is done by first *defining* these new types in terms of already existing types and then deriving properties about the new types by formal proof. This guarantees that adding a new type to the logic will not introduce inconsistency. Sections 3 through 6 of this paper explain the formal mechanism for defining new types in higher order logic and give a series of detailed examples illustrating this mechanism.

In general, defining a new type in higher order logic can be tricky; the details of the definition have to be got just right to yield a type with the desired properties. But certain kinds of types can be defined systematically, and the process of defining them and proving that they have the required properties can therefore be automated. However, for this to be of *practical* value in a theorem prover such as HOL, it is essential that the automated tools for defining new types be reasonably efficient. To derive the properties of a defined type in HOL, all the logical inferences involved must be actually carried out in the system. To automate the definition of new types in HOL, it is therefore desirable to reduce to a minimum the amount of inference that must be done. Section 7 of this paper shows how a certain class of widely-used concrete recursive types can be defined by a method which requires relatively little logical inference, and can therefore be efficiently automated in HOL.

All the theorems shown in this paper have been proved completely formally in the HOL system. And the method for automating recursive type definitions described in Section 7 has been fully implemented and is included in the latest release of HOL.

The Organization of the Paper

The organization of the paper is as follows. Section 1 contains an introduction to the version of higher order logic that is used in the paper. Section 2 describes how proofs in this formulation of higher order logic are mechanized in the HOL theorem prover. It is not possible to give more than a sketch of the HOL approach to theorem proving in this section; but a full description of HOL can be found in Gordon's paper [7]. In Section 3, a method is described by which new logical types can be defined as conservative extensions of higher order logic. Sections 4 through 6 consist of a series of examples which illustrate this method for defining new types. In Section 4, three simple logical types are defined: the 'trivial' type with only one value, the cartesian product type, and the disjoint sum type. In Section 5, two simple recursive types are defined: the type of natural numbers, and the type of lists. And in Section 6, the construction of two recursive types of trees is described. Finally, Section 7 outlines an efficient method for automating the definition of arbitrary concrete recursive types in higher order logic. This method uses types previously defined in Sections 4, 5, and 6. The implementation of the algorithm in the HOL system is also discussed in this section.

Note: Type constructions of the kind described in this paper are well-known in set theory (and logic), and no new theory of type constructions is presented here. The contribution of this paper consists rather in: (1) working out the details of defining these types in the particular logic implemented by the HOL theorem prover, and (2) building a logical basis for the *efficient* automation of recursive type definitions in HOL.

1 Introduction to Higher Order Logic

The version of higher order logic supported by the HOL system is based on Church's type theory [3], extended with the type discipline of the LCF logic PPλ [8]. This formulation of higher order logic was developed by Mike Gordon at the University of Cambridge, and is described in detail in [6]. This section gives a brief and informal introduction to the notation and some of the important features of this logic.

1.1 Notation

The syntax of higher order logic used in the HOL theorem prover includes terms corresponding to the conventional notation of predicate calculus. A term of the form $P\,x$ expresses the proposition that x has property P, and a term of the form $R(x, y)$ means that the relation R holds between x and y. The usual logical operators $\neg$, $\wedge$,

$\vee$, $\supset$ and $\equiv$ denote negation, conjunction, disjunction, implication, and equivalence respectively. The syntax of terms in HOL also includes the conventional notation for universal and existential quantifiers: $\forall x.P\,x$ means that P holds for every value of x, and $\exists x.P\,x$ means that P holds for at least one value of x. The additional quantifier $\exists!$ denotes unique existence: $\exists!x.P\,x$ means that P holds for exactly one value of x. Nested quantifiers of the form $\forall v_1.\forall v_2.\cdots\forall v_n.tm$ can also be written $\forall v_1\,v_2\,\cdots\,v_n.tm$. Other notation includes $(c\Rightarrow t_1\,|\,t_2)$ to denote the conditional 'if c then t_1 else t_2', and $f\circ g$ to denote the composition of the functions f and g. The constants T and F denote the truth values *true* and *false* respectively.

Higher order logic extends the notation of predicate calculus in three important ways: (1) variables are allowed to range over functions and predicates, (2) functions can take functions as arguments and yield functions as results, and (3) the notation of the λ-calculus can be used to write terms which denote functions.

The first two of these notational extensions are illustrated by the theorem of higher order logic shown below:

$$\vdash \forall x f.\,((\mathsf{rec}\ f)\,0 = x)\wedge\forall n.\,(\mathsf{rec}\ f)\,(n{+}1) = f\,((\mathsf{rec}\ f)\,n)$$

This theorem states that functions can be defined on the natural numbers such that they satisfy simple primitive recursive equations. It asserts that for any value x and any function f, the term $(\mathsf{rec}\ f)$ denotes a function that yields x when applied to 0 and satisfies the recursive equation $(\mathsf{rec}\ f)\,(n{+}1) = f\,((\mathsf{rec}\ f)\,n)$ for all n. The universally quantified variable f in this theorem is an example of a higher-order variable: it ranges over functions. And the constant rec is an example of a higher order function: it both takes a function as an argument and yields a function as a result. Conventional practice is that function application in higher order logic associates to the left. So, for example, the term $(\mathsf{rec}\ f)\,n$ can also be written rec $f\,n$.

The syntax of higher order logic also includes terms of the (typed) λ-calculus. If tm is a term and v is a variable, then the expression '$\lambda v.tm$' is also a term. It denotes the function whose value for an argument x is given by substituting x for v in tm. The term $\lambda n.n{+}1$, for example, denotes the successor function on natural numbers; and the term $(\lambda n.n{+}1)\,7$ can be simplified to $7{+}1$ by substituting 7 for n in $n{+}1$. Simplifications of this kind are called β-reductions.

1.2 Types in Higher Order Logic

Higher order logic is a *typed* logic; every syntactically well-formed term of the logic must have a type that is consistent with the types of its subterms. Informally, types can be thought of as denoting sets of values and terms as denoting elements of these sets.[1] As a syntactic device, types are necessary in higher order logic to eliminate certain paradoxes (e.g. Russell's paradox) which would otherwise arise because variables are allowed to range over functions and predicates.

Writing $tm{:}ty$ indicates explicitly that the term tm has type ty. Such explicit type information will usually be omitted, however, when it is clear from the form or context of the term what its type must be. The HOL mechanization of higher order

---

[1] Because of the polymorphism introduced by type variables, the notion of types as sets is inadequate for a formal semantics of the logic. But it will do for the purposes of this paper.

logic uses Milner's elegant algorithm for type inference [11] to assign consistent types to logical terms entered by the user. The user of HOL therefore only occasionally has to give the types of terms explicitly.

1.2.1 The Syntax of Types

There are three syntactic classes of types in higher order logic: type constants, type variables, and compound types.

Type constants are identifiers that name sets of values. Examples are the two primitive types *bool* and *ind*, which denote the set of booleans and the set of 'individuals' (an infinite set) respectively. Another example is the type constant *num*, which denotes the set of natural numbers. The type *num* is not primitive but is defined in terms of *ind*; its definition is given in Section 5.1.

Type variables are used to stand for 'any type'; they are written α, β, γ, etc. Types that contain type variables are called *polymorphic* types. A *substitution instance* of a polymorphic type *ty* is a type obtained by substituting types for all occurrences of one or more of the type variables in *ty*. Theorems of higher order logic that contain polymorphic types are also true for any substitution instance of these types.

Compound types are expressions built from other types using *type operators*. They have the form: $(ty_1, ty_2, \ldots, ty_n)op$, where ty_1 through ty_n are types and *op* is the name of an *n*-ary type operator. An example is the binary type operator *fun*, which denotes the function space operation on types. The compound type $(ind, bool)fun$, for instance, is the type of all total functions from *ind* to *bool*. Types constructed using the type operator *fun* can also be written in a special infix form: $ty_1 \rightarrow ty_2$. The infix type operator $\rightarrow$ associates to the right; so the type $ind \rightarrow bool \rightarrow bool$, for example, is the same as $(ind, (bool, bool)fun)fun$.

In principle, every type needed for doing proofs in higher order logic can be written using type variables, the primitive type constants *bool* and *ind*, and the type operator *fun*. In practice, however, it is desirable to extend the syntax of types to help make theorems and proofs more concise and intelligible than would otherwise be possible. Section 3 shows how this can be done by adding new type constants and type operators to the logic using type 'definitions'.

1.3 Hilbert's ε-operator

An important primitive constant of higher order logic, which will be used frequently in this paper, is Hilbert's ε-operator. Its syntax and informal semantics are as follows. If P[*x*] is a boolean term involving a variable *x* of type *ty* then $\varepsilon x.\, P[x]$ denotes some value, *v* say, of type *ty* such that P[*v*] is true. If there is no such value (i.e. P[*v*] is false for each value *v* of type *ty*) then $\varepsilon x.\, P[x]$ denotes some fixed but arbitrarily chosen value of type *ty*. Thus, for example, '$\varepsilon n.\, 4 < n \wedge n < 6$' denotes the value 5, '$\varepsilon n.\, (\exists m.\, n = 2 \times m)$' denotes an unspecified even natural number, and '$\varepsilon n.\, n < n$' denotes an arbitrary natural number.

The informal semantics of Hilbert's ε-operator outlined above is formalized in higher order logic by the following theorem:

$$\vdash \forall P.\, (\exists x.\, P\, x) \supset P(\varepsilon x.\, P\, x)$$

It follows that if P is a predicate and $\vdash \exists x.\, Px$ is a theorem of the logic, then so is $\vdash P(\varepsilon x.\, Px)$. The ε-operator can therefore be used to obtain a logical term which provably *denotes* a value with a given property P from a theorem merely stating that such a value *exists*. This property of ε is used extensively in the proofs outlined in this paper. For further discussion of the ε-operator, see [10].

An immediate consequence of the semantics of ε described above is that all logical types must denote non-empty sets. For any type ty, the term $\varepsilon x{:}ty.\mathsf{T}$ denotes an element of the set denoted by ty. Thus the set denoted by ty must have at least one element. This will be important when the method for adding new types to the logic is discussed in Section 3.

2 The HOL Theorem Proving System

The HOL system [7] is a mechanized proof-assistant developed by Mike Gordon at the University of Cambridge for conducting proofs in the version of higher order logic described in the previous section. It has been primarily used to reason about the correctness of digital hardware. But much of what has been developed in HOL for hardware verification—the theory of arithmetic, for example—is also fundamental to many other applications. The underlying logic and basic facilities of the system are completely general and can in principle be used to support reasoning in any area that can be formalized in higher order logic.

HOL is based on the LCF approach to interactive theorem proving and has many features in common with the LCF theorem provers developed at Cambridge [12] and Edinburgh [8]. Like LCF, the HOL system supports secure theorem proving by representing its logic in the strongly-typed functional programming language ML [4]. Propositions and theorems of the logic are represented by ML abstract data types, and interaction with the theorem prover takes place by executing ML procedures that operate on values of these data types. Because HOL is built on top of a general-purpose programming language, the user can write arbitrarily complex programs to implement proof strategies. Furthermore, because of the way the logic is represented using ML abstract data types, such user-defined proof strategies are guaranteed to perform only valid logical inferences.

The HOL system has a special ML abstract data type thm whose values are theorems of higher order logic. There are no literals of type thm; that is, it is not possible to obtain an object of type thm by simply typing one in. There are, however, certain predefined ML identifiers which are given values of type thm when the system is built. These values correspond to the axioms of higher order logic. In addition, HOL makes available several predefined ML procedures that take theorems as arguments and return theorems as results. Each of these procedures corresponds to one of the primitive inference rules of the logic and returns only theorems that logically follow from its input theorems using the corresponding inference rule. Since ML is a strongly-typed language, the type checker ensures that values of type thm can be generated only by using these predefined functions. In HOL, therefore, every value of type thm must either be an axiom or have been obtained by computation using the predefined functions that represent the primitive inference rules of the logic. Thus every theorem in HOL must be generated from the axioms using the inference rules. In this way, the ML type checker guarantees the soundness of the HOL theorem prover.

In addition to the primitive inference rules, there are many *derived* inference rules available in HOL. These are ML procedures which perform commonly-used sequences of primitive inferences by applying the appropriate sequence of primitive inference rules. Derived inference rules relieve the HOL user of the need to explicitly give the all primitive inferences required in a proof. The ML code for a derived rule can be arbitrarily complex; but it will never return a theorem that does not follow by valid logical inference, since the type checker ensures that derived rules can only return theorems if they have been obtained by a series of calls to the primitive inference rules.

The approach to theorem proving described above ensures the soundness of the HOL theorem prover—but it is computationally expensive. Formal proofs of even simple theorems in higher order logic can take thousands of primitive inferences. And when these proofs are done in HOL, all the inferences must actually be carried out by executing the corresponding ML procedures.

There are, however, two important features of HOL which together allow *efficient* proof strategies to be programmed. The first of these is merely this: theorems proved in HOL can be saved on disk and therefore do not have to be generated each time they are needed in future proofs. The second feature is the expressive power of higher order logic itself, which allows useful and very general 'lemmas' to be stated in the logic. The amount of inference that a programmed proof rule must do can therefore be reduced by pre-proving general theorems from which the desired results follow by a relatively small amount of deduction. These theorems can then be saved and used by the derived inference rule in future proofs. This strategy of replacing 'run time' inference by pre-proved theorems is possible in HOL because type polymorphism and higher-order variables make the logic expressive enough to yield theorems of sufficient generality. This is illustrated in Section 7.4 of this paper, where a single general theorem is given from which the 'axiomatization' of any concrete recursive type can be efficiently deduced.

3 Defining New Logical Types

The primary function of types in higher order logic is to eliminate the potential for inconsistency that comes with allowing higher order variables. The type expressions needed to prevent inconsistency have a very simple and economical syntax; all that is needed are the types that can be constructed from type variables, the two primitive types *bool* and *ind*, and the type operator →. In principle, every type needed for doing proofs in higher order logic can be written using only these primitive types. But in practice it is desirable to extend the syntax of types to include more kinds of types than are strictly necessary to prevent inconsistency.

Extending the syntax of type is of practical importance; it makes it possible to formulate propositions in logic in a more natural and concise way than can be done with only the primitive types. This pragmatic motivation for a rich syntax of types is similar to the motivation for the use of abstract data types in high-level programming languages; using higher level data types helps to control the size and complexity of proofs. This is essential in a theorem proving system (such as HOL) intended to be used as a practical tool for generating large formal proofs.

This section shows how new types can be consistently added to higher order logic by *defining* them in terms of already existing types. This is done in a way that allows theorems which 'axiomatize' these new types to be derived by formal proof from their definitions. The motivation for first defining a type and then deriving abstract 'axioms' for it is that this process guarantees consistency. Simply postulating axioms to describe the properties of new types may introduce inconsistency into the logic. But defining new types in terms of already existing types and then deriving axioms for them amounts to giving a consistency proof of these axioms.

3.1 Outline of the Method for Defining a New Type

The approach to defining new a logical type used in this paper involves the following three distinct steps:

1. finding an appropriate subset of an existing type to represent the new type;

2. extending the syntax of logical types to include a new type symbol, and using a *type definition axiom* to relate this new type to its representation; and

3. deriving from the type definition axiom and the properties of the representing type a set of theorems that serves as an 'axiomatization' of the new type.

In the first of these steps, a model for the new type is given by specifying a set of values that will be used to represent it. This is done formally by defining a predicate P on an existing type such that the set of values satisfying P has exactly the properties that the new type is expected to have. In general, finding representations for new types and defining predicates that specify them can be difficult; but, as will be shown in Section 7.3, the representations of a certain class of recursive types can be constructed systematically.

In the second step, the syntax of types is extended to include a new type constant (or type operator) which denotes the set of values of the new type. This is done by adding a *type definition axiom* to the logic that serves to relate values of the new type to the corresponding values of the existing type that represent them. Type definition axioms are explained below in Section 3.2.

In the last step, a collection of theorems is proved that abstractly characterizes the new type. These theorems state the essential properties of the new type without reference to the way its values are represented and therefore act as an abstract 'axiomatization' of it. They are not, however, axioms in the sense that they are postulated without proof, but are *derived* by formal proof from the definition of the subset predicate given in step (1) and the type definition axiom postulated in step (2). This final step therefore amounts to giving a consistency proof of the axioms for the new type by showing that there is a model for them. Several examples of the derivation of axioms for new types are given in Sections 4–6; and, in Section 7, a method is described whereby the proof of the axioms for concrete recursive types can be efficiently automated.

3.2 Type Definition Axioms

The syntax of types in higher order logic can be extended to include new type constants as well as new type operators, by means of *type definition axioms*. This type definition mechanism is based on a suggestion by Mike Fourman which was formalized by Mike Gordon in [6]. The idea is that a new type is defined by adding an axiom to the logic which asserts that it is isomorphic to an appropriate 'subset' of an existing type:

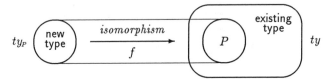

Suppose, for example, that ty is a type of the logic and $P{:}ty{\to}bool$ is a predicate on values of type ty that defines some useful subset of the set denoted by ty. A type definition axiom defines a new type constant ty_P which denotes a set having exactly the same properties as the subset defined by P. This is done by extending the syntax of types to include the new type constant ty_P and then adding an *axiom* to the logic asserting that the set of values denoted by the new type is isomorphic to the set specified by P:

$$\vdash \exists f{:}ty_P{\to}ty.\,(\forall a_1\,a_2.f\;a_1 = f\;a_2 \supset a_1 = a_2) \wedge (\forall r.\,P\;r = (\exists a.\,r = f\;a)) \quad (1)$$

This axiom states that there is a function f from the new type ty_P to the existing type ty which is one-to-one and onto the subset defined by P. The function f can be thought of as a representation function that maps a value of the new type ty_P to the value of type ty that represents it. Because f is an isomorphism, it can be shown that the set denoted by ty_P has the same properties as the subset of ty defined by P. By adding this axiom to the logic, the new type ty_P is therefore defined in terms of the existing type ty.

As was discussed in Section 1.3, the semantics of ε requires all types of the logic to denote non-empty sets. This means that the predicate P used in the type definition above must be true of at least one value of the representing type; i.e. it must be the case that $\vdash \exists x{:}ty.\,P\;x$. This existence theorem must be proved before the type definition axiom can be added to the logic. In the HOL theorem prover, the system requires the user to supply such an existence theorem before allowing a type definition axiom to be created.

If the subset defined by P is non-empty, then adding the type definition axiom (1) shown above is a *conservative extension* of the logic.[2] That is, for all boolean terms tm not containing the new type, $\vdash tm$ is a theorem of the extended logic if and only if it is a theorem of the original logic. In particular, $\vdash$ F is a theorem of the extended logic if and only if it is a theorem of the original logic. Thus adding type definition axioms to the logic will not introduce inconsistency; adding type definition axioms is 'safe'.

In addition to type constants, new type operators can also be defined by adding axioms of the form shown above. For example, if $ty[\alpha, \beta]$ is an existing type that contains type variables α and β, and $P{:}ty[\alpha, \beta]{\to}bool$ is a predicate where $\vdash \exists x.\,P\;x$, then a new binary type operator $(\alpha, \beta)op$ can be defined by asserting the axiom:

$$\vdash \exists f{:}(\alpha, \beta)op{\to}ty[\alpha, \beta].$$
$$(\forall a_1\,a_2.f\;a_1 = f\;a_2 \supset a_1 = a_2) \wedge (\forall r.\,P\;r = (\exists a.\,r = f\;a))$$

---

[2]The term P in the type definition axiom must also satisfy certain syntactic conditions (which will not be discussed here) having to do with free variables and polymorphism.

Detailed examples of type operator definitions are given in Sections 4.2 and 4.3 below, where definitions are described for the binary type operators *prod* (cartesian product) and *sum* (disjoint sum).

3.3 Defining Representation and Abstraction Functions

A type definition axiom of the form shown above merely asserts the *existence* of an isomorphism between a new type and the corresponding subset of an existing type. To formulate abstract axioms for a new type, it is convenient to have logical constants which in fact *denote* such an isomorphism and its inverse. These mappings are used to define operations on the values of a new type in terms of operations on values of the representing type. These operations can then be used to formulate the abstract axioms for the new type. Using the primitive constant ε described in Section 1.3, constants denoting isomorphisms between new types and the subsets of existing types which represent them are easily defined as follows.

Given a type definition axiom stating the existence of an isomorphism between a new type ty_P and a subset of an existing type ty defined by a predicate P:

$$\vdash \exists f{:}ty_P{\rightarrow}ty. (\forall a_1 a_2. f\ a_1 = f\ a_2 \supset a_1 = a_2) \land (\forall r. P\ r = (\exists a.\ r = f\ a))$$

a corresponding *representation* function $\mathsf{REP}{:}ty_P{\rightarrow}ty$ can be defined which maps a value of type ty_P to the value of type ty which represents it. Using the ε-operator, the function REP is defined by:

$$\vdash \mathsf{REP} = \varepsilon f. (\forall a_1 a_2. f\ a_1 = f\ a_2 \supset a_1 = a_2) \land (\forall r. P\ r = (\exists a.\ r = f\ a)).$$

From the property of ε discussed in Section 1.3, it follows immediately that the function REP is one-to-one and onto the subset of ty given by P:

$$\vdash \forall a_1 a_2.\ \mathsf{REP}\ a_1 = \mathsf{REP}\ a_2 \supset a_1 = a_2$$
$$\vdash \forall r. P\ r = (\exists a.\ r = \mathsf{REP}\ a)$$

Once the representation function REP is defined, the ε-operator can be used to define the inverse *abstraction* function $\mathsf{ABS}{:}ty{\rightarrow}ty_P$ as follows:

$$\vdash \forall r.\ \mathsf{ABS}\ r = (\varepsilon a.\ r = \mathsf{REP}\ a).$$

It is straightforward to prove that the abstraction function ABS is one-to-one for values of type ty satisfying P and that ABS is onto the new type ty_P:

$$\vdash \forall r_1 r_2. P\ r_1 \supset (P\ r_2 \supset (\mathsf{ABS}\ r_1 = \mathsf{ABS}\ r_2 \supset r_1 = r_2))$$
$$\vdash \forall a. \exists r.\ (a = \mathsf{ABS}\ r) \land P\ r$$

It also follows from the definitions of the abstraction and representation functions that ABS is the left inverse of REP and, for values of type ty satisfying P, REP is the left inverse of ABS:

$$\vdash \forall a.\ \mathsf{ABS}(\mathsf{REP}\ a) = a$$
$$\vdash \forall r. P\ r = (\mathsf{REP}(\mathsf{ABS}\ r) = r)$$

Abstraction and representation functions of the kind illustrated by ABS and REP are used in every new type definition described in this paper. In each case, these functions are defined formally using the corresponding type definition axiom in the way shown above for ABS and REP. Theorems corresponding to those shown above for ABS and REP are used in the proofs of abstract axioms for each new type defined.

4   Three Simple Type Definitions

Three simple examples are given in this section to illustrate the method for defining new types described above in Section 3. In each example, a new type is defined using the three steps described in Section 3.1. First, an appropriate subset of an existing type is found to represent the values of the new type, and a predicate is defined to specify this subset. A type definition axiom for the new type is then postulated, and abstraction and representation functions are defined as described in Section 3.3. An abstract axiomatization is then formulated for the new type, which describes its properties without reference to the way it is represented and defined. This axiomatization follows by formal proof from the properties of the new type's representation. Some basic theorems about the new type are then derived from its abstract axiomatization.

The three types defined in this section will be used as basic 'building blocks' in the general method outlined in Section 7.3 for finding appropriate representations for arbitrary concrete recursive types.

4.1   The Type Constant *one*

This section describes the definition and axiomatization of the simplest (and the smallest) type possible in higher order logic: the type constant *one*, which denotes a set having exactly one element.

4.1.1   The Representation

To represent the type *one*, any singleton subset of an existing type will do. In the type definition given below, the subset of *bool* containing only the truth-value T will be used. This subset can be specified by the predicate $\lambda b{:}bool.\, b$, which denotes the identity function on *bool*. The set of booleans satisfying this predicate clearly has the property that the new type *one* is expected to have, namely the property of having exactly one element.

4.1.2   The Type Definition

As discussed in Section 3.2, a type definition axiom cannot be added to the logic unless the representing subset is non-empty. In the present case, the representing subset is specified by the predicate $\lambda b.\, b$. It is trivial to prove that this predicate specifies a non-empty set of booleans; the theorem $\vdash \exists x.\,(\lambda b.b)x$ follows immediately from $\vdash (\lambda b.b)\mathsf{T}$, which is itself equivalent to $\vdash \mathsf{T}$. Once it has been shown that $\lambda b.b$ specifies a non-empty set of booleans, the type constant *one* can be defined by postulating the type definition axiom shown below.

$$\vdash \exists f{:}one{\to}bool.\,(\forall a_1\, a_2.f\ a_1 = f\ a_2 \supset a_1 = a_2) \wedge (\forall r.\,(\lambda b.b)\ r = (\exists a.\, r = f\ a))$$

Using this type definition axiom, a representation function $\mathsf{REP\_one}{:}one{\to}bool$ can be defined to map the single value of type *one* to the boolean value T which represents it. As described in Section 3.3, this representation function can be defined such that it is one-to-one:

$$\vdash \forall a_1\, a_2.\, \mathsf{REP\_one}\ a_1 = \mathsf{REP\_one}\ a_2 \supset a_1 = a_2 \tag{2}$$

and onto the subset of *bool* defined by $\lambda b.b$:

$$\vdash \forall r.\,(\lambda b.b)\, r = (\exists a.\, r = \mathsf{REP\_one}\ a)$$

which, by the β-reduction $\vdash (\lambda b.b)\, r = r$, immediately yields the following theorem:

$$\vdash \forall r.\, r = (\exists a.\, r = \mathsf{REP\_one}\ a) \tag{3}$$

Theorems (2) and (3) about the representation function $\mathsf{REP\_one}$ will be used in the proof given in the following section of the abstract axiomatization of *one*. The inverse abstraction function $\mathsf{ABS\_one}$:*bool→one* will not be needed in this proof.[3]

4.1.3   Deriving the Axiomatization of *one*

The axiomatization of the type *one* will consist of the following single theorem:

$$\vdash \forall f{:}\alpha{\rightarrow}one.\ \forall g{:}\alpha{\rightarrow}one.\ (f = g)$$

This theorem states that any two functions f and g mapping values of type α to values of type *one* are equal. From this it follows that there is only one value of type *one*, since if there were more than one such value it would be possible to define two different functions of type $\alpha{\rightarrow}one$. This theorem is therefore an abstract characterization of the type *one*; it expresses the essential properties of the type, but does so without reference to the way the type is represented.

The proof of the axiom for *one* uses the properties of $\mathsf{REP\_one}$ given by theorems (2) and (3) above. Specializing the variable r in (3) to the term $\mathsf{REP\_one}(f\ x)$ yields:

$$\vdash \mathsf{REP\_one}(f\ x) = (\exists a.\ \mathsf{REP\_one}(f\ x) = \mathsf{REP\_one}\ a)$$

The right hand side of this equation is equal to T; this theorem can therefore be simplified to $\vdash \mathsf{REP\_one}(f\ x)$. Similar reasoning yields the theorem $\vdash \mathsf{REP\_one}(g\ x)$, from which it follows that:

$$\vdash \mathsf{REP\_one}(f\ x) = \mathsf{REP\_one}(g\ x)$$

From this theorem and theorem (2) stating that the function $\mathsf{REP\_one}$ is one-to-one, it follows that $\vdash f\ x = g\ x$ and therefore that $\vdash \forall f\, g.\,(f = g)$, as desired.

4.1.4   A Theorem about *one*

Once the axiom for *one* has been proved, it is straightforward to prove a theorem which states explicitly that there is only one value of type *one*. This is done by defining a constant **one** to denote the single value of type *one*. Using the ε-operator, the definition of **one** can be written:

$$\vdash \mathbf{one} = \varepsilon x{:}one.\mathsf{T}$$

From the axiom for *one*, it follows that $\vdash \lambda x{:}\alpha.\, v = \lambda x{:}\alpha.\,\mathbf{one}$. Applying both sides of this equation to $x{:}\alpha$, and doing a β-reduction, gives $\vdash v = \mathbf{one}$. Generalizing v yields $\vdash \forall v{:}one.\, v = \mathbf{one}$, which states that every value v of type *one* is equal to the constant **one**, i.e. there is only one value of type *one*.

---

[3]In fact, the axiomatization of *one* can be derived directly from its type definition theorem; the constant $\mathsf{REP\_one}$ is defined here merely to simplify the presentation of the proof that follows.

4.2   The Type Operator $prod$

In this section, a binary type operator $prod$ is defined to denote the cartesian product operation on types. If ty_1 and ty_2 are types, then the type $(ty_1, ty_2)prod$ will be the type of ordered pairs whose first component is of type ty_1 and whose second component is of type ty_2.

4.2.1   The Representation

The type $(\alpha, \beta)prod$ can be represented by a subset of the polymorphic primitive type $\alpha \rightarrow \beta \rightarrow bool$. The idea is that an ordered pair $\langle a{:}\alpha,\ b{:}\beta \rangle$ will be represented by the function

$$\lambda x\, y.\, (x{=}a) \wedge (y{=}b)$$

which yields the truth-value T when applied to the two components a and b of the pair, and yields F when applied to any other two values of types α and β.

Every pair can be represented by a function of the form shown above; but not every function of type $\alpha \rightarrow \beta \rightarrow bool$ represents a pair. The functions that do represent pairs are those which satisfy the predicate Is_pair_REP defined by:

$$\vdash \text{Is\_pair\_REP } f = \exists v_1\, v_2.\, f = \lambda x\, y.\, (x{=}v_1) \wedge (y{=}v_2),$$

i.e. those functions f which have the form $\lambda x\, y.\, (x{=}v_1) \wedge (y{=}v_2)$ for some pair of values v_1 and v_2. This will be the subset predicate for the representation of $(\alpha, \beta)prod$. As will be shown below, the set of functions satisfying Is_pair_REP has exactly the standard properties of the cartesian product of types α and β.

4.2.2   The Type Definition

To introduce a type definition axiom for $prod$, one must first show that the predicate Is_pair_REP defines a non-empty subset of $\alpha \rightarrow \beta \rightarrow bool$. This is easy, since it is the case that $\vdash \forall a\, b.\, \text{Is\_pair\_REP}(\lambda x\, y.\, (x{=}a) \wedge (y{=}b))$ and therefore $\vdash \exists f.\, \text{Is\_pair\_REP}\, f$. Once this theorem has been proved, a type definition axiom of the usual form can be introduced for the type operator $prod$:

$$\vdash \exists f{:}(\alpha, \beta)prod \rightarrow (\alpha \rightarrow \beta \rightarrow bool).$$
$$(\forall a_1\, a_2.\, f\, a_1 = f\, a_2 \supset a_1 = a_2) \wedge (\forall r.\, \text{Is\_pair\_REP } r = (\exists a.\, r = f\, a))$$

This theorem defines the compound type $(\alpha, \beta)prod$ to be isomorphic to the subset of $\alpha \rightarrow \beta \rightarrow bool$ defined by Is_pair_REP. Since the type variables α and β in this theorem can be instantiated to any two types, it has the effect of giving a representation not only for the particular type '$(\alpha, \beta)prod$', but also for the product of *any* two types. For example, instantiating both α and β to $bool$ yields a type definition axiom for the cartesian product $(bool, bool)prod$. As will be shown below, the abstract axiomatization of $prod$ derived from the type definition axiom given above is also formulated in terms of the compound type $(\alpha, \beta)prod$. It therefore also holds for any substitution instance of $(\alpha, \beta)prod$—i.e. for the product of any two types.

The abstract axiomatization of *prod* derived in the following section will use the abstraction and representation functions:

ABS_pair:$(\alpha \to \beta \to bool) \to (\alpha, \beta)prod$ and
REP_pair:$(\alpha, \beta)prod \to (\alpha \to \beta \to bool)$

which relate pairs to the functions of type $\alpha \to \beta \to bool$ which represent them. These representation and abstraction functions are defined formally as described above in Section 3.3. A set of theorems stating that Abs_pair and Rep_pair are isomorphisms can also be proved as outlined in Section 3.3. These theorems will be used in the proof of the axiom for *prod* given in the next section.

For notational convenience, an infix type operator '×' will be used in the remainder of this paper for the product of two types. Type expressions of the form $ty_1 \times ty_2$ will be simply syntactic abbreviations for $(ty_1, ty_2)prod$.

4.2.3   Deriving the Axiomatization of *prod*

To formulate the axiomatization of $(\alpha \times \beta)$, two constants will be defined:

Fst:$(\alpha \times \beta) \to \alpha$ and   Snd:$(\alpha \times \beta) \to \beta$.

These denote the usual *projection* functions on pairs; the function Fst extracts the first component of a pair, and the function Snd extracts the second component of a pair. The definitions of these functions are:

$\vdash$ Fst $p = \varepsilon x. \exists y. (\text{REP\_pair } p) \, x \, y$
$\vdash$ Snd $p = \varepsilon y. \exists x. (\text{REP\_pair } p) \, x \, y$

These definitions first use the representation function REP_pair to map a pair p to the function that represents it. They then 'select' the required component of the pair using the ε-operator. From the definitions of Fst and Snd, it is possible to show that

$\vdash$ Fst(ABS_pair($\lambda x \, y. (x{=}a) \wedge (y{=}b)$)) $= a$
$\vdash$ Snd(ABS_pair($\lambda x \, y. (x{=}a) \wedge (y{=}b)$)) $= b$

$$(4)$$

by using the fact that Rep_pair is the left inverse of ABS_pair for functions that satisfy the subset predicate Is_pair_REP. Once these two theorems have been proved, the axiomatization of the cartesian product of two types can be derived without further reference to the way Fst and Snd are defined.

Using the functions Fst and Snd, the axiomatization of the cartesian product of two types can be formulated based on the notion of a *product* in category theory. The following theorem will be the single axiom for the product of two types:

$\vdash \forall f{:}\gamma \to \alpha. \, \forall g{:}\gamma \to \beta. \, \exists! h{:}\gamma \to (\alpha \times \beta). (\text{Fst o } h = f) \wedge (\text{Snd o } h = g)$

This theorem states that for all functions f and g, there is a unique function h such

that the diagram

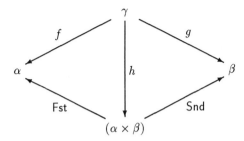

$$(\alpha \times \beta)$$

is commutative, i.e. $\forall x.\,\mathsf{Fst}(h\ x) = f\ x$ and $\forall x.\,\mathsf{Snd}(h\ x) = g\ x$. As noted above, this theorem is proved for the polymorphic type $(\alpha \times \beta)$. It therefore characterizes the product of any two types, since the type variables α and β in this theorem can be instantiated to any two types of the logic to yield an axiom for their product.

An outline of the proof of the axiom shown above is as follows. Given two functions $f{:}\gamma{\to}\alpha$ and $g{:}\gamma{\to}\beta$, define the function $h{:}\gamma{\to}(\alpha \times \beta)$ as follows:

$$h\ v = \mathsf{ABS\_pair}(\lambda x\, y.\,(x{=}f\ v) \wedge (y{=}g\ v))$$

Using the theorems (4) above, it follows that $\mathsf{Fst\ o}\ h = f$ and $\mathsf{Snd\ o}\ h = g$. To show that h is unique, suppose that there is also a function h' such that $\mathsf{Fst\ o}\ h' = f$ and $\mathsf{Snd\ o}\ h' = g$. Suppose v is some value of type γ. Since $\mathsf{ABS\_pair}$ is onto $(\alpha \times \beta)$, there exist a and b such that $h'\ v = \mathsf{ABS\_pair}(\lambda x\, y.\,(x{=}a) \wedge (y{=}b))$. Thus,

$$
\begin{aligned}
f\ v &= \mathsf{Fst}(h'\ v) &= \mathsf{Fst}(\mathsf{ABS\_pair}(\lambda x\, y.\,(x{=}a) \wedge (y{=}b))) &= a \qquad \text{and} \\
g\ v &= \mathsf{Snd}(h'\ v) &= \mathsf{Snd}(\mathsf{ABS\_pair}(\lambda x\, y.\,(x{=}a) \wedge (y{=}b))) &= b
\end{aligned}
$$

which means that

$$h'\ v = \mathsf{ABS\_pair}(\lambda x\, y.\,(x{=}f\ v) \wedge (y{=}g\ v)) = h\ v$$

and therefore that $h' = h$.

4.2.4 Theorems about *prod*

Using the axiom for products proved in the previous section, an infix operator $\otimes$ can be defined such that for all functions $f{:}\gamma{\to}\alpha$ and $g{:}\gamma{\to}\beta$ the expression $f \otimes g$ denotes the unique function of type $\gamma{\to}(\alpha \times \beta)$ which the axiom asserts to exist. This operator can be defined using the ε-operator as follows:

$$\vdash \forall f\, g.\,(f \otimes g) = \varepsilon h.\,(\mathsf{Fst\ o}\ h = f) \wedge (\mathsf{Snd\ o}\ h = g)$$

It follows from the axiom for products and the property of ε shown in Section 1.3 that $(f \otimes g)$ denotes a function which makes the diagram shown above commute:

$$\vdash \mathsf{Fst\ o}\ (f \otimes g) = f \quad \text{and} \quad \vdash \mathsf{Snd\ o}\ (f \otimes g) = g.$$

It can also be shown that for all f and g, the term $f \otimes g$ denotes the unique function with this property:

$$\vdash \forall f\, g\, h.\, (\mathsf{Fst}\; \mathsf{o}\; h = f) \wedge (\mathsf{Snd}\; \mathsf{o}\; h = g) \supset (h = (f \otimes g)).$$

Using the operator $\otimes$, an infix pairing function ',' can be defined to give the usual syntax for pairs, with (a, b) denoting the ordered pair having first component a and second component b. The definition is:

$$\vdash \forall a\, b.(a, b) = ((\mathsf{K}\; a) \otimes \mathsf{I})\; b \qquad \text{where } \mathsf{K} = \lambda a\, b.a \text{ and } \mathsf{I} = \lambda a.a.$$

The projection functions Fst and Snd and the constructor ',' defined above satisfy three theorems shown below, which are commonly used to characterize pairs.

$$\vdash \forall a\, b.\, \mathsf{Fst}(a, b) = a$$
$$\vdash \forall a\, b.\, \mathsf{Snd}(a, b) = b$$
$$\vdash \forall p.\, p = (\mathsf{Fst}\; p, \mathsf{Snd}\; p)$$

The first two of these theorems follow from the definition of the infix pairing operator ',' and the fact that $\vdash \mathsf{Fst}\; \mathsf{o}\; ((\mathsf{K}\; a) \otimes \mathsf{I}) = \mathsf{K}\; a$ and $\vdash \mathsf{Snd}\; \mathsf{o}\; ((\mathsf{K}\; a) \otimes \mathsf{I}) = \mathsf{I}$. The third theorem follows from the uniqueness of functions defined using $\otimes$.

4.3 The Type Operator *sum*

The final example in this section is the definition and axiomatization of a binary type operator *sum* to denote the disjoint sum operation on types. The set that will denoted by the compound type $(ty_1, ty_2)sum$ can be thought of as the union of two disjoint sets: a copy of the set denoted by ty_1, in which each element is labelled as coming from ty_1; and a copy of the set denoted by ty_2, in which each element is labelled as coming from ty_2. Thus each value of type $(ty_1, ty_2)sum$ will correspond either to a value of type ty_1 or to a value of type ty_2. Furthermore, each value of type ty_1 and each value of type ty_2 will correspond to a unique value of type $(ty_1, ty_2)sum$.

4.3.1 The Representation

One way of representing a value v of type $(\alpha, \beta)sum$ would be to use a triple (a, b, f) of type $\alpha \times \beta \times bool$, where f is a boolean 'flag' stating whether v corresponds to the value a of type α or the value b of type β. With this representation, each value a of type α would correspond to a triple (a, d_β, T) in the representation, where d_β is some fixed 'dummy' value of type β. Likewise, each value b of type β would have a corresponding triple $(d_\alpha, b, \mathsf{F})$ in the representation, where d_α is a dummy value of type α. Using this representation, every value in the representing subset of $\alpha \times \beta \times bool$ would correspond either to a value of type α labelled by T or to a value of type β labelled by F.

The representation of values of type $(\alpha, \beta)sum$ can be both simplified and made independent of the product type operator by noting that a triple (a, d_β, T), for example, can itself be represented by the function:

$$\lambda x\, y\, fl.\, (x{=}a) \wedge (y{=}d_\beta) \wedge (fl{=}\mathsf{T})$$

This function is true exactly when applied to the value a, the dummy value d_β and the truth-value T. Every function of this form corresponds to unique value of type α, and every value of type α corresponds to a function of this form. But the same can be said of functions of the form:

$$\lambda x\, y\, fl.\,(x{=}a) \wedge (fl{=}\mathsf{T})$$

The dummy value d_β is therefore not necessary. A value of type $(\alpha, \beta)sum$ that corresponds to a value b of type β can likewise be represented by a function of the form:

$$\lambda x\, y\, fl.\,(y{=}b) \wedge (fl{=}\mathsf{F}).$$

The type $(\alpha, \beta)sum$ can therefore be represented by the subset of functions of type $\alpha{\to}\beta{\to}bool{\to}bool$ that satisfy the predicate $\mathsf{Is\_sum\_REP}$ defined by:

$$\vdash \mathsf{Is\_sum\_REP}\ f = (\exists v_1.f = \lambda x\, y\, fl.\,(x{=}v_1) \wedge (fl{=}\mathsf{T})) \vee$$
$$(\exists v_2.f = \lambda x\, y\, fl.\,(y{=}v_2) \wedge (fl{=}\mathsf{F}))$$

The set of functions satisfying $\mathsf{Is\_sum\_REP}$ contains exactly one function for each value of type α and exactly one function for each value of type β. It therefore represents the disjoint sum of the set of values of type α and the set of values of type β.

4.3.2   The Type Definition

The type definition axiom for sum is introduced in exactly the same way as the defining axioms for one and $prod$. The first step is to prove a theorem stating that $\mathsf{Is\_sum\_REP}$ is true of at least one value in the representing set: $\vdash \exists f.\,\mathsf{Is\_sum\_REP}\ f$. A type definition axiom of the usual form can then be introduced:

$$\vdash \exists f{:}(\alpha, \beta)sum{\to}(\alpha{\to}\beta{\to}bool).$$
$$(\forall a_1\, a_2.f\ a_1 = f\ a_2 \supset a_1 = a_2) \wedge (\forall r.\,\mathsf{Is\_sum\_REP}\ r = (\exists a.\,r = f\ a))$$

and the abstraction and representation functions

$$\mathsf{ABS\_sum}{:}(\alpha{\to}\beta{\to}bool{\to}bool){\to}(\alpha, \beta)sum \quad \text{and}$$
$$\mathsf{REP\_sum}{:}(\alpha, \beta)sum{\to}(\alpha{\to}\beta{\to}bool{\to}bool)$$

defined in the usual way. As outlined in Section 3.3, the definitions of $\mathsf{Abs\_sum}$ and $\mathsf{REP\_sum}$ and the type definition axiom for sum yield the usual isomorphism theorems about such abstraction and representation functions. These theorems will be used in the derivation of the abstract axiom for sum.

For notational clarity, an infix type operator '$+$' will now be used for the disjoint sum of two types. In what follows, the syntactic abbreviation $ty_1 + ty_2$ will be used instead of the form $(ty_1, ty_2)sum$.

4.3.3 Deriving the Axiomatization of *sum*

The axiomatization of $(\alpha + \beta)$ will use two constants:

$$\text{Inl}:\alpha \rightarrow (\alpha + \beta) \quad \text{and} \quad \text{Inr}:\beta \rightarrow (\alpha + \beta)$$

defined by:

$$\vdash \text{Inl } a = \text{ABS\_sum}(\lambda x\, y\, fl.\,(x{=}a) \wedge (fl{=}\text{T}))$$
$$\vdash \text{Inr } b = \text{ABS\_sum}(\lambda x\, y\, fl.\,(y{=}b) \wedge (fl{=}\text{F}))$$

The constants Inl and Inr denote the left and right *injection* functions for sums. Every value of type $(\alpha + \beta)$ is either a left injection Inl a for some value $a{:}\alpha$ or a right injection Inr b for some value $b{:}\beta$.

The form of the axiom for $(\alpha{+}\beta)$ is based on the categorical notion of a *coproduct*. The axiom for $(\alpha + \beta)$ is:

$$\vdash \forall f{:}\alpha{\rightarrow}\gamma.\ \forall g{:}\beta{\rightarrow}\gamma.\ \exists!\, h{:}(\alpha + \beta){\rightarrow}\gamma.\,(h \circ \text{Inl} = f) \wedge (h \circ \text{Inr} = g)$$

This theorem asserts that for all functions f and g there is a unique function h such that the diagram shown below is commutative.

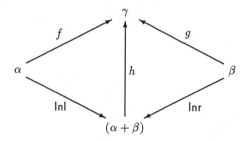

The proof of the axiom for sums is similar to the one outlined in the previous section for products. The proof will therefore not be given in full here. The existence of h follows simply by defining

$$h\, s = ((\exists v_1.\, x = \text{Inl } v_1) \Rightarrow f(\varepsilon v_1.\, x = \text{Inl } v_1) \mid g(\varepsilon v_2.\, x = \text{Inr } v_2))$$

for given f and g. The uniqueness of h follows from the fact that Inl and Inr are one-to-one, and from the fact that ABS_sum is onto.

4.3.4 Theorems about *sum*

Using the axiom for sums, it is possible to define an operator $\oplus$ which is analogous to the operator $\otimes$ defined above for products. The definition of $\oplus$ is:

$$\vdash \forall f\, g.\,(f \oplus g) = \varepsilon h.\,(h \circ \text{Inl} = f) \wedge (h \circ \text{Inr} = g)$$

From the axiom for sums, it follows that for all functions f and g the term $(f \oplus g)$ denotes a function that makes the diagram for sums commute:

$$\vdash (f \oplus g) \circ \text{Inl} = f \quad \text{and} \quad \vdash (f \oplus g) \circ \text{Inr} = g$$

and that $(f \oplus g)$ denotes the unique function with this property:

$$\vdash \forall f\, g\, h.\, (h \circ \mathsf{Inl} = f) \land (h \circ \mathsf{Inr} = g) \supset (h = (f \oplus g)).$$

Using $\oplus$, it is possible to define two *discriminator* functions $\mathsf{Isl}{:}(\alpha + \beta){\to}bool$ and $\mathsf{Isr}{:}(\alpha + \beta){\to}bool$ as follows:

$$\vdash \mathsf{Isl} = (\mathsf{K}\ \mathsf{T}) \oplus (\mathsf{K}\ \mathsf{F}) \quad \text{and} \quad \vdash \mathsf{Isr} = (\mathsf{K}\ \mathsf{F}) \oplus (\mathsf{K}\ \mathsf{T})$$

From these definitions, and the properties of $\oplus$ shown above, it follows that every value of type $(\alpha + \beta)$ satisfies either Isl or Isr:

$$\vdash \forall s{:}(\alpha + \beta).\, \mathsf{Isl}\ s \lor \mathsf{Isr}\ s$$

and that Isl is true of left injections and Isr is true of right injections:

$$\vdash \forall a.\, \mathsf{Isl}(\mathsf{Inl}\ a) \qquad \vdash \forall b.\, \neg\mathsf{Isl}(\mathsf{Inr}\ b)$$
$$\vdash \forall b.\, \mathsf{Isr}(\mathsf{Inr}\ b) \qquad \vdash \forall a.\, \neg\mathsf{Isr}(\mathsf{Inl}\ a)$$

The operator $\oplus$ can also be used to define *projection* functions $\mathsf{Outl}{:}(\alpha + \beta){\to}\alpha$ and $\mathsf{Outr}{:}(\alpha + \beta){\to}\beta$ that map values of type $(\alpha + \beta)$ to the corresponding values of type α or β. Their definitions are:

$$\vdash \mathsf{Outl} = \mathsf{I} \oplus (\mathsf{K}\ \varepsilon b.\ \mathsf{F}) \quad \text{and} \quad \vdash \mathsf{Outr} = (\mathsf{K}\ \varepsilon a.\ \mathsf{F}) \oplus \mathsf{I}$$

where $\varepsilon a.\ \mathsf{F}$ and $\varepsilon b.\ \mathsf{F}$ denote 'arbitrary' values of type α and β respectively. From these definitions, it follows that the projection functions Outl and Outr have the properties:

$$\vdash \forall a.\, \mathsf{Outl}(\mathsf{Inl}\ a) = a \qquad \vdash \forall s.\, \mathsf{Isl}\ s \supset \mathsf{Inl}(\mathsf{Outl}\ s) = s$$
$$\vdash \forall a.\, \mathsf{Outr}(\mathsf{Inr}\ a) = a \qquad \vdash \forall s.\, \mathsf{Isr}\ s \supset \mathsf{Inr}(\mathsf{Outr}\ s) = s$$

5 Two Recursive Types: Numbers and Lists

This section outlines the definition of two recursive types: *num* (the type natural numbers) and $(\alpha)list$ (the polymorphic type of lists). Both *num* and $(\alpha)list$ are simple examples of the kind of recursive types which can be defined using the general method that will be described in Section 7. Their definitions are given here as examples to introduce the idea of defining recursive types in higher order logic. They also provide examples of the general form of abstract axiomatization that will be used in Section 7 for such types.

Both *num* and $(\alpha)list$ will be used in Section 6 to construct representations for two logical types of trees. Along with the basic building blocks: *one*, *prod* and *sum*, these types of trees will then be used in Section 7.3 to construct representations for arbitrary concrete recursive types.

5.1 The Natural Numbers

The construction of the natural numbers described in this section is based on the definition of the type *num* outlined by Gordon in [6]. The type *num* of natural numbers is defined using a subset of the primitive type *ind* of individuals. This primitive type is characterized by a single axiom, the 'axiom of infinity' shown below:

$$\vdash \exists f{:}ind{\rightarrow}ind.\,(\forall x_1\, x_2.\,(f\ x_1 = f\ x_2) \supset (x_1 = x_2)) \wedge \neg(\forall y.\,\exists x.\,y = fx) \qquad (5)$$

This theorem is one of the basic axioms of higher order logic. It asserts the existence of a function f from *ind* to *ind* which is one-to-one but not onto.

From this axiom, it follows that there are at least a countably infinite number of distinct values of type *ind*. Informally, this follows by observing that there is at least one value of type *ind* which is not in the image of f. Call this value i_0. Now define i_1 to be $f(i_0)$. Since i_1 is in the image of the function f and i_0 is not, it follows that they are distinct values of type *ind*. Now, define i_2 to be $f(i_1)$. By the same argument as given above for i_1, it is clear that i_2 is not equal to i_0. Furthermore, i_2 is also not equal to i_1, since from the fact that f is one-to-one it follows that if $i_2 = i_1$ then $f(i_1) = f(i_0)$ and so $i_1 = i_0$. So i_2 is distinct from both i_1 and i_0. Defining i_3 to be $f(i_2)$, i_4 to be $f(i_3)$, etc. gives—by the same reasoning—an infinite sequence of distinct values of type *ind*. This infinite sequence can be used to represent the natural numbers.

5.1.1 The Representation and Type Definition

As was outlined informally above, it follows from the axiom of infinity (5) that there exists a function which can be used to 'generate' an infinite sequence of distinct values of type *ind*. The axiom of infinity merely asserts the existence of this function; the first step in representing the natural numbers is therefore to define a constant $S{:}ind{\rightarrow}ind$ which in fact *denotes* this function. Using the ε-operator, the definition of S is simply:

$$\vdash S = \varepsilon f{:}ind{\rightarrow}ind.\,(\forall x_1\, x_2.\,(f\ x_1 = f\ x_2) \supset (x_1 = x_2)) \wedge \neg(\forall y.\,\exists x.\,y = fx)$$

Once S has been defined, a constant $Z{:}ind$ can be defined which denotes a value not in the image of S. From this value Z, an infinite sequence of distinct individuals can then be generated by repeated application of S. The definition of Z simply uses the ε-operator to choose an arbitrary value not in the image of S:

$$\vdash Z = \varepsilon y{:}ind.\,\forall x.\,\neg(y = S\ x)$$

From the definitions of S and Z, the semantics of ε, and the axiom of infinity, it follows immediately that Z is not in the image of S and that S is one-to-one. Formally:

$$\vdash \forall i.\,\neg(S\ i = Z)$$
$$\vdash \forall i_1\, i_2.\,(S\ i_1 = S\ i_2) \supset (i_1 = i_2) \qquad (6)$$

By the informal argument given in the introduction to this section, these two theorems imply that the individuals denoted by Z, S(Z), S(S(Z)), S(S(S(Z))), …

form an infinite sequence of distinct values, and can therefore be used to represent the type *num* of natural numbers. To make a type definition for *num*, a predicate N:*ind*→*bool* must be defined which is true of just those individuals in this infinite sequence. This can be done by defining N to be true of the values of type *ind* in the smallest subset of individuals which contains Z and is closed under S. The formal definition of N in higher order logic is:

$$\vdash N\, i = \forall P{:}ind{\rightarrow}bool.\, P\, Z \land (\forall x.\, P\, x \supset P(S\, x)) \supset P\, i$$

This definition states that N is true of a value *i*:*ind* exactly when *i* is an element of *every* subset of *ind* which contains Z and is closed under S. This means that the subset of *ind* defined by N is the *smallest* such set and therefore contains just those individuals obtainable from Z by zero or more applications of S.

From the definition of N, it is easy to prove the following three theorems:

$$\vdash N\, Z$$
$$\vdash \forall i.\, N\, i \supset N(S\, i) \tag{7}$$
$$\vdash \forall P.\, (P\, Z \land \forall i.\, (P\, i \supset P(S\, i))) \supset \forall i.\, N\, i \supset P\, i$$

The first two of these theorems state that the subset of *ind* defined by N contains Z and is closed under the function S. The third theorem states that the subset of *ind* defined by N is the smallest such set. That is, any set of individuals containing Z and closed under S has the set of individuals specified by N as a subset.

Using the predicate N, the type constant *num* can be defined by introducing a type definition axiom of the usual form. From the theorem ⊢ N Z, it follows immediately that ⊢ ∃*i*. N *i*. The following type definition axiom for the type *num* can therefore be introduced:

$$\vdash \exists f{:}num{\rightarrow}ind.(\forall a_1\, a_2.f\, a_1 = f\, a_2 \supset a_1 = a_2) \land (\forall r.\, N\, r = (\exists a.\, r = f\, a))$$

and the usual abstraction and representation functions

$$\text{ABS\_num:}ind{\rightarrow}num \quad \text{and} \quad \text{REP\_num:}num{\rightarrow}ind$$

for mapping between values of type *num* and their representations of type *ind* can defined as described in Section 3.3.

5.1.2 Deriving the Axiomatization of *num*

The natural numbers are conventionally axiomatized by Peano's postulates. The five theorems labelled (6) and (7) in the previous section amount to a formulation of the Peano postulates for the natural numbers represented by individuals. It is therefore easy to derive Peano's postulates for the type *num* of natural numbers from these corresponding theorems about the subset of *ind* specified by N.

The first step in deriving the Peano postulates for *num* is to define the two constants:

$$0{:}num \quad \text{and} \quad \text{Suc:}num{\rightarrow}num,$$

which denote the number zero and the successor function on natural numbers. Using the abstraction and representation functions ABS_num and REP_num, the constants 0 and Suc can be defined as follows:

$$\vdash 0 = \mathsf{ABS\_num}\ \mathsf{Z}$$
$$\vdash \mathsf{Suc}\ n = \mathsf{ABS\_num}(\mathsf{S}(\mathsf{REP\_num}\ n))$$

From these definitions, the five theorems labelled (6) and (7), and the fact that the abstraction and representation functions ABS_num and REP_num are isomorphisms, it is easy to prove the abstract axiomatization of *num*, consisting of the three Peano postulates shown below:

$$\vdash \forall n.\ \neg(\mathsf{Suc}\ n = 0)$$
$$\vdash \forall n_1\ n_2.\ \mathsf{Suc}\ n_1 = \mathsf{Suc}\ n_2 \supset n_1 = n_2$$
$$\vdash \forall P.\ (P\ 0 \wedge \forall n.\ P\ n \supset P(\mathsf{Suc}\ n)) \supset \forall n.\ P\ n$$

The first of Peano's postulates shown above states that zero is not the successor of any natural number. This theorem follows immediately from the corresponding theorem $\vdash \forall i.\ \neg(\mathsf{S}\ i = \mathsf{Z})$ derived in the previous section for the representing values of type *ind*. Likewise, the second of Peano's postulates, which states that Suc is one-to-one, follows from the corresponding theorem about S. The third postulate states the validity of mathematical induction on natural numbers; it follows from the last of three theorems (7) derived in the previous section.

5.1.3 The Primitive Recursion Theorem

Once Peano's postulates have been proved, all the usual properties of the natural numbers can be derived from them. One important property is that functions can be uniquely defined on the natural numbers by primitive recursion. This is stated by the primitive recursion theorem, shown below:

$$\vdash \forall x f.\ \exists! fn.\ (fn\ 0 = x) \wedge \forall n.\ fn\ (\mathsf{Suc}\ n) = f\ (fn\ n)\ n \tag{8}$$

This theorem states that a function $fn:num{\to}\alpha$ can be *uniquely* defined by primitive recursion—i.e. by specifying a value for x to define the value of $fn(0)$ and an expression f to define the value of $fn(\mathsf{Suc}\ n)$ recursively in terms of $fn(n)$ and n. The proof of this theorem will not be given here, but an outline of the proof can be found in Gordon's paper [6]. The proof of a similar theorem for a logical type of *trees* is given in Section 6.1.3.

An important fact about the primitive recursion theorem is that it is equivalent to the three Peano postulates for *num* derived in Section 5.1.2. The single theorem (8) can therefore be used as the abstract axiomatization of the defined type *num*, instead of the three separate theorems expressing Peano's postulates. In Section 7.2, it will be shown how *any* concrete recursive type can be axiomatized in higher order logic by a similar 'primitive recursion' theorem.

Any function definition by primitive recursion on natural numbers can be justified formally in logic by appropriately specializing x and f in theorem (8). For example, specializing x and f to:

$$\lambda n.\ n \quad \text{and} \quad \lambda f x.\ \lambda m.\mathsf{Suc}(f\ m)$$

in a suitably type-instantiated version of the primitive recursion theorem yields (after some simplification) the theorem:

$$\vdash \exists! fn. (fn\ 0\ n = n) \wedge \forall n\ m. (fn\ (\mathsf{Suc}\ n)\ m = \mathsf{Suc}(fn\ n\ m))$$

which asserts the (unique) existence of an *addition* function on natural numbers. Primitive recursive definitions of other standard arithmetic operations (e.g. $+$, $\times$, and exponentiation) can also be formally justified using theorem (8).

5.2   Finite-Length Lists

This section describes the definition of a recursive type $(\alpha)list$ of lists containing values of type α. In principle, it is possible to represent this type by a subset of some *primitive* compound type. But in practice, it is easier to use the defined type constant num and the type operator $\times$ (defined above in Section 4.2). The representation using num and $\times$ described below is based on Gordon's construction of lists in [6].

5.2.1   The Representation and Type Definition

Lists are simply finite sequences of values, all of the same type. A list with n values of type α will be represented by a pair (f, n), where f is a function of type $num \rightarrow \alpha$ and n is a value of type num. The idea is that the function f will give the sequence of values in the list; $f(0)$ will be the first value, $f(1)$ will be the second value, and so on. The second component of a pair (f, n) representing a list will be a number n giving the length of the list represented.

The set of values used to represent lists can not be simply the set of all pairs of type $(num \rightarrow \alpha) \times num$. The pairs used must be restricted so that each list has a *unique* representation. The one-element list [42], for example, will be represented by a pair $(f, 1)$, where $f(0)=42$. But there are an infinite number of different functions $f:num \rightarrow num$ that satisfy the equation $f(0)=42$. To make the representation of [42] unique, some 'standard' value must be chosen for the value of $f(m)$ when $m > 0$. The predicate Is_list_REP defined below uses the standard value $\varepsilon x:\alpha.\mathsf{T}$ to specify a set of pairs containing a unique representation for each list:

$$\vdash \mathsf{Is\_list\_REP}(f, n) = \forall m. m \geq n \supset (f\ m = \varepsilon x:\alpha.\mathsf{T})$$

If a pair (f, n) satisfies Is_list_REP, then for $m < n$ the value of $f(m)$ will be the corresponding element of the list represented. For $m \geq n$, the value of $f(m)$ will be the standard value $\varepsilon x.\mathsf{T}$. With this representation, there is exactly one pair (f, n) for each finite-length list of values of type α.

It is easy to prove that $\vdash \exists f\ n.\ \mathsf{Is\_list\_REP}(f, n)$, since Is_list_REP holds of the pair $(\lambda n.\ \varepsilon x.\mathsf{T}, 0)$. A type definition axiom of the usual form can therefore be introduced for the type $(\alpha)list$:

$$\vdash \exists f:(\alpha)list \rightarrow ((num \rightarrow \alpha) \times num).$$
$$(\forall a_1\ a_2. f\ a_1 = f\ a_2 \supset a_1 = a_2) \wedge (\forall r.\ \mathsf{Is\_list\_REP}\ r = (\exists a.\ r = f\ a))$$

and the abstraction and representation functions:

ABS_list:$((num \rightarrow \alpha) \times num) \rightarrow (\alpha)list$ and
REP_list:$(\alpha)list \rightarrow ((num \rightarrow \alpha) \times num)$

can be defined based on the type definition axiom in the usual way.

5.2.2 Deriving the Axiomatization of $(\alpha)list$

The abstract axiomatization of lists will be based on two constructors:

$$\text{Nil} : (\alpha)list \qquad \text{and} \qquad \text{Cons} : \alpha \rightarrow (\alpha)list \rightarrow (\alpha)list.$$

The constant Nil denotes the empty list. The function Cons constructs lists in the usual way: if h is a value of type α and t is a list then Cons h t denotes the list with head h and tail t.

The definition of Nil is

$$\vdash \text{Nil} = \text{ABS\_list}((\lambda n{:}num.\, \varepsilon x{:}\alpha.\, \mathsf{T}), 0)$$

This equation simply defines Nil to be the list whose representation is the pair $(f, 0)$, where $f(n)$ has the value $\varepsilon x.\mathsf{T}$ for all n.

The constructor Cons can be defined by first defining a corresponding function Cons_REP which performs the Cons-operation on list representations. The definition is:

$$\vdash \text{Cons\_REP } h \ (f, n) = ((\lambda m.(m{=}0 \Rightarrow h \,|\, f(m-1))),\ n+1)$$

The function Cons_REP takes a value h and pair (f, n) representing a list and yields the representation of the result of inserting h at the head of the represented list. This result is a pair whose first component is a function yielding value h when applied to 0 (the head of the resulting list) and the value given by $f(m-1)$ when applied to m for all $m>0$ (the tail of the resulting list). The second component is the length $n+1$, one greater than the length of the input list representation.

Once Cons_REP has been defined, it is easy to define Cons. The definition is:

$$\vdash \text{Cons } h \ t = \text{ABS\_list}(\text{Cons\_REP } h \ (\text{REP\_list } t))$$

The function Cons defined by this equation simply takes a value h and a list t, maps t to its representation, computes the representation of the desired result using Cons_REP, and then maps that result back to the corresponding abstract list.

Once Nil and Cons have been defined, the following abstract axiom for lists can be derived by formal proof:

$$\vdash \forall x\, f.\, \exists !\, fn.\, (fn(\text{Nil}) = x) \wedge (\forall h\, t.\, fn(\text{Cons } h\, t) = f\, (fn\, t)\, h\, t) \tag{9}$$

This axiom is analogous to the primitive recursion theorem for natural numbers, and is an example of the general form of the theorems which will be used in Section 7 to characterize all recursive types. Like the primitive recursion theorem, the abstract axiom for lists asserts that functions can be uniquely defined by primitive recursion. Once this theorem has been derived from the type definition axiom for lists and the definitions of Cons and Nil, all the usual properties of lists follow without further reference to the way lists are defined.

The axiom (9) for lists can be proved formally from the type definition for $(\alpha)list$. Full details will not be given here, but the proof is comparatively simple. The existence of the function fn in theorem (9) follows by demonstrating the existence of a corresponding function on list representations. This function can be defined by primitive recursion on the length component of the representation by using the primitive recursion theorem (8) for natural numbers. The uniqueness of the function fn in the abstract axiom for lists can then be proved by mathematical induction on the length component of list representations.

5.2.3 Theorems about $(\alpha)list$

Once the abstract axiom (9) for lists has been proved, the following three theorems can be derived from it:

$$\vdash \forall h\, t.\, \neg(\mathsf{Nil} = \mathsf{Cons}\ h\ t)$$
$$\vdash \forall h_1\, h_2\, t_1\, t_2.\ (\mathsf{Cons}\ h_1\ t_1 = \mathsf{Cons}\ h_2\ t_2) \supset ((h_1 = h_2) \wedge (t_1 = t_2))$$
$$\vdash \forall P.\ (P(\mathsf{Nil}) \wedge \forall t.\, P\ t \supset \forall h.\, P(\mathsf{Cons}\ h\ t)) \supset \forall l.\, P\ l$$

These three theorems are analogous to the Peano postulates for the natural numbers derived in Section 5.1.2. The first theorem states that Nil is not equal to any list constructed by Cons. The second theorem states that Cons is one-to-one. And the third theorem asserts the validity of structural induction on lists.

6 Two Recursive Types of Trees

This section describes the formal definitions of two different logical types which denote sets of trees. First, a type $tree$ is defined which denotes the set of all trees whose nodes can branch any (finite) number of times. This type is then used to define a second logical type of trees, $(\alpha)Tree$, which denotes the set of $labelled$ trees. These have the same sort of structure as values of type $tree$, but they also have a label of type α associated with each node.

The type $(\alpha)Tree$ defined in this section is of interest because each logical type in the class of recursive types discussed in Section 7 can be represented by some subset of it. Once the type of labelled trees has been defined, it can be used (along with the type one and the type operators $\times$ and $+$) to construct systematically a representation for any concrete recursive type. This avoids the problem of having to find an $ad\ hoc$ representation for each recursive type, and so makes it possible to mechanize efficiently the formal definition of such types.

6.1 The Type of Trees: $tree$

Values of the logical type $tree$ defined in this section will be finite trees whose internal nodes can branch any finite number of times. These trees will be $ordered$. That is, the relative order of each node's immediate subtrees will be important; and two similar trees which differ only in the order of their subtrees will be considered to be different trees.

6.1.1 The Representation and Type Definition

Trees will be represented by coding them as natural numbers; each tree will be represented by a unique value of type num. The smallest possible tree consists of a single leaf node with no subtrees; it will be represented by the number 0. To represent a tree with one or more subtrees, a function $\mathsf{node\_REP}:(num)list{\rightarrow}num$ will be defined which computes the natural number representing such a tree from a list of the numbers which represent its subtrees. The function $\mathsf{node\_REP}$ will take as an argument a list l of numbers. If each of the numbers in the list represents a tree, then $\mathsf{node\_REP}\ l$ will represent the tree whose subtrees are represented by the numbers in l.

Consider, for example, a tree with three subtrees: t_1, t_2, and t_3. Suppose that the three subtrees t_1, t_2, and t_3 are represented by the natural numbers i, j, and k respectively:

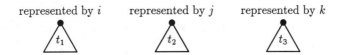

The number representing the tree which has t_1, t_2, and t_3 as subtrees will then be denoted by node_REP$[i; j; k]$:

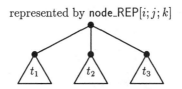

where the conventional list notation $[i; j; k]$ is a syntactic abbreviation for the list denoted by Cons i (Cons j (Cons k Nil)).

Since node_REP takes a *list* of numbers as arguments, it can be used to compute the code for a tree with any finite number of immediate subtrees. Thus, using node_REP, the natural number representing a tree of any shape can be computed recursively from the natural numbers representing its subtrees. The only property that node_REP must have for this to work is the property of being a one-to-one function on lists of numbers:

$$\vdash \forall l_1\, l_2.\,(\text{node\_REP } l_1 = \text{node\_REP } l_2) \supset (l_1 = l_2) \tag{10}$$

This theorem asserts that if node_REP computes the same natural number from two lists l_1 and l_2, then these lists must be equal and therefore must consist of the same finite sequence of numbers. If node_REP has this property, then it can be used to compute a *unique* numerical representation for every possible tree. It remains to define the function node_REP such that theorem (10) holds.

One way of formally defining node_REP is to use the well-known coding function $(n, m) \longmapsto (2n + 1) \times 2^m$ which codes a pair of natural numbers by a single natural number. Using this coding function, node_REP can be defined by recursion on lists such that the following two theorems hold:

$$\vdash \text{node\_REP Nil} \quad\ \ = 0$$
$$\vdash \text{node\_REP (Cons } n\ t) = ((2 \times n) + 1) \times (2 \text{ Exp (node\_REP } t)) \tag{11}$$

These two equations define the value of node_REP l by 'primitive recursion' on the list l. When l is the empty list Nil, the result is 0. When l is a non-empty list with head n and tail t, the result is computed by coding as a single natural number the pair consisting of n and the result of applying node_REP recursively to t. Primitive recursive definitions of this kind can be justified by formal proof using the abstract axiom (9) for lists derived in Section 5.2.2; the two theorems (11) can be derived from an appropriate instance of this axiom and a non-recursive definition of the constant node_REP.

Theorem (10) stating that node_REP is one-to-one can be derived from the two theorems (11) which define node_REP by primitive recursion. The proof is done by structural induction on the lists l_1 and l_2 using the theorem shown in Section 5.2.3 stating the validity of proofs by induction on lists.

The function node_REP can be used to compute a natural number to represent any finitely branching tree. To make a type definition for the type constant *tree*, a predicate on natural numbers Is_tree_REP:*num→bool* must be defined which is true of just those numbers representing trees. This predicate will be defined in the same way as the corresponding predicate was defined in Section 5.1.1 for the representation of numbers by individuals: Is_tree_REP n will be true if the number n is in the smallest set of natural numbers closed under node_REP.

The formal definition of Is_tree_REP uses the auxiliary function Every, defined recursively on lists as follows:

$$\vdash \text{Every } P \text{ Nil} \qquad = \text{ T}$$
$$\vdash \text{Every } P \text{ (Cons } h \text{ } t) = (P \text{ } h) \wedge \text{Every } P \text{ } t$$

These two theorems define Every P l to mean that the predicate P holds of every element of the list l. Using Every, the predicate Is_tree_REP is defined as follows:

$$\vdash \text{Is\_tree\_REP } n = \forall P. (\forall tl. \text{ Every } P \text{ } tl \supset P(\text{node\_REP } tl)) \supset P \text{ } n$$

This definition states that a number n represents a tree exactly when it is an element of every subset of *num* which is closed under node_REP. It follows that the set of numbers for which Is_tree_REP is true is the smallest set closed under node_REP. This set contains just those natural numbers which can be computed using node_REP and therefore contains only those numbers which represent trees.

To use Is_tree_REP to define a new type, the theorem $\vdash \exists n. \text{ Is\_tree\_REP } n$ must first be proved. This theorem follows immediately from the fact that Is_tree_REP is true of 0, i.e. the number denoted by node_REP Nil. Once this theorem has been proved, a type definition axiom of the usual form can be introduced:

$$\vdash \exists f : tree \to num.$$
$$(\forall a_1 a_2. f \text{ } a_1 = f \text{ } a_2 \supset a_1 = a_2) \wedge (\forall r. \text{ Is\_tree\_REP } r = (\exists a. r = f \text{ } a))$$

along with the usual abstraction and representation functions:

$$\text{ABS\_tree}:num \to tree \quad \text{and} \quad \text{REP\_tree}:tree \to num.$$

6.1.2 The Axiomatization of *tree*

The abstract axiom for *tree* will be based on the constructor:

$$\text{node}:(tree)list \to tree$$

The function node builds trees from smaller trees. If tl:$(tree)list$ is a list of trees, then the term node tl denotes the tree whose immediate subtrees are the trees in the list tl. If tl is the empty list of trees, then node tl denotes the tree consisting of a single leaf node. Using node, it is possible to construct a tree of any shape. For example, the tree:

is denoted by the expression: node[node Nil; node Nil; node[node Nil; node Nil]].

An auxiliary function Map will be used in the formal definition of the constructor node. The function Map is the usual mapping function for lists; it takes a function $f{:}\alpha{\to}\beta$ and a list $l{:}(\alpha)list$ and yields the result of applying f to each member of l in turn. The recursive definition of Map is:

$$\vdash \text{Map } f \text{ Nil} = \text{Nil}$$
$$\vdash \text{Map } f \text{ (Cons } h \ t) = \text{Cons } (f \ h) \text{ (Map } f \ t)$$

Using Map and the function node_REP:$(num)list{\to}num$ defined in the previous section, the formal definition in logic of node is:

$$\vdash \text{node } tl = (\text{ABS\_tree}(\text{node\_REP}(\text{Map REP\_tree } tl)))$$

The constructor node defined by this equation takes a list of trees tl, applies node_REP to the corresponding list of numbers representing the trees in tl, and then maps the result to the corresponding abstract tree.

The following two important theorems follow from the formal definition of node given above; they are analogous to the Peano postulates for the natural numbers, and are used to prove the abstract axiom for the type $tree$:

$$\vdash \forall tl_1 \ tl_2. \,(\text{node } tl_1 = \text{node } tl_2) \supset (tl_1 = tl_2)$$
$$\vdash \forall P. \,(\forall tl. \text{ Every } P \ tl \supset P \text{ (node } tl)) \supset \forall t. P \ t$$

The first of these theorems states that the constructor node is one-to-one. This follows directly from theorem (10), which states that the corresponding function node_REP is one-to-one. The second theorem shown above asserts the validity of induction on trees, and can be used to justify proving properties of trees by structural induction. This theorem can be proved from the definitions of node and Is_tree_REP and the fact that ABS_tree and REP_tree are isomorphisms relating trees and the numbers that represent them.

The abstract axiomatization of the defined type $tree$ consists of the single theorem shown below:

$$\vdash \forall f. \exists! fn. \forall tl. \ fn(\text{node } tl) = f \text{ (Map } fn \ tl) \ tl \qquad (12)$$

This theorem is analogous to the primitive recursion theorem (8) for natural numbers and the abstract axiom (9) for lists. It asserts the unique existence of functions defined recursively on trees. The universally quantified variable f ranges over functions that map a list of values of type α and a list of trees to a value of type α. For any such function, there is a unique function $fn{:}tree{\to}\alpha$ that satisfies the equation $fn(\text{node } tl) = f \text{ (Map } fn \ tl) \ tl$. For any tree (node tl), this equation defines the value of $fn(\text{node } tl)$ recursively in terms of the result of applying fn to each of the immediate subtrees in the list tl.

6.1.3 An Outline of the Proof of the Axiom for *tree*

It is straightforward to prove the uniqueness part of the abstract axiom for trees; the uniqueness of the function fn in theorem (12) follows by structural induction on trees using the induction theorem for the defined type *tree*. The existence part of theorem (12) is considerably more difficult to prove. It follows from a slightly weaker theorem in which the list of subtrees tl is not an argument to the universally quantified function f:

$$\vdash \forall f.\, \exists fn.\, \forall tl.\, fn(\text{node } tl) = f\,(\text{Map } fn\ tl) \tag{13}$$

This weaker theorem can be proved by first defining a height function $\text{Ht:}tree{\rightarrow}num$ on trees and then proving that, for any number n, there exists a function fun which satisfies the desired recursive equation for trees whose height is bounded by n:

$$\vdash \forall f\, n.\, \exists fun.\, \forall tl.\, (\text{Ht}(\text{node } tl) \le n) \supset (fun(\text{node } tl) = f\,(\text{Map } fun\ tl)) \tag{14}$$

The main step in the proof of this theorem is an induction on the natural number n.

Theorem (14) can be used to define a higher order function fun which yields approximations of the function fn whose existence is asserted by theorem (13). For any n and f, the term $(\text{fun } n\ f)$ denotes an approximation of fn which satisfies the recursive equation in theorem (13) for trees whose height is no greater than n. This is stated formally by the following theorem:

$$\vdash \forall f\, n\, tl.\, (\text{Ht}(\text{node } tl) \le n) \supset (\text{fun } n\ f\ (\text{node } tl) = f\,(\text{Map } (\text{fun } n\ f)\ tl)) \tag{15}$$

The approximations of fn constructed by fun have the following important property: for any two numbers n and m, the corresponding functions constructed by fun behave the same for trees whose height is bounded by both n and m. This property follows by structural induction on trees, and is expressed formally by the theorem:

$$\vdash \forall t\, n\, m\, f.\, (\text{Ht } t){<}n \wedge (\text{Ht } t){<}m \supset (\text{fun } n\ f\ t = \text{fun } m\ f\ t) \tag{16}$$

Theorem (13) asserts the existence of a function fn for any given f; the higher order function fun can be used to explicitly construct this function fn from the given function f. For any f, the term $\lambda t.\, \text{fun }(\text{Ht}(\text{node } [t]))\ f\ t$ denotes the function which satisfies the desired recursive equation. An outline of the proof of this is as follows. Specializing f, n, and tl in theorem (15) to f, $\text{Ht}(\text{node}[\text{node } tl])$, and tl respectively yields the following implication:

$$\vdash \text{Ht}(\text{node } tl) \le \text{Ht}(\text{node}[\text{node } tl]) \supset$$
$$\text{fun }(\text{Ht}(\text{node}[\text{node } tl]))\ f\ (\text{node } tl) = f(\text{Map }(\text{fun }(\text{Ht}(\text{node}[\text{node } tl]))\ f)\ tl)$$

The height function Ht has the property: $\vdash \forall t.\, \text{Ht } t \le \text{Ht}(\text{node } [t])$. The antecedent of the implication shown above is therefore always true, and the theorem can be simplified to:

$$\vdash \text{fun }(\text{Ht}(\text{node}[\text{node } tl]))\ f\ (\text{node } tl) = f(\text{Map }(\text{fun }(\text{Ht}(\text{node}[\text{node } tl]))\ f)\ tl)$$

The property of fun expressed by theorem (16) implies that the above theorem is equivalent to:

$$\vdash \mathsf{fun}\ (\mathsf{Ht}(\mathsf{node}[\mathsf{node}\ tl]))\ f\ (\mathsf{node}\ tl) = f(\mathsf{Map}\ (\lambda t.\,\mathsf{fun}\ (\mathsf{Ht}(\mathsf{node}[t]))\ f\ t)\ tl)$$

which is itself equivalent (by β-reduction) to:

$$\vdash (\lambda t.\,\mathsf{fun}\ (\mathsf{Ht}(\mathsf{node}[t]))f\ t)(\mathsf{node}\ tl) = f(\mathsf{Map}\ (\lambda t.\,\mathsf{fun}\ (\mathsf{Ht}(\mathsf{node}[t]))\ f\ t)\ tl)$$

Theorem (13) follows immediately from this last result. The stronger theorem (12), which axiomatizes the defined type *tree*, then follows from theorem (13) by a relatively straightforward formal proof.

6.2 The Type of Labelled Trees: $(\alpha)Tree$

This section outlines the definition of the type $(\alpha)Tree$ which denotes the set of labelled trees. Labelled trees of the kind defined in this section have the same sort of general structure as values of the logical type *tree* defined in the previous section. The only difference is that a tree of type $(\alpha)Tree$ has a value or 'label' of type α associated with each of its nodes. It is therefore comparatively simple to define the type $(\alpha)Tree$, since the values of the structurally similar type *tree* can be used in its representation.

6.2.1 The Representation and Type Definition

The representation of a labelled tree of type $(\alpha)Tree$ will be a pair (t, l), where t is a value of type *tree* giving the shape of the tree being represented and l is a list of type $(\alpha)list$ containing the values associated with its nodes. The values in the list l will occur in the sequence which corresponds to a *preorder traversal* of the labelled tree being represented. Consider, for example, the labelled tree shown below:

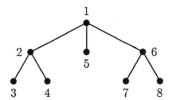

This tree has a natural number associated with each node and can be represented by a pair (t, l) of type *tree* $\times$ $(num)list$. The first component t of this pair will be the value of type *tree* whose structure corresponds to the above picture. The second component l will be a list of length eight containing the numbers associated with the nodes of the corresponding labelled tree. The numbers in this list will occur in the order $[1; 2; 3; 4; 5; 6; 7; 8]$, corresponding to a preorder traversal of the labelled tree being represented.

Any α-labelled tree can be similarly represented by a pair of type *tree* $\times$ $(\alpha)list$; but not every such pair represents a tree. For a pair (t, l) to represent a labelled tree, the length of the list l must be the same as the number of nodes in the tree t. This can be expressed in logic by defining two functions:

$$\mathsf{Length}{:}(\alpha)list{\rightarrow}num \quad \text{and} \quad \mathsf{Size}{:}tree{\rightarrow}num$$

which compute the length of a list and the number of nodes in a tree, respectively.

The function Length can be defined recursively by using the abstract axiom (9) for lists to derive the following two equations:

$$\vdash \text{Length Nil} \quad = 0$$
$$\vdash \text{Length (Cons } h\ t) = (\text{Length } t) + 1$$

The function Size can be defined by first defining a recursive function on lists Sum:$(num)list \rightarrow num$ which computes the sum of a list of natural numbers:

$$\vdash \text{Sum Nil} \quad = 0$$
$$\vdash \text{Sum (Cons } n\ l) = n + (\text{Sum } l)$$

and then using the abstract axiom (12) for the defined type $tree$ to derive the following recursive definition of Size:

$$\vdash \text{Size(node } tl) = (\text{Sum(Map Size } tl)) + 1$$

Using the functions Length and Size, the values of type $tree \times (\alpha)list$ that represent labelled trees can be specified by the predicate Is_Tree_REP defined as follows:

$$\vdash \text{Is\_Tree\_REP}(t, l) = (\text{Length } l = \text{Size } t)$$

This predicate is true of just those pairs (t, l) where the number of nodes in the tree t equals the length of the list l. It is therefore true of precisely those values of type $tree \times (\alpha)list$ which can be used to represent labelled trees.

For any value $v{:}\alpha$, the predicate Is_Tree_REP holds of the pair: (node Nil, $[v]$). From this, it immediately follows that $\vdash \exists p.\ \text{Is\_Tree\_REP}\ p$. The following type definition axiom can therefore be introduced to define $(\alpha)Tree$:

$$\vdash \exists f{:}(\alpha)Tree \rightarrow (tree \times (\alpha)list).$$
$$(\forall a_1\, a_2.f\ a_1 = f\ a_2 \supset a_1 = a_2) \wedge (\forall r.\ \text{Is\_Tree\_REP}\ r = (\exists a.\ r = f\ a))$$

The associated abstraction and representation functions:

$$\text{ABS\_Tree:}(tree \times (\alpha)list) \rightarrow (\alpha)Tree \quad \text{and}$$
$$\text{REP\_Tree:}(\alpha)Tree \rightarrow (tree \times (\alpha)list)$$

can then be defined in the usual way (as described in Section 3.3).

6.2.2 Deriving the Axiomatization of $(\alpha)Tree$

The abstract axiom for $(\alpha)Tree$ is based on the constructor

$$\text{Node:}\alpha \rightarrow ((\alpha)Tree)list \rightarrow (\alpha)Tree$$

which is analogous to the constructor node for $tree$. If v is a value of type α, and l is a list of labelled trees, then the term (Node $v\ l$) denotes the labelled tree whose immediate subtrees are those occurring in l and whose root node is labelled by the value v. The function Node can be used to construct labelled trees of any shape.

For example, the tree:

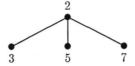

$$2$$
$$3 \quad 5 \quad 7$$

is denoted by the term: Node 2 [Node 3 Nil; Node 5 Nil; Node 7 Nil].

The formal definition of Node uses an auxiliary function $\mathsf{Flat}{:}((\alpha)list)list{\rightarrow}(\alpha)list$ which takes a list of lists and yields the result of appending them all together into a single list. The recursive definition of Flat is:

$$\vdash \mathsf{Flat\ Nil} \qquad\quad = \mathsf{Nil}$$
$$\vdash \mathsf{Flat\ (Cons}\ h\ t) = \mathsf{Append}\ h\ (\mathsf{Flat}\ t)$$

where Append is defined (also recursively) by:

$$\vdash \mathsf{Append\ Nil}\ l \qquad\quad = l$$
$$\vdash \mathsf{Append\ (Cons}\ h\ l_1)\ l_2 = \mathsf{Cons}\ h\ (\mathsf{Append}\ l_1\ l_2)$$

Using Flat, and the mapping function Map defined above in Section 6.1.2, the formal definition of the constructor Node is given by the following theorem:

$$\vdash \mathsf{Node}\ v\ l = \mathsf{ABS\_Tree}((\mathsf{node}(\mathsf{Map}\ (\mathsf{Fst\ o\ REP\_Tree})\ l)),$$
$$((\mathsf{Cons}\ v\ (\mathsf{Flat}(\mathsf{Map}\ (\mathsf{Snd\ o\ REP\_Tree})\ l))))))$$

This definition uses REP_Tree to obtain the representation of each labelled tree in the list l. This yields a list of pairs representing labelled trees. The function node is then used to construct a new tree whose subtrees are the tree components in this list of pairs, and Flat is used to construct the corresponding list of node-values. The result is then mapped back to an abstract labelled tree using the abstraction function ABS_Tree.

Using the constructor Node v l defined above, the abstract axiom for $(\alpha)Tree$ can be written:

$$\vdash \forall f.\, \exists!fn.\, \forall v\, tl.\, fn(\mathsf{Node}\ v\ tl) = f\ (\mathsf{Map}\ fn\ tl)\ v\ tl \tag{17}$$

This theorem is of the same general form as theorem (12), the abstract axiom for the defined type *tree*. It states the uniqueness of functions defined by 'primitive recursion' on labelled trees. The proof of this theorem is straightforward, but it requires some tricky (and uninteresting) lemmas involving the partitioning of lists. Details of the proof will therefore not be given here. The general strategy of the proof is to use the abstract axiom for values of type *tree* to define a recursive function on *representations* which 'implements' the function fn asserted to exist by the axiom (17).

7 Automating Recursive Type Definitions

This section outlines a method for formally defining any simple concrete recursive type in higher order logic. This method has been used to implement an efficient derived inference rule in HOL which defines such recursive types automatically. The input to this derived rule is a user-supplied informal[4] specification of the recursive type to be defined. This type specification is written in a notation which resembles a data type declaration in functional programming languages like Standard ML [9]. It simply states the names of the new type's constructors and the logical types of their arguments. The output is a theorem of higher order logic which abstractly characterizes the properties of the desired recursive type—i.e. a derived 'abstract axiomatization' of the type.

An overview of the algorithm used by this programmed inference rule to define a new recursive type is shown in the diagram below. The algorithm follows the three steps for defining a new logical type described in Section 3.1.

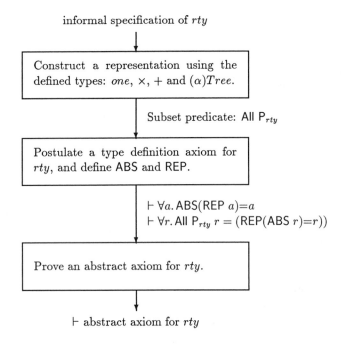

informal specification of rty

Construct a representation using the defined types: one, $\times$, $+$ and $(\alpha)Tree$.

Subset predicate: All P_{rty}

Postulate a type definition axiom for rty, and define ABS and REP.

$\vdash \forall a.\, \text{ABS}(\text{REP}\; a)=a$
$\vdash \forall r.\, \text{All}\; P_{rty}\; r = (\text{REP}(\text{ABS}\; r)=r))$

Prove an abstract axiom for rty.

$\vdash$ abstract axiom for rty

In the first step, an appropriate representation is found for the values of the recursive type rty to be defined. This representation is always some subset of a substitution instance of $(\alpha)Tree$—i.e. a subset of some type $(ty)Tree$ of general trees labelled by values of type ty. The type ty of labels for these trees is built up systematically using the type constant one and the type operators $\times$ and $+$. The output of this stage is a 'subset predicate' which defines the set of labelled trees used to represent values of the new type rty. This predicate has the standard form: 'All P_{rty}', where P_{rty} is a predicate whose exact form is determined by the specification of the type to be defined. (The meaning of 'All' is explained below in

---

[4]In this context, *informal* means not in the language of higher order logic.

Section 7.3.2.) No logical inference needs to be done in this step; so the ML code which implements it in the HOL system is quite fast.

In the second step, a type definition axiom is introduced for the new type, based on the subset predicate All P_{rty}. The associated abstraction and representation functions ABS and REP are then defined and proved to be isomorphisms between the new type rty and the set of values specified by All P_{rty}. The output of this stage consists of the two theorems about ABS and REP shown in the diagram above. The proofs done in this step are easy and routine (see Section 3.3), and their mechanization in HOL is therefore efficient and straightforward.

In the final step, an abstract axiom for the new type rty is derived by formal proof from the definition of the subset predicate All P_{rty} and the two theorems about ABS and REP proved in the previous stage. This is the only step in the algorithm where a non-trivial amount of logical inference has to be done. The ML implementation of this step therefore uses the 'optimization' strategy for HOL derived inference rules discussed in Section 2: a pre-proved general theorem about recursive types is used to reduce to a minimum the amount of inference that has to be done at 'run time' to derive the desired result. This pre-proved theorem has the form shown below:

$$\vdash \forall P. \cdots \; \langle \beta \text{ is isomorphic to 'All } P \text{'} \rangle \supset \langle \text{abstract axiom for } \beta \rangle$$

Informally, this theorem states that *any* type β which is represented by a set of labelled trees 'All P' satisfies an abstract axiomatization of the required form. By specializing P in this theorem to the predicate $P_{r}ty$ constructed in the first step, the abstract axiom for rty follows simply by modus ponens (using the theorems about ABS and REP derived in the second step) and a relatively small amount of straightforward simplification.

A detailed description of the HOL implementation of this algorithm for defining recursive types is beyond the scope of this paper; but the sections which follow give an overview of the logical basis of this implementation. In Section 7.1, the syntax of informal type specifications is described, and some simple examples are given of type specifications written in this notation. Section 7.2 then describes the general form of the abstract axioms that are derived by the system. Section 7.3 explains how appropriate representations for these types can be systematically constructed from their informal type specifications. Finally, Section 7.4 gives the general theorem stating that any recursive type represented in the way described in Section 7.3 satisfies an abstract axiom of the form shown in Section 7.2. An example of the application of this theorem is also given.

7.1 Informal Type Specifications[5]

Every logical type which can be defined by the method outlined in the following sections can be described informally by a *type specification* of the following general form:

$$(\alpha_1, \ldots, \alpha_n)rty \quad ::= \quad C_1 \, ty_1^1 \, \ldots \, ty_1^{k_1} \quad | \quad \cdots \quad | \quad C_m \, ty_m^1 \, \ldots \, ty_m^{k_m} \quad (18)$$

where each ty_i^j is either an existing logical type (not containing rty) or is the type expression $(\alpha_1, \ldots, \alpha_n)rty$ itself. This equation specifies a type $(\alpha_1, \ldots, \alpha_n)rty$ with

---

[5]Some of the notation used in this section is adapted from Bird and Wadler's clear description of the syntax of type definitions in their excellent book [1] on functional programming.

n type variables $\alpha_1, \ldots, \alpha_n$ where $n \geq 0$. If $n=0$ then rty is a type constant; otherwise rty is an n-ary type operator. The type specified has m distinct constructors $C_1, \ldots, C_m$ where $m \geq 1$. Each constructor C_i takes k_i arguments, where $k_i \geq 0$; and the types of these arguments are given by the type expressions ty_i^j for $1 \leq j \leq k_i$. If one or more of the type expressions ty_i^j is the type $(\alpha_1, \ldots, \alpha_n)rty$ itself, then the equation specifies a *recursive* type. In any specification of a recursive type, at least one constructor must be non-recursive—i.e. all its arguments must have types which already exist in the logic.

The logical type specified by equation (18) denotes the set of all values which can be finitely generated using the constructors $C_1, \ldots, C_m$, where each constructor is one-to-one and any two different constructors yield different values. I.e. the type specified by (18) is the *initial algebra* [2] with constructors $C_1, \ldots, C_m$. Every value of this logical type is denoted by some term of the form:

$$C_i \, x_i^1 \, \ldots \, x_i^{k_i}$$

where x_i^j is a term of logical type ty_i^j for $1 \leq j \leq k_i$. In addition, any two terms:

$$C_i \, x_i^1 \, \ldots \, x_i^{k_i} \qquad \text{and} \qquad C_j \, x_j^1 \, \ldots \, x_j^{k_j}$$

denote equal values exactly when their constructors are the same (i.e. $i = j$) and these constructors are applied to equal arguments (i.e. $x_i^n = x_j^n$ for $1 \leq n \leq k_i$).

7.1.1 Some Examples of Type Specifications

The two simple recursive types num and $(\alpha)list$ which were defined in Section 5 are both examples of types that can be described by type specifications of the general form illustrated by (18) above.

The specification of the type num of natural numbers is the simple equation shown below:

$$num \quad ::= \quad 0 \quad | \quad \mathsf{Suc} \ num$$

This equation specifies the type constant num to have two constructors: $0{:}num$ and $\mathsf{Suc}{:}num{\to}num$. The type num which is described by this type specification denotes the set of values generated from the constant 0 by zero or more applications of the constructor Suc—i.e. the set of values denoted by terms of the form: 0, $\mathsf{Suc}(0)$, $\mathsf{Suc}(\mathsf{Suc}(0))$, $\ldots$ etc.

The type specification for the type $(\alpha)list$ of finite lists is similar to the one given above for num. It is:

$$(\alpha)list \quad ::= \quad \mathsf{Nil} \quad | \quad \mathsf{Cons} \ \alpha \ (\alpha)list$$

This equation states that the type $(\alpha)list$ denotes the set of all values generated by the two constructors: $\mathsf{Nil}{:}(\alpha)list$ and $\mathsf{Cons}{:}\alpha{\to}(\alpha)list{\to}(\alpha)list$.

A slightly more complex example is the recursive type $btree$, described by the type specification shown below:

$$btree \quad ::= \quad \mathsf{Leaf} \ num \quad | \quad \mathsf{Tree} \ btree \ btree$$

This equation specifies a type of *binary trees* whose leaf nodes (but not internal nodes) are labelled by natural numbers. When defined formally in higher order logic, this type has two constructors: $\mathsf{Leaf}{:}num{\to}btree$ and $\mathsf{Tree}{:}btree{\to}btree{\to}btree$. The function Leaf constructs leaf nodes; if n is a value of type num, then $(\mathsf{Leaf}\ n)$ denotes a leaf node labelled by n. The constructor Tree builds binary trees from smaller binary trees; if t_1 and t_2 are binary trees then $(\mathsf{Tree}\ t_1\ t_2)$ denotes the binary tree with left subtree t_1 and right subtree t_2.

In addition to recursive types, simple enumerated and 'record' types can also be specified by equations of the form given by (18). For example, the type constant one and the two type operators $prod$ and sum, whose formal definitions were given in Section 4, can be informally specified by the three equations shown below:

$$
\begin{aligned}
one &::= \quad \mathsf{one} \\
(\alpha, \beta)prod &::= \quad \mathsf{pair}\ \alpha\ \beta \\
(\alpha, \beta)sum &::= \quad \mathsf{Inl}\ \alpha \quad | \quad \mathsf{Inr}\ \beta
\end{aligned}
$$

The first of these specifications simply states that one is the enumerated type with exactly one value: the value denoted by the constant one. The second specification states that every value of type $(\alpha, \beta)prod$ is denoted by some term of the form $(\mathsf{pair}\ a\ b)$, i.e. an ordered pair with first component $a{:}\alpha$ and second component $b{:}\beta$. The third equation states that every value of type $(\alpha, \beta)sum$ is either a left injection constructed by Inl or a right injection constructed by Inr.

Many more examples of types—both recursive and non-recursive—which can be specified by equations of the form given by (18) can be found in books on functional programming. (See, for example, chapter 8 of [1].)

7.2 Formulating Abstract Axioms for Recursive Types

The input to the HOL programmed inference rule which defines types is, in general, an informal specification of the form:

$$
(\alpha_1, \ldots, \alpha_n)rty \quad ::= \quad \mathsf{C}_1\ ty_1^1\ \ldots\ ty_1^{k_1} \quad | \quad \cdots \quad | \quad \mathsf{C}_m\ ty_m^1\ \ldots\ ty_m^{k_m}
$$

Each type $(\alpha_1, \ldots, \alpha_n)rty$ specified by an equation of this form can be abstractly characterized by a single theorem of higher order logic. This theorem is the output of the HOL derived rule for defining types and has the following general form:

$$
\vdash \forall f_1\ \cdots\ f_m.\ \exists! fn{:}(\alpha_1, \ldots, \alpha_n)rty{\to}\beta.
$$
$$
\forall x_1^1 \cdots x_1^{k_1}.\ fn(\mathsf{C}_1\ x_1^1\ \ldots\ x_1^{k_1}) = f_1\ (fn\ x_1^1) \ldots (fn\ x_1^{k_1})\ x_1^1\ \ldots\ x_1^{k_1}\ \wedge
$$
$$
\vdots \tag{19}
$$
$$
\forall x_m^1 \cdots x_m^{k_m}.\ fn(\mathsf{C}_m\ x_m^1\ \ldots\ x_m^{k_m}) = f_m\ (fn\ x_m^1) \ldots (fn\ x_m^{k_m})\ x_m^1\ \ldots\ x_m^{k_m}
$$

where the right hand sides of the equations include recursive applications $(fn\ x_i^j)$ of the function fn only for variables x_i^j of type $(\alpha_1, \ldots, \alpha_n)rty$.

Theorem (19) states that for any m functions $f_1, \ldots, f_m$ there is a $unique$ function fn which satisfies a 'primitive recursive' definition whose form is determined by the given functions $f_1, \ldots, f_m$. This is an abstract characterization of the type $(\alpha_1, \ldots, \alpha_n)rty$: it states the essential properties of the type, but does so without reference to the way it is represented. It follows from this theorem that every value of type $(\alpha_1, \ldots, \alpha_n)rty$ is constructed by one of the constructors $\mathsf{C}_1, \ldots, \mathsf{C}_m$, that each of these constructors is one-to-one, and that different constructors yield different values. The proof that theorem (19) implies these properties of $(\alpha_1, \ldots, \alpha_n)rty$ and the constructors $C_1, \ldots, C_m$ can be outlined as follows.

The fact that every value of type $(\alpha_1, \ldots, \alpha_n)rty$ is constructed by one of the functions $C_1, \ldots, C_m$ follows from the uniqueness part of theorem (19). Suppose there is some value, v say, such that $v \neq (C_i\ x_i^1\ \ldots\ x_i^{k_i})$ for $1 \leq i \leq m$. I.e. v is not constructed by any C_i. One could then define two functions f and g of type $(\alpha_1, \ldots, \alpha_n)rty \rightarrow bool$ which yield the boolean T for all values constructed by any constructor C_i:

$$\forall x_i^1 \cdots x_i^{k_i}.\, f(C_i\ x_i^1\ \ldots\ x_i^{k_i}) = g(C_i\ x_i^1\ \ldots\ x_i^{k_i}) = \mathsf{T} \qquad \text{for } 1 \leq i \leq m \qquad (20)$$

and when applied to v yield different results: $f\ v = \mathsf{T}$ and $g\ v = \mathsf{F}$. If f and g are defined this way then $f \neq g$, since $f\ v \neq g\ v$. But from the uniqueness part of theorem (19) it follows that if f and g satisfy (20) then $f = g$. Therefore no such value v exists, and every value of type $(\alpha_1, \ldots, \alpha_n)rty$ is constructed by some C_i.

The fact that the constructors $C_1, \ldots, C_m$ are one-to-one can be proved by using theorem (19) to define a 'destructor' function D_i for each C_i such that:

$$\vdash D_i(C_i\ x_i^1\ \ldots\ x_i^{k_i}) = (x_i^1, \ldots, x_i^{k_i})$$

For each constructor C_i, such a function can be defined by appropriately specializing the corresponding quantified variable f_i in theorem (19). From the property of the destructor D_i shown above, it is then easy to prove that:

$$\vdash (C_i\ x_i^1\ \ldots\ x_i^{k_i}) = (C_i\ y_i^1\ \ldots\ y_i^{k_i}) \supset (x_i^1 = y_i^1 \wedge \cdots \wedge x_i^{k_i} = y_i^{k_i})$$

which states that C_i is one-to-one, as desired.

Finally, the fact that different constructors yield different values can be proved by appropriately specializing the universally quantified functions $f_1, \ldots, f_m$ in theorem (19) to obtain a theorem asserting the proposition shown below:

$$\vdash \exists fn.\, \forall x_i^1 \cdots x_i^{k_i}.\, fn(C_i\ x_i^1\ \ldots\ x_i^{k_i}) = i \qquad \text{for } 1 \leq i \leq m$$

This states the existence of a function fn which yields the natural number i when applied to values constructed by the ith constructor. This means that any two different constructors C_i and C_j yield different values of type $(\alpha_1, \ldots, \alpha_n)rty$, since applying fn to these values gives different natural numbers.

Using theorems of the form illustrated by (19) to axiomatize recursive types is closely related to the initial algebra approach to the theory of abstract data types [2,5]. This approach is very elegant from a theoretical point of view, but it is also of *practical* value in the HOL mechanization of recursive type definitions. Each recursive type is characterized by a single theorem, and all the theorems which characterize such types have the same general form. This uniform treatment of recursive types is the basis for the *efficient* automation of their construction in HOL. It allows the axiom for any recursive type to be quickly derived from a pre-proved theorem stating that axioms of this kind hold for *all* such types. Furthermore, it makes it possible to derive useful standard properties of recursive types (e.g. structural induction) in a uniform way, with relatively short formal proofs and therefore by efficient programmed inference rules.

7.3 Constructing Representations for Recursive Types

This section outlines a method by which a representation can be found for any type specified by an equation of the form described in Section 7.1. Each representation is an appropriately-defined subset of a type constructed using the type constant *one*, the type operators $\times$ and $+$, and the type $(\alpha)Tree$. A simple example is first given in Sections 7.3.1 and 7.3.2; the method for finding representations in general is then outlined in Section 7.3.3.

7.3.1 An Example: The Representation of Binary Trees

Consider the type *btree* described above in Section 7.1.1. This type was specified informally by:

$$btree \quad ::= \quad \textbf{Leaf} \ num \quad | \quad \textbf{Tree} \ btree \ btree$$

The type *btree* specified by this equation can be represented in higher order logic by a subset of the set denoted by the compound type $(num + one)Tree$. This type denotes the set of all trees (of any shape) whose nodes are labelled either by a value of type *num* or by the single value **one** of type *one*. The idea of this representation is that each binary tree t of type *btree* is represented by a corresponding tree of type $(num + one)Tree$ which has both the same shape as t and the same labels on its nodes as t.

Consider, for example, the binary tree ($\textbf{Leaf} \ n$), consisting of a single leaf node labelled by the natural number n. This binary tree will be represented by a leaf node of type $(num + one)Tree$ labelled by the left injection ($\textbf{Inl} \ n$):

A binary tree ($\textbf{Tree} \ t_1 \ t_2$) which is not a leaf node, but has two subtrees t_1 and t_2, will be represented by a tree of type $(num + one)Tree$ which also has two subtrees and is labelled by the right injection ($\textbf{Inr} \ \textbf{one}$):

where r_1 and r_2 are the representations of the two binary trees t_1 and t_2 respectively. The 'dummy' value ($\textbf{Inr} \ \textbf{one}$) is used in this case to label the root node of the representation, since the corresponding binary tree being represented has no value associated with its root node.

7.3.2 Defining the Subset Predicate for *btree*

To introduce a type definition axiom for *btree*, a predicate Is_btree_REP must first be defined which is true of just those values of type $(num + one)Tree$ which represent binary trees using the scheme outlined above. This predicate is defined formally by building it up from two auxiliary predicates: Is_Leaf and Is_Tree. These two auxiliary predicates correspond to the two kinds of binary trees which will be represented, and each one states what the representation of the corresponding kind of binary tree looks like.

The predicates Is_Leaf and Is_Tree are defined as follows. Every value in the representation is a tree of the form (Node v tl), where v is a label of type $(num+one)$ and tl is a list of subtrees. If such a tree represents a leaf node (Leaf n), then the label v must be the value (Inl n) and the list tl must be empty. These conditions are expressed formally by the predicate Is_Leaf, defined as follows:

$$\vdash \text{Is\_Leaf } v\ tl = (\exists n.\, v = \text{Inl } n \wedge \text{Length } tl = 0)$$

If (Node v tl) represents a binary tree (Tree t_1 t_2) with two subtrees, then the list of subtrees tl must have length two, and the label v must be the value (Inr one). The definition of Is_Tree is therefore:

$$\vdash \text{Is\_Tree } v\ tl = (v = \text{Inr one} \wedge \text{Length } tl = 2)$$

The two predicates Is_Leaf and Is_Tree state what kind of values v and tl must be for the tree (Node v tl) to be the *root* node of legal binary-tree representation. But if a general tree of type $(num + one)Tree$ in fact represents a binary tree, then not only its root node but every node it contains (i.e. all its subtrees) must also satisfy either Is_Leaf or Is_Tree. This can be expressed formally in logic by first defining a higher order function All recursively on trees as follows:

$$\vdash \text{All } P\ (\text{Node } v\ tl) = P\ v\ tl \wedge \text{Every } (\text{All } P)\ tl$$

Using All, the predicate Is_btree_REP can then be defined such that it is true of a tree t exactly when the label and subtree list of every node in t satisfies either Is_Leaf or Is_Tree. The definition of Is_btree_REP is simply:

$$\vdash \text{Is\_btree\_REP } t = \text{All } (\lambda v.\, \lambda tl.\, \text{Is\_Leaf } v\ tl \vee \text{Is\_Tree } v\ tl)\ t$$

This predicate exactly specifies the subset of $(num + one)Tree$ whose values represent binary trees, and can therefore be used to introduce a type definition axiom for the new type *btree* in the usual way. All the predicates which specify representations of recursive type are defined using All in exactly the way shown above for Is_btree_REP.

7.3.3 Finding Representations in General

The representation of binary trees by a subset of $(num + one)Tree$ illustrates the general method for finding representations of any type specified by an equation of the form described in Section 7.1. In general, a recursive type specified by an equation of this kind denotes a set of labelled *trees* with a fixed number of different kinds of nodes. Any such type can therefore be represented by a subset of values denoted by some instance of the defined type $(\alpha)Tree$ of general trees.

Suppose, for example, that $(\alpha_1, \ldots, \alpha_n)rty$ is specified by:

$$(\alpha_1, \ldots, \alpha_n)rty \quad ::= \quad C_1\, ty_1^1 \ldots ty_1^{k_1} \quad | \quad \cdots \quad | \quad C_m\, ty_m^1 \ldots ty_m^{k_m}$$

This equation specifies a type with m different kinds of values, corresponding to the m constructors $C_1, \ldots, C_m$. When this type is defined formally in higher order logic, each of its values will be denoted by some term of the form:

$$C_i\, x_i^1 \ldots x_i^{k_i}$$

where C_i is a constructor and each argument x_i^j is a value of type ty_i^j for $1 \leq j \leq k_i$. In the general case of a recursive type, some of the k_i arguments to C_i will have existing logical types and some will have the type $(\alpha_1, \ldots, \alpha_n)rty$ itself. Let p_i be the number of arguments which have existing logical types and let q_i be the number of arguments which have type $(\alpha_1, \ldots, \alpha_n)rty$, where $k_i = p_i + q_i$. The abstract value of type $(\alpha_1, \ldots, \alpha_n)rty$ denoted by $C_i\, x_i^1 \ldots x_i^{k_i}$ can be represented by a tree which has q_i subtrees and p_i values associated with its root node. This is illustrated by the diagram shown below:

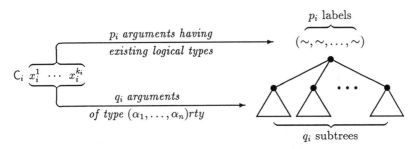

In the general case illustrated by this diagram, the tree representing $C_i\, x_i^1 \ldots x_i^{k_i}$ is labelled by p_i-tuple of values. Each of these values is one of the p_i arguments to C_i which are not of type $(\alpha_1, \ldots, \alpha_n)rty$ but have types which already exist in the logic. When $p_i = 0$, the representing tree is labelled not by a tuple but by the constant **one** (as was done for the constructor **Tree** of $btree$). And when $p_i = 1$ the representing tree is labelled simply by a single value of the appropriate type (as was done for the constructor **Leaf** of $btree$). The q_i subtrees shown in the diagram are the representations of the arguments to C_i which have the type $(\alpha_1, \ldots, \alpha_n)rty$. If $q_i = 0$ then the representing tree has no subtrees.

Each of the m kinds of values constructed by $C_1, \ldots, C_m$ can be represented by a tree using the scheme outlined above. In general, a value obtained using the ith constructor C_i will be represented by a tree labelled by a tuple of p_i values. The representing type for $(\alpha_1, \ldots, \alpha_n)rty$ will therefore be a type expression of the form:

$$(\overbrace{(\underbrace{ty \times \cdots \times ty}_{\text{product of } p_1 \text{ types}}) + \cdots + (\underbrace{ty \times \cdots \times ty}_{\text{product of } p_m \text{ types}})}^{\text{sum of } m \text{ products}})Tree$$

where the ty's are the existing logical types occurring in the equation which specifies the new type $(\alpha_1, \ldots, \alpha_n)rty$ being defined.

Using this scheme, a predicate Is_rty_REP can be defined to specify a set of trees to represent $(\alpha_1, \ldots, \alpha_n)rty$ in exactly the same way as the predicate Is_btree_REP was defined for the representation of $btree$. The definition of Is_rty_REP will have the form:

$$\vdash \text{Is\_rty\_REP } t = \text{All } (\lambda v. \lambda tl. \text{Is\_C}_1 \ v \ tl \ \vee \ \cdots \ \vee \ \text{Is\_C}_m \ v \ tl) \ t$$

where each Is_C_i is an auxiliary predicate specifying which trees represent values constructed by the corresponding constructor C_i. The ith auxiliary predicate Is_C_i is defined as follows. When $i \neq m$, the definition is:

$$\vdash \text{Is\_C}_i \ v \ tl = \exists x_1 \ldots x_{p_i}. \ v = \text{Inl}(\underbrace{\text{Inr} \cdots (\text{Inr}}_{i-1 \text{ Inr's}}(x_1, \ldots, x_{p_i})) \cdots) \wedge \text{Length } tl = q_i$$

where p_i is the number of arguments to C_i which have existing logical types, and q_i is the number of arguments of type $(\alpha_1, \ldots, \alpha_n)rty$. This definition states that if a tree (Node $v \ tl$) represents a value $C_i \ x_i^1 \ \ldots \ x_i^{k_i}$ then it must have the right number subtrees in tl and its label v must be an appropriate injection of some p_i-tuple (of the right logical type, of course). When $i = m$, the definition is similar:

$$\vdash \text{Is\_C}_m \ v \ tl = \exists x_1 \ldots x_{p_m}. \ v = (\underbrace{\text{Inr} \cdots (\text{Inr}}_{m-1 \text{ Inr's}}(x_1, \ldots, x_{p_m})) \cdots) \wedge \text{Length } tl = q_m$$

The only difference is that the last injection applied is Inr, not Inl.

7.4   Deriving Abstract Axioms for Recursive Types

The *uniform* treatment of representations for recursive types makes it possible to write an efficient HOL derived inference rule which proves abstract axioms for them efficiently. Every representation is some subset 'All P' of an instance of $(\alpha)Tree$. A general theorem can therefore be formulated stating that an abstract axiom of the required form holds for *any* recursive type represented this way. This theorem can then be simply 'instantiated' to obtain an abstract axiom for any particular recursive type.

The theorem stating that every recursive type satisfies an abstract axiom of the desired form is shown below:

$$\vdash \forall P. \ \forall Abs{:}(\alpha)Tree{\rightarrow}\beta. \ \forall Rep{:}\beta{\rightarrow}(\alpha)Tree.$$
$$(\forall a. \ Abs(Rep \ a){=}a \wedge \forall r. \ \text{All } P \ r = (Rep(Abs \ r){=}r)) \supset$$
$$\forall f. \ \exists! \ fn. \ \forall v \ tl. \ P \ v \ (\text{Map } Rep \ tl) \supset$$
$$fn(Abs(\text{Node } v \ (\text{Map } Rep \ tl))) = f \ (\text{Map } fn \ tl) \ v \ tl \tag{21}$$

Informally, this theorem states that any type β which is represented by (i.e. is isomorphic to) a set 'All P' of trees satisfies an abstract axiom of the form described in Section 7.2. Theorem 21 makes this assertion in form of an implication:

$$\vdash \forall P. \cdots \ \langle \beta \text{ is isomorphic to 'All } P' \rangle \supset \langle \text{abstract axiom for } \beta \rangle$$

where the antecedent of this implication is written formally as follows:

$$\forall a. \ Abs(Rep \ a){=}a \ \wedge \ \forall r. \ \text{All } P \ r = (Rep(Abs \ r){=}r)$$

This simply says that β is isomorphic to the set of trees of type $(\alpha)Tree$ which satisfy All P. The type variable β stands for the new recursive type which is represented by All P, and the variables Abs and Rep are the abstraction and representation functions for β.

The conclusion of theorem (21) states that functions can be uniquely defined by 'primitive recursion' on the structure which β inherits from All P. That is, for any f, there is a unique function $fn:\beta \rightarrow \gamma$ which satisfies the recursive equation:

$$fn(Abs(\text{Node } v \text{ (Map } Rep \text{ } tl))) = f \text{ (Map } fn \text{ } tl) \text{ } v \text{ } tl$$

whenever the condition P v (Map Rep tl) holds of v and tl. This condition on v and tl restricts the recursive equation shown above to apply only to 'well-constructed' values of type β. If P v (Map Rep tl) holds, then All P is true of the value Node v (Map Rep tl) on the left hand side of the equation. The corresponding *abstract* value, denoted by:

$$Abs(\text{Node } v \text{ (Map } Rep \text{ } tl)),$$

will then be a correctly-represented value of type β. The example in Section 7.4.1 below shows how the form of the predicate P in the condition P v (Map Rep tl) determines the final 'shape' of the resulting axiom.

Theorem (21) illustrates the expressive power which higher-order variables and type polymorphism give to higher order logic. The variable P in this theorem ranges (essentially) over all predicates on $(\alpha)Tree$. And the two type variables α and β can be instantiated to any two logical types. Theorem (21) therefore asserts that an abstract axiom holds for *any* recursive type, since any such type is isomorphic to an appropriate subset All P of some instance of (α)Tree. Because general results like theorem (21) can be formulated as theorems in the logic, they can be used to make programmed inference rules in HOL efficient. Derived inference rules can use such pre-proved general theorems to avoid having to do costly 'run time' inference. Theorem (21) is used in this way by the derived rule which automates recursive type definitions.

The example given in the following section shows how this derived rule uses the general theorem (21) to prove the abstract axiom for a particular recursive type.

7.4.1 Example: Deriving the Axiom for *btree*

The example given in this section is the proof of the abstract axiom for *btree*, the type whose representation was described in Section 7.3.2. The following is the sequence of main steps which the HOL system carries out to define *btree* and derive an abstract axiom for it:

(1) Define the subset predicate Is_btree_REP, introduce a type definition axiom for *btree*, and define the associated abstraction and representation functions ABS:$(num + one)Tree \rightarrow btree$ and REP:$btree \rightarrow (num + one)Tree$.

This is done as outlined in Sections 7.3.2 and 3.3. The result of this step is the two theorems shown below:

$\vdash \forall a. \text{ABS}(\text{REP } a) = a$
$\vdash \forall r. \text{All Is\_btree\_REP } r = (\text{REP}(\text{ABS } r) = r))$

These theorems simply state that the newly-introduced type constant *btree* denotes a set of values which is isomorphic to the subset of $(num + one)Tree$ defined by All Is_btree_REP.

(2) Use theorem (21) to obtain an (unsimplified) abstract axiom for *btree*.

If the type variables α and β in theorem (21) are instantiated to $(num+one)$ and *btree* respectively, then the universally quantified variables P, *Abs*, and *Rep* can be specialized to Is_btree_REP, ABS, and REP. The resulting instance of theorem (21) is an implication whose antecedent matches the two theorems about ABS and REP derived in the previous step. The theorem shown below therefore follows simply by modus ponens (and rewriting, with the definition of Is_btree_REP):

$$\vdash \forall f.\, \exists!fn.\, \forall v\, tl.\, (\text{Is\_Leaf}\ v\ (\text{Map REP}\ tl) \lor \text{Is\_Tree}\ v\ (\text{Map REP}\ tl)) \supset$$
$$fn(\text{ABS}(\text{Node}\ v\ (\text{Map REP}\ tl))) = f\ (\text{Map}\ fn\ tl)\ v\ tl$$

This theorem expresses the essence of the desired abstract axiom for *btree*. The remaining steps carried out by the system are sequence of straightforward simplifications of this theorem which put it into the desired final form.

(3) Remove the disjunction: Is_Leaf v (Map REP tl) $\lor$ Is_Tree v (Map REP tl).

The theorem derived in the previous step contains a term which has the form $\forall v\, tl.\, (P \lor Q) \supset R$. By a simple proof in predicate calculus, this term is equivalent to the conjunction: $(\forall v\, tl.\, P \supset R) \land (\forall v\, tl.\, Q \supset R)$. The theorem derived in the previous step is therefore equivalent to:

$$\vdash \forall f.\, \exists!fn.\, \forall v\, tl.\, \text{Is\_Leaf}\ v\ (\text{Map REP}\ tl) \supset$$
$$fn(\text{ABS}(\text{Node}\ v\ (\text{Map REP}\ tl))) = f\ (\text{Map}\ fn\ tl)\ v\ tl\ \land$$
$$\forall v\, tl.\, \text{Is\_Tree}\ v\ (\text{Map REP}\ tl) \supset$$
$$fn(\text{ABS}(\text{Node}\ v\ (\text{Map REP}\ tl))) = f\ (\text{Map}\ fn\ tl)\ v\ tl$$

In the general case of a type with m constructors, the subset predicate contains a disjunction of the general form:

$$\text{Is\_C}_1\ v\ (\text{Map}\ Rep\ tl) \lor \cdots \lor \text{Is\_C}_m\ v\ (\text{Map}\ Rep\ tl)$$

When this step is done, it will introduce a conjunction of m implications in the body of the abstract axiom, each of which corresponds to one of the m constructors $C_1, \ldots, C_m$.

(4) Rewrite with the definitions of Is_Leaf and Is_Tree. This yields:

$$\vdash \forall f.\, \exists!fn.\, \forall v\, tl.\, (\exists n.\, v = \text{Inl}\ n \land \text{Length}(\text{Map REP}\ tl) = 0) \supset$$
$$fn(\text{ABS}(\text{Node}\ v\ (\text{Map REP}\ tl))) = f\ (\text{Map}\ fn\ tl)\ v\ tl\ \land$$
$$\forall v\, tl.\, (v = \text{Inr}\ one \land \text{Length}(\text{Map REP}\ tl) = 2) \supset$$
$$fn(\text{ABS}(\text{Node}\ v\ (\text{Map REP}\ tl))) = f\ (\text{Map}\ fn\ tl)\ v\ tl$$

Note: In the HOL implementation, the predicates Is_Leaf and Is_Tree are not actually defined as new constants; they are instead written using λ-terms. This step therefore does not need to be done in the HOL implementation.

(5) Simplify terms of the form: $\mathsf{Length}(\mathsf{Map\ REP}\ tl) = m$.

A term of the form $\mathsf{Length}(\mathsf{Map\ REP}\ tl) = m$ is equivalent to a simplified term of the form $\mathsf{Length}\ tl = m$. This in turn is equivalent to saying that tl is equal to some list of m values: $\exists t_1 \dots t_m.\, tl = [t_1; \dots; t_m]$. The terms involving Length in the previous theorem can therefore be simplified, resulting in the following theorem:

$$\vdash \forall f.\, \exists!fn.\, \forall v\ tl.\, (\exists n.\, v = \mathsf{Inl}\ n \wedge tl = \mathsf{Nil}) \supset$$
$$fn(\mathsf{ABS}(\mathsf{Node}\ v\ (\mathsf{Map\ REP}\ tl))) = f\ (\mathsf{Map}\ fn\ tl)\ v\ tl\ \wedge$$
$$\forall v\ tl.\, (v = \mathsf{Inr}\ one \wedge \exists t_1\ t_2.\, tl = [t_1; t_2]) \supset$$
$$fn(\mathsf{ABS}(\mathsf{Node}\ v\ (\mathsf{Map\ REP}\ tl))) = f\ (\mathsf{Map}\ fn\ tl)\ v\ tl$$

This step introduces the variables t_1 and t_2. They range over values of type *btree* and occur in the axiom for *btree* in its final form.

(6) Remove equations of the form: $v = \cdots$ and $tl = \cdots$.

The antecedents of the two logical implications in the previous theorem both contain equations giving values for v and tl. These can be removed by using (a generalization of) the fact that in predicate calculus a term of the form $\forall y.\, (\exists x.\, y = tm_1[x]) \supset tm_2[y]$ is equivalent to $\forall x.\, tm_2[tm_1[x]]$. The result of removing the equations for v and tl is:

$$\vdash \forall f.\, \exists!fn.\, \forall n.\, fn(\mathsf{ABS}(\mathsf{Node}\ (\mathsf{Inl}\ n)(\mathsf{Map\ REP\ Nil})))$$
$$= f\ (\mathsf{Map}\ fn\ \mathsf{Nil})\ (\mathsf{Inl}\ n)\ \mathsf{Nil}\ \ \wedge$$
$$\forall t_1\ t_2.\, fn(\mathsf{ABS}(\mathsf{Node}\ (\mathsf{Inr}\ one)\ (\mathsf{Map\ REP}\ [t_1; t_2])))$$
$$= f\ (\mathsf{Map}\ fn\ [t_1; t_2])\ (\mathsf{Inr}\ one)\ [t_1; t_2]$$

The body of the theorem now consists of two equations. These define the value of fn for the two different kinds of binary trees.

(7) Rewrite with the definition of Map. This yields:

$$\vdash \forall f.\, \exists!fn.\, \forall n.\, fn(\mathsf{ABS}(\mathsf{Node}\ (\mathsf{Inl}\ n)\ \mathsf{Nil}))$$
$$= f\ \mathsf{Nil}\ (\mathsf{Inl}\ n)\ \mathsf{Nil}\ \ \wedge$$
$$\forall t_1\ t_2.\, fn(\mathsf{ABS}(\mathsf{Node}\ (\mathsf{Inr}\ one)\ [\mathsf{REP}\ t_1; \mathsf{REP}\ t_2]))$$
$$= f\ [fn\ t_1; fn\ t_2]\ (\mathsf{Inr}\ one)\ [t_1; t_2]$$

(8) Define the abstract constructors Leaf and Tree as follows:

$$\vdash \mathsf{Leaf}\ n\ \ = \ \mathsf{ABS}(\mathsf{Node}\ (\mathsf{Inl}\ n)\ \mathsf{Nil})$$
$$\vdash \mathsf{Tree}\ t_1\ t_2\ = \ \mathsf{ABS}(\mathsf{Node}\ (\mathsf{Inr}\ one)\ [\mathsf{REP}\ t_1; \mathsf{REP}\ t_2])$$

The constructors Leaf and Tree defined by these equations first use Node to construct the representations of the required values and then use ABS to obtain the corresponding values of type *btree*. Rewriting the theorem derived in the previous step with these definitions yields:

$$\vdash \forall f.\, \exists!fn.\, \forall n.\, fn(\mathsf{Leaf}\ n) = f\ \mathsf{Nil}\ (\mathsf{Inl}\ n)\ \mathsf{Nil}\ \ \wedge$$
$$\forall t_1\ t_2.\, fn(\mathsf{Tree}\ t_1\ t_2) = f\ [fn\ t_1; fn\ t_2]\ (\mathsf{Inr}\ one)\ [t_1; t_2]$$

(9) Introduce two functions f_1 and f_2 in place of f.

With an appropriate choice of value for the universally quantified variable f, two functions f_1 and f_2 can be introduced for the right hand sides of the two equations. These define the value of fn separately for the two constructors Leaf and Tree. Specializing f to the appropriate function, and simplifying, gives:

$$\vdash \forall f_1\, f_2.\, \exists! fn.\, \forall n.\, fn(\mathsf{Leaf}\ n) = f_1\ n\ \wedge$$
$$\forall t_1\, t_2.\, fn(\mathsf{Tree}\ t_1\ t_2) = f_2\ (fn\ t_1)\ (fn\ t_2)\ t_1\ t_2$$

This theorem is the abstract axiom for *btree*—in its final form.

The HOL derived rule which automates recursive type definitions carries out the sequence of steps shown above for each informal type specification entered by the user. An appropriate instance of theorem (21) yields an 'unsimplified' abstract axiom for the type being defined. This axiom is then systematically transformed into the form described in Section 7.2 by the sequence of simple equivalence-preserving steps shown above. The amount of actual logical inference that must be carried out is relatively small, and each step is a straightforward transformation of the theorem derived in the previous step. The HOL implementation of this procedure is therefore both efficient and robust.

8 Concluding Remarks

The method for defining recursive types described in Section 7 is the logical basis for a set of efficient theorem-proving tools the HOL system. In addition to the derived inference rule which automates recursive type definitions, a number of related tools have been implemented in HOL for generating proofs involving recursive types. These include:

- an inference rule which derives structural induction for recursive types, and related tools for interactively generating proofs by structural induction (e.g. a general structural induction *tactic*),

- a set of rules which automate the inference necessary to define functions by 'primitive recursion' on recursive types,

- derived rules which prove that the constructors of recursive types are one-to-one and yield distinct values, and

- tools for generating interactive proofs by case analysis on the constructors of recursive types.

Preliminary work is underway to extend these tools to deal with mutually recursive types, and types with equational constraints.

Defining a logical type in HOL is rarely the primary goal of the user of the system, but often a necessary part of some more interesting proof. The efficient automation of type definitions in HOL is therefore of significant practical value, since defining types 'by hand' in the system is tedious and tricky. The mechanization of type definitions described in this paper allows new recursive types to be introduced by the

HOL user quickly and easily. This is made possible by the systematic construction of representations for these types, the uniform treatment of abstract axioms for them (using essentially the initial algebra approach to type specifications), and the expressive power of higher order logic itself.

Acknowledgements

Thanks are due to Albert Camilleri, Inder Dhingra and Mike Gordon for helpful comments on drafts of this paper, and to Thomas Forster for useful discussions about the construction of trees. I am grateful to Gonville and Caius College Cambridge for support in the form of an unofficial fellowship, during which the work described in this paper was done.

References

[1] Bird, R., and Wadler, P, *Introduction to Functional Programming*, Prentice Hall International Series in Computer Science (Prentice Hall, 1988).

[2] Burstall, R., and Goguen, J., 'Algebras, theories and freeness: an introduction for computer scientists', in: *Theoretical Foundations of Programming Methodology*, edited by M. Wirsing and G. Schmidt, Proceedings of the 1981 Marktoberdorf NATO Summer School, NATO ASI Series, Vol. C91 (Reidel, 1982), pp. 329–350.

[3] Church, A., 'A Formulation of the Simple Theory of Types', *Journal of Symbolic Logic*, Vol. 5 (1940), pp. 56–68.

[4] Cousineau, G., G. Huet, and L. Paulson, *The ML Handbook*, INRIA, (1986).

[5] Goguen, J.A., J.W. Thatcher, and E.G. Wagner, 'An initial algebra approach to the specification, correctness, and implementation of abstract data types', in: *Current Trends in Programming Methodology*, edited by R.T. Yeh (Prentice-Hall, New Jersey, 1978), IV, pp. 80–149.

[6] Gordon, M., 'HOL: A Machine Oriented Formulation of Higher Order Logic', Technical Report No. 68, Computer Laboratory, The University of Cambridge, Revised version (July 1985).

[7] Gordon, M.J.C., 'HOL: A Proof Generating System for Higher-Order Logic', in: *VLSI Specification, Verification and Synthesis*, edited by G. Birtwistle and P.A. Subrahmanyam, Kluwer International Series in Engineering and Computer Science, SECS35 (Kluwer Academic Publishers, Boston, 1988), pp. 73–128.

[8] Gordon, M.J., R. Milner, and C.P. Wadsworth, 'Edinburgh LCF: A Mechanised Logic of Computation', Lecture Notes in Computer Science, Vol. 78 (Springer-Verlag, Berlin, 1979).

[9] Harper, R., D. MacQueen, and R. Milner, Standard ML, Report No. ECS-LFCS-86-2, Laboratory for Foundations of Computer Science, Department of Computer Science, The University of Edinburgh (March 1986).

[10] Leisenring, A.C., *Mathematical Logic and Hilbert's ε-Symbol*, University Mathematical Series (Macdonald & Co., London, 1969).

[11] Milner, R., 'A Theory of Type Polymorphism in Programming', *Journal of Computer and System Sciences*, No. 17 (1978).

[12] Paulson, L.C., *Logic and Computation: Interactive Proof with Cambridge LCF*, Cambridge Tracts in Theoretical Computer Science 2 (Cambridge University Press, Cambridge, 1987).

Mechanizing Programming Logics
in Higher Order Logic

Michael J.C. Gordon

Computer Laboratory   SRI International
New Museums Site   Suite 23
Pembroke Street   Millers Yard
Cambridge CB2 3QG   Cambridge CB2 1RQ

Abstract:

Formal reasoning about computer programs can be based directly on the semantics of the programming language, or done in a special purpose logic like Hoare logic. The advantage of the first approach is that it guarantees that the formal reasoning applies to the language being used (it is well known, for example, that Hoare's assignment axiom fails to hold for most programming languages). The advantage of the second approach is that the proofs can be more direct and natural.

In this paper, an attempt to get the advantages of both approaches is described. The rules of Hoare logic are *mechanically* derived from the semantics of a simple imperative programming language (using the HOL system). These rules form the basis for a simple program verifier in which verification conditions are generated by LCF-style tactics whose validations use the derived Hoare rules. Because Hoare logic is derived, rather than postulated, it is straightforward to mix semantic and axiomatic reasoning. It is also straightforward to combine the constructs of Hoare logic with other application-specific notations. This is briefly illustrated for various logical constructs, including termination statements, VDM-style 'relational' correctness specifications, weakest precondition statements and dynamic logic formulae .

The theory underlying the work presented here is well known. Our contribution is to propose a way of mechanizing this theory in a way that makes certain practical details work out smoothly.

Contents

1 Introduction

The work described here is part of a long term project on verifying combined hardware/software systems by mechanized formal proof. The ultimate goal is to develop techniques and tools for modelling and analysing systems like computer controlled chemical plants, fly-by-wire aircraft and microprocessor based car brake controllers. These typically contain both software and hardware and must respond in real time to asynchronous inputs.

This paper concentrates on software verification. A mechanization of Hoare logic is constructed via a representation of it in higher order logic. The main experiment described here is the implementation, in the HOL system [13], of a program verifier for a simple imperative language. This translates correctness questions formulated in Hoare logic [16] to theorem proving problems (called 'verification conditions') in pure predicate logic. This is a standard technique [17], the only novelty in our approach is the way rules of the programming logic are mechanically derived from the semantics of the programming language in which programs are written. This process of generating verification conditions is implemented as a *tactic*[1] in HOL whose validation part uses the derived Hoare rules; it is thus guaranteed to be sound. This way of implementing a verifier ensures that theorems proved with it are logical consequences of the underlying programming language semantics.

Work is already in progress [20] to prove that the semantics used by the verifier (which is described in Section 5 below) is the same as the semantics determined by running the programs on a simple microprocessor. When this proof is completed, it will follow that theorems proved using the verifier are true statements about the actual behaviour of programs when they are executed on hardware.

To explore the flexibility of representing programming logics in higher order logic, three well-known programming logics are examined. Although we do not construct mechanizations of these here, it is shown that doing so would be straightforward.

There is no new mathematical theory in this paper. It is purely a contribution to the methodology of doing proofs mechanically. We hope that our example verifier demonstrates that it is possible to combine both the slickness of a special purpose logic with the rigor of reasoning directly from the reference semantics of a programming language. We hope also that it demonstrates the expressiveness of higher order logic and the flexibility of Milner's LCF approach [11,30] to interactive proof. The contents of the remaining sections of the paper are as follows:

Section 2 is a description of the example programming language that will be used.

Section 3 is a brief review of Hoare logic. The presentation is adapted from the book *Programming Language Theory and its Implementation* [14].

Section 4 outlines the version of predicate logic used in this paper.

Section 5 gives the semantics of the language described in Section 2 and the semantics of Hoare-style partial correctness specifications.

Section 6 discusses how partial correctness specifications can be regarded as abbreviations for logic formulae. The axioms and rules in Section 3 then become derivable from the semantics in Section 5.

---

[1]Tactics are described in Section 8 below. The idea of implementing verification condition generation with tactics was developed jointly with Tom Melham.

Section 7 is an account of how the derived axioms and rules of Hoare logic can be mechanized in the HOL system.

Section 8 is a brief introduction to tactics and tacticals.

Section 9 shows how verification conditions can be generated using tactics.

Section 10 explains how a weak form of termination statements can be added to Hoare logic and the necessary additional axioms and inference rules derived from suitable definitions. Additional mechanized proof generating tools (in HOL) are also described.

Section 11 explains how VDM-style specifications, weakest preconditions and dynamic logic can be represented in higher order logic.

Section 12 contains concluding remarks and a brief discussion of future work.

2 A Simple Imperative Programming Language

The syntax of the little programming language used in this paper is specified by the BNF given below. In this specification, the variable $\mathcal{N}$ ranges over the *numerals* 0, 1, 2 etc, the variable $\mathcal{V}$ ranges over *program variables*[2] X, Y, Z etc, the variables $\mathcal{E}$, $\mathcal{E}_1$, $\mathcal{E}_2$ etc. range over *integer expressions*, the variables $\mathcal{B}$, $\mathcal{B}_1$, $\mathcal{B}_2$ etc. range over *boolean expressions* and the variables $\mathcal{C}$, $\mathcal{C}_1$, $\mathcal{C}_2$ etc. range over *commands*.

$$\mathcal{E} \ ::= \ \mathcal{N} \ | \ \mathcal{V} \ | \ \mathcal{E}_1 + \mathcal{E}_2 \ | \ \mathcal{E}_1 - \mathcal{E}_2 \ | \ \mathcal{E}_1 \times \mathcal{E}_2 \ | \ \ldots$$

$$\mathcal{B} \ ::= \ \mathcal{E}_1 {=} \mathcal{E}_2 \ | \ \mathcal{E}_1 \leq \mathcal{E}_2 \ | \ \ldots$$

$$
\begin{aligned}
\mathcal{C} \ ::= \ &\textbf{skip} \\
| \ &\mathcal{V} := \mathcal{E} \\
| \ &\mathcal{C}_1 \ ; \ \mathcal{C}_2 \\
| \ &\textbf{if } \mathcal{B} \textbf{ then } \mathcal{C}_1 \textbf{ else } \mathcal{C}_2 \\
| \ &\textbf{while } \mathcal{B} \textbf{ do } \mathcal{C}
\end{aligned}
$$

Note that the BNF syntax above is ambiguous: it does not specify, for example, whether **if** $\mathcal{B}$ **then** $\mathcal{C}_1$ **else** $\mathcal{C}_2$; $\mathcal{C}_3$ means (**if** $\mathcal{B}$ **then** $\mathcal{C}_1$ **else** $\mathcal{C}_2$) ; $\mathcal{C}_3$ or **if** $\mathcal{B}$ **then** $\mathcal{C}_1$ **else** $(\mathcal{C}_2$; $\mathcal{C}_3)$. We will clarify, whenever necessary, using brackets. Here, for example, is a command $\mathcal{C}$ to compute the quotient Q and remainder R that results from dividing Y into X.

$$
\left.
\begin{aligned}
&R := X; \\
&Q := 0; \\
&\textbf{while } Y \leq R \textbf{ do} \\
&\quad (R := R - Y; \ Q := Q + 1)
\end{aligned}
\right\} \mathcal{C}
$$

---

[2]To distinguish program variables from logical variables, the convention is adopted here that the former are upper case and the latter are lower case. The need for such a convention is explained in Section 5.

3 Hoare Logic

In a seminal paper, C.A.R. Hoare [16] introduced the notation $\{\mathcal{P}\}\ \mathcal{C}\ \{\mathcal{Q}\}$ for specifying what a program does[3]. In this notation, $\mathcal{C}$ is a program from the programming language whose programs are being specified (the language in Section 2 in our case); and $\mathcal{P}$ and $\mathcal{Q}$ are conditions on the program variables used in $\mathcal{C}$.

The semantics of $\{\mathcal{P}\}\ \mathcal{C}\ \{\mathcal{Q}\}$ is now described informally. A formal semantics in higher order logic is given in Section 5 below.

The effect of executing a command $\mathcal{C}$ is to change the *state*, where a state is simply an assignment of values to program variables. If $\mathcal{F}[X_1, \ldots, X_n]$ is a formula containing free[4] program variables $X_1, \ldots, X_n$, then we say $\mathcal{F}[X_1, \ldots, X_n]$ is true in a state s if $\mathcal{F}[X_1, \ldots, X_n]$ is true when $X_1, \ldots, X_n$ have the values assigned to them by s. The $\mathcal{P}$ and $\mathcal{Q}$ in $\{\mathcal{P}\}\ \mathcal{C}\ \{\mathcal{Q}\}$ are logic formulae containing the program variables used in $\mathcal{C}$ (and maybe other variables also).

$\{\mathcal{P}\}\ \mathcal{C}\ \{\mathcal{Q}\}$ is said to be true, if whenever $\mathcal{C}$ is executed in a state in which $\mathcal{P}$ is true, and if the execution of $\mathcal{C}$ terminates, then $\mathcal{Q}$ is true in the state in which $\mathcal{C}$'s execution terminates.

Example

$\{X = 1\}\ X := X + 1\ \{X = 2\}$. Here $\mathcal{P}$ is the condition that the value of X is 1, $\mathcal{Q}$ is the condition that the value of X is 2 and $\mathcal{C}$ is the assignment command $X := X + 1$ (i.e. 'X becomes $X + 1$'). $\square$

An expression $\{\mathcal{P}\}\ \mathcal{C}\ \{\mathcal{Q}\}$ is called a *partial correctness specification*; $\mathcal{P}$ is called its *precondition* and $\mathcal{Q}$ its *postcondition*.

These specifications are 'partial' because for $\{\mathcal{P}\}\ \mathcal{C}\ \{\mathcal{Q}\}$ to be true it is *not* necessary for the execution of $\mathcal{C}$ to terminate when started in a state satisfying $\mathcal{P}$. It is only required that *if* the execution terminates, *then* $\mathcal{Q}$ holds.

Example

$$
\left.
\begin{array}{l}
\{\mathsf{T}\} \\
\quad R := X; \\
\quad Q := 0; \\
\quad \textbf{while } Y \leq R \textbf{ do} \\
\quad\quad (R := R - Y;\ Q := Q + 1) \\
\{X = R + (Y \times Q)\ \wedge\ R < Y\}
\end{array}
\right\} \mathcal{C}
$$

This is $\{\mathsf{T}\}\ \mathcal{C}\ \{X = R + (Y \times Q)\ \wedge\ R < Y\}$ where $\mathcal{C}$ is the command indicated by the braces above. The formula T making up the precondition is the universally true formula; the symbol $\wedge$ denotes logical conjunction (i.e. 'and'). The specification is true if whenever the execution of $\mathcal{C}$ halts, then Q is the quotient and R is the remainder that results from dividing Y into X. It is true, even if X is initially negative. $\square$

A stronger kind of specification is a *total correctness specification*. There is no standard notation for these. We will use $[\mathcal{P}]\ \mathcal{C}\ [\mathcal{Q}]$.

---

[3] Actually, Hoare's original notation was $\mathcal{P}\ \{\mathcal{C}\}\ \mathcal{Q}$ not $\{\mathcal{P}\}\ \mathcal{C}\ \{\mathcal{Q}\}$, but the latter form is now more widely used. Note that some authors (e.g. [15]) use $\{\mathcal{P}\}\ \mathcal{C}\ \{\mathcal{Q}\}$ for total correctness rather than partial correctness.

[4] A *free* occurrence of a variable is one that is not bound by $\forall$, $\exists$ or λ etc.

A total correctness specification $[\mathcal{P}]\ \mathcal{C}\ [\mathcal{Q}]$ is true if and only if the following conditions apply:

(i) Whenever $\mathcal{C}$ is executed in a state satisfying $\mathcal{P}$, then the execution of $\mathcal{C}$ terminates.

(ii) After termination $\mathcal{Q}$ holds.

The relationship between partial and total correctness can be informally expressed by the equation:

$$\text{Total correctness} = \text{Termination} + \text{Partial correctness}.$$

It is usually easier to prove total correctness by establishing termination and partial correctness separately. We show how this separation of concerns is supported by our program verifier in Section 10.

3.1   Axioms and Rules of Hoare Logic

In his 1969 paper, Hoare gave a deductive system for partial correctness specifications. The axioms and rules that follow are minor variants of these[5]. We write $\vdash \{\mathcal{P}\}\ \mathcal{C}\ \{\mathcal{Q}\}$ if $\{\mathcal{P}\}\ \mathcal{C}\ \{\mathcal{Q}\}$ is either an instance of one of the axiom schemes A1 or A2 below, or can be deduced by a sequence of applications of the rules R1, R2, R3, R4 or R5 below from such instances. We write $\vdash \mathcal{P}$, where $\mathcal{P}$ is a formula of predicate logic, if $\mathcal{P}$ can be deduced from the laws of logic and arithmetic. We shall not give an axiom system for either predicate logic or arithmetic here. In the little program verifier that we describe in Section 9, such logical and arithmetical theorems are deduced using the existing proof infrastructure of the HOL system [13]. The goal of the work described in this paper is to gracefully embed Hoare logic in higher order logic, so that the HOL system can also be used to verify progams via the axioms and rules below.

If $\vdash \mathcal{P}$, where $\mathcal{P}$ is a formula of predicate calculus or arithmetic, then we say ' $\vdash \mathcal{P}$ is a theorem of pure logic'; if $\vdash \{\mathcal{P}\}\ \mathcal{C}\ \{\mathcal{Q}\}$ we say ' $\vdash \{\mathcal{P}\}\ \mathcal{C}\ \{\mathcal{Q}\}$ is a theorem of Hoare logic'.

A1: The skip-axiom. For any formula $\mathcal{P}$:

$$\vdash \{\mathcal{P}\}\ \textbf{skip}\ \{\mathcal{P}\}$$

A2: The assignment-axiom. For any formula $\mathcal{P}$, program variable $\mathcal{V}$ and integer expression $\mathcal{E}$:

$$\vdash \{\mathcal{P}[\mathcal{E}/\mathcal{V}]\}\ \mathcal{V} := \mathcal{E}\ \{\mathcal{P}\}$$

where $\mathcal{P}[\mathcal{E}/\mathcal{V}]$ denotes the result of substituting $\mathcal{E}$ for all free occurrences of $\mathcal{V}$ in $\mathcal{P}$ (and free variables are renamed, if necessary, to avoid capture).

Rules R1 to R5 below are stated in standard notation: the hypotheses of the rule above a horizontal line and the conclusion below it. For example, R1 states that if

---

[5]The presentation of Hoare logic in this paper is based on the one in *Programming Language Theory and its Implementation* [14].

$\vdash \mathcal{P}' \Rightarrow \mathcal{P}$ is a theorem of pure logic and $\vdash \{\mathcal{P}\}\, \mathcal{C}\, \{\mathcal{Q}\}$ is a theorem of Hoare logic, then $\vdash \{\mathcal{P}'\}\, \mathcal{C}\, \{\mathcal{Q}\}$ can be deduced by R1.

R1: The rule of precondition strengthening. For any formulae $\mathcal{P}$, $\mathcal{P}'$ and $\mathcal{Q}$, and command $\mathcal{C}$:

$$\frac{\vdash \mathcal{P}' \Rightarrow \mathcal{P} \qquad \vdash \{\mathcal{P}\}\, \mathcal{C}\, \{\mathcal{Q}\}}{\vdash \{\mathcal{P}'\}\, \mathcal{C}\, \{\mathcal{Q}\}}$$

R2: The rule of postcondition weakening. For any formulae $\mathcal{P}$, $\mathcal{Q}$ and $\mathcal{Q}'$, and command $\mathcal{C}$:

$$\frac{\vdash \{\mathcal{P}\}\, \mathcal{C}\, \{\mathcal{Q}\} \qquad \vdash \mathcal{Q} \Rightarrow \mathcal{Q}'}{\vdash \{\mathcal{P}\}\, \mathcal{C}\, \{\mathcal{Q}'\}}$$

Notice that in R1 and R2, one hypothesis is a theorem of ordinary logic whereas the other hypothesis is a theorem of Hoare logic. This shows that proofs in Hoare logic may require subproofs in pure logic; more will be said about the implications of this later.

R3: The sequencing rule. For any formulae $\mathcal{P}$, $\mathcal{Q}$ and $\mathcal{R}$, and commands $\mathcal{C}_1$ and $\mathcal{C}_2$:

$$\frac{\vdash \{\mathcal{P}\}\, \mathcal{C}_1\, \{\mathcal{Q}\} \qquad \vdash \{\mathcal{Q}\}\, \mathcal{C}_2\, \{\mathcal{R}\}}{\vdash \{\mathcal{P}\}\, \mathcal{C}_1;\, \mathcal{C}_2\, \{\mathcal{R}\}}$$

R4: The if-rule. For any formulae $\mathcal{P}$, $\mathcal{Q}$ and $\mathcal{B}$, and commands $\mathcal{C}_1$ and $\mathcal{C}_2$:

$$\frac{\vdash \{\mathcal{P} \wedge \mathcal{B}\}\, \mathcal{C}_1\, \{\mathcal{Q}\} \qquad \vdash \{\mathcal{P} \wedge \neg\mathcal{B}\}\, \mathcal{C}_2\, \{\mathcal{Q}\}}{\vdash \{\mathcal{P}\}\, \text{if } \mathcal{B} \text{ then } \mathcal{C}_1 \text{ else } \mathcal{C}_2\, \{\mathcal{Q}\}}$$

Notice that in this rule (and also in R5 below) it is assumed that $\mathcal{B}$ is both a boolean expression of the programming language and a formula of predicate logic. We shall only assume that the boolean expressions of the language are a *subset* of those in predicate logic. This assumption is reasonable since we are the designers of our example language and can design the language so that it is true; it would not be reasonable if we were claiming to provide a logic for reasoning about an existing language like Pascal. One consequence of this assumption is that the semantics of boolean expressions must be the usual logical semantics. We could not, for example, have 'sequential' boolean operators in which the boolean expression $\mathsf{T} \vee (1/0 = 0)$ evaluates to T, but $(1/0 = 0) \vee \mathsf{T}$ causes an error (due to division by 0).

R5: The while-rule. For any formulae $\mathcal{P}$ and $\mathcal{B}$, and command $\mathcal{C}$:

$$\frac{\vdash \{\mathcal{P} \wedge \mathcal{B}\}\, \mathcal{C}\, \{\mathcal{P}\}}{\vdash \{\mathcal{P}\}\, \text{while } \mathcal{B} \text{ do } \mathcal{C}\, \{\mathcal{P} \wedge \neg\mathcal{B}\}}$$

A formula $\mathcal{P}$ such that $\vdash \{\mathcal{P} \wedge \mathcal{B}\}\, \mathcal{C}\, \{\mathcal{P}\}$ is called an *invariant* of $\mathcal{C}$ for $\mathcal{B}$.

3.2   An Example Proof in Hoare Logic

The simple proof that follows will be used later to illustrate the mechanization of Hoare logic in the HOL system.

By the assignment axiom:

Th1 : $\vdash \{X = X + (Y \times 0)\}\ R := X\ \{X = R + (Y \times 0)\}$

Th2 : $\vdash \{X = R + (Y \times 0)\}\ Q := 0\ \{X = R + (Y \times Q)\}$

Hence by the sequencing rule:

Th3 : $\vdash \{X = X + (Y \times 0)\}\ R := X;\ Q := 0\ \{X = R + (Y \times Q)\}$

By a similar argument consisting of two instances of the assignment axiom followed by a use of the sequencing rule:

Th4 : $\vdash \{X = (R - Y) + (Y \times (Q + 1))\}$
$R := R - Y$
$\{X = R + (Y \times (Q + 1))\}$

Th5 : $\vdash \{X = R + (Y \times (Q + 1))\}\ Q := Q + 1\ \{X = R + (Y \times Q)\}$

Th6 $\vdash \{X = (R - Y) + (Y \times (Q + 1))\}$
$R := R - Y;\ Q := Q + 1$
$\{X = R + (Y \times Q)\}$

The following is a trivial theorem of arithmetic:

Th7 : $\vdash (X = R + (Y \times Q)) \wedge Y \leq R \Rightarrow (X = (R - Y) + (Y \times (Q + 1)))$

hence by precondition strengthening applied to Th7 and Th6:

Th8 : $\vdash \{(X = R + (Y \times Q)) \wedge Y \leq R\}$
$R := R - Y;\ Q := Q + 1$
$\{X = R + (Y \times Q)\}$

and so by the **while**-rule

Th9 : $\vdash \{X = R + (Y \times Q)\}$
while $Y \leq R$ **do** $(R := R - Y;\ Q := Q + 1)$
$\{(X = R + (Y \times Q)) \wedge \neg Y \leq R\}$

By the sequencing rule applied to Th3 and Th9

Th10 : $\vdash \{X = X + (Y \times 0)\}$
$R := X;$
$Q := 0;$
while $Y \leq R$ **do** $(R := R - Y;\ Q := Q + 1)$
$\{(X = R + (Y \times Q)) \wedge \neg Y \leq R\}$

The next two theorems are trivial facts of arithmetic:

Th11 : $\vdash$ T $\Rightarrow$ $X = X + (Y \times 0)$

Th12 : $\vdash$ $(X = R + (Y \times Q)) \land \neg Y \le R \Rightarrow (X = R + (Y \times Q)) \land R < Y$

Finally, combining the last three theorems using precondition strengthening and postcondition weakening:

Th13 : $\vdash$ $\{T\}$
$R := X;$
$Q := 0;$
while $Y \le R$ **do** $(R := R - Y; Q := Q + 1)$
$\{(X = R + (Y \times Q)) \land R < Y\}$

In the example just given, it was shown how to prove $\{\mathcal{P}\}\mathcal{C}\{\mathcal{Q}\}$ by proving properties of the components of $\mathcal{C}$ and then putting these together (with the appropriate proof rules) to get the desired property of $\mathcal{C}$ itself. For example, Th3 and Th6 both had the form $\vdash \{\mathcal{P}\}\mathcal{C}_1;\mathcal{C}_2\{\mathcal{Q}\}$ and to prove them we first proved $\vdash \{\mathcal{P}\}\mathcal{C}_1\{\mathcal{R}\}$ and $\vdash \{\mathcal{R}\}\mathcal{C}_2\{\mathcal{Q}\}$ (for suitable $\mathcal{R}$), and then deduced $\vdash \{\mathcal{P}\}\mathcal{C}_1;\mathcal{C}_2\{\mathcal{Q}\}$ by the sequencing rule.

This process is called *forward proof*, because one moves forward from axioms via rules to conclusions. In practice, it is much more natural to work backwards: starting from the goal of showing $\{\mathcal{P}\}\mathcal{C}\{\mathcal{Q}\}$, one generates subgoals, subsubgoals etc. until the problem is solved. For example, suppose one wants to show:

$$\{X = x \land Y = y\} \ R := X; \ X := Y; \ Y := R \ \{Y = x \land X = y\}$$

then by the assignment axiom and sequencing rule it is sufficient to show the subgoal

$$\{X = x \land Y = y\} \ R := X; \ X := Y \ \{R = x \land X = y\}$$

(because $\vdash \{R = x \land X = y\} \ Y := R \ \{Y = x \land X = y\}$). By a similar argument this subgoal can be reduced to

$$\{X = x \land Y = y\} \ R := X \ \{R = x \land Y = y\}$$

which clearly follows from the assignment axiom.

In Section 9 we describe how LCF style tactics (which are described in Section 8) can be used to implement a *goal oriented* method of proof based on verification conditions [17]. The user supplies a partial correctness specification annotated with mathematical statements describing relationships between variables (e.g. **while** invariants). HOL tactics can then be used to generate a set of purely mathematical statements, called *verification conditions* (or vcs). The validation of the vc-generating tactic (see Section 8) ensures that if the verification conditions are provable, then the original specification can be deduced from the axioms and rules of Floyd-Hoare logic. The following diagram (adapted from [14]) gives an overview of this approach.

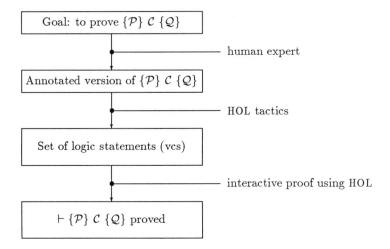

The next section explains the version of predicate calculus underlying the HOL system; the section after that is a quick introduction to Milner's ideas on tactics and tacticals.

4 Higher Order Logic

The table below shows the logical notations used in this paper.

Predicate calculus notation	
Notation	*meaning*
T	truth
F	falsity
$P(x)$ (or $P\ x$)	x has property P
$\neg t$	not t
$t_1 \vee t_2$	t_1 or t_2
$t_1 \wedge t_2$	t_1 and t_2
$t_1 \Rightarrow t_2$	t_1 implies t_2
$t_1 \equiv t_2$	t_1 if and only if t_2
$t_1 = t_2$	t_1 equals t_2
$\forall x.\ t[x]$	for all x it is the case that $t[x]$
$\exists x.\ t[x]$	for some x it is the case that $t[x]$
$(t \rightarrow t_1 \mid t_2)$	if t is true then t_1 else t_2

Higher order logic extends first-order logic by allowing higher order variables (i.e. variables whose values are functions) and higher order functions (i.e. functions whose arguments and/or results are other functions). For example[6], partial correctness specifications can be represented by defining a predicate (i.e. a function whose result is a truth value) Spec by:

$$\mathsf{Spec}(p, c, q) \ = \ \forall s_1 \ s_2. \ p \ s_1 \ \wedge \ c(s_1, s_2) \ \Rightarrow \ q \ s_2$$

Spec is a predicate on triples (p, c, q) where p and q are unary predicates and c is a binary predicate. To represent command sequencing we can define a constant Seq by:

$$\mathsf{Seq}(c_1, c_2)(s_1, s_2) \ = \ \exists s. \ c_1(s_1, s) \ \wedge \ c_2(s, s_2)$$

The sequencing rule in Hoare logic (which was explained in Section 3) can be stated directly in higher order logic as:

$$\vdash \ \forall p \ q \ r \ c_1 \ c_2. \ \mathsf{Spec}(p, c_1, q) \ \wedge \ \mathsf{Spec}(q, c_2, r) \ \Rightarrow \ \mathsf{Spec}(p, \mathsf{Seq}(c_1, c_2), r)$$

These examples make essential use of higher order variables; they can't be expressed in first-order logic.

The version of higher order logic that we use[7] has function-denoting terms called λ-expressions. These have the form $\lambda x. \ t$ where x is a variable and t is an expression. Such a λ-term denotes the function $a \mapsto t[a/x]$ where $t[a/x]$ is the result of substituting a for x in t. For example, $\lambda x. \ x{+}3$ denotes the function $a \mapsto a{+}3$ which adds 3 to its argument. The simplification of $(\lambda x. \ t)t'$ to $t[t'/x]$ is called β-reduction. λ-expressions of the form $\lambda(x_1, \ldots, x_n). \ t$ will also be used; these denote functions defined on n-tuples. For example, $\lambda(m, n). \ m + n$ denotes the function $(m, n) \mapsto m + n$.

To save writing brackets, function applications can be written as $f \ x$ instead of $f(x)$. More generally, we adopt the standard convention that $t_1 \ t_2 \ t_3 \ \cdots \ t_n$ abbreviates $(\ \cdots \ ((t_1 \ t_2) \ t_3) \ \cdots \ t_n)$ i.e. application associates to the left.

The notation $\lambda x_1 \ x_2 \ \cdots \ x_n. \ t$ abbreviates $\lambda x_1. \ \lambda x_2. \ \cdots \ \lambda x_n. \ t$. The scope of a λ extends as far to the right as possible. Thus, for example, $\lambda b. \ b = \lambda x. \ \mathsf{T}$ means $\lambda b. \ (b = (\lambda x. \ \mathsf{T}))$ not $(\lambda b. \ b) = (\lambda x. \ \mathsf{T})$.

4.1 Types

Higher order variables can be used to formulate Russell's paradox: define the predicates P and Ω by:

$$\begin{aligned} \mathsf{P} \ x \ &= \ \neg \ (x \ x) \\ \Omega \ &= \ \mathsf{P} \ \mathsf{P} \end{aligned}$$

Then it immediately follows that $\Omega = \neg \ \Omega$. Russell invented his theory of types to prevent such inconsistencies. Church simplified Russell's idea; HOL uses a slight extension of Church's type system[8].

Types are expressions that denote sets of values, they are either *atomic* or *compound*. Examples of atomic types are:

$$bool, \qquad num, \qquad real, \qquad string$$

---

[6]The examples here are explained in more detail in Sections 5.1 and 5.2 below.

[7]The version of higher order logic used in this paper is a slight extension of a system invented by Church [5]. Church's system is sometimes called 'simple type theory'; an introductory textbook on this has recently been written by Andrews [1].

[8]The type system of the HOL logic is the type system used in LCF's logic PPλ [11,30].

these denote the sets of booleans, natural numbers, real numbers and character strings respectively. Compound types are built from atomic types (or other compound types) using *type operators*. For example, if σ, σ_1 and σ_2 are types then so are:

$$\sigma \ list, \qquad \sigma_1 \times \sigma_2, \qquad \sigma_1 \rightarrow \sigma_2$$

where *list* is a postfixed unary type operator and $\rightarrow$ and $\times$ are infixed binary type operators. The type σ *list* denotes the set of lists of values of elements from the set denoted by σ; the type $\sigma_1 \times \sigma_2$ denotes the set of pairs (x_1, x_2) where x_1 is in the set denoted by σ_1 and x_2 is in the set denoted by σ_2; the type $\sigma_1 \rightarrow \sigma_2$ denotes the set of total functions with domain denoted by σ_1 and range denoted by σ_2. Lower case *slanted* identifiers will be used for particular types, and greek letters (mostly σ) to range over arbitrary types.

Terms of higher order logic must be *well-typed* in the sense that each subterm can be assigned a type 'in a consistent way'. More precisely, it must be possible to assign a type to each subterm such that both 1 and 2 below hold.

1. For every subterm of the form $t_1 \ t_2$ there are types σ and σ' such that:

   (a) t_1 is assigned $\sigma' \rightarrow \sigma$

   (b) t_2 is assigned σ'

   (c) $t_1 \ t_2$ is assigned the type σ.

2. Every subterm of the form $\lambda x. \ t$ is assigned a type $\sigma_1 \rightarrow \sigma_2$ where:

   (a) x is assigned σ_1

   (b) t is assigned σ_2.

Variables with the same name can be assigned different types, but then they are regarded as different variables.

Writing $t{:}\sigma$ indicates that a term t has type σ. Thus $x{:}\sigma_1$ is the same variable as $x{:}\sigma_2$ if and only if $\sigma_1 = \sigma_2$. In Church's original notation (which is also used by Andrews) $t{:}\sigma$ would be written t_σ.

In some formulations of higher-order logic, the types of variables have to be written down explicitly. For example, $\lambda x. \ \cos(\sin(x))$ would not be allowed in Church's system, instead one would have to write:

$$\lambda x_{real}. \ \cos_{real \rightarrow real}(\sin_{real \rightarrow real}(x_{real}))$$

We allow the types of variables to be omitted if they can be inferred from the context. There is an algorithm, due to Robin Milner [27], for doing such type inference.

We adopt the standard conventions that $\sigma_1 \rightarrow \sigma_2 \rightarrow \sigma_3 \rightarrow \ \cdots \ \sigma_n \rightarrow \sigma$ is an abbreviation for $\sigma_1 \rightarrow (\sigma_2 \rightarrow (\sigma_3 \rightarrow \ \cdots \ (\sigma_n \rightarrow \sigma) \ \cdots \ ))$ *i.e.* $\rightarrow$ associates to the right. This convention blends well with the left associativity of function application, because if f has type $\sigma_1 \rightarrow \ \cdots \ \sigma_n \rightarrow \sigma$ and $t_1, \ \ldots \ , \ t_n$ have types $\sigma_1, \ \ldots \ , \ \sigma_n$ respectively, then $f \ t_1 \ \cdots \ t_n$ is a well-typed term of type σ. We also assume $\times$ is more tightly binding than $\rightarrow$; for example, *state* $\times$ *state* $\rightarrow$ *bool* means (*state* $\times$ *state*) $\rightarrow$ *bool*.

4.2 Definitions

In addition to the 'built-in' constants like $\wedge$, $\vee$, $\neg$ etc, new constants can be introduced by *definitions*. The definition of a constant, c say, is an axiom of the form:

$$\vdash c = t$$

where t is a closed[9] term. For example, the definition of Seq given in Section 5 below is: .

$$\vdash \text{Seq} = \lambda(C_1, C_2).\ \lambda(s_1, s_2).\ \exists s.\ C_1(s_1, s) \wedge C_2(s, s_2)$$

which is logically equivalent to:

$$\vdash \text{Seq}(C_1, C_2)(s_1, s_2) = \exists s.\ C_1(s_1, s) \wedge C_2(s, s_2)$$

New types can also be defined as names for subsets of existing types; see Melham's paper for details [25]. A particular collection of constants, types and definitions is called a *theory*. Theories can be hierarchically structured and stored on disk [13].

5 Semantics in Logic

The traditional denotation of a command $\mathcal{C}$ is a function, Meaning($\mathcal{C}$) say, from machine states to machine states. The idea is:

$$\text{Meaning}(\mathcal{C})(s) = \text{'the state resulting from executing } \mathcal{C} \text{ in state } s\text{'}$$

Since **while**-commands need not terminate, the functions denoted by commands will be *partial*. For example, for any state s and command $\mathcal{C}$

$$\text{Meaning}(\textbf{while T do } \mathcal{C})(s)$$

will be undefined. Since functions in conventional predicate calculus are total,[10] we cannot use them as command denotations. Instead we will take the meaning of commands to be predicates on pairs of states (s_1, s_2); the idea being that if $\mathcal{C}$ denotes c then:

$$c(s_1, s_2) \equiv (\text{Meaning}(\mathcal{C})(s_1) = s_2)$$

i.e.

$$c(s_1, s_2) = \begin{cases} \text{T} & \text{if executing } \mathcal{C} \text{ in state } s_1 \text{ results in state } s_2 \\ \text{F} & \text{otherwise} \end{cases}$$

If c_{while} is the predicate denoted by **while T do** $\mathcal{C}$, we will simply have:

$$\forall s_1\ s_2.\ c_{while}(s_1, s_2) = \text{F}$$

---

[9]A term is closed if it has no free variables; i.e. all variables are bound by $\forall$, $\exists$ or λ.

[10]There are versions of predicate calculus than can handle partial functions; for example, the Scott/Milner 'Logic of Computable Functions' (LCF) [30] and the Scott/Fourman formulation of intuitionisitic higher order logic [24]. The HOL system used in this paper is actually a version of Milner's proof assistant for LCF [11] which has been modified to support higher order logic. Although programming language semantics are particularly easy to represent in LCF, other kinds of semantics (e.g. the behaviour of hardware) are more straightforward in classical logic.

Formally, the type *state* of states that we use is defined by:

$$state = string \rightarrow num$$

The notation 'XYZ' will be used for the string consisting of the three characters X, Y and Z; thus 'XYZ' : *string*. A state s in which the strings 'X', 'Y' and 'Z' are bound to 1, 2 and 3 respectively, and all other strings are bound to 0, is defined by:

$$s = \lambda x. (x = `X` \rightarrow 1 \mid (x = `Y` \rightarrow 2 \mid (x = `Z` \rightarrow 3 \mid 0)))$$

If e, b and c are the denotations of $\mathcal{E}$, $\mathcal{B}$ and $\mathcal{C}$ respectively, then:

$$e : state \rightarrow num$$
$$b : state \rightarrow bool$$
$$c : state \times state \rightarrow bool$$

For example, the denotation of $X + 1$ would be $\lambda s. \, s`X` + 1$ and the denotation of $(X + Y) > 10$ would be $\lambda s. \, (s`X` + s`Y`) > 10$.

It is convenient to introduce the notations $[\![\mathcal{E}]\!]$ and $[\![\mathcal{B}]\!]$ for the logic terms representing the denotations of $\mathcal{E}$ and $\mathcal{B}$. For example:

$$[\![X + 1]\!] \quad = \quad \lambda s. \, s`X` + 1$$
$$[\![(X + Y) > 10]\!] \quad = \quad \lambda s. \, (s`X` + s`Y`) > 10$$

Note that $[\![\mathcal{E}]\!]$ and $[\![\mathcal{B}]\!]$ are *terms*, i.e. syntactic objects. In traditional denotational semantics, the meanings of expressions are represented by abstract mathematical entities, like functions. Such abstract entities are the 'meanings' of $[\![\mathcal{E}]\!]$ and $[\![\mathcal{B}]\!]$; but because we are concerned with mechanical reasoning, we work with the terms and formulae themselves, rather than with the intangible abstract entities they denote.

Sometimes it is necessary for pre and postconditions to contain logical variables that are not program variables. An example is:

$$\{X = x \wedge Y = y\} \, Z := X; \, X := Y; \, Y := Z \, \{X = y \wedge Y = x\}$$

Here x and y are logical variables whereas X and Y (and Z) are program variables. The formulae representing the correct semantics of the pre and post conditions of this specification are:

$$[\![X = x \wedge Y = y]\!] \quad = \quad \lambda s. \, s`X` = x \wedge s`Y` = y$$
$$[\![X = y \wedge Y = x]\!] \quad = \quad \lambda s. \, s`X` = y \wedge s`Y` = x$$

The convention adopted in this paper is that upper case variables are program variables and lower case variables are logical variables (as in the example just given). Logical variables occurring in pre and post conditions are sometimes called *ghost variables* or *auxiliary variables*. In our little programming language the only data type is numbers, hence program variables will have type *num*. The definition of $[\![\cdots]\!]$ can now be stated more precisely: if $T[X_1, \ldots, X_n]$ is a term of higher order logic whose upper case free variables of type *num* are $X_1, \ldots, X_n$ then

$$[\![T[X_1, \ldots, X_n]]\!] \quad = \quad \lambda s. \, T[s`X_1`, \ldots, s`X_n`]$$

In other words if T is a term of type σ then the term $[\![T]\!]$ of type $state \rightarrow \sigma$ is obtained as follows:

(i) Each free upper case variable $\mathcal{V}$ of type *num* is replaced by the term $s`\mathcal{V}`$, where s is a variable of type *state* not occurring in $\mathcal{P}$.

(ii) The result of (i) is prefixed by '$\lambda s.$'.

5.1 Semantics of Commands

To represent the semantics of our little programming language, predicates in higher order logic that correspond to the five kinds of commands are defined. For each command $\mathcal{C}$, a term $[\![\mathcal{C}]\!]$ of type $state \times state \rightarrow bool$ is defined as follows:

1. $[\![\text{skip}]\!] = \text{Skip}$

   where the constant Skip is defined by:

   $$\text{Skip}(s_1, s_2) = (s_1 = s_2)$$

2. $[\![\mathcal{V} := \mathcal{E}]\!] = \text{Assign}(`\mathcal{V}`, [\![\mathcal{E}]\!])$

   where the constant Assign is defined by:

   $$\text{Assign}(v, e)(s_1, s_2) = (s_2 = \text{Bnd}(e, v, s_1))$$

   where:

   $$\text{Bnd}(e, v, s) = \lambda x.\ (x = v \rightarrow e\ s_1\ |\ s\ x)$$

3. $[\![\mathcal{C}_1;\ \mathcal{C}_2]\!] = \text{Seq}([\![\mathcal{C}_1]\!], [\![\mathcal{C}_2]\!])$

   where the constant Seq is defined by:

   $$\text{Seq}(c_1, c_2)(s_1, s_2) = \exists s.\ c_1(s_1, s)\ \wedge\ c_2(s, s_2)$$

4. $[\![\text{if } \mathcal{B} \text{ then } \mathcal{C}_1 \text{ else } \mathcal{C}_2]\!] = \text{If}([\![\mathcal{B}]\!], [\![\mathcal{C}_1]\!], [\![\mathcal{C}_2]\!])$

   where the constant If is defined by:

   $$\text{If}(b, c_1, c_2)(s_1, s_2) = (b\ s_1 \rightarrow c_1(s_1, s_2)\ |\ c_2(s_1, s_2))$$

5. $[\![\text{while } \mathcal{B} \text{ do } \mathcal{C}]\!] = \text{While}([\![\mathcal{B}]\!], [\![\mathcal{C}]\!])$

   where the constant While is defined by:

   $$\text{While}(b, c)(s_1, s_2) = \exists n.\ \text{Iter}(n)(b, c)(s_1, s_2)$$

   where $\text{Iter}(n)$ is defined by primitive recursion as follows:

   $$\begin{aligned}
   \text{Iter}(0)(b, c)(s_1, s_2) &= \text{F} \\
   \text{Iter}(n{+}1)(b, c)(s_1, s_2) &= \text{If}(b, \text{Seq}(c, \text{Iter}(n)(b, c)), \text{Skip})(s_1, s_2)
   \end{aligned}$$

Example

$R := X;$
$Q := 0;$
while $Y \le R$
   do $(R := R - Y;\ Q := Q + 1)$

denotes:

```
Seq
 (Assign('R', [[X]]),
  Seq
  (Assign('Q', [[0]]),
   While
   ([[Y ≤ R]],
    Seq
    (Assign('R', [[R − Y]],
     Assign('Q', [[Q + 1]])))))
```

Expanding the $[\![\cdots]\!]$s results in:

```
Seq
 (Assign('R', λs. s'X'),
  Seq
  (Assign('Q', λs. 0),
   While
   ((λs. s'Y' ≤ s'R'),
    Seq
    (Assign('R', λs. s'R' − s'Y'),
     Assign('Q', λs. s'Q' + 1)))))
```
□

It might appear that by representing the meaning of commands with relations, we can give a semantics to nondeterministic constructs. For example, if $\mathcal{C}_1 \parallel \mathcal{C}_2$ is the nondeterministic choice 'either do $\mathcal{C}_1$ or do $\mathcal{C}_2$', then one might think that a satisfactory semantics would be given by:

$$[\![\mathcal{C}_1 \parallel \mathcal{C}_2]\!] \;=\; \mathsf{Choose}([\![\mathcal{C}_1]\!], [\![\mathcal{C}_2]\!])$$

where the constant Choose is defined by:

$$\mathsf{Choose}(c_1, c_2)(s_1, s_2) \;=\; c_1(s_1, s_2) \;\vee\; c_2(s_1, s_2)$$

Unfortunately this semantics has some undesirable properties. For example, if c_{while} is the predicate denoted by the non-terminating command **while** T **do skip**, then

$$\forall s_1 \; s_2. \; c_{while}(s_1, s_2) = \mathsf{F}$$

and hence, because $\forall t. \; t \vee \mathsf{F} = t$, it follows that:

$$\mathbf{skip} \parallel c_{while} \;=\; \mathbf{skip}$$

Thus the command that does nothing is equivalant to a command that *either* does nothing *or* loops! It is well known how to distinguish guaranteed termination from possible termination [31]; the example above shows that the relational semantics used in this paper does not do it. This problem will appear again in connection with Dijkstra's theory of weakest preconditions in Section 11.2.

5.2   Semantics of Partial Correctness Specifications

A partial correctness specification $\{\mathcal{P}\}\; \mathcal{C}\; \{\mathcal{Q}\}$ denotes:

$$\forall s_1\; s_2.\; [\![\mathcal{P}]\!]\; s_1\; \wedge\; [\![\mathcal{C}]\!](s_1, s_2)\; \Rightarrow\; [\![\mathcal{Q}]\!]\; s_2$$

To abbreviate this formula, we define a constant Spec by:

$$\mathsf{Spec}(p, c, q)\; =\; \forall s_1\; s_2.\; p\; s_1\; \wedge\; c(s_1, s_2)\; \Rightarrow\; q\; s_2$$

Note that the denotation of pre and postconditions $\mathcal{P}$ and $\mathcal{Q}$ are not just the logical formulae themselves, but are $[\![\mathcal{P}]\!]$ and $[\![\mathcal{Q}]\!]$. For example, in the specification $\{X = 1\}\; \mathcal{C}\; \{\mathcal{Q}\}$, the precondition $X = 1$ asserts that the value of the string 'X' in the initial state is 1. The precondition thus denotes $[\![\mathcal{P}]\!]$, i.e. $\lambda s.\; s\text{'}X\text{'} = 1$. Thus:

$$\{X = 1\}\; X := X + 1\; \{X = 2\}$$

denotes

$$\mathsf{Spec}([\![X = 1]\!],\; \mathsf{Assign}('X', [\![X + 1]\!]),\; [\![X = 2]\!])$$

i.e.

$$\mathsf{Spec}((\lambda s.\; s\text{'}X\text{'} = 1),\; \mathsf{Assign}('X', \lambda s.\; s\text{'}X\text{'} + 1),\; \lambda s.\; s\text{'}X\text{'} = 2)$$

Example

In the specification below, x and y are logical variables whereas X and Y (and Z) are program variables.

$$\{X = x\; \wedge\; Y = y\}\; Z := X;\; X := Y;\; Y := Z\; \{X = y\; \wedge\; Y = x\}$$

The semantics of this is thus represented by the term:

$$\begin{aligned}
\mathsf{Spec}(&[\![X = x\; \wedge\; Y = y]\!], \\
&\mathsf{Seq}(\mathsf{Assign}('Z', [\![X]\!]), \\
&\quad\mathsf{Seq}(\mathsf{Assign}('X', [\![Y]\!]), \mathsf{Assign}('Y', [\![Z]\!]))), \\
&[\![X = y\; \wedge\; Y = x]\!])
\end{aligned}$$

which abbreviates:

$$\begin{aligned}
\mathsf{Spec}(&(\lambda s.\; s\text{'}X\text{'} = x\; \wedge\; s\text{'}Y\text{'} = y), \\
&\mathsf{Seq}(\mathsf{Assign}('Z', \lambda s.\; s\text{'}X\text{'}), \\
&\quad\mathsf{Seq}(\mathsf{Assign}('X', \lambda s.\; s\text{'}Y\text{'}), \mathsf{Assign}('Y', \lambda s.\; s\text{'}Z\text{'}))), \\
&\lambda s.\; s\text{'}X\text{'} = y\; \wedge\; s\text{'}Y\text{'} = x)
\end{aligned}$$

$\square$

6   Hoare Logic as Higher Order Logic

Hoare logic can be embedded in higher order logic simply by regarding the concrete syntax given in Sections 2 and 3 as an *abbreviation* for the corresponding semantic formulae described in Section 5. For example:

$$\{X = x\}\; X := X + 1\; \{X = x + 1\}$$

can be interpreted as as abbreviating:

$$\mathsf{Spec}(\llbracket X = x \rrbracket, \ \mathsf{Assign}('X', \llbracket X + 1 \rrbracket), \ \llbracket X = x + 1 \rrbracket)$$

i.e.

$$\mathsf{Spec}((\lambda s. \ s'X' = x), \ \mathsf{Assign}('X', \lambda s. \ s'X' + 1), \ \lambda s. \ s'X' = x + 1)$$

The translation between the syntactic 'surface structure' and the semantic 'deep structure' is straightforward; it can easily be mechanized with a simple parser and pretty-printer. Section 7 contains examples illustrating this in a version of the HOL system.

If partial correctness specifications are interpreted this way then, as shown in the rest of this section, the axioms and rules of Hoare logic described in Section 3 become derived rules of higher order logic.

The first step in this derivation is to prove the following seven theorems from the definitions of the constants Spec, Skip, Assign, Bnd, Seq, If, While and Iter.

H1. $\vdash \ \forall p. \ \mathsf{Spec}(p, \mathsf{Skip}, p)$

H2. $\vdash \ \forall p \ v \ e. \ \mathsf{Spec}((\lambda s. \ p(\mathsf{Bnd}(e \ s, v, s))), \mathsf{Assign}(v, e), p)$

H3. $\vdash \ \forall p \ p' \ q \ c. \ (\forall s. \ p' \ s \ \Rightarrow \ p \ s) \ \wedge \ \mathsf{Spec}(p, c, q) \ \Rightarrow \ \mathsf{Spec}(p', c, q)$

H4. $\vdash \ \forall p \ q \ q' \ c. \ \mathsf{Spec}(p, c, q) \ \wedge \ (\forall s. \ q \ s \ \Rightarrow \ q' \ s) \ \Rightarrow \ \mathsf{Spec}(p, c, q')$

H5. $\vdash \ \forall p \ q \ r \ c_1 \ c_2. \ \mathsf{Spec}(p, c_1, q) \ \wedge \ \mathsf{Spec}(q, c_2, r) \ \Rightarrow \ \mathsf{Spec}(p, \mathsf{Seq}(c_1, c_2), r)$

H6. $\vdash \ \forall p \ q \ c_1 \ c_2 \ b.$
$\mathsf{Spec}((\lambda s. \ p \ s \ \wedge \ b \ s), c_1, q) \ \wedge \ \mathsf{Spec}((\lambda s. \ p \ s \ \wedge \ \neg(b \ s)), c_2, q)$
$\Rightarrow$
$\mathsf{Spec}(p, \mathsf{If}(b, c_1, c_2), q)$

H7. $\vdash \ \forall p \ c \ b.$
$\mathsf{Spec}((\lambda s. \ p \ s \ \wedge \ b \ s), c, p)$
$\Rightarrow$
$\mathsf{Spec}(p, \mathsf{While}(b, c), (\lambda s. \ p \ s \ \wedge \ \neg(b \ s)))$

The proofs of H1 to H7 are routine. All the axioms and rules of Hoare logic, *except* for the assignment axiom, can be implemented in a uniform way from H1 – H7. The derivation of the assignment axiom from H2, although straightforward, is a bit messy; it is thus explained last (in Section 6.7).

6.1   Derivation of the skip-axiom

To derive the **skip**-axiom it must be shown for arbitrary $\mathcal{P}$ that:

$$\vdash \ \{\mathcal{P}\} \ \mathbf{skip} \ \{\mathcal{P}\}$$

which abbreviates:

$$\vdash \ \mathsf{Spec}(\llbracket \mathcal{P} \rrbracket, \mathsf{Skip}, \llbracket \mathcal{P} \rrbracket)$$

This follows by specializing p to $\llbracket \mathcal{P} \rrbracket$ in H1.

6.2 Derivation of Precondition Strengthening

To derive the rule of precondition strengthening it must be shown that for arbitrary $\mathcal{P}$, $\mathcal{P}'$, $\mathcal{C}$ and $\mathcal{Q}$ that:

$$\frac{\vdash\ \mathcal{P}' \Rightarrow\ \mathcal{P} \qquad \vdash\ \{\mathcal{P}\}\ \mathcal{C}\ \{\mathcal{Q}\}}{\vdash\ \{\mathcal{P}'\}\ \mathcal{C}\ \{\mathcal{Q}\}}$$

Expanding abbreviations converts this to:

$$\frac{\vdash\ \mathcal{P}' \Rightarrow\ \mathcal{P} \qquad \vdash\ \mathsf{Spec}([\![\mathcal{P}]\!], [\![\mathcal{C}]\!], [\![\mathcal{Q}]\!])}{\vdash\ \mathsf{Spec}([\![\mathcal{P}']\!], [\![\mathcal{C}]\!], [\![\mathcal{Q}]\!])}$$

Specializing H3 yields:

$$\vdash\ (\forall s.\ [\![\mathcal{P}']\!]\ s \ \Rightarrow\ [\![\mathcal{P}]\!]\ s) \wedge \mathsf{Spec}([\![\mathcal{P}]\!], [\![\mathcal{C}]\!], [\![\mathcal{Q}]\!]) \ \Rightarrow\ \mathsf{Spec}([\![\mathcal{P}']\!], [\![\mathcal{C}]\!], [\![\mathcal{Q}]\!])$$

The rule of precondition strengthening will follow if $\vdash\ \forall s.\ [\![\mathcal{P}']\!]\ s\ \Rightarrow\ [\![\mathcal{P}]\!]\ s$ can be deduced from $\vdash\ \mathcal{P}' \Rightarrow \mathcal{P}$. To see that this is indeed the case, let us make explicit the program variables $X_1, \ldots, X_n$ occurring in $\mathcal{P}$ and $\mathcal{P}'$ by writing $\mathcal{P}[X_1, \ldots, X_n]$ and $\mathcal{P}[X_1, \ldots, X_n]'$. Then $\vdash\ \mathcal{P}' \Rightarrow \mathcal{P}$ becomes

$$\vdash\ \mathcal{P}'[X_1, \ldots, X_n] \ \Rightarrow\ \mathcal{P}[X_1, \ldots, X_n]$$

Since $X_1, \ldots, X_n$ are free variables in this theorem, they are implicitly universally quantified, and hence each X_i can be instantiated to $s`X_i`$ to get:

$$\vdash\ \mathcal{P}'[s`X_1`, \ldots, s`X_n`] \ \Rightarrow\ \mathcal{P}[s`X_1`, \ldots, s`X_n`]$$

Generalizing on the free variable s yields:

$$\vdash\ \forall s.\ \mathcal{P}'[s`X_1`, \ldots, s`X_n`] \ \Rightarrow\ \mathcal{P}[s`X_1`, \ldots, s`X_n`]$$

which is equivalent (by β-reduction) to

$$\vdash\ \forall s.\ (\lambda s.\ \mathcal{P}'[s`X_1`, \ldots, s`X_n`])\ s \ \Rightarrow\ (\lambda s.\ \mathcal{P}[s`X_1`, \ldots, s`X_n`])\ s$$

i.e.

$$\vdash\ \forall s.\ [\![\mathcal{P}'[X_1, \ldots, X_n]]\!]\ s \ \Rightarrow\ [\![\mathcal{P}[X_1, \ldots, X_n]]\!]\ s$$

The derivation sketched above can be done via higher order matching applied to H3. This is, in fact, how the HOL derived rule PRE_STRENGTH_RULE described in Section 7 is programmed. Systems like *Isabelle* [29], which have higher order unification built in, would handle such rules very smoothly.

6.3 Derivation of Postcondition Weakening

To derive the rule of postcondition weakening, it must be shown that for arbitrary $\mathcal{P}$, $\mathcal{C}$, and $\mathcal{Q}$ and $\mathcal{Q}'$ that:

$$\frac{\vdash\ \{\mathcal{P}\}\ \mathcal{C}\ \{\mathcal{Q}\} \qquad \vdash\ \mathcal{Q} \Rightarrow\ \mathcal{Q}'}{\vdash\ \{\mathcal{P}\}\ \mathcal{C}\ \{\mathcal{Q}'\}}$$

The derivation of this from H4 is similar to the derivation of precondition strengthening from H3.

6.4 Derivation of the Sequencing Rule

To derive the sequencing rule, it must be shown that for arbitrary $\mathcal{P}$, $\mathcal{C}_1$, $\mathcal{R}$, $\mathcal{C}_2$ and and $\mathcal{Q}$ that:

$$\frac{\vdash \{\mathcal{P}\}\,\mathcal{C}_1\,\{\mathcal{Q}\} \qquad \vdash \{\mathcal{Q}\}\,\mathcal{C}_2\,\{\mathcal{R}\}}{\vdash \{\mathcal{P}\}\,\mathcal{C}_1;\,\mathcal{C}_2\,\{\mathcal{R}\}}$$

Expanding the abbreviations yields:

$$\frac{\vdash \mathsf{Spec}([\![\mathcal{P}]\!], [\![\mathcal{C}_1]\!], [\![\mathcal{Q}]\!]) \qquad \vdash \mathsf{Spec}([\![\mathcal{Q}]\!], [\![\mathcal{C}_2]\!], [\![\mathcal{R}]\!])}{\vdash \mathsf{Spec}([\![\mathcal{P}]\!], \mathsf{Seq}([\![\mathcal{C}_1]\!], [\![\mathcal{C}_2]\!]), [\![\mathcal{R}]\!])}$$

The validity of this rule follows directly from H5.

6.5 Derivation of the if-rule

To derive the **if**-rule, it must be shown that for arbitrary $\mathcal{P}$, $\mathcal{B}$, $\mathcal{C}_1$, $\mathcal{C}_2$ and and $\mathcal{Q}$ that:

$$\frac{\vdash \{\mathcal{P} \wedge \mathcal{B}\}\,\mathcal{C}_1\,\{\mathcal{Q}\} \qquad \vdash \{\mathcal{P} \wedge \neg\mathcal{B}\}\,\mathcal{C}_2\,\{\mathcal{Q}\}}{\vdash \{\mathcal{P}\}\ \text{if } \mathcal{B} \text{ then } \mathcal{C}_1 \text{ else } \mathcal{C}_2\,\{\mathcal{Q}\}}$$

Expanding abbreviations yields:

$$\frac{\vdash \mathsf{Spec}([\![\mathcal{P} \wedge \mathcal{B}]\!], [\![\mathcal{C}_1]\!], [\![\mathcal{Q}]\!]) \qquad \vdash \mathsf{Spec}([\![\mathcal{P} \wedge \neg\mathcal{B}]\!], [\![\mathcal{C}_2]\!], [\![\mathcal{Q}]\!])}{\vdash \mathsf{Spec}([\![\mathcal{P}]\!], \mathsf{If}([\![\mathcal{B}]\!], [\![\mathcal{C}_1]\!], [\![\mathcal{C}_2]\!]), [\![\mathcal{Q}]\!])}$$

This follows from H6 in a similar fashion to the way precondition strenthening follows from H3.

6.6 Derivation of the while-rule

To derive the **while**-rule, it must be shown that for arbitrary $\mathcal{P}$, $\mathcal{B}$ and $\mathcal{C}$ that:

$$\frac{\vdash \{\mathcal{P} \wedge \mathcal{B}\}\,\mathcal{C}\,\{\mathcal{P}\}}{\vdash \{\mathcal{P}\}\ \text{while } \mathcal{B} \text{ do } \mathcal{C}\,\{\mathcal{P} \wedge \neg\mathcal{B}\}}$$

Expanding abbreviations yields:

$$\frac{\vdash \mathsf{Spec}([\![\mathcal{P} \wedge \mathcal{B}]\!], [\![\mathcal{C}]\!], [\![\mathcal{P}]\!])}{\vdash \mathsf{Spec}([\![\mathcal{P}]\!], \mathsf{While}([\![\mathcal{B}]\!], [\![\mathcal{C}]\!]), [\![\mathcal{P} \wedge \neg\mathcal{B}]\!])}$$

This follows from H7.

6.7 Derivation of the Assignment Axiom

To derive the assignment axiom, it must be shown that for arbitrary $\mathcal{P}$, $\mathcal{E}$ and $\mathcal{V}$:

$$\vdash \{\mathcal{P}[\mathcal{E}/\mathcal{V}]\} \; \mathcal{V} := \mathcal{E} \; \{\mathcal{P}\}$$

This abbreviates:

$$\vdash \; \mathsf{Spec}([\![\mathcal{P}[\mathcal{E}/\mathcal{V}]]\!], \mathsf{Assign}(`\mathcal{V}`, [\![\mathcal{E}]\!]), [\![\mathcal{P}]\!])$$

By H2:

$$\vdash \; \forall p \; x \; e. \; \mathsf{Spec}((\lambda s. \; p(\mathsf{Bnd}(e \; s, x, s))), \mathsf{Assign}(x, e), p)$$

Specializing p, x and e to $[\![P]\!]$, `$\mathcal{V}$` and $[\![\mathcal{E}]\!]$ yields:

$$\vdash \; \mathsf{Spec}((\lambda s. \; [\![P]\!](\mathsf{Bnd}([\![\mathcal{E}]\!]s, `\mathcal{V}`, s))), \mathsf{Assign}(`\mathcal{V}`, [\![\mathcal{E}]\!]), [\![P]\!])$$

Thus, to derive the assignment axiom it must be shown that:

$$\vdash \; [\![\mathcal{P}[\mathcal{E}/\mathcal{V}]]\!] \;=\; \lambda s. \; [\![P]\!](\mathsf{Bnd}([\![\mathcal{E}]\!]s, `\mathcal{V}`, s))$$

To see why this holds, let us make explicit the free program variables in $\mathcal{P}$ and $\mathcal{E}$ by writing $\mathcal{P}[\mathcal{V}, X_1, \ldots, X_n]$ and $\mathcal{E}[\mathcal{V}, X_1, \ldots, X_n]$, where $X_1, \ldots, X_n$ are the free program variables that are not equal to $\mathcal{V}$. Then, for example, $\mathcal{P}[1, \ldots, n]$ would denote the result of substituting $1, \ldots, n$ for $X_1, \ldots, X_n$ in $\mathcal{P}$ respectively. The equation above thus becomes:

$$[\![\mathcal{P}[\mathcal{V}, X_1, \ldots, X_n][\mathcal{E}[\mathcal{V}, X_1, \ldots, X_n]/\mathcal{V}]]\!]$$
$$=$$
$$\lambda s. \; [\![\mathcal{P}[\mathcal{V}, X_1, \ldots, X_n]]\!](\mathsf{Bnd}([\![\mathcal{E}[\mathcal{V}, X_1, \ldots, X_n]]\!]s, `\mathcal{V}`, s))$$

Performing the substitution in the left hand side yields:

$$[\![\mathcal{P}[\mathcal{E}[\mathcal{V}, X_1, \ldots, X_n], X_1, \ldots, X_n]]\!]$$
$$=$$
$$\lambda s. \; [\![\mathcal{P}[\mathcal{V}, X_1, \ldots, X_n]]\!](\mathsf{Bnd}([\![\mathcal{E}[\mathcal{V}, X_1, \ldots, X_n]]\!]s, `\mathcal{V}`, s))$$

Replacing expressions of the form $[\![\mathcal{P}[\cdots]]\!]$ by their meaning yields:

$$(\lambda s. \; \mathcal{P}[\mathcal{E}[s`\mathcal{V}`, s`X_1`, \ldots, s`X_n`], s`X_1`, \ldots, s`X_n`])$$
$$=$$
$$\lambda s. \; (\lambda s. \; \mathcal{P}[s`\mathcal{V}`, s`X_1`, \ldots, s`X_n`])(\mathsf{Bnd}([\![\mathcal{E}[\mathcal{V}, X_1, \ldots, X_n]]\!]s, `\mathcal{V}`, s))$$

Performing a β-reduction on the right hand side, and then simplifying with the following easily derived properties of Bnd (the second of which assumes `$\mathcal{V}` \neq X_i$):

$$\vdash \; \mathsf{Bnd}([\![\mathcal{E}[\mathcal{V}, X_1, \ldots, X_n]]\!]s, `\mathcal{V}`, s) \; `\mathcal{V}` \;=\; [\![\mathcal{E}[\mathcal{V}, X_1, \ldots, X_n]]\!]s$$

$$\vdash \; \mathsf{Bnd}([\![\mathcal{E}[\mathcal{V}, X_1, \ldots, X_n]]\!]s, `\mathcal{V}`, s) \; X_i \;=\; s \; X_i$$

results in:

$$(\lambda s. \; \mathcal{P}[\mathcal{E}[s`\mathcal{V}`, s`X_1`, \ldots, s`X_n`], s`X_1`, \ldots, s`X_n`])$$
$$=$$
$$\lambda s. \; \mathcal{P}[[\![\mathcal{E}[\mathcal{V}, X_1, \ldots, X_n]]\!]s, s`X_1`, \ldots, s`X_n`]$$

which is true since:

$$[\![\mathcal{E}[\mathcal{V}, X_1, \ldots, X_n]\!]s \ = \ \mathcal{E}[s\raisebox{0.3ex}{`}\mathcal{V}\raisebox{0.3ex}{`}, s\raisebox{0.3ex}{`}X_1\raisebox{0.3ex}{`}, \ldots, s\raisebox{0.3ex}{`}X_n\raisebox{0.3ex}{`}]$$

Although this derivation might appear tricky at first sight, it is straightforward and easily mechanized. The HOL derived rule ASSIGN_AX described in Section 7 performs this deduction for each $\mathcal{P}$, $\mathcal{E}$ and $\mathcal{V}$.

It is tempting to try to formulate the assignment axiom as a theorem of higher order logic looking something like:

$$\forall p \ e \ v. \ \mathsf{Spec}(p[e/v], \mathsf{Assign}(v, e), p)$$

Unfortunately, the expression $p[e/v]$ does not make sense when p is a variable. $\mathcal{P}[\mathcal{E}/\mathcal{V}]$ is a meta notation and consequently the assignment axiom can only be stated as a meta theorem. This elementary point is nevertheless quite subtle. In order to prove the assignment axiom as a theorem within higher order logic it would be necessary to have types in the logic corresponding to formulae, variables and terms. One could then prove something like:

$$\forall P \ E \ V. \ \mathsf{Spec}(\mathsf{Truth}(\mathsf{Subst}(P, E, V)), \ \mathsf{Assign}(V, \mathsf{Value} \ E), \ \mathsf{Truth} \ P)$$

It is clear that working out the details of this would be lots of work. This sort of embedding of a subset of a logic within itself has been explored in the context of the Boyer-Moore theorem prover [3].

7 Hoare Logic in HOL

In this section, the mechanization of the axioms and rules of Hoare logic using the HOL system [13] will be illustrated. We will try to make this comprehensible to readers unfamiliar with HOL, but it would help to have some prior exposure to Milner's LCF approach to interactive proof [10,11].

The HOL system is a version of Cambridge LCF [30] with higher order logic as object language[11]. Cambridge LCF evolved from Edinburgh LCF [11]. Some of the material in this and following sections has been taken from the paper *A Proof generating System for Higher Order Logic* [13].

The boxes below contain a session with a version of the HOL system that has been extended to support Hoare logic using the approach described in Section 6. The actual code implementing the extensions is not described here; it is mostly just straightforward ML (but the parser and pretty-printer are implemented in Lisp). To help the reader, the transcripts of the sessions with HOL have been edited so that proper logical symbols appear instead of their ASCII representations[12].

The user interface to HOL (and LCF) is the interactive programming language ML. At top level, expressions can be evaluated and declarations performed. The former results in the value of the expression and it ML type being printed; the latter

---

[11]LCF has the Scott/Milner 'Logic of Computable Functions' as object language.

[12]An interface to HOL that supports proper logical characters has been implemented by Roger Jones of ICL Defence Systems. This interface is also used to support a surface syntax based on the Z notation [33]. ICL use HOL for the specification and verification of security properties of both hardware and software.

in a value being bound to a name. The interactions in the boxes that follow should be understood as occurring in sequence. For example, variable bindings made in earlier boxes are assumed to persist to later ones. The ML prompt is #, so lines beginning with # are typed by the user and other lines are the system's response.

The ML language has a similar type system to the one used by the HOL logic. It is very common to confuse the two type systems. In this paper small teletype font will be used for ML types and *slanted* font for logical types. For example, the ML expression 1 has ML type int, whereas the HOL constant 1 has logical type *num*.

The first box below illustrates how the parser and pretty-printer have been modified to translate between 'surface' and 'deep' structure. The curly brackets { and } function like $[\![$ and $]\!]$ of Section 5; i.e. "$\{\mathcal{F}[X_1,\ldots,X_n]\}$" parses to "\s.$\mathcal{F}[s`X_1`,\ldots,s`X_n`]$". Evaluating pretty_on() switches the pretty-printer on; it can be switched off by evaluating pretty_off().

```
#"{(R=x) ∧ (Y=y)}" ;;
"λs. (s `R` = x) ∧ (s `Y` = y)" : term

#"R:=X" ;;
"Assign(`R`,(λs. s `X`))" : term

#pretty_on();;
() : void

#"R:=X" ;;
"R := X" : term
```

The ML function ASSIGN_AX has ML typeterm -> term -> thm and implements the assignment axiom.

$$\text{ASSIGN\_AX } "\{\mathcal{P}\}" \ "\mathcal{V} := \mathcal{E}" \ \mapsto \ \vdash \ \{\mathcal{P}[\mathcal{E}/\mathcal{V}]\} \ \mathcal{V} := \mathcal{E} \ \{\mathcal{P}\}$$

```
#let hth1 = ASSIGN_AX "{(R=x) ∧ (Y=y)}" "R:=X" ;;
hth1 = ⊢ {(X = x) ∧ (Y = y)} R := X {(R = x) ∧ (Y = y)}

#let hth2 = ASSIGN_AX "{(R=x) ∧ (X=y)}" "X:=Y" ;;
hth2 = ⊢ {(R = x) ∧ (Y = y)} X := Y {(R = x) ∧ (X = y)}

#let hth3 = ASSIGN_AX "{(Y=x) ∧ (X=y)}" "Y:=R" ;;
hth3 = ⊢ {(R = x) ∧ (X = y)} Y := R {(Y = x) ∧ (X = y)}
```

The ML function SEQ_RULE has ML type thm # thm -> thm and implements the sequencing rule.

$$\text{SEQ\_RULE } (\vdash \ \{\mathcal{P}\} \ C_1 \ \{\mathcal{Q}\}, \ \vdash \ \{\mathcal{Q}\} \ C_2 \ \{\mathcal{R}\}) \ \mapsto \ \vdash \ \{\mathcal{P}\} \ C_1; \ C_2 \ \{\mathcal{R}\}$$

```
#let hth4 = SEQ_RULE (hth1,hth2);;
hth4 = ⊢ {(X = x) ∧ (Y = y)} R := X; X := Y {(R = x) ∧ (X = y)}

#let hth5 = SEQ_RULE (hth4,hth3);;
hth5 =
⊢ {(X = x) ∧ (Y = y)}
    R := X; X := Y; Y := R
  {(Y = x) ∧ (X = y)}
```

If the pretty printing is switched off, the actual terms being manipulated become visible.

```
#pretty_off();;
() : void

#hth5;;
⊢ Spec
  ((λs. (s 'X' = x) ∧ (s 'Y' = y)),
   Seq
   (Seq(Assign('R',(λs. s 'X')),Assign('X',(λs. s 'Y'))),
    Assign('Y',(λs. s 'R'))),(λs. (s 'Y' = x) ∧ (s 'X' = y)))

#pretty_on();;
() : void
```

Using ML it is easy to define a function `SEQL_RULE` of type `thm list -> thm` that implements a derived rule generalizing `SEQ_RULE` from two arguments to a list of arguments.

> SEQL_RULE
> $$[\; \vdash \{\mathcal{P}\} \, \mathcal{C}_1 \, \{\mathcal{Q}_1\}; \quad \vdash \{\mathcal{Q}_1\} \, \mathcal{C}_2 \, \{\mathcal{Q}_2\}; \; \dots \; ; \quad \vdash \{\mathcal{Q}_{n-1}\} \, \mathcal{C}_n \, \{\mathcal{R}\}]$$
> $$\longmapsto$$
> $$\vdash \{\mathcal{P}\} \, \mathcal{C}_1; \dots ; \mathcal{C}_n \, \{\mathcal{R}\}$$

For readers familiar with ML, here is the definition of `SEQL_RULE`.

```
#letrec SEQL_RULE thl =
# if null(tl thl) then hd thl
#                 else SEQL_RULE
#                      (SEQ_RULE(hd thl,hd(tl thl)).tl(tl thl))
SEQL_RULE = - : proof

#let hth6 = SEQL_RULE[hth1;hth2;hth3];;
hth6 =
⊢ {(X = x) ∧ (Y = y)}
   R := X; X := Y; Y := R
  {(Y = x) ∧ (X = y)}
```

The ML function `PRE_STRENGTH_RULE` has type `thm # thm -> thm` and implements the rule of precondition strengthening.

$$\text{PRE\_STRENGTH\_RULE} \; (\; \vdash \mathcal{P}' \Rightarrow \mathcal{P}, \vdash \{\mathcal{P}\} \, \mathcal{C} \, \{\mathcal{Q}\}) \; \longmapsto \; \vdash \{\mathcal{P}'\} \, \mathcal{C} \, \{\mathcal{Q}\}$$

The ML function `POST_WEAK_RULE` has type `thm # thm -> thm` implements the rule of postcondition weakening.

$$\text{POST\_WEAK\_RULE} \; (\; \vdash \{\mathcal{P}\} \, \mathcal{C} \, \{\mathcal{Q}\}, \vdash \mathcal{Q} \Rightarrow \mathcal{Q}') \; \longmapsto \; \vdash \{\mathcal{P}'\} \, \mathcal{C} \, \{\mathcal{Q}\}$$

`POST_WEAK_RULE` is illustrated in the sessions with HOL below.

In the box below, the predefined constant `MAX` and lemma `MAX_LEMMA1` are used.

```
#MAX;;
⊢ MAX(m,n) = (m > n → m | n)

#let hth7 = ASSIGN_AX "{R = MAX(x,y)}" "R := Y" ;;
hth7 = ⊢ {Y = MAX(x,y)} R := Y {R = MAX(x,y)}

#MAX_LEMMA1;;
⊢ ((X = x) ∧ (Y = y)) ∧ Y > X ⇒ (Y = MAX(x,y))

#let hth8 = PRE_STRENGTH_RULE(MAX_LEMMA1,hth7);;
hth8 = ⊢ {((X = x) ∧ (Y = y)) ∧ Y > X} R := Y {R = MAX(x,y)}
```

The ML function IF_RULE has type thm # thm -> thm and implements the **if**-rule.

$$\text{IF\_RULE}\,(\,\vdash\,\{\mathcal{P} \wedge \mathcal{B}\}\,\mathcal{C}_1\,\{\mathcal{Q}\},\ \vdash\,\{\mathcal{P} \wedge \neg\mathcal{B}\}\,\mathcal{C}_2\,\{\mathcal{Q}\})$$
$$\mapsto$$
$$\vdash\,\{\mathcal{P}\}\ \text{if}\ \mathcal{B}\ \text{then}\ \mathcal{C}_1\ \text{else}\ \mathcal{C}_2\ \{\mathcal{Q}\}$$

MAX_LEMMA2 is used in the next box; it is a pre-proved lemma about MAX.

```
#let hth9 = ASSIGN_AX "{R = MAX(x,y)}" "R := X" ;;
hth9 = ⊢ {X = MAX(x,y)} R := X {R = MAX(x,y)}

#MAX_LEMMA2;;
⊢ ((X = x) ∧ (Y = y)) ∧ ¬ Y > X ⇒ (X = MAX(x,y))

#let hth10 = PRE_STRENGTH_RULE(MAX_LEMMA2,hth9);;
hth10 = ⊢ {((X = x) ∧ (Y = y)) ∧ ¬ Y > X} R := X {R = MAX(x,y)}

#let hth11 = IF_RULE(hth8,hth10);;
hth11 =
⊢ {(X = x) ∧ (Y = y)}
    if Y > X then R := Y else R := X
  {R = MAX(x,y)}
```

The ML function WHILE_RULE has type thm -> thm and implements the **while**-rule.

$$\text{WHILE\_RULE} \vdash \{\mathcal{P} \wedge \mathcal{B}\}\,\mathcal{C}\,\{\mathcal{P}\} \mapsto \vdash \{\mathcal{P}\}\ \text{while}\ \mathcal{B}\ \text{do}\ \mathcal{C}\ \{\mathcal{P} \wedge \neg\mathcal{B}\}$$

To illustrate the **while**-rule, a HOL transcript of the example proof in Section 3.2 is now given. The **while**-rule is used to prove Th9 below. For completeness, all the proofs of the pure logic theorems that are needed are included. These are performed using tactics; readers unfamiliar with tactics should either skip the proofs of Th7, Th11 and Th12, or read Section 8 below.

```
#let Th1 = ASSIGN_AX "{X = R + (Y × 0)}" "R := X" ;;
Th1 = ⊢ {X = X + (Y × 0)} R := X {X = R + (Y × 0)}

#let Th2 = ASSIGN_AX "{X = R + (Y × Q)}" "Q := 0" ;;
Th2 = ⊢ {X = R + (Y × 0)} Q := 0 {X = R + (Y × Q)}

#let Th3 = SEQ_RULE(Th1,Th2);;
Th3 = ⊢ {X = X + (Y × 0)} R := X; Q := 0 {X = R + (Y × Q)}
```

```
#let Th4 = ASSIGN_AX "{X = R + (Y × (Q + 1))}" "R := (R - Y)" ;;
Th4 =
⊢ {X = (R - Y) + (Y × (Q + 1))}
    R := R - Y
  {X = R + (Y × (Q + 1))}

#let Th5 = ASSIGN_AX "{X = R + (Y × Q)}" "Q := (Q + 1)" ;;
Th5 = ⊢ {X = R + (Y × (Q + 1))} Q := Q + 1 {X = R + (Y × Q)}

#let Th6 = SEQ_RULE(Th4,Th5);;
Th6 =
⊢ {X = (R - Y) + (Y × (Q + 1))}
    R := R - Y; Q := Q + 1
  {X = R + (Y × Q)}
```

In the next box, a simple arithmetical fact called Th7 is proved. Some systems can prove such facts fully automatically [2]; unfortunately this is not so with HOL and the user must supply a proof outline expressed as a *tactic*. Tactics are described in Section 8 below, and readers unfamiliar with them might want to read that section now. To make this paper less dependent on detailed knowledge of HOL's particular repertoire of tactics, some of the ML code used in the sessions that follow has been replaced by English descriptions. For example, the actual code for proving Th7 is:

```
#let Th7 =
# TAC_PROOF
#  (([], "(X = R + (Y * Q)) /\ (Y <= R)
#          ==> (X = (R - Y) + (Y * (Q + 1)))"),
#    REPEAT STRIP_TAC
#      THEN REWRITE_TAC[LEFT_ADD_DISTRIB;MULT_CLAUSES]
#      THEN ONCE_REWRITE_TAC[SPEC "Y*Q" ADD_SYM]
#      THEN ONCE_REWRITE_TAC[ADD_ASSOC]
#      THEN IMP_RES_TAC SUB_ADD
#      THEN ASM_REWRITE_TAC[]);;

Th7 = |- (X = R + (Y * Q)) /\ Y <= R ==> (X = (R - Y) + (Y * (Q + 1)))
```

which can be read informally as:

```
#let Th7 =
# TAC_PROOF
#  (([], "(X = R + (Y × Q)) ∧ (Y ≤ R)
#          ⇒ (X = (R - Y) + (Y × (Q + 1)))"),
#    'Move conjuncts of antecedent of implication to assumption list'
#    THEN 'Simplify using lemmas about + and ×'
#    THEN 'Expand assumptions using ⊢ n ≤ m ⇒ (m - n) + n = m '
#    THEN 'Simplify using assumptions' );;

Th7 = ⊢ (X = R + (Y × Q)) ∧ Y ≤ R ⇒ (X = (R - Y) + (Y × (Q + 1)))
```

Subsequent sessions will contain similar informal English descriptions of tactics, rather than exact ML code.

Continuing our session: if Th7 is used to strengthen the precondition of Th6, the result is then a suitable hypothesis for the **while**-rule.

```
#let Th8 = PRE_STRENGTH_RULE(Th7,Th6);;
Th8 =
⊢ {(X = R + (Y × Q)) ∧ Y ≤ R}
    R := R − Y; Q := Q + 1
  {X = R + (Y × Q)}

#let Th9 = WHILE_RULE Th8;;
Th9 =
⊢ {X = R + (Y × Q)}
    while Y ≤ R do R := R − Y; Q := Q + 1
  {(X = R + (Y × Q)) ∧ ¬ Y ≤ R}

#let Th10 = SEQ_RULE(Th3,Th9);;
Th10 =
⊢ {X = X + (Y × 0)}
    R := X; Q := 0; while Y ≤ R do R := R − Y; Q := Q + 1
  {(X = R + (Y × Q)) ∧ ¬ Y ≤ R}
```

The proof of Th10 could have been done in a single complicated step.

```
#SEQL_RULE
# [ASSIGN_AX "{X = R + (Y × 0)}" "R := X";
#  ASSIGN_AX "{X = R + (Y × Q)}" "Q := 0";
#  WHILE_RULE
#  (PRE_STRENGTH_RULE
#    (DISTRIB_LEMMA,
#      SEQL_RULE
#      [ASSIGN_AX "{X = R + (Y × (Q + 1))}" "R := (R − Y)";
#        ASSIGN_AX "{X = R + (Y × Q)}" "Q := (Q + 1)"]))];;
⊢ {X = X + (Y × 0)}
    R := X; Q := 0; while Y ≤ R do R := R − Y; Q := Q + 1
  {(X = R + (Y × Q)) ∧ ¬ Y ≤ R}
```

Two lemmas are now proved that will be used to simplify the precondition and postcondition of Th10.

```
#let Th11 =
# TAC_PROOF
#  (([],"T ⇒ (X = X + (Y × 0))"),
#   'Simplify' );;
Th11 = ⊢ T ⇒ (X = X + (Y × 0))

#let Th12 =
# TAC_PROOF
#  (([],"(X = R + (Y × Q)) ∧ ¬ Y ≤ R
#    ⇒ (X = R + (Y × Q)) ∧ R < Y"),
#   'Move antecedent of implication to assumption list'
#   THEN 'Simplify using assumptions and ⊢ ¬(m < n = n ≤ m)'
Th12 = ⊢ (X = R + (Y × Q)) ∧ ¬ Y ≤ R ⇒ (X = R + (Y × Q)) ∧ R < Y
```

The pre and postconditions of Th10 can now be simplified in a single step.

```
#let Th13 = POST_WEAK_RULE(PRE_STRENGTH_RULE(Th11,Th10),Th12);;
Th13 =
⊢ {T}
    R := X; Q := 0; while Y ≤ R do R := R − Y; Q := Q + 1
  {(X = R + (Y × Q)) ∧ R < Y}
```

This completes the mechanical generation of the proof in Section 3.2.

The proof of Th13 could be done by rewriting Th10 without recourse to Hoare logic at all. First the lemma $\vdash \neg(Y \leq R) = (R < Y)$ is proved:

```
#let Th14 =
# TAC_PROOF
#   ((□,"¬ (Y ≤ R) = (R < Y)"),
#    'Simplify using ⊢ ¬(m < n = n ≤ m' );;
Th14 = ⊢ ¬ Y ≤ R = R < Y
```

Then Th10 is rewritten using Th14 and elementary properties of addition and multiplication.

```
# 'Simplify Th10 using Th14 and properties of + and ×' ;;
⊢ {T}
    R := X; Q := 0; while Y ≤ R do R := R − Y; Q := Q + 1
  {(X = R + (Y × Q)) ∧ R < Y}
```

To see how this works let us look at the 'deep structure' of Th10.

```
#pretty_off();;
() : void

#Th10;;
⊢ Spec
   ((λs. s 'X' = (s 'X') + ((s 'Y') × 0)),
    Seq(Seq(Assign('R',(λs. s 'X')),Assign('Q',(λs. 0))),
        While
        ((λs. (s 'Y') ≤ (s 'R')),
         Seq
         (Assign('R',(λs. (s 'R') − (s 'Y'))),
          Assign('Q',(λs. (s 'Q') + 1))))),
    (λs. (s 'X' = (s 'R') + ((s 'Y') × (s 'Q'))) ∧
         ¬ (s 'Y') ≤ (s 'R')))
```

Rewriting Th10 with $\vdash m \times 0 = 0$ will replace (s 'Y') × 0 by 0. Rewriting with $\vdash m + 0 = m$ and Th14 works similarly.

```
# 'Simplify using Th14 and properties of + and ×' ;;
⊢ Spec
   ((λs. T),
    Seq(Seq(Assign('R',(λs. s 'X')),Assign('Q',(λs. 0))),
        While
        ((λs. (s 'Y') ≤ (s 'R')),
         Seq
         (Assign('R',(λs. (s 'R') − (s 'Y'))),
          Assign('Q',(λs. (s 'Q') + 1))))),
    (λs. (s 'X' = (s 'R') + ((s 'Y') × (s 'Q'))) ∧
         (s 'R') < (s 'Y')))
```

Thus although the pretty-printer makes it look as though Y ×0 is rewritten to 0, what actually happens is that (s 'Y') × 0 is rewritten to 0. This direct application of a HOL theorem proving tool to the semantics of a partial correctness specification illustrates how reasoning with HOL tools can be mixed with reasoning in Hoare logic. We suspect such mixtures of 'axiomatic' and 'semantic' reasoning to be quite powerful.

In Section 9 it is shown how tactics can be formulated that correspond to reasoning based on verification conditions. This enables all HOL's infrastructure for goal directed proof to be brought to bear on partial correctness specifications. Before describing this, here is a quick review of tactics and tacticals abridged from [13]. Readers familiar with Cambridge LCF or HOL should skip this section.

8 Introduction to Tactics and Tacticals

A *tactic* is an ML function which is applied to a goal to reduce it to subgoals. A *tactical* is a (higher-order) ML function for combining tactics to build new tactics[13].

For example, if T_1 and T_2 are tactics, then the ML expression T_1 THEN T_2 evaluates to a tactic which first applies T_1 to a goal and then applies T_2 to each subgoal produced by T_1. The tactical THEN is an infixed ML function.

8.1 Tactics

It simplifies the description of tactics if the following ML type abbreviations are used:

```
proof       = thm list -> thm
subgoals    = goal list # proof
tactic      = goal -> subgoals
```

If T is a tactic and g is a goal, then applying T to g (*i.e.* evaluating the ML expression T g) will result in an object of ML type subgoals, *i.e.* a pair whose first component is a list of goals and whose second component has ML type proof.

Suppose T g = $([g_1;\ldots;g_n],p)$. The idea is that g_1 , $\ldots$, g_n are subgoals and p is a 'validation' of the reduction of goal g to subgoals g_1 , $\ldots$, g_n. Suppose further that the subgoals g_1 , $\ldots$, g_n have been solved. This would mean that theorems th_1 , $\ldots$, th_n have been proved such that each th_i 'achieves' the goal g_i. The validation p (produced by applying T to g) is an ML function which when applied to the list $[th_1;\ldots;th_n]$ returns a theorem, th, which 'achieves' the original goal g. Thus p is a function for converting a solution of the subgoals to a solution of the original goal. If p does this successfully, then the tactic T is called *valid*. Invalid tactics cannot result in the proof of invalid theorems; the worst they can do is result in insolvable goals or unintended theorems being proved. If T were invalid and were used to reduce goal g to subgoals g_1 , $\ldots$, g_n, then a lot of effort might be put into into proving theorems th_1 , $\ldots$, th_n achieving g_1 , $\ldots$, g_n, only to find that these theorems are useless because $p[th_1;\ldots;th_n]$ doesn't achieve g (*i.e.* it fails, or else it achieves some other goal).

A theorem *achieves* a goal if the assumptions of the theorem are included in the assumptions of the goal *and* if the conclusion of the theorem is equal (up to renaming of bound variables) to the conclusion of the goal. More precisely, a theorem $t_1, \ldots$, t_m |- t achieves a goal $([u_1;\ldots;u_n],u)$ if and only if t_1 , $\ldots$, t_m are included among u_1 , $\ldots$, u_n and t is equal to u (up to renaming of bound variables). For

---

[13]The terms 'tactic' and 'tactical' are due to Robin Milner, who invented the concepts.

example, the goal (["x=y";"y=z";"z=w"],"x=z") is achieved by the theorem x=y, y=z ⊢ x=z (the assumption "z=w" is not needed).

A tactic *solves* a goal if it reduces the goal to the empty list of subgoals. Thus T solves g if T g = ([],p). If this is the case, and if T is valid, then p[] will evaluate to a theorem achieving g. Thus if T solves g then the ML expression snd(T g)[] evaluates to a theorem achieving g.

Tactics are specified using the following notation:

$$\frac{goal}{goal_1 \quad goal_2 \quad \ldots \quad goal_n}$$

For example, the tactic CONJ_TAC is specified by

$$\frac{t_1 \;/\backslash\; t_2}{t_1 \qquad t_2}$$

CONJ_TAC reduces a goal of the form (Γ,"$t_1/\backslash t_2$") to subgoals (Γ,"t_1") and (Γ,"t_2"). The fact that the assumptions of the top-level goal are propagated unchanged to the two subgoals is indicated by the absence of assumptions in the notation.

Tactics generally 'fail' (in the ML sense [13]) if they are applied to inappropriate goals. For example, CONJ_TAC will fail if it is applied to a goal whose conclusion is not a conjunction.

8.2 Using Tactics to Prove Theorems

Suppose a goal g is to be solved. If g is simple it might be possible to think up a tactic, T say, which reduces it to the empty list of subgoals. If this is the case then executing

$$\text{let } gl,p = T \; g$$

will bind p to a function which when applied to the empty list of theorems yields a theorem th achieving g. The declaration above will also bind gl to the empty list of goals. Executing

$$\text{let } th = p \; []$$

will thus bind th to a theorem achieving g.

To simplify the use of tactics, there is a standard function called TAC_PROOF with type goal # tactic -> thm such that evaluating

$$\text{TAC\_PROOF}(g,T)$$

proves the goal g using tactic T and returns the resulting theorem (or fails, if T does not solve g).

When conducting a proof that involves many subgoals and tactics, it is necessary to keep track of all the validations and compose them in the correct order. While this is feasible even in large proofs, it is tedious. HOL provides a package for building and traversing the tree of subgoals, stacking the validations and applying them when subgoals are solved; this package was originally implemented for LCF by Larry Paulson [30].

The subgoal package implements a simple framework for interactive proof in which proof trees can be created and traversed in a top-down fashion. Using a tactic, the current goal is expanded into subgoals and a validation, which are automatically pushed onto the goal stack. Subgoals can be attacked in any order. If the tactic solves the goal (i.e. returns an empty subgoal list), then the package proceeds to the next goal in the tree.

The function `goal` has type `term -> void` and takes a term t and then sets up the goal $([],t)$.

The function `expand` has type `tactic -> void` and applies a tactic to the top goal on the stack, then pushes the resulting subgoals onto the stack, then prints the resulting subgoals. If there are no subgoals, the validation is applied to the theorems solving the subgoals that have been proved and the resulting theorems are printed.

8.3 Tacticals

A *tactical* is an ML function that returns a tactic (or tactics) as result. Tacticals may take various parameters; this is reflected in the various ML types that the built-in tacticals have. The tacticals used in this paper are:

```
ORELSE : tactic -> tactic -> tactic
```

The tactical `ORELSE` is an ML infix. If T_1 and T_2 are tactics, then the ML expression T_1 `ORELSE` T_2 evaluates to a tactic which first tries T_1 and then if T_1 fails it tries T_2.

```
THEN : tactic -> tactic -> tactic
```

The tactical `THEN` is an ML infix. If T_1 and T_2 are tactics, then the ML expression T_1 `THEN` T_2 evaluates to a tactic which first applies T_1 and then applies T_2 to all the subgoals produced by T_1.

```
THENL : tactic -> tactic list -> tactic
```

If T is a tactic which produces n subgoals and $T_1, \ldots, T_n$ are tactics, then T `THENL` $[T_1;\ldots;T_n]$ is a tactic which first applies T and then applies T_i to the ith subgoal produced by T. The tactical `THENL` is useful for doing different things to different subgoals.

```
REPEAT : tactic -> tactic
```

If T is a tactic then `REPEAT` T is a tactic which repeatedly applies T until it fails.

9 Verification Conditions via Tactics

If one wants to prove the partial correctness specification $\{\mathcal{P}\}\ \mathcal{V}{:=}\mathcal{E}\ \{\mathcal{Q}\}$ then, by the assignment axiom and precondition strengthening, it is clearly sufficient to prove the pure logic theorem $\vdash \mathcal{P} \Rightarrow \mathcal{Q}[\mathcal{E}/\mathcal{V}]$. The formula $\mathcal{P} \Rightarrow \mathcal{Q}[\mathcal{E}/\mathcal{V}]$ is called the *verification condition* for $\{\mathcal{P}\}\ \mathcal{V}:=\mathcal{E}\ \{\mathcal{Q}\}$.

More generally, the verification conditions for $\{\mathcal{P}\}\ \mathcal{C}\ \{\mathcal{Q}\}$ are a set of pure logic formulae $\mathcal{F}_1, \ldots, \mathcal{F}_n$ such that if $\vdash \mathcal{F}_1, \ldots, \vdash \mathcal{F}_n$ are theorems of pure logic then $\vdash \{\mathcal{P}\}\ \mathcal{C}\ \{\mathcal{Q}\}$ is a theorem of Hoare logic. Verification conditions are related to Dijkstra's weakest liberal preconditions [6], see the definition of Wlp in Section 11.2. For example, the weakest liberal precondition of $\mathcal{Q}$ for the assignment command $\mathcal{V}:=$ $\mathcal{E}$ is $\mathcal{Q}[\mathcal{E}/\mathcal{V}]$. The verification conditions for $\{\mathcal{P}\}\ \mathcal{C}\ \{\mathcal{Q}\}$ is $\mathcal{P} \Rightarrow \mathsf{Wlp}(\mathcal{C}, \mathcal{Q})$ and it is possible that the treatment of verification conditions that follows could be improved by formulating it in terms of Wlp.

The generation of verification conditions can be represented by a tactic

$$\frac{\{\mathcal{P}\}\ \mathcal{C}\ \{\mathcal{Q}\}}{\mathcal{F}_1 \quad \mathcal{F}_2 \quad \cdots \quad \mathcal{F}_n}$$

where the validation (proof) part of the tactic is a composition of the functions representing the axioms and rules of Hoare logic. For example, the tactic `ASSIGN_TAC` is:

$$\frac{\{\mathcal{P}\}\ \mathcal{V}:=\mathcal{E}\ \{\mathcal{Q}\}}{\mathcal{P} \Rightarrow \mathcal{Q}[\mathcal{E}/\mathcal{V}]}$$

Here is a little session illustrating `ASSIGN_TAC`.

```
#goal "{X=x} X:=X+1 {X=x+1}" ;;
"{X = x} X := X + 1 {X = x + 1}"

#expand ASSIGN_TAC;;
OK..
"(X = x) ⇒ (X + 1 = x + 1)"

#expand 'Move antecedent to assumptions and then simplify' ;;
OK..
goal proved
⊢ (X = x) ⇒ (X + 1 = x + 1)
⊢ {X = x} X := X + 1 {X = x + 1}

Previous subproof:
goal proved
```

Each of the Hoare axioms and rules can be 'inverted' into a tactic which accomplishes one step in the process of verification condition generation. The composition of these tactics then results in a complete verification condition generator.

The specification of these tactics is as follows.

`SKIP_TAC : tactic`

$$\frac{}{\{\mathcal{P}\}\ \mathbf{skip}\ \{\mathcal{P}\}}$$

`SKIP_TAC` solves all goals of the form $\{\mathcal{P}\}\ \mathbf{skip}\ \{\mathcal{P}\}$.

```
ASSIGN_TAC : tactic
```

$$\frac{\{\mathcal{P}\}\ \mathcal{V}:=\ \mathcal{E}\ \{\mathcal{Q}\}}{\mathcal{P}\Rightarrow\mathcal{Q}[\mathcal{E}/\mathcal{V}]}$$

```
SEQ_TAC : tactic
```

To make generating verification conditions for sequences simple, it is convenient to require annotations to be inserted in sequences $C_1;C_2$ when C_2 is *not* an assignment [14]. Such an annotation will be of the form assert$\{\mathcal{R}\}$, where $\mathcal{R}$ is a pure logic formula. For such sequences SEQ_TAC is as follows:

$$\frac{\{\mathcal{P}\}\ C_1;\ \text{assert}\{\mathcal{R}\};\ C_2\ \{\mathcal{Q}\}}{\{\mathcal{P}\}C_1\{\mathcal{R}\}\qquad\{\mathcal{R}\}C_2\{\mathcal{Q}\}}$$

In the case that C_2 is an assignment $\mathcal{V}:=\mathcal{E}$, the postcondition $\mathcal{Q}$ can be 'passed through' C_2, using the assignment axiom, to automatically generate the assertion $\mathcal{Q}[\mathcal{E}/\mathcal{V}]$. Thus in this case SEQ_TAC simplifies to:

$$\frac{\{\mathcal{P}\}\ C;\ \mathcal{V}:=\ \mathcal{E}\ \{\mathcal{Q}\}}{\{\mathcal{P}\}\ C\ \{\mathcal{Q}[\mathcal{E}/\mathcal{V}]\}}$$

```
IF_TAC : tactic
```

$$\frac{\{\mathcal{P}\}\ \text{if}\ \mathcal{B}\ \text{then}\ C_1\ \text{else}\ C_2\ \{\mathcal{Q}\}}{\{\mathcal{P}\wedge\mathcal{B}\}\ C_1\ \{\mathcal{Q}\}\qquad\{\mathcal{P}\wedge\neg\mathcal{B}\}\ C_2\ \{\mathcal{Q}\}}$$

```
WHILE_TAC : tactic
```

In order for verification conditions to be generated from **while**-commands, it is necessary to specify an invariant [14] by requiring that an *annotation* invariant$\{\mathcal{R}\}$, where $\mathcal{R}$ is the invariant, be inserted just after the do.

$$\frac{\{\mathcal{P}\}\ \text{while}\ \mathcal{B}\ \text{do invariant}\ \{\mathcal{R}\};\ C\ \{\mathcal{Q}\}}{\mathcal{P}\Rightarrow\mathcal{R}\qquad\{\mathcal{R}\wedge\mathcal{B}\}\ C\ \{\mathcal{R}\}\qquad\mathcal{R}\wedge\neg\mathcal{B}\Rightarrow\mathcal{Q}}$$

Here now is a continuation of the session started above in which tactics for Hoare logic are illustrated. First the individual tactics are illustrated, then it is shown how they can be combined, using tacticals, into a single verification condition generator.

The first step is to set up the specification of the simple division program as a goal.

```
#pretty_on();;
() : void

#goal "{T}
#       R:=X;
#       Q:=0;
#       assert{(R = X) ∧ (Q = 0)};
#       while Y ≤ R
#         do (invariant{X = (R + (Y × Q))};
#            R := R - Y; Q := Q + 1)
#       {(X = (R + (Y × Q))) ∧ (R < Y)}" ;;
"{T}
  R := X;
  Q := 0;
  assert{(R = X) ∧ (Q = 0)};
  while Y ≤ R do invariant{X = R + (Y × Q)}; R := R - Y; Q := Q + 1
  {(X = R + (Y × Q)) ∧ R < Y}"
```

The command in this goal is a sequence, so SEQ_TAC can be applied.

```
#expand SEQ_TAC;;
OK..
2 subgoals
"{(R = X) ∧ (Q = 0)}
  while Y ≤ R do invariant{X = R + (Y × Q)};
   R := R - Y; Q := Q + 1
  {(X = R + (Y × Q)) ∧ R < Y}"

"{T} R := X; Q := 0 {(R = X) ∧ (Q = 0)}"
```

The top goal is printed last; SEQ_TAC can be applied to it.

```
#expand SEQ_TAC;;
OK..
"{T} R := X {(R = X) ∧ (0 = 0)}"

#expand ASSIGN_TAC;;
OK..
"T ⇒ (X = X) ∧ (0 = 0)"
```

The goal T⇒(X =X) ∧(0 =0) is solved by rewriting with standard facts. The subgoal package prints out the theorems produced and backs up to the next pending subgoal.

```
#expand 'Simplify' ;;
OK..
goal proved
⊢ T ⇒ (X = X) ∧ (0 = 0)
⊢ {T} R := X {(R = X) ∧ (0 = 0)}
⊢ {T} R := X; Q := 0 {(R = X) ∧ (Q = 0)}

Previous subproof:
"{(R = X) ∧ (Q = 0)}
  while Y ≤ R do invariant{X = R + (Y × Q)};
   R := R - Y; Q := Q + 1
  {R < Y ∧ (X = R + (Y × Q))}"
```

The remaining subgoal is a **while**-command, so WHILE_TAC is applied to get three subgoals.

```
#expand WHILE_TAC;;
OK..
3 subgoals
"(X = R + (Y × Q)) ∧ ¬ Y ≤ R ⇒ (X = R + (Y × Q)) ∧ R < Y"

"{(X = R + (Y × Q)) ∧ Y ≤ R}
  R := R - Y; Q := Q + 1
 {X = R + (Y × Q)}"

"(R = X) ∧ (Q = 0) ⇒ (X = R + (Y × Q))"
```

The first subgoal (i.e. the one printed last) is quickly solved by moving the antecedent of the implication to the assumptions, and then rewriting using standard properties of + and ×.

```
#expand   'Move antecedent to assumptions' ;;
OK..
"X = R + (Y × Q)"
    [ "R = X" ]
    [ "Q = 0" ]

#expand   'Simplify using assumptions' ;;
OK..
goal proved
.. ⊢ X = R + (Y × Q)
⊢ (R = X) ∧ (Q = 0) ⇒ (X = R + (Y × Q))

Previous subproof:
2 subgoals
"(X = R + (Y × Q)) ∧ ¬ Y ≤ R ⇒ (X = R + (Y × Q)) ∧ R < Y"

"{(X = R + (Y × Q)) ∧ Y ≤ R}
  R := R - Y; Q := Q + 1
 {X = R + (Y × Q)}"
```

The first subgoal is a sequence, so SEQ_TAC is applied. This results in an assignment, so ASSIGN_TAC is then applied. The resulting purely logical subgoal is the already proved Th7.

```
#expand SEQ_TAC;;
OK..
"{(X = R + (Y × Q)) ∧ Y ≤ R} R := R - Y {X = R + (Y × (Q + 1))}"

#expand ASSIGN_TAC;;
OK..
"(X = R + (Y × Q)) ∧ Y ≤ R ⇒ (X = (R - Y) + (Y × (Q + 1)))"
```

Th7 is used to solve the first subgoal.

```
#expand 'Use Th7 ' ;;
OK..
goal proved
⊢ (X = R + (Y × Q)) ∧ Y ≤ R ⇒ (X = (R − Y) + (Y × (Q + 1)))
⊢ {(X = R + (Y × Q)) ∧ Y ≤ R}
    R := R − Y
  {X = R + (Y × (Q + 1))}
⊢ {(X = R + (Y × Q)) ∧ Y ≤ R}
    R := R − Y; Q := Q + 1
  {X = R + (Y × Q)}

Previous subproof:
"(X = R + (Y × Q)) ∧ ¬ Y ≤ R ⇒ (X = R + (Y × Q)) ∧ R < Y"
```

The previously proved theorem Th12 solves the remaining subgoal.

```
#expand 'Use Th12 ' ;;
OK..
goal proved
⊢ (X = R + (Y × Q)) ∧ ¬ Y ≤ R ⇒ (X = R + (Y × Q)) ∧ R < Y
⊢ {(R = X) ∧ (Q = 0)}
    while Y ≤ R do R := R − Y; Q := Q + 1
  {(X = R + (Y × Q)) ∧ R < Y}
⊢ {T}
    R := X; Q := 0; while Y ≤ R do R := R − Y; Q := Q + 1
  {(X = R + (Y × Q)) ∧ R < Y}

Previous subproof:
goal proved
```

The tactics ASSIGN_TAC, SEQ_TAC, IF_TAC and WHILE_TAC can be combined into a single tactic, called VC_TAC below, which generates the verification conditions in a single step.

The definition of VC_TAC in ML is given in the next box; it uses the tacticals REPEAT and ORELSE described in Section 8.3.

```
#let VC_TAC = REPEAT(ASSIGN_TAC
#                     ORELSE SEQ_TAC
#                     ORELSE IF_TAC
#                     ORELSE WHILE_TAC);;
VC_TAC = - : tactic
```

This compound tactic is illustrated by using it to repeat the proof just done. The original goal is made the top goal, and then this is expanded using VC_TAC. The result is four verification conditions.

```
#goal "{T}
#       R:=X;
#       Q:=0;
#       assert{(R = X) ∧ (Q = 0)};
#       while Y ≤ R
#         do (invariant{X = (R + (Y × Q))};
#            R := R − Y; Q := Q + 1)
#      {(X = (R + (Y × Q))) ∧ (R < Y)}" ;;
"{T}
  R := X;
  Q := 0;
  assert{(R = X) ∧ (Q = 0)};
  while Y ≤ R do invariant{X = R + (Y × Q)};
   R := R − Y; Q := Q + 1
 {(X = R + (Y × Q)) ∧ R < Y}"

#expand VC_TAC;;
OK..
4 subgoals
"(X = R + (Y × Q)) ∧ ¬ Y ≤ R ⇒ (X = R + (Y × Q)) ∧ R < Y"

"(X = R + (Y × Q)) ∧ Y ≤ R ⇒ (X = (R − Y) + (Y × (Q + 1)))"

"(R = X) ∧ (Q = 0) ⇒ (X = R + (Y × Q))"

"T ⇒ (X = X) ∧ (0 = 0)"
```

Notice how `VC_TAC` converts the goal of proving a Hoare specification into pure logic subgoals. These can be solved as above.

```
#expand 'Simplify' ;;
OK..
goal proved
⊢ T ⇒ (X = X) ∧ (0 = 0)

Previous subproof:
3 subgoals
"(X = R + (Y × Q)) ∧ ¬ Y ≤ R ⇒ (X = R + (Y × Q)) ∧ R < Y"

"(X = R + (Y × Q)) ∧ Y ≤ R ⇒ (X = (R − Y) + (Y × (Q + 1)))"

"(R = X) ∧ (Q = 0) ⇒ (X = R + (Y × Q))"
```

```
#expand 'Move antecedent to assumptions and then rewrite' ;;
OK..
goal proved
⊢ (R = X) ∧ (Q = 0) ⇒ (X = R + (Y × Q))

Previous subproof:
2 subgoals
"(X = R + (Y × Q)) ∧ ¬ Y ≤ R ⇒ (X = R + (Y × Q)) ∧ R < Y"

"(X = R + (Y × Q)) ∧ Y ≤ R ⇒ (X = (R − Y) + (Y × (Q + 1)))"
```

```
#expand 'Use Th7 ' ;;
OK..
goal proved
⊢ (X = R + (Y × Q)) ∧ Y ≤ R ⇒ (X = (R − Y) + (Y × (Q + 1)))

Previous subproof:
"(X = R + (Y × Q)) ∧ ¬ Y ≤ R ⇒ (X = R + (Y × Q)) ∧ R < Y"
```

```
#expand 'Use Th12 ' ;;
OK..
goal proved
⊢ (X = R + (Y × Q)) ∧ ¬ Y ≤ R ⇒ (X = R + (Y × Q)) ∧ R < Y
⊢ {T}
    R := X; Q := 0; while Y ≤ R do R := R − Y; Q := Q + 1
  {(X = R + (Y × Q)) ∧ R < Y}

Previous subproof:
goal proved
```

Finally, here is the proof in one step.

```
#prove
# ("{T},
#        R:=X;
#        Q:=0;
#        assert{(R = X) ∧ (Q = 0)};
#        while Y ≤ R
#        do (invariant{X = (R + (Y × Q))};
#             R:=R − Y; Q:=Q + 1)
#   {(X = (R + (Y × Q))) ∧ R < Y}",
#   VC_TAC
#     THENL
#     [ 'Simplify' ;
#       'Move antecedent to assumptions and then simplify' ;
#       'Use Th7 ' ;
#       'Use Th12 ' ]);;
⊢ {T}
    R := X; Q := 0; while Y ≤ R do R := R − Y; Q := Q + 1
  {(X = R + (Y × Q)) ∧ R < Y}
```

10   Termination and Total Correctness

Hoare logic is usually presented as a self-contained calculus. However, if it is re-garded as a derived logic, as it is here, then it's easy to add extensions and modi-fications without fear of introducing unsoundness. To illustrate this, we will sketch how termination assertions can be added, and how these can be used to prove total correctness.

A *termination assertion* is a formula $\mathsf{Halts}(\llbracket \mathcal{P} \rrbracket, \llbracket \mathcal{C} \rrbracket)$, where the constant Halts is defined by:

$$\mathsf{Halts}(p, c) \ = \ \forall s_1. \ p \ s_1 \ \Rightarrow \ \exists s_2. \ c(s_1, s_2)$$

Notice that this says that c 'halts' under precondition p if there is *some* final state for each initial state satisfying p. For example, although **while** T **do skip** does not terminate, the definition above suggests that (**while** T **do skip**) ∥ **skip** does, since:

$$\vdash \ \mathsf{Halts}(\llbracket\mathsf{T}\rrbracket, \ \mathsf{Choose}(\llbracket\textbf{while T do skip}\rrbracket, \ \llbracket\textbf{skip}\rrbracket))$$

(∥ and **Choose** are described in Section 5). The meaning of $\mathsf{Halts}(\llbracket\mathcal{P}\rrbracket, \llbracket\mathcal{C}\rrbracket)$ is 'some computation of $\mathcal{C}$ starting from a state satisfying $\mathcal{P}$ terminates' this is quite different from 'every computation of $\mathcal{C}$ starting from a state satisfying $\mathcal{P}$ terminates'. The latter stronger kind of termination requires a more complex kind of semantics for its formalization (e.g. one using powerdomains [31]). If commands are deterministic, then termination is adequately formalized by Halts. It is intuitively clear (and can be proved using the methods described in Melham's paper [25]) that the relations denoted by commands in our little language (*not* including ∥) are partial functions. If **Det** is defined by:

$$\mathsf{Det} \ c \ = \ \forall s \ s_1 \ s_2. \ c(s, s_1) \ \wedge \ c(s, s_2) \ \Rightarrow \ (s_1 = s_2)$$

then for any command $\mathcal{C}$ it can be proved that $\vdash$ **Det** $\llbracket\mathcal{C}\rrbracket$. This fact will be needed to show that the formalization of weakest preconditions in Section 11.2 is correct.

The informal equation

$$\text{Total correctness} = \text{Termination} + \text{Partial correctness.}$$

can be implemented by defining:

$$\mathsf{Total\_Spec}(p, c, q) \ = \ \mathsf{Halts}(p, c) \ \wedge \ \mathsf{Spec}(p, c, q)$$

Then $[\mathcal{P}] \ \mathcal{C} \ [\mathcal{Q}]$ is represented by $\mathsf{Total\_Spec}(\llbracket\mathcal{P}\rrbracket, \llbracket\mathcal{C}\rrbracket, \llbracket\mathcal{Q}\rrbracket)$.

From the definition of Halts it is straightforward to prove the following theorems:

T1. $\quad \vdash \ \forall p. \ \mathsf{Halts}(p, \mathsf{Skip})$

T2. $\quad \vdash \ \forall p \ v \ e. \ \mathsf{Halts}(p, \mathsf{Assign}(v, e))$

T3. $\quad \vdash \ \forall p \ p' \ c. \ (\forall s. \ p' \ s \ \Rightarrow \ p \ s) \ \wedge \ \mathsf{Halts}(p, c) \ \Rightarrow \ \mathsf{Halts}(p', c)$

T4. $\quad \vdash \ \forall p \ c_1 \ c_2 \ q. \ \mathsf{Halts}(p, c_1) \ \wedge \ \mathsf{Spec}(p, c_1, q) \ \wedge \ \mathsf{Halts}(q, c_2)$
$$\Rightarrow \mathsf{Halts}(p, \mathsf{Seq}(c_1, c_2))$$

T5. $\quad \vdash \ \forall p \ c_1 \ c_2 \ b. \ \mathsf{Halts}(p, c_1) \ \wedge \ \mathsf{Halts}(p, c_2) \ \Rightarrow \ \mathsf{Halts}(p, \mathsf{If}(b, c_1, c_2))$

T6. $\quad \vdash \ \forall b \ c \ x.$
$$(\forall n. \ \mathsf{Spec}((\lambda s. \ p \ s \ \wedge \ b \ s \ \wedge \ (s \ x = n)), c, (\lambda s. \ p \ s \ \wedge \ s \ x < n)))$$
$$\wedge \ \mathsf{Halts}((\lambda s. \ p \ s \ \wedge \ b \ s), c)$$
$$\Rightarrow \mathsf{Halts}(p, \mathsf{While}(b, c))$$

Although these theorems are fairly obvious, when I first wrote them down I got a few details wrong. These errors soon emerged when the proofs were done using the HOL system.

T6 shows that if x is a *variant*, i.e. a variable whose value decreases each time 'around the loop', then the **while**-command halts. Proving this in HOL was much harder than any of the other theorems (but was still essentially routine).

10.1   Derived Rules for Total Correctness

Using T1 – T6 above and H1 – H7 of Section 6, it is straightforward to apply the methods described in Section 6 to implement the derived rules for total correctness shown below. These are identical to the corresponding rules for partial correctness except for having '[' and ']' instead of '{' and '}' respectively.

$$\vdash [\mathcal{P}] \text{ skip } [\mathcal{P}]$$

$$\frac{\vdash \mathcal{P}' \Rightarrow \mathcal{P} \qquad \vdash [\mathcal{P}] \, \mathcal{C} \, [\mathcal{Q}]}{\vdash [\mathcal{P}'] \, \mathcal{C} \, [\mathcal{Q}]}$$

$$\frac{\vdash [\mathcal{P}] \, \mathcal{C} \, [\mathcal{Q}] \qquad \vdash \mathcal{Q} \Rightarrow \mathcal{Q}'}{\vdash [\mathcal{P}] \, \mathcal{C} \, [\mathcal{Q}']}$$

$$\frac{\vdash [\mathcal{P}] \, \mathcal{C}_1 \, [\mathcal{Q}] \qquad \vdash [\mathcal{Q}] \, \mathcal{C}_2 \, [\mathcal{R}]}{\vdash [\mathcal{P}] \, \mathcal{C}_1; \, \mathcal{C}_2 \, [\mathcal{R}]}$$

$$\frac{\vdash [\mathcal{P} \wedge \mathcal{B}] \, \mathcal{C}_1 \, [\mathcal{Q}] \qquad \vdash [\mathcal{P} \wedge \neg \mathcal{B}] \, \mathcal{C}_2 \, [\mathcal{Q}]}{\vdash [\mathcal{P}] \text{ if } \mathcal{B} \text{ then } \mathcal{C}_1 \text{ else } \mathcal{C}_2 \, [\mathcal{Q}]}$$

The total correctness rule for **while**-commands needs a stronger hypothesis than the corresponding one for partial correctness. This is to ensure that the command terminates. For this purpose, a variant is needed in addition to an invariant.

$$\frac{\vdash [\mathcal{P} \wedge \mathcal{B} \wedge (\mathbb{N} = n)] \, \mathcal{C} \, [\mathcal{P} \wedge (\mathbb{N} < n)]}{\vdash [\mathcal{P}] \text{ while } \mathcal{B} \text{ do } \mathcal{C} \, [\mathcal{P} \wedge \neg \mathcal{B}]}$$

Notice that since

$$\mathsf{Total\_Spec}(p, c, q) \;=\; \mathsf{Halts}(p, c) \,\wedge\, \mathsf{Spec}(p, c, q)$$

it is clear that the following rule is valid

$$\frac{\vdash [\mathcal{P}] \, \mathcal{C} \, [\mathcal{Q}]}{\vdash \{\mathcal{P}\} \, \mathcal{C} \, \{\mathcal{Q}\}}$$

The converse to this is only valid if $\mathcal{C}$ contains no **while**-commands. It would be straightforward to implement a HOL derived rule

$$\frac{\vdash \{\mathcal{P}\} \, \mathcal{C} \, \{\mathcal{Q}\}}{\vdash [\mathcal{P}] \, \mathcal{C} \, [\mathcal{Q}]}$$

that would fail (in the ML sense) if $\mathcal{C}$ contained **while**-commands.

10.2   Tactics for Total Correctness

Tactics for total correctness can be implemented that use the derived rules in the previous section as validations. The tactics for everything except **while**-commands are obtained by replacing '[' and ']' by '{' and '}'. Namely:

`SKIP_T_TAC : tactic`

$$\overline{[\mathcal{P}]\ \textbf{skip}\ [\mathcal{P}]}$$

`ASSIGN_T_TAC : tactic`

$$\frac{[\mathcal{P}]\ \mathcal{V}:=\ \mathcal{E}\ [\mathcal{Q}]}{\mathcal{P}\Rightarrow\mathcal{Q}[\mathcal{E}/\mathcal{V}]}$$

`SEQ_T_TAC : tactic`

$$\frac{[\mathcal{P}]\ \mathcal{C}_1;\ \texttt{assert}\{\mathcal{R}\};\ \mathcal{C}_2\ [\mathcal{Q}]}{[\mathcal{P}]\mathcal{C}_1[\mathcal{R}]\qquad[\mathcal{R}]\mathcal{C}_2[\mathcal{Q}]}$$

$$\frac{[\mathcal{P}]\ \mathcal{C};\ \mathcal{V}:=\ \mathcal{E}\ [\mathcal{Q}]}{[\mathcal{P}]\ \mathcal{C}\ [\mathcal{Q}[\mathcal{E}/\mathcal{V}]]}$$

`IF_T_TAC : tactic`

$$\frac{[\mathcal{P}]\ \textbf{if}\ \mathcal{B}\ \textbf{then}\ \mathcal{C}_1\ \textbf{else}\ \mathcal{C}_2\ [\mathcal{Q}]}{[\mathcal{P}\wedge\mathcal{B}]\ \mathcal{C}_1\ [\mathcal{Q}]\qquad[\mathcal{P}\wedge\neg\mathcal{B}]\ \mathcal{C}_2\ [\mathcal{Q}]}$$

`WHILE_T_TAC : tactic`

To enable verification conditions to be generated from **while**-commands they must be annotated with a variant as well as an invariant.

$$\frac{[\mathcal{P}]\ \textbf{while}\ \mathcal{B}\ \textbf{do invariant}\ \{\mathcal{R}\};\ \textbf{variant}\ \{\texttt{N}\};\ \mathcal{C}\ [\mathcal{Q}]}{\mathcal{P}\Rightarrow\mathcal{R}\qquad[\mathcal{R}\wedge\mathcal{B}\wedge(\texttt{N}=n)]\ \mathcal{C}\ [\mathcal{R}\wedge(\texttt{N}<n)]\qquad\mathcal{R}\wedge\neg\mathcal{B}\Rightarrow\mathcal{Q}}$$

To illustrate these tactics, here is a session in which the total correctness of the division program is proved. First suppose that a verification condition generator is defined by:

```
#let VC_T_TAC =
# REPEAT(ASSIGN_T_TAC
#        ORELSE SEQ_T_TAC
#        ORELSE IF_T_TAC
#        ORELSE WHILE_T_TAC);;
VC_T_TAC = - : tactic
```

and the following goal is set up:

```
##pretty_on();;
() : void

#goal "[0 < Y]
#        R := X;
#        Q := 0;
#        assert{(0 < Y) ∧ (R = X) ∧ (Q = 0)};
#        while Y ≤ R
#          do (invariant{(0 < Y) ∧ (X = R + (Y × Q))}; variant{R};
#          R := R − Y; Q := Q + 1)
#      [(X = R + (Y × Q)) ∧ R < Y]" ;;
"[0 < Y]
  R := X;
  Q := 0;
  assert{0 < Y ∧ (R = X) ∧ (Q = 0)};
  while Y ≤ R do
    invariant{0 < Y ∧ (X = R + (Y × Q))};
    variant{R};
    R := R − Y; Q := Q + 1
  [(X = R + (Y × Q)) ∧ R < Y]"

() : void
```

then applying `VC_T_GEN` results in the following four verification conditions:

```
#expand VC_T_TAC;;
OK..
4 subgoals
"(0 < Y ∧ (X = R + (Y × Q))) ∧ ¬ Y ≤ R ⇒
 (X = R + (Y × Q)) ∧ R < Y"

"(0 < Y ∧ (X = R + (Y × Q))) ∧ Y ≤ R ∧ (R = r) ⇒
 (0 < Y ∧ (X = (R − Y) + (Y × (Q + 1)))) ∧ (R − Y) < r"

"0 < Y ∧ (R = X) ∧ (Q = 0) ⇒ 0 < Y ∧ (X = R + (Y × Q))"

"0 < Y ⇒ 0 < Y ∧ (X = X) ∧ (0 = 0)"

() : void
```

These are routine to prove using HOL. Notice that `WHILE_T_TAC` has been implemented so that it automatically generates a logical (or ghost) variable by lowering the case of the variant. In the example above, r is a logical variable generated from the program variable R that is given as the variant.

11 Other Programming Logic Constructs

In this section, three variants on Hoare logic are described.

(i) VDM-style specifications.

(ii) Weakest preconditions.

(iii) Dynamic logic.

None have these have been fully mechanized in HOL, but it is hoped that enough detail is given to show that doing so should be straightforward.

11.1 VDM-Style Specifications

The Vienna Development Method (VDM) [19]) is a formal method for program development which uses a variation on Hoare-style specifications. The VDM notation reduces the need for auxiliary logical variables by providing a way of refering to the initial values of variables in postconditions. For example, the following Hoare-style partial correctness specification:

$$\{X = x \wedge Y = y\}\ R\text{:= } X;\ X\text{:= } Y;\ Y\text{:= } R\ \{Y = x \wedge X = y\}$$

could be written in a VDM-style as:

$$\{\mathsf{T}\}\ R\text{:= } X;\ X\text{:= } Y;\ Y\text{:= } R\ \{Y = \overleftarrow{X} \wedge X = \overleftarrow{Y}\}$$

where $\overleftarrow{X}$ and $\overleftarrow{Y}$ denote the values X and Y had before the three assignments were executed. More generally,

$$\{\mathcal{P}[X_1, \ldots, X_n]\}\ \mathcal{C}\ \{\mathcal{Q}[X_1, \ldots, X_n, \overleftarrow{X_1}, \ldots, \overleftarrow{X_n}]\}$$

can be thought of as an abbreviation for

$$\{\mathcal{P}[X_1, \ldots, X_n] \wedge X_1 = \overleftarrow{X_1} \wedge \ldots \wedge X_n = \overleftarrow{X_n}\}\ \mathcal{C}\ \{\mathcal{Q}[X_1, \ldots, X_n, \overleftarrow{X_1}, \ldots, \overleftarrow{X_n}]\}$$

where $\overleftarrow{X_1}, \ldots, \overleftarrow{X_n}$ are distinct logical variables not occurring in $\mathcal{C}$. It should be straightforward to build a parser and pretty-printer that supports this interpretation of VDM specifications.

It is claimed that VDM specifications are more natural than convential Hoare-style ones. I have not worked with them enough to have an opinion on this, but the point I hope to make here is that there is no problem mechanizing a VDM-style programming logic using the methods in this paper.

Although the meaning of individual VDM specifications is clear, it is not so easy to see what the correct Hoare-like rules of inference are. For example, the sequencing rule must somehow support the deduction of

$$\{\mathsf{T}\}\ X\text{:= } X + 1;\ X\text{:= } X + 1\ \{X = \overleftarrow{X} + 2\}$$

from

$$\{\mathsf{T}\}\ X\text{:= } X + 1\ \{X = \overleftarrow{X} + 1\}$$

There is another semantics of VDM specifications, which Jones attributes to Peter Aczel [19]. This semantics avoids the need for hidden logical variables and also makes it easy to see what the correct rules of inference are. The idea is to regard the postcondition as a binary relation on the initial and final states. This can be formalized by regarding

$$\{\mathcal{P}[X_1, \ldots, X_n]\}\ \mathcal{C}\ \{\mathcal{Q}[X_1, \ldots, X_n, \overleftarrow{X_1}, \ldots, \overleftarrow{X_n}]\}$$

as an abbreviation for

$$\mathsf{VDM\_Spec}(\llbracket \mathcal{P}[X_1, \ldots, X_n] \rrbracket,\ \llbracket \mathcal{C} \rrbracket,\ \llbracket \mathcal{Q}[X_1, \ldots, X_n, \overleftarrow{X_1}, \ldots, \overleftarrow{X_n}] \rrbracket_2)$$

where VDM_Spec is defined by:

$$\text{VDM\_Spec}(p, c, r) = \forall s_1\ s_2.\ p\ s_1\ \wedge\ c(s_1, s_2)\ \Rightarrow\ r(s_1, s_2)$$

and the notation $[\![\ \cdots\ ]\!]_2$ is defined by:

$$[\![\mathcal{Q}[X_1, \ldots, X_n, \overleftarrow{X_1}, \ldots, \overleftarrow{X_n}]]\!]_2 = \\ \lambda(s_1, s_2).\ \mathcal{Q}[s_2{}^\prime X_1{}^\prime, \ldots, s_2{}^\prime X_n{}^\prime, s_1{}^\prime X_1{}^\prime, \ldots, s_1{}^\prime X_n{}^\prime]$$

It is clear that $[\![\ \cdots\ ]\!]_2$ could be supported by a parser and pretty-printer in the same way that $[\![\ \cdots\ ]\!]$ is supported.

The sequencing rule now corresponds to the theorem:

$$\vdash\ \forall p_1\ p_2\ r_1\ r_2\ c_1\ c_2. \\ \text{VDM\_Spec}(p_1,\ c_1,\ \lambda(s_1, s_2).\ p_2\ s_2\ \wedge\ r_1(s_1, s_2))\ \wedge \\ \text{VDM\_Spec}(p_2,\ c_2,\ r_2)\ \Rightarrow \\ \text{VDM\_Spec}(p_1,\ \text{Seq}(c_1, c_2),\ \text{Seq}(r_1, r_2))$$

Example

If $\{T\}\ X := X + 1\ \{X = \overleftarrow{X} + 1\}$ is interpreted as:

$$\text{VDM\_Spec}([\![T]\!],\ [\![X := X + 1]\!],\ [\![X = \overleftarrow{X} + 1]\!]_2)$$

which (since $\vdash\ \forall x.\ T\ \wedge\ x\ =\ x$) implies:

$$\text{VDM\_Spec}([\![T]\!],\ [\![X := X + 1]\!],\ \lambda(s_1, s_2).\ [\![T]\!]s_2\ \wedge\ [\![X = \overleftarrow{X} + 1]\!]_2(s_1, s_2))$$

and hence it follows by the sequencing theorem above that:

$$\text{VDM\_Spec}([\![T]\!],\ [\![X := X + 1;\ X := X + 1]\!],\ \text{Seq}([\![X = \overleftarrow{X} + 1]\!]_2,\ [\![X = \overleftarrow{X} + 1]\!]_2))$$

By the definition of Seq in Section 5:

$$
\begin{aligned}
&\text{Seq}([\![X = \overleftarrow{X} + 1]\!]_2,\ [\![X = \overleftarrow{X} + 1]\!]_2)(s_1, s_2) \\
&= \exists s.\ [\![X = \overleftarrow{X} + 1]\!]_2(s_1, s)\ \wedge\ [\![X = \overleftarrow{X} + 1]\!]_2(s, s_2) \\
&= \exists s.\ (\lambda(s_1, s_2).\ s_2{}^\prime X^\prime = s_1{}^\prime X^\prime + 1)(s_1, s)\ \wedge\ (\lambda(s_1, s_2).\ s_2{}^\prime X^\prime = s_1{}^\prime X^\prime + 1)(s, s_2) \\
&= \exists s.\ (s^\prime X^\prime = s_1{}^\prime X^\prime + 1)\ \wedge\ (s_2{}^\prime X^\prime = s^\prime X^\prime + 1) \\
&= \exists s.\ (s^\prime X^\prime = s_1{}^\prime X^\prime + 1)\ \wedge\ (s_2{}^\prime X^\prime = (s_1{}^\prime X^\prime + 1) + 1) \\
&= \exists s.\ (s^\prime X^\prime = s_1{}^\prime X^\prime + 1)\ \wedge\ (s_2{}^\prime X^\prime = s_1{}^\prime X^\prime + 2) \\
&= (\exists s.\ s^\prime X^\prime = s_1{}^\prime X^\prime + 1)\ \wedge\ (\exists s.\ s_2{}^\prime X^\prime = s_1{}^\prime X^\prime + 2) \\
&= T\ \wedge\ (s_2{}^\prime X^\prime = s_1{}^\prime X^\prime + 2) \\
&= (s_2{}^\prime X^\prime = s_1{}^\prime X^\prime + 2) \\
&= [\![X = \overleftarrow{X} + 2]\!]_2(s_1, s_2)
\end{aligned}
$$

Hence:

$$\vdash\ \{T\}\ X := X + 1;\ X := X + 1\ \{X = \overleftarrow{X} + 2\}$$

$\square$

An elegant application of treating postconditions as binary relations is Aczel's version of the **while**-rule [19]:

$$\frac{\vdash \{\mathcal{P} \wedge \mathcal{B}\}\ \mathcal{C}\ \{\mathcal{P} \wedge \mathcal{R}\}}{\vdash \{\mathcal{P}\}\ \textbf{while}\ \mathcal{B}\ \textbf{do}\ \mathcal{C}\ \{\mathcal{P} \wedge \neg\mathcal{B} \wedge \mathcal{R}^*\}}$$

Where $\mathcal{R}^*$ is the reflexive closure of $\mathcal{R}$ defined by

$$\mathcal{R}^*(s_1, s_2)\ =\ \exists n.\ \mathcal{R}^n(s_1, s_2)$$

and $\mathcal{R}^n$ is definable in higher order logic by the following primitive recursion:

$$\mathcal{R}^0\ =\ \lambda(s_1, s_2).\ (s_1 = s_2)$$

$$\mathcal{R}^{n+1}\ =\ \mathsf{Seq}(\mathcal{R}, \mathcal{R}^n)$$

Aczel pointed out that his version of the **while**-rule can be converted into a rule of total correctness simply by requiring $\mathcal{R}$ to be transitive and well-founded:

$$\frac{\vdash [\mathcal{P} \wedge \mathcal{B}]\ \mathcal{C}\ [\mathcal{P} \wedge \mathcal{R}] \qquad \vdash \mathsf{Transitive}\ \mathcal{R} \qquad \vdash \mathsf{Well\_Founded}\ \mathcal{R}}{\vdash [\mathcal{P}]\ \textbf{while}\ \mathcal{B}\ \textbf{do}\ \mathcal{C}\ [\mathcal{P} \wedge \neg\mathcal{B} \wedge \mathcal{R}^*]}$$

where:

$$\mathsf{Transitive}\ r\ =\ \forall s_1\ s_2\ s_3.\ r(s_1, s_2) \wedge r(s_2, s_3) \Rightarrow r(s_1, s_3)$$

$$\mathsf{Well\_Founded}\ r\ =\ \neg\exists f : num{\rightarrow}state.\ \forall n.\ r(f(n), f(n+1))$$

Notice how it is straightforward to define notions like $\mathsf{Transitive}$ and $\mathsf{Well\_Founded}$ in higher order logic; these cannot be defined in first order logic.

11.2   Dijkstra's Weakest Preconditions

Dijkstra's theory of weakest preconditions, like VDM, is primarily a theory of rigorous program construction rather than a theory of post hoc verification. As will be shown, it is straightforward to define weakest preconditions for deterministic programs in higher order logic[14].

In his book [6], Dijkstra introduced both 'weakest liberal preconditions' (Wlp) and 'weakest preconditions' (Wp); the former for partial correctness and the latter for total correctness. The idea is that if $\mathcal{C}$ is a command and $\mathcal{Q}$ a predicate, then:

- $\mathsf{Wlp}(\mathcal{C}, \mathcal{Q})\ =\ $ '*The weakest predicate $\mathcal{P}$ such that $\{\mathcal{P}\}\ \mathcal{Q}\ \{\mathcal{Q}\}$*'

- $\mathsf{Wp}(\mathcal{C}, \mathcal{Q})\ =\ $ '*The weakest predicate $\mathcal{P}$ such that $[\mathcal{P}]\ \mathcal{Q}\ [\mathcal{Q}]$*'

Before defining these notions formally, it is necessary to first define the general notion of the 'weakest predicate' satisfying a condition. If p and q are predicates on states (i.e. have type $state{\rightarrow}bool$), then define $p{\Leftarrow}q$ to mean p is weaker (i.e. 'less constraining') than q, in the sense that everything satisfying q also satisfies p. Formally:

$$p{\Leftarrow}q\ =\ \forall s.\ q\ s \Rightarrow p\ s$$

The weakest predicate satisfying a condition can be given a general definition using Hilbert's ε-operator. This is an operator that chooses an object satisfying a

---

[14]Dijkstra's semantics of nondeterministic programs can also be formalized in higher order logic, but not using the simple methods described in this paper (see the end of Section 5.1).

predicate. If P is a predicate on predicates, then $\varepsilon p.\ P\ p$ is the predicate defined by the property:

$$(\exists p.\ P\ p) \;\Rightarrow\; P(\varepsilon p.\ P\ p)$$

Thus, if there exists a p such that $P\ p$, then $\varepsilon p.\ P\ p$ denotes such a p; if no such p exists, then $\varepsilon p.\ P\ p$ denotes an arbitrary predicate. Hilbert invented ε and showed that it could be consistently added to first order logic. Allowing the use of ε in higher order logic is equivalent to assuming the Axiom of Choice. The weakest predicate satisfying P can be defined using ε:

$$\mathsf{Weakest}\ P \;=\; \varepsilon p.\ P\ p \;\wedge\; \forall p'.\ P\ p' \;\Rightarrow\; (p\Leftarrow p')$$

Dijkstra's two kinds of weakest preconditions can be defined by:

$$\mathsf{Wlp}(c, q) \;=\; \mathsf{Weakest}(\lambda p.\ \mathsf{Spec}(p, c, q))$$

$$\mathsf{Wp}(c, q) \;=\; \mathsf{Weakest}(\lambda p.\ \mathsf{Total\_Spec}(p, c, q))$$

These definitions seems to formalize the intuitive notions described by Dijkstra, but are cumbersome to work with. The theorems shown below are easy consequences of the definitions above, and are much more convenient to use in formal proofs.

$$\vdash\ \mathsf{Wlp}(c, q) \;=\; \lambda s.\ \forall s'.\ c(s, s') \;\Rightarrow\; q\ s'$$

$$\vdash\ \mathsf{Wp}(c, q) \;=\; \lambda s.\ (\exists s'.\ c(s, s')) \;\wedge\; \forall s'.\ c(s, s') \;\Rightarrow\; q\ s'$$

The relationship between Hoare's notation and weakest preconditions is given by:

$$\vdash\ \mathsf{Spec}(p, c, q) \;\;=\;\; \forall s.\ p\ s \;\Rightarrow\; \mathsf{Wlp}(c, q)\ s$$

$$\vdash\ \mathsf{Total\_Spec}(p, c, q) \;=\; \forall s.\ p\ s \;\Rightarrow\; \mathsf{Wp}(c, q)\ s$$

The statement of the last two theorems, as well as other results below, can be improved if 'big' versions of the logical operators $\wedge$, $\vee$ $\Rightarrow$ and $\neg$, and constants T and F are introduced which are 'lifted' to predicates. These are defined in the table below, together with the operator $\models$ which tests whether a predicate is always true. These lifted predicates will also be useful in connection with dynamic logic.

Operators on predicates	
$p \wedge q$	$= \lambda s.\ p\ s \wedge q\ s$
$p \vee q$	$= \lambda s.\ p\ s \vee q\ s$
$p \Rightarrow q$	$= \lambda s.\ p\ s \Rightarrow q\ s$
$\neg p$	$= \lambda s.\ \neg p\ s$
T	$= \lambda s.\ \mathsf{T}$
F	$= \lambda s.\ \mathsf{F}$
$\models p$	$= \forall s.\ p\ s$

The last two theorems can now be reformulated more elegantly as:

$$\vdash\ \mathsf{Spec}(p, c, q) \quad = \quad \models p \Rightarrow \mathsf{Wlp}(c, q)$$

$$\vdash\ \mathsf{Total\_Spec}(p, c, q) \ = \ \models p \Rightarrow \mathsf{Wp}(c, q)$$

In Dijkstra's book, various properties of weakest preconditions are stated as axioms, for example:

Property 1. $\vdash\ \forall c. \models \mathsf{Wp}(c, \mathsf{F}) = \mathsf{F}$

Property 2. $\vdash\ \forall q\ r\ c.\ (q \Rightarrow r) \quad \Rightarrow \quad \models \mathsf{Wp}(c, q) \Rightarrow \mathsf{Wp}(c, r)$

Property 3. $\vdash\ \forall q\ r\ c.\ \models \mathsf{Wp}(c, q) \wedge \mathsf{Wp}(c, r) \ = \ \mathsf{Wp}(c,\ q \wedge r)$

Property 4. $\vdash\ \forall q\ r\ c.\ \models \mathsf{Wp}(c, q) \vee \mathsf{Wp}(c, r) \Rightarrow \mathsf{Wp}(c,\ q \vee r)$

Property 4'. $\vdash\ \forall q\ r\ c.\ \mathsf{Det}\ c \ \Rightarrow\ \models \mathsf{Wp}(c, q) \vee \mathsf{Wp}(c, r) \ = \ \mathsf{Wp}(c,\ q \vee r)$

These all follow easily from the definition of Wp given above (Det is the determinacy predicate defined in Section 10). It is also straightforward to derive analogous properties of weakest liberal preconditions:

$$\vdash\ \forall c. \models \mathsf{Wlp}(c, \mathsf{F}) \ = \ \lambda s.\ \neg \exists s'.\ c(s, s')$$

$$\vdash\ \forall q\ r\ c.\ (q \Rightarrow r) \ \Rightarrow\ \models \mathsf{Wlp}(c, q) \Rightarrow \mathsf{Wlp}(c, r)$$

$$\vdash\ \forall q\ r\ c.\ \models \mathsf{Wlp}(c, q) \wedge \mathsf{Wlp}(c, r) \ = \ \mathsf{Wlp}(c,\ q \wedge r)$$

$$\vdash\ \forall q\ r\ c.\ \models \mathsf{Wlp}(c, q) \vee \mathsf{Wlp}(c, r) \Rightarrow \mathsf{Wlp}(c,\ q \vee r)$$

$$\vdash\ \forall q\ r\ c.\ \mathsf{Det}\ c \ \Rightarrow\ \models \mathsf{Wlp}(c, q) \vee \mathsf{Wlp}(c, r) \ = \ \mathsf{Wlp}(c,\ q \vee r)$$

Many of the properties of programming constructs given in Dijkstra's book [6] are straightforward to verify for the constructs of our little language. For example:

$$\vdash\ \mathsf{Wp}([\![\mathbf{skip}]\!],\ q) = q$$

$$\vdash\ \mathsf{Wlp}([\![\mathbf{skip}]\!],\ q) = q$$

$$\vdash\ \mathsf{Wp}([\![\mathcal{V} := \mathcal{E}]\!],\ q) = \lambda s.\ q(\mathsf{Bnd}\ ([\![\mathcal{E}]\!]s)\ `\mathcal{V}`\ s)$$

$$\vdash\ \mathsf{Wlp}([\![\mathcal{V} := \mathcal{E}]\!],\ q) = \lambda s.\ q(\mathsf{Bnd}\ ([\![\mathcal{E}]\!]s)\ `\mathcal{V}`\ s)$$

$$\vdash\ \mathsf{Wp}([\![\mathbf{if}\ \mathcal{B}\ \mathbf{then}\ \mathcal{C}_1\ \mathbf{else}\ \mathcal{C}_2]\!],\ q) = \lambda s.\ ([\![\mathcal{B}]\!]s \to \mathsf{Wp}([\![\mathcal{C}_1]\!], s)\ |\ \mathsf{Wp}([\![\mathcal{C}_2]\!], s))$$

$$\vdash\ \mathsf{Wlp}([\![\mathbf{if}\ \mathcal{B}\ \mathbf{then}\ \mathcal{C}_1\ \mathbf{else}\ \mathcal{C}_2]\!],\ q) = \lambda s.\ ([\![\mathcal{B}]\!]s \to \mathsf{Wlp}([\![\mathcal{C}_1]\!], s)\ |\ \mathsf{Wlp}([\![\mathcal{C}_2]\!], s))$$

The inadequacy of the relational model reveals itself when we try to derive Dijkstra's Wp-law for sequences. This law is:

$$\mathsf{Wp}(\llbracket \mathcal{C}_1; \mathcal{C}_2 \rrbracket, q) = \mathsf{Wp}(\llbracket \mathcal{C}_1 \rrbracket, \mathsf{Wp}(\llbracket \mathcal{C}_2 \rrbracket, q))$$

which is not true with our semantics. For example, taking:

$$
\begin{aligned}
s_1 &= \lambda x.\ 0 \\
s_2 &= \lambda x.\ 1 \\
c_1(s_1, s_2) &= (s_1 = s_1) \ \vee \ (s_2 = s_2) \\
c_2(s_1, s_2) &= (s_1 = s_1) \ \wedge \ (s_2 = s_2)
\end{aligned}
$$

results in:

$$\mathsf{Wp}(\mathsf{Seq}(c_1, c_2), \mathsf{T}) = \mathsf{T}$$

but

$$\mathsf{Wp}(c_1, \ \mathsf{Wp}(c_2, \mathsf{T})) \ = \mathsf{F}$$

The best that can be proved using the relational semantics is the following:

$$\vdash \ \mathsf{Det} \ \llbracket \mathcal{C}_1 \rrbracket \ \Rightarrow \ \mathsf{Wp}(\llbracket \mathcal{C}_1; \mathcal{C}_2 \rrbracket, q) = \mathsf{Wp}(\llbracket \mathcal{C}_1 \rrbracket, \mathsf{Wp}(\llbracket \mathcal{C}_2 \rrbracket, q))$$

As discussed in Section 5.1, the problem lies in the definition of Halts. For partial correctness there is no problem; the following sequencing law for weakest liberal preconditions can be proved from the relational semantics.

$$\vdash \ \mathsf{Wlp}(\llbracket \mathcal{C}_1; \mathcal{C}_2 \rrbracket, q) = \mathsf{Wlp}(\llbracket \mathcal{C}_1 \rrbracket, \mathsf{Wlp}(\llbracket \mathcal{C}_2 \rrbracket, q))$$

With relational semantics, the Wp-law for **while**-commands also requires a determinacy assumption:

$$\vdash \ \mathsf{Det} \ c \ \Rightarrow \ \mathsf{Wp}(\llbracket \textbf{while} \ \mathcal{B} \ \textbf{do} \ \mathcal{C} \rrbracket, q) \ s \ = \ \exists n. \ \mathsf{Iter\_Wp} \ n \ \llbracket \mathcal{B} \rrbracket \ \llbracket \mathcal{C} \rrbracket \ q \ s$$

where

$$
\begin{aligned}
\mathsf{Iter\_Wp} \ 0 \ b \ c \ q &= \ \neg b \wedge p \\
\mathsf{Iter\_Wp} \ (n{+}1) \ b \ c \ q &= \ b \wedge \mathsf{Wp}(c, \ \mathsf{Iter\_Wp} \ n \ b \ c \ p)
\end{aligned}
$$

However, the Wlp-law for **while**-commands does not require a determinacy assumption:

$$\vdash \ \mathsf{Wlp}(\llbracket \textbf{while} \ \mathcal{B} \ \textbf{do} \ \mathcal{C} \rrbracket, q) \ s \ = \ \forall n. \ \mathsf{Iter\_Wlp} \ n \ \llbracket \mathcal{B} \rrbracket \ \llbracket \mathcal{C} \rrbracket \ q \ s$$

where

$$
\begin{aligned}
\mathsf{Iter\_Wlp} \ 0 \ b \ c \ q &= \ \neg b \ \Rightarrow \ p \\
\mathsf{Iter\_Wlp} \ (n{+}1) \ b \ c \ q &= \ b \ \Rightarrow \ \mathsf{Wlp}(c, \ \mathsf{Iter\_Wlp} \ n \ b \ c \ p)
\end{aligned}
$$

The Wlp-law for **while**-commands given above was not the first one I thought of. Initially I tried the same tactic that was used to prove the Wp-law for **while**-commands on the goal obtained from it by deleting the determinacy assumption and replacing Wp by Wlp. It soon became clear that this would not work, and after some 'proof hacking' with HOL I came up with the theorem above. A possible danger of powerful proof assistants is that they will encourage the generation of theorems without much understanding of their significance. I tend to use HOL for simple mathematics rather like I use a pocket calculator for arithmetic. I have already forgotten how to do some arithmetic operations by hand; I hope HOL will not cause me to forget how to do mathematical proofs manually!

11.3 Dynamic Logic

Dynamic logic is a programming logic which emphasizes an analogy between Hoare logic and modal logic; it was invented by V.R. Pratt based on an idea of R.C. Moore [32,9]. In dynamic logic, states of computation are thought of as *possible worlds*, and if a command C transforms an initial state s to a final state s' then s' is thought of as *accessible* from s (the preceding phrases in italics are standard concepts from modal logic).

Modal logic is characterized by having formulae $\Box q$ and $\Diamond q$ with the following interpretations.

- $\Box q$ is true in s if q is true in all states accessible from s.

- $\Diamond q$ is true in s if $\neg\Box\neg q$ is true in s.

Instead of a single $\Box$ and $\Diamond$, dynamic logic has operators $[C]$ and $<C>$ for each command C. These can be defined on the relation c denotated by C as follows:

$$[c]q \quad = \quad \lambda s.\ \forall s'.\ c(s,s') \ \Rightarrow \ q\ s'$$

$$<c>q \quad = \quad \neg([c](\neg q))$$

where $\neg$ is negation lifted to predicates (see preceding section).

A typical theorem of dynamic logic is:

$$\vdash \ \forall c\ q.\ \mathsf{Det}\ c \ \Rightarrow \ \models <c>q \Rightarrow [c]q$$

This is a version of the modal logic principle that says that if the accessibility relation is functional then $\Diamond q \ \Rightarrow \ \Box q$ [9].

From the definitions of $[c]q$ and $<c>q$ it can be easily deduced that:

$$\vdash \ (\models [c]q) \ = \ \mathsf{Spec}(\mathsf{T},c,q)$$

$$\vdash \ \mathsf{Det}\ c \ \Rightarrow \ ((\models <c>q) \ = \ \mathsf{Total\_Spec}(\mathsf{T},c,q))$$

$$\vdash \ \mathsf{Spec}(p,c,q) \ = \ (\models p \Rightarrow [c]q)$$

$$\vdash \ \mathsf{Det}\ c \ \Rightarrow \ (\mathsf{Total\_Spec}(p,c,q) \ = \ (\models p \Rightarrow <c>q))$$

Where $\models$, $\Rightarrow$ and T were defined in the preceding section. Using these relationships, theorems of dynamic logic can be converted to theorems of Hoare logic (and vice versa).

Dynamic logic is closely related to weakest preconditions as follows:

$$\vdash \ \mathsf{Wlp}(c,q) \ = \ [c]\,q$$

$$\vdash \ \mathsf{Det} \ c \ \Rightarrow \ (\mathsf{Wp}(c,q) \ = \ \langle c \rangle q)$$

These theorems can be used to translate results from one system to the other.

12   Conclusions and Future Work

The examples in the previous section show that it is straightforward to define the semantic content of diverse programming logics directly in higher order logic. In earlier sections it is shown how, with a modest amount of parsing and pretty printing, this semantic representation can be made syntactically palatable. If a general purpose system like HOL is used, the choice of specification constructs can be optimized to the problem at hand and to the tastes of the specifier; a particular choice need not be hard-wired into the verifier. For example, Hoare-style and VDM-style correctness specifications can be freely mixed. A significant benefit of working in a single logical system is that only a single set of theorem proving tools is needed. Typical software verifications require some general mathematical reasoning, as well as specialized manipulations in a programming logic. If everything is embedded in a single logic, then there is no need to interface a special purpose program verifier to a separate theorem prover for handling verification conditions. This benefit is even greater if both software and hardware are being simultaneously reasoned about, because hardware verification tools can also be embedded in systems like HOL [12].

Although the methods presented here seem to work smoothly, the examples done so far are really too trivial to permit firm conclusions to be drawn. The next step in this research is to try to develop a *practical* program verifier on top of the HOL system. We plan to extend the methods of this paper to a programming language containing at least procedures, functions, arrays and some input/output. Various possibilities are under consideration, ranging from adopting an existing language like Tempura, Occam or Vista, to designing our own verification-oriented language. It is intended that whatever language we choose will be supported by a verified compiler generating code for a verified processor. Preliminary work on this has already started. Eventually it is planned to verify a reasonably non-trivial program using our tools. It is expected that this case study will be some kind of simple real-time system. Our goal is to show the *possibility* of totally verified systems, and to give a preliminary idea of their practicability.

One unsatisfactory aspect of the verifier described here is that the parser and pretty-printer were implemented by descending from ML into Lisp (the HOL system is implemented in Lisp). This could be avoided if ML were augmented with a general purpose parser and pretty-printer generator similar to the one in Mosses' semantics implementation system SIS [26]. Providing these tools would be quite a lot of work, but fortunately some progress in this direction has already been made. For example:

(i) Huet's group at INRIA have interfaced CAML (a version of ML that extends the ML used by HOL) to the Unix YACC parser generator.

(ii) It is planned that a parser generator (from Edinburgh) will be distributed with *Standard ML of New Jersey*, a high performance implementation of Standard ML from AT&T Bell Laboratories.

(iii) The Esprit project *GIPE* (*Generating Interactive Programming Environments*) is producing a powerful general purpose interface for manipulating formal languages. This supports a user-specifiable parser and pretty-printer which is closely integrated with a mouse-driven syntax-directed editor. For example, the pretty-printer adapts its line breaks to the width of the window in which it is being used.

These three projects suggest that tools will soon be available to enable syntactic interfaces to logical systems to be smoothly implemented without the low-level hacking I had to use.

13 Acknowledgements

The style of interactive proof supported by the HOL system was invented by Robin Milner and extended by Larry Paulson. The idea of representing verification condition generation by a tactic was developed jointly with Tom Melham; he also spotted errors in an earlier version of this paper. Job Zwiers, of the Philips Research Laboratories in Eindhoven, explained to me the connection between determinacy, relational semantics, dynamic logic and Dijkstra's weakest preconditions. Avra Cohn provided techical advice, as well as general support and encouragement. Finally, the past and present members of the Cambridge hardware verification group generated the exciting intellectual environment in which the research reported here was conducted.

References

[1] Andrews, P.B., *An Introduction to Mathematical Logic and Type Theory*, Academic Press, 1986.

[2] Boyer, R.S. and Moore, J S., *A Computational Logic*, Academic Press, 1979.

[3] Boyer, R.S., and Moore, J S., 'Metafunctions: proving them correct and using them efficiently as new proof procedures' in Boyer, R.S. and Moore, J S. (eds), *The Correctness Problem in Computer Science*, Academic Press, New York, 1981.

[4] Clarke, E.M. Jr., 'The characterization problem for Hoare logics', in Hoare, C.A.R. and Shepherdson, J.C. (eds), *Mathematical Logic and Programming Languages*, Prentice Hall, 1985.

[5] A. Church, 'A Formulation of the Simple Theory of Types', Journal of Symbolic Logic **5**, 1940.

[6] Dijkstra, E.W., *A Discipline of Programming*, Prentice-Hall, 1976.

[7] Floyd, R.W., 'Assigning meanings to programs', in Schwartz, J.T. (ed.), *Mathematical Aspects of Computer Science, Proceedings of Symposia in Applied Mathematics* **19** (American Mathematical Society), Providence, pp. 19-32, 1967.

[8] Good, D.I., 'Mechanical proofs about computer programs', in Hoare, C.A.R. and Shepherdson, J.C. (eds), *Mathematical Logic and Programming Languages*, Prentice Hall, 1985.

[9] Goldblatt, R., *Logics of Time and Computation*, CSLI Lecture Notes **7**, CSLI/Stanford, Ventura Hall, Stanford, CA 94305, USA, 1987.

[10] Gordon, M.J.C., 'Representing a logic in the LCF metalanguage', in Néel, D. (ed.), *Tools and Notions for Program Construction*, Cambridge University Press, 1982.

[11] Gordon, M.J.C., Milner, A.J.R.G. and Wadsworth, C.P., *Edinburgh LCF: a mechanized logic of computation*, Springer Lecture Notes in Computer Science **78**, Springer-Verlag, 1979.

[12] M. Gordon, 'Why Higher-order Logic is a Good Formalism for Specifying and Verifying Hardware', in G. Milne and P. A. Subrahmanyam (eds), *Formal Aspects of VLSI Design*, North-Holland, 1986.

[13] Gordon, M.J.C., 'HOL: A Proof Generating System for Higher-Order Logic', University of Cambridge, Computer Laboratory, Tech. Report No. 103, 1987; Revised version in G. Birtwistle and P.A. Subrahmanyam (eds), *VLSI Specification, Verification and Synthesis*, Kluwer, 1987.

[14] Gordon, M.J.C., *Programming Language Theory and its Implementation*, Prentice-Hall, 1988.

[15] Gries, D., *The Science of Programming*, Springer-Verlag, 1981.

[16] Hoare, C.A.R., 'An axiomatic basis for computer programming', *Communications of the ACM* **12**, pp. 576-583, October 1969.

[17] Igarashi, S., London, R.L., Luckham, D.C., 'Automatic program verification I: logical basis and its implementation', Acta Informatica **4**, 1975, pp. 145-182.

[18] INMOS Limited, 'Occam Programming Language', Prentice-Hall.

[19] Jones, C.B., 'Systematic Program Development' in Gehani, N. & McGettrick, A.D. (eds), *Software Specification Techniques*, Addison-Wesley, 1986.

[20] Joyce, J.J., Forthcoming Ph.D. thesis, University of Cambridge Computer Laboratory, expected 1989.

[21] Ligler, G.T., 'A mathematical approach to language design', in *Proceedings of the Second ACM Symposium on Principles of Programming Languages*, pp. 41-53.

[22] Loeckx, J. and Sieber, K., *The Foundations of Program Verification*, John Wiley & Sons Ltd. and B.G. Teubner, Stuttgart, 1984.

[23] London, R.L., et al. 'Proof rules for the programming language Euclid', *Acta Informatica* **10**, No. 1, 1978.

[24] Fourman, M.P., 'The Logic of Topoi', in Barwise, J. (ed.), *Handbook of Mathematical Logic*, North-Holland, 1977.

[25] Melham. T.F., 'Automating Recursive Type Definitions in Higher Order Logic', Proceedings of the *1988 Banff Conference on Hardware Verification*, this volume.

[26] Mosses, P.D., 'Compiler Generation using Denotational Semantics', in *Mathematical Foundations of Computer Science*, Lecture Notes in Computer Science **45**, Springer-Verlag, 1976.

[27] Milner, A.R.J.G., 'A Theory of Type Polymorphism in Programming', *Journal of Computer and System Sciences* **17**, 1978.

[28] Paulson, L.C., 'A higher-order implementation of rewriting', *Science of Computer Programming* **3**, pp 143-170, 1985.

[29] Paulson, L.C., 'Natural deduction as higher-order resolution', *Journal of Logic Programming* **3**, pp 237-258, 1986.

[30] Paulson, L.C., *Logic and Computation: Interactive Proof with Cambridge LCF*, Cambridge University Press, 1987.

[31] Plotkin, G.D., 'Dijkstra's Predicate Transformers and Smyth's Powerdomains', in Bjørner, D. (ed.), *Abstract Software Specifications*, Lecture Notes in Computer Science **86**, Springer-Verlag, 1986.

[32] Pratt, V.R., 'Semantical Considerations on Floyd-Hoare Logic', *Proceedings of the 17th IEEE Symposium on Foundations of Computer Science*, 1976.

[33] Hayes, I. (ed.), *Specification Case Studies*, Prentice-Hall.

Automated Theorem Proving for Analysis and Synthesis of Computations

David R. Musser
Rensselaer Polytechnic Institute
Computer Science Department
Troy, NY 12180

Abstract

This paper explores two aspects of automated theorem proving for support of computer software and hardware development: rewriting techniques for equational and inductive reasoning; and construction of proof trees representing key steps of a proof, as an aid to interactive proof search and as a basis for synthesis of computations.

1 Introduction

Automated theorem proving is a subject with a rapidly growing literature and a wide variety of research efforts on theory development, implementations, and applications. One of the prime application areas, and a stimulus for much theoretical and implementational work, has been the goal of automating reasoning about computations. For a more than two decades there has been substantial activity in the areas of program verification and program synthesis, with related efforts toward design of programming languages and methodologies based at least in part on the goal of producing software whose correctness is supported by mathematical arguments rather than just syntactic analysis and testing. More recently, some of the attention has been shifted to the hardware side, attempting to formalize the techniques used to design VLSI circuits and to develop techniques for constructing mathematical proofs that particular circuit designs implement some specified behavior. The aim of this paper is to present a brief account of a selection of automated theorem proving techniques that are useful in reasoning about computations, either in the software or the hardware area. While there are certainly some techniques that are more useful for software than hardware, or vice versa, I believe the techniques I discuss in this paper are fundamental for both aspects of computation.

I will concentrate on two main topics:

- use of rewriting techniques for deciding equality between computational objects, including the case when some form of inductive reasoning is required; and

- interactive construction of proofs as data structures that help provide feedback to a proof system user, enable the user to provide key steps of direction and manage the large numbers of theorems and subgoals that typically arise in proofs about computations, and potentially serve as a basis for *synthesis of computations*.

Rather than attempting to be comprehensive, I focus on a few of the techniques that I regard as being both of a fundamental nature and as having been demonstrated to be highly effective. For rewriting techniques I draw mainly on experience with RRL, a Rewrite Rule Laboratory [22], and for interactive proof construction I use examples developed with the Affirm program verification system [40]. Only software examples are discussed; for examples of applications of Affirm and RRL to hardware designs, see [34], [35], [36]. Most of the material presented in this paper is not new, although some of the rewriting-related methods are recent, and the particular proof construction exercise discussed in Section 5.3 has not been previously documented in the literature.

Although I draw mainly on RRL and Affirm for examples, many of the techniques discussed in this paper have also been widely used in other systems and will, I believe, continue to be of fundamental importance. Although one might wish to be able to make effective use of automated theorem provers without having to learn the fundamental theory underlying them, the present state of the art makes this impossible. I have tried to organize the discussion in this paper so that it will provide the reader who is a novice in this area with many of the key ideas which can serve as a foundation for a thorough understanding the work of others, including other automated theorem provers. Of course, one is unlikely to obtain a complete understanding of the issues from a brief paper such as this. My coverage of the various topics is not even; I go into depth only on some of the main issues in equational inference, and only try to provide some motivation and guide to literature on other topics. The interested reader certainly should delve further into the literature and, if possible, use one or more automated theorem provers on nontrivial examples.

In focusing on rewriting methods and interactive proof construction, it may appear that I am leaving out many important topics in automation of reasoning about computations. There is partial compensation in the fact that rewriting methods themselves have been continually expanded to subsume or substitute for many other approaches to deduction; e.g., see [12], [20], [42]. In particular, many of the methods associated with resolution-based theorem proving techniques can now also be performed with rewriting techniques; the same is true for mathematical induction techniques. I won't attempt to argue that rewriting techniques make the older methods obsolete, or even that they are likely to do so in the future, but there is at least a good case for making the study of rewriting techniques a starting point for a broad understanding of many different approaches to automated deduction. There are, nevertheless, a number of important topics to which the rewriting framework has little to offer, such as dealing with integer inequalities. For an introduction and references to this topic, see, for example, [3].

As for interactive proof construction, there has been significant disagreement among automated theorem prover researchers on the role of user interaction, with one side arguing that the only feasible role of computer reasoning systems is as an automated mathematical assistant, and thus that interaction is intrinsically necessary, while the other side argues for minimizing interaction and concentrating on totally automated proofs. I believe it is not necessary to settle this argument in order to accept the value of the concepts sketched in this paper relating to construction of proofs as data structures, for such proof structures have uses other than aiding in man-machine interaction:

- As will be discussed, proof structures can serve a basis for synthesis of programs or hardware designs, an idea that several research groups are pursuing.

- Proof structures can serve as an internal communication medium between decision and semi-decision procedures on the one hand and heuristic guidance mechanisms on the other.

- Proof structures can help solve the problem of "how do we verify the theorem prover?"

On the last point, even if it were possible to achieve *today* the total automation of the search for a proof (within the bounds imposed by undecidability), we would be unlikely to have much confidence in the answers returned by a theorem prover unless the implementation of the prover itself had been subjected to an extremely rigorous analysis; perhaps even a formal proof of correctness. But direct attack on this problem for a program of the complexity of a theorem prover would be a task of enormous magnitude (even if we had a totally automatic prover to help with it, since just *specifying* the internal workings in sufficient detail would be very difficult). But with a theorem prover that produces a sufficiently detailed proof structure, one could require that each proof found be put through a *proof checker*, and this proof checker could be relatively simple compared to the mechanisms required to search for a proof in the main theorem prover, and thus much more amenable to a correctness proof.

2   Equational Reasoning

Programming languages and hardware design languages generally provide only a small part of the formalism we need to reason about programs (and usually what they do provide comes with a great deal more complexity than we would like to contend with). What we ultimately need for this purpose is a logic; i.e, a language with a formalized syntax and semantics, both of which support logical deduction; moreover, both should be simple enough to make the reasoning process understandable to humans and feasible for machines, and rich enough to express the kinds of computations we want to carry out in real applications. Needless to say, we do not presently have such a logic. However, some admirable progress has been made for limited areas of computation, e.g., Boyer and Moore's logic [2] related to pure Lisp. Another important area of an ideal computational logic, where it is both reasonably clear what is needed and in which substantial success has been achieved in satisfying these needs, is reasoning about equality.

2.1   Validity versus Proof

A problem that continually arises in reasoning about computations is the question of when objects denoted by distinct terms are equal. This is of course a fundamental problem of many other mathematical domains aside from computation, and much of the discussion that follows applies more generally. Nevertheless, the goal of a computational logic has stimulated much of the development of this topic over the past two decades.

Given a finite set of equations $E = \{t_1 = u_1, \ldots, t_n = u_n\}$ and another equation $t = u$, the fundamental problem is to determine when $t = u$ is a *valid* consequence of E; that is, whether $t = u$ is true in all models of the equations. Here, as in most cases, validity is a concept involving infinitely many cases, and we usually resort instead to the finitary notion of a *proof* of $t = u$ from the equations E and some set of *inference rules*. A classic theorem of Birkhoff tells us that $t = u$ is a valid consequence of E if and only if u can be obtained from t by a finite number of applications of the equality rules of inference (equivalence axioms and substitution property) to the equations in E: that is, there is a finite chain of terms $s_0, s_2, \ldots, s_n$ such that $t = s_0$, $u = s_n$, and each $s_i = s_{i+1}$ is obtained by applying one of the equality rules of inference to one of the equations in E.

With validity, we could sometimes succeed in showing that $t = u$ is *not* a valid consequence of E by exhibiting a model in which $t = u$ is false although all the equations in E are true, but we cannot directly deal with the infinitely many cases that are required to show an equation *is* a valid consequence of E. With proofs of equations, the situation appears to be reversed: we can prove $t = u$ by exhibiting a finite chain of equality steps connecting t and u, but it would appear to require an infinitary argument to establish that no such chain can exist. Even when such an equality chain does exist, it is still a formidable computational problem to find one if a brute force approach to the search is taken.

Fortunately, the situation with proofs is much better than this in many important cases of equality in computational domains, for we can use well-established techniques of *Church-Rosser rewriting relations* and *canonical forms* to eliminate entirely the need for searching for an equality chain. The foundations for automating the use of Church-Rosser techniques were laid in the seminal paper by Knuth and Bendix in 1969 [24], although much of the theory on which they based their automation was known earlier; e.g., [37], [10]. Although Knuth and Bendix did not consider the problem in their paper, their methods have also turned out to constitute an important foundation for one approach to automating proofs about inductive properties of computational objects, the so-called inductionless induction method (see Section 3.2).

2.2 Terms, Substitutions, and Rewriting

Although Knuth and Bendix worked in an algebraic setting, their treatment of equations can be equally well described in a logical framework. The language of the logic contains a set of *terms*, which are finite tree structures built from a given finite set F of function symbols and a denumerably infinite set V of variable symbols; the set of all such terms is denoted $T(F, V)$. An equation is a pair of such terms, say (t_1, u_1), usually written $t_1 = u_1$, and the equality rules of inference are captured in the notion of *rewriting* a subterm of a term using an equation as a *rewrite rule*. Specifically, a pair of terms (l, r) is a rewrite rule if l is not just a variable and the variables that appear in r also appear in l. We usually write the rule as $l \to r$, and l is called the left-hand side and r the right-hand side of the rule. Note that in some cases an equation $t = u$ could be used as a rewrite rule as either $t \to u$ or $u \to t$.

A *substitution* is a mapping from σ from terms to terms which is determined entirely by its value on a finite number of variables; a substitution is denoted by an expression of the form $\{t_1/v_1, \ldots, t_k/v_k\}$, read "substitute t_1 for v_1, ..., t_k for v_k." The $k \geq 0$ variable symbols $v_1, \ldots, v_k$ must be distinct, and the case $k = 0$ is the identity substitution ι such that $\iota(t) = t$ for all terms t. Following convention we write an application of a substitution as $t\sigma$ rather than $\sigma(t)$.

To define rewriting precisely we also need some notion of position of an occurrence of a subterm s within a term t. One way to do this to introduce an extra variable symbol $\Box$ and the concept of a "box term": a term in $T(F, V \cup \{\Box\})$ with a single occurrence of $\Box$. Then an ordinary term t in $T(F, V)$ can be described as some box term t_1 with a subterm s replacing the box, which we make precise as an application of a substitution: $t = t_1\{s/\Box\}$.

For a given rewrite rule $l \to r$, a relation on pairs of terms, t *rewrites to* u, can be defined as: for some subterm s of t and box term t_1 such that $t = t_1\{s/\Box\}$, there is a substitution σ such that $s = l\sigma$ and $u = t_1\{r\sigma/\Box\}$. We write this as $t \to u$ using $l \to r$, overloading the use of the symbol $\to$.

For a given set of rewrite rules R, we say $t \to u$ using R if $t \to u$ for some rule $l \to r$ in R; and for a given set of equations E, we say $t \leftrightarrow u$ using E if there is an equation $t_i = u_i$ in E such that $t \to u$ for either $t_i \to u_i$ or $u_i \to t_i$.

Thus term rewriting is a binary relation on the term set $T(F, V)$, determined by a given set of rules or equations. This is a rather restricted notion of rewriting, but much of the basic Church-Rosser theory can be developed in a more abstract setting, and therefore can be applied much more widely. In the following subsection, we derive many of the most basic principles in the abstract, though the reader may find it helpful to relate each result back to the particular case of rewriting as just defined for terms. This abstract framework is helpful also in considering other cases such as conditional rewriting systems.

2.3 Abstract Rewriting Relations

We now let the rewriting relation $\to$ be any binary relation on a set S. The irreflexive, transitive closure of $\to$ is denoted by $\to^+$, and the reflexive, transitive closure is denoted by $\to^*$; the latter is referred to as *reduction*. It is helpful to think of $\to$ as a directed graph on S and reduction as the reflexive, transitive closure of the graph.

The inverse of $\to$ is denoted $\leftarrow$, and the union of $\to$ and $\leftarrow$ is denoted $\leftrightarrow$. Thus $x \leftrightarrow y$ if and only if either $x \to y$ or $y \to x$. This relation is an undirected graph on S. The reflexive, transitive closure of $\leftrightarrow$, denoted $\leftrightarrow^*$, is an equivalence relation. The graph $\leftrightarrow$ is thus partitioned into subgraphs, the equivalence classes of $\leftrightarrow^*$.

Often our main interest is in this equivalence relation $\leftrightarrow^*$: for any given pair of elements x and y, we want to know whether $x \leftrightarrow^* y$ or not; that is, whether it is possible to connect x and y with a finite chain of $\leftrightarrow$ links. In the general case we must deal with a very large or infinite graph, so searching for such a chain directly is too expensive; so also would be obtaining the equivalence relation $\leftrightarrow^*$ by a transitive closure computation, although in a few cases such techniques can work. Instead, we seek a more efficient approach that works with the $\to$ and $\to^*$ relations rather than $\leftrightarrow^*$. This approach requires $\to$ to have some special properties.

2.4 Joinability and the Church-Rosser Property

Elements p and q are *joinable* if and only if there exists an r such that $p \to^* r$ and $q \to^* r$.

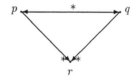

Joinability Relation   Church-Rosser Property

A rewriting relation $\to$ is said to have the *Church-Rosser property* if and only if for all p, q such that $p \leftrightarrow^* q$, it is the case that p and q are joinable.

Thus when a rewriting relation has the Church-Rosser property, it possible to check for equivalence, $\leftrightarrow^*$, of two elements p and q by searching for a common term to which they are both reducible. If a rewriting relation is *locally finite*, i.e., at most finitely many elements are reachable in one step from any element, then such a search could be carried out by simultaneously following all possible reduction paths from both p and q, which would be more efficient than following all $\leftrightarrow$ paths, but still very costly. Further improvement is possible if reduction paths are of at most finite length.

2.5 Normal Forms

An element p of S is said to be *terminal* if and only if there is no q such that $p \rightarrow q$. If $p \rightarrow^* q$ and q is terminal, then q is said to be a *normal form* of p.

Obviously if two distinct elements are terminal they are not joinable. With a Church-Rosser rewriting relation, therefore, an element can have at most one normal form. However, some elements might not have any normal form.

2.6 The Noetherian Property

A rewriting relation $\rightarrow$ is said to be *finitely terminating*, or *Noetherian*, if and only if there is no infinite sequence of the form $x_1 \rightarrow x_2 \rightarrow x_3 \rightarrow \dots$. Thus with a Noetherian relation every reduction path is finite and every element has a normal form. An element might have several distinct normal forms, however.

2.7 An Efficient Equivalence Test

Now suppose a rewriting relation $\rightarrow$ is both Church-Rosser and Noetherian. Then every element has a unique normal form, called its *canonical form*, and we can decide whether $p \leftrightarrow^* q$ without any searching: just follow a reduction path from p to its canonical form p_1 and from q to its canonical form q_1, and check whether p_1 and q_1 are identical. If they are indeed identical, then clearly $p \leftrightarrow^* q$. (See Figure A.) Conversely, if $p \leftrightarrow^* q$, they are joinable to a common element r, by the Church-Rosser property. Letting r_1 be the canonical form of r, it must be the case that $p_1 = r_1$ and $q_1 = r_1$, by the uniqueness of normal forms under a Church-Rosser relation, and thus $p_1 = q_1$. (See Figure B.)

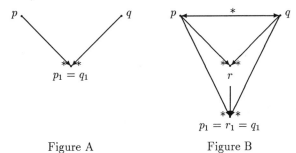

Figure A   Figure B

To be able to use this efficient equivalence test, we must be able to determine whether a rewriting relation is Church-Rosser and Noetherian.

2.8 Confluence Test for the Church-Rosser Property

Testing a rewriting relation for the Church-Rosser property can be done by testing for a "confluence" condition.

A rewriting relation $\rightarrow$ is *confluent* if and only if for all elements p, q_1, q_2, if $p \rightarrow^* q_1$ and $p \rightarrow^* q_2$ then q_1 and q_2 are joinable.

Theorem 1 *A rewriting relation is Church-Rosser if and only if it is confluent.*

Remark: Some authors *define* the Church-Rosser property as confluence.

Proof: If $\rightarrow$ is Church-Rosser, it is easily seen to be confluent: if $p \rightarrow^* q_1$ and $p \rightarrow^* q_2$ then $q_1 \leftrightarrow^* q_2$ and thus by the Church-Rosser property q_1 and q_2 are joinable.

If $\rightarrow$ is confluent, we can prove that it is Church-Rosser, by induction on the length N of a $\leftrightarrow$ chain connecting two elements p and q. If $N = 0$, p and q are identical and thus trivially joinable. Let $N > 0$, so that there is an element p_1 such that $p \leftrightarrow p_1 \leftrightarrow^* q$, where the length of the chain from p_1 to q is $N - 1$. By the induction hypothesis, p_1 and q are joinable, to r, say.

Either $p \rightarrow p_1$ or $p \leftarrow p_1$. If $p \rightarrow p_1$, then p and q are joinable (to r). (See Figure C.) If $p \leftarrow p_1$, then by confluence p and r are joinable, to r_1, say, and thus p and q are joinable (to r_1). (See Figure D.) $\square$

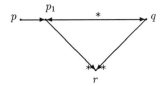

Figure C

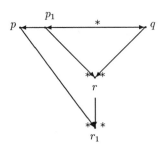

Figure D

Thus instead of directly checking for the Church-Rosser property, we can check for confluence. It is possible in some cases to use a more localized test, as is discussed next.

2.9 Local Confluence Test for the Church-Rosser Property

A rewriting relation $\rightarrow$ is *locally confluent* if and only if for all elements p, q_1, q_2, if $p \rightarrow q_1$ and $p \rightarrow q_2$ then q_1 and q_2 are joinable. (Note the difference from confluence: only one step is taken from p rather than arbitrarily many.)

A locally confluent relation may fail to be confluent, as shown by the following diagram:

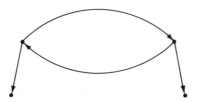

However, if a rewriting relation is Noetherian and locally confluent, then it is confluent. This is called the "Diamond Lemma," because of the structure of its proof; see [15]. The Church-Rosser property then follows from its equivalence to confluence. One can also prove the Church-Rosser property directly, using the "mountain-range" ideas in [19]. In preparation for this proof we need some additional machinery relating to the Noetherian property, which is also useful in many other contexts.

2.10 Well-Founded Partial Orderings

Let $>$ be a strict partial ordering on a set S; i.e. a transitive binary relation on S without any cycles. A partial ordering $>$ is said to be *well-founded* if there is no infinite sequence of the form $x_1 > x_2 > a_3 > \ldots$ consisting of elements from S. A rewriting relation $\to$ is Noetherian if and only if $\to^+$ is Noetherian. The relation $\to^+$ is a strict partial ordering on S, and the Noetherian property for $\to^+$ is equivalent to $\to^+$ being well-founded. We extend this ordering to sets of elements; more precisely, to multisets of elements.

2.11 Multiset Orderings

A multiset is like a set but with elements possibly repeated a finite number of times; formally it is a set S and a mapping m from the set to the nonnegative integers; the value given by $m(x)$ is called the multiplicity, or number of occurrences, of x. The multiset is said to be on the set S. Multiset union is computed by adding multiplicities and multiset difference by subtracting them (a negative result being replaced by 0).

A well-founded partial ordering $>$ can be extended to a well-founded partial order on the set of all finite multisets on S as follows: $M_1 > M_2$ if and only if for all x in $M_2 - M_1$ there is a y in $M_1 - M_2$ such that $y > x$ (see [8]). A natural illustration of this ordering is given by the following lemma, which follows immediately from the definitions, and is used in the following subsection.

Lemma 1 *Let $p \to q_1$, $p \to q_2$, $q_1 \to^* r$ and $q_2 \to^* r$. Also let M_1 be the multiset $\{p, q_1, q_2\}$ and M_2 be the multiset containing all of the elements in the chains $q_1 \to^* r$ and $q_2 \to^* r$. Then $M_1 > M_2$ according to the multiset extension of the $\to^+$ partial ordering.*

2.12 Noetherian and Locally Confluent Rewriting Relations

Now suppose $\to$ is Noetherian and locally confluent, and $t \leftrightarrow^* u$. We will show that t and u are joinable. Joinability will already hold unless the chain $\leftrightarrow^*$ contains one or more triples t_{i-1}, t_i, t_{i+1} such that $t_i \to t_{i-1}$ and $t_i \to t_{i+1}$. We can call such a triple a "peak" since by the Noetherian property t_i is "higher" than t_{i-1} and t_{i+1}, and the whole chain can be consistently thought of as a "mountain range"; our objective is to transform it into a "valley." By local confluence and the preceding lemma, a peak can be replaced by a valley, giving a new mountain range that is "lower" than the original range; formally, the multiset of terms in the new chain is less than that of the original chain in the multiset ordering that extends the $\to^+$ ordering. Thus, as long as there are peaks, we can find keep extending the sequence of $M_1 > M_2 > \ldots$ of multisets of elements in mountain ranges connecting t and u. By the well-foundedness of this multiset ordering, this sequence must eventually terminate, so we must eventually obtain a range with no peaks; i.e., a valley between t and u.

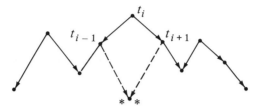

Thus we have proved:

Theorem 2 *If a rewriting relation is Noetherian and locally confluent, then it is Church-Rosser.*

(Note that in the above proof an induction on the number of peaks in a chain would not work, since replacing a peak by a valley can increase the number of peaks.)

2.13   Local Confluence of a Term Rewriting Relation

With a rewriting relation determined by a finite set of rewrite rules, the test for local confluence can be reduced to a finite number of cases, based on an analysis of how the left-hand sides of the rules "overlap" each other, producing "critical pairs" to be checked for joinability. The definition of overlap, or "superposition," as it is usually called, depends on the important concept of *unification* of terms.

Two terms t and u are *unifiable* if there is a substitution σ such that $t\sigma = u\sigma$. σ is called a *unifier* of t and u, and whenever σ is chosen so that it is a factor of any other unifier σ_1 (i.e., σ_1 can be written as a composition $\sigma \circ \sigma_2$), then it is called a *most general unifier* of t and u. It can be shown that the most general unifier of two terms, if it exists, is unique up to variable renaming.

Let $l_1 \to r_1$ and $l_2 \to r_2$ be two rewrite rules (possibly the same rule). If l_1 unifies with a nonvariable subterm s of l_2, with most general unifier σ, then we say that the term $l_2\sigma$ is a *superposition* of the rules, and that the corresponding *critical pair* of terms is $(b_2\sigma\{r_1\sigma/\square\}, r_2\sigma)$, where b_2 is the box term such that $l_2 = b_2\{s/\square\}$. Thus $l_2\sigma$ is a term to which both rules apply, and $b_2\sigma\{r_1\sigma/\square\}$ is the term obtained using rule 1 while $r_2\sigma$ is obtained using rule 2.

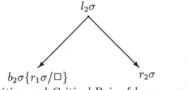

A Superposition and Critical Pair of $l_1 \to r_1$ and $l_2 \to r_2$

A superposition and critical pair for a pair of rules essentially represents a most general case of how a term can be rewritten in two different ways using the rules. Note that other superpositions and corresponding critical pairs might exist for the same pair of rules, with the overlap occurring in different positions.

Theorem 3 *For a given set of rewrite rules, there is a finite number of pairs of terms (p, q), called critical pairs, such that if each (p, q) is joinable, then the rewriting relation is locally confluent.*

In order to use this theorem for a practical test for local confluence, one must be able to test for joinability. When a set of rewrite rules has the Noetherian property, we define for each critical pair (p, q) a corresponding *reduced critical pair* (p_1, q_1) where p_1 is any normal form of p and q_1 any normal form of q.

Corollary 1 *If a set of rewrite rules has the Noetherian property, then the rewriting relation is locally confluent (and therefore confluent) if and only if for every critical pair there is a corresponding reduced critical pair whose terms are identical.*

Thus, for Noetherian set of rules, an algorithmic test for confluence is to form all critical pairs and for each pair rewrite each member to a normal form, checking the resulting reduced critical pair for identity of their terms.

2.14 Knuth-Bendix Completion Procedure

When a reduced critical pair has non-identical terms, one can try to extend the rule set by forming the reduced critical pair terms into a new rule, such that when added to the existing rule set the Noetherian property still holds. If this can be done, one must then consider any additional superpositions and critical pairs resulting from pairing the new rule with itself or one of the old rules, which may in turn cause additional rules to be added, and so on. Eventually, this procedure may terminate with a set of rules that are both Noetherian and confluent; by the way the resulting rules are constructed from critical pairs, they determine the same equalities as the original set.

On the other hand, the procedure may fail at some point because a reduced critical pair cannot be formed into a rewrite rule that preserves the Noetherian property, or it may continue indefinitely, producing an infinite set of rules.

The procedure just described is known as the Knuth-Bendix Completion Procedure, since a procedure of this nature was first described in [24]. Actually, the procedure in [24] was somewhat less general, being given in terms of a particular way of establishing the Noetherian property. The more general version sketched here was first given by Lankford [25].

2.15 Proving the Noetherian Property

The discussion up to this point shows that it is very important to the goal of reasoning efficiently about equality to be able to establish the Noetherian property of a term rewriting relation. Although the problem is undecidable in the general case [14], there are a number of approaches to establishing sufficient conditions for the Noetherian property. I will not attempt to discuss these approaches in this paper; fortunately, Dershowitz ([6], see also [7]) has already provided an excellent introduction to research on this question, with references to most of the main work.

The Rewrite Rule Laboratory, RRL[1], implements two methods called *recursive path ordering (rpo)* and *lexicographic recursive path ordering (lrpo)*, both of which extend a precedence relation on function symbols to terms. With *lrpo*, the *status* of associative binary function symbols is also used, with values *left-to-right* or *right-to-left*, indicating in which way operands should be associated in normal forms. The user can supply precedences and statuses with declarations, or interactively and incrementally in response to questions posed by RRL as it tries to convert equations into rules. Learning how to provide this ordering information is one of the main skills a user must learn in order to use RRL effectively.

2.16 Conditional Rewriting

An important variation on the kind of term rewriting discussed in the foregoing sections is *conditional rewriting*, in which the rules have the form $(t \rightarrow u \text{ if } P)$, where P is a term of type boolean. There are a number of ways to define the rewriting relation associated with a set of such rules depending what kind of rewriting is applied to P; see [9]. One of the most useful ways appears to be *contextual rewriting* [42] (which differs from an earlier definition in [44]); for example, a theorem proving method for full first order logic can be formulated in terms of contextual rewriting [42]. Zhang's implementation of contextual rewriting in RRL plays an important role in many of the theorem proving experiments carried out with that system.

2.17 Associative-Commutative Rewriting

Another important variation is *associate and commutative (AC)* rewriting [28], [39], [11], in which the associative and commutative properties of designated function symbols are taken into account in reduction and in an extension of the Knuth-Bendix procedure. Some such approach is necessary since the equation defining the commutative law, e.g., $x + y = y + x$ cannot be converted into a rewrite rule with the Noetherian property. RRL implements AC-rewriting by using unification and matching algorithms that work with AC operators.

Using AC unification in the Knuth-Bendix procedure tends to produce a large number of superpositions and critical pairs, causing significant degrading of efficiency of the procedure. However, a modification described in [19] regains some efficiency by detecting that many of the superpositions generated are unnecessary (so that the critical pair generation and joinability test do not have to be performed).

3 Inductive Proof

Many kinds of computational objects cannot be characterized purely by a finite set of equations; although the objects may satisfy some finite set of equations E, one can always find models of E for which some properties of the objects do not hold.

[1]RRL [22] is a software environment for experimenting with automated deduction methods based on term rewriting techniques. It was developed under the direction of D. Kapur at General Electric Research and Development Center and Rensselaer Polytechnic Institute. The implementation, primarily by G. Sivakumar and H. Zhang, consists of about 18,000 lines of Lisp code that runs on Digital Equipment VAX computers, SUN Workstations, or Symbolics Lisp Machines. The REVE system [29], developed at MIT and the University of Nancy, provides a software laboratory with capabilities similar to RRL, but implemented in CLU. Both REVE and RRL were developed with partial support from the National Science Foundation.

The domains of these objects are defined by inductive definitions; e.g., the natural numbers, finite sequences, trees, etc. Operations on these domains are also defined inductively, such as defining $+$ on natural numbers by $0+x = x$ and $s(x)+y = s(x+y)$, where s is the successor function. Most properties of $+$, such as the associative law, $x+(y+z) = (x+y)+z$, are not provable by purely equational inference, so how do we know they are true? The reason is that, first, the natural numbers are representable by the ground terms $0, s(0), s(s(0)), \ldots$, i.e., every value of a variable in an equation to be proved, such as $x+0 = x$, must correspond to one of these ground terms. Secondly, *every* substitution of these ground terms for the variables yields an equation which is true; e.g., in the case of $x + 0 = x$ the equations $0 + 0 = 0$, $s(0) + 0 = s(0)$, $s(s(0)) + 0 = s(s(0))$, ... are all true (and in fact provable by equational inference).

The first of these claims, that there are no natural numbers other than those corresponding to $0, s(0), s(s(0)), \ldots$, can be taken essentially as part of the definition of the natural numbers. The second claim involves infinitely many equations and must somehow be proved; usually this is done with some Peano-like rule of inference: e.g., as a proof of $P(x)$ it suffices to prove $P(0)$ and that $P(x)$ implies $P(s(x))$. For example to prove $x + 0 = x$, it suffices to prove $0 + 0 = 0$ and that $x + 0 = x$ implies $s(x) + 0 = s(x)$, both of which are provable by equational inference from $0 + x = x$ and $s(x) + y = s(x + y)$.

Burstall [4] showed how to generalize the Peano type of inductive inference rule to more general classes of computational objects than the natural numbers, through *structural induction* principles. The main limitation to such an approach is that one or more of the proof subgoals produced may turn out not to be provable by equational inference, in which case one is no closer to having a proof of the original goal.

Examples of this problem occur even with the natural numbers. Consider, for example [43], what happens if one defines a greatest common divisor function with $\gcd(x, 0) = x$, $\gcd(0, y) = y$, $\gcd(x + y, y) = \gcd(x, y)$, along with $0 + x = x$ and $s(x) + y = s(x + y)$, and then tries to prove

$$P(x) = \forall y[\gcd(x, y) = gcd(y, x)]$$

One of the subgoals, $\gcd(0, y) = \gcd(y, 0)$ is provable equationally from the definitions, but the other,

$$\gcd(x, y) = \gcd(y, x) \text{ implies } \gcd(s(x), y) = \gcd(y, s(x)),$$

is not.

The Boyer-Moore theorem prover [2] overcomes some of the limitations of structural induction, in a setting where functions are defined recursively rather than by a set of equations. Instead of having a fixed inductive inference rule or set of such rules, the Boyer-Moore prover identifies the functions involved in a statement to be proved and generates new inference rules based on the pattern of recursion in definition of those functions. Because the structure of these generated rules mirrors the structure of the recursive function definitions, they often yield provable subgoals.

3.1 Cover Set Induction

The *cover set induction* principle recently developed by Zhang, Kapur and Krishnamoorthy [43] is similar to the Boyer-Moore approach, but works with functions defined by sets of equations rather than recursive definitions. Again, instead of having a fixed inductive inference rule or set of such rules, one generates new inference rules based on the structure of the equational axioms used to define the functions of the language. Because the structure of these generated rules mirrors the structure of the defining axioms, they often yield subgoals that are provable with equational inference. The cover set approach actually is more powerful than the Boyer-Moore approach in two important respects: it works even when some functions are incompletely defined and when there are relations between constructors, as is discussed in [43].

For example, from the defining equations for the gcd function given above, one identifies the following set of pairs of terms as a cover set:

$$(x, 0), (0, y), (x + y, y), (x, x + y)$$

This set of pairs of terms has the property that every pair of ground terms of type natural number is equal (by the equations defining $+$) to an instance of one of the pairs in the set. (This is one of the requirements in the definition of cover set given in [43].) From this cover set, one obtains the following induction principle:

$$\begin{array}{ccc} & P(x, 0) & \\ & P(0, y) & \\ P(x, y) & \text{implies} & P(x + y, y) \\ \underline{P(x, y)} & \text{implies} & P(x, x + y) \\ & P(x, y) & \end{array}$$

Using this induction principle, the proof of a statement such as $P(x) = \forall y[\gcd(x, y) = \gcd(y, x)]$ does result in subgoals that are provable by equational inference.

For further details about how cover sets and corresponding induction principles are defined, see [43]. An example of the power of this approach, let us consider a proof found by Zhang of the correctness of the quick sort algorithm, using his implementation of cover-set induction in RRL. Quick sort is formulated as a function from lists to lists, where lists are axiomatized as being generated by constructor functions nil and cons. The complete formulation of the problem is shown in Figure 1, just as it is input to RRL. Note the form of the definition of the predicate sortedp; the cover set method produces an inductive rule of inference adapted to this form.

As with the Boyer-Moore prover, the proof is not completely automatic; one must provide extensive guidance in the form of intermediate lemmas. In this case sixteen lemmas are required, some of which are treated by RRL as conditional rewrite rules. It takes 90 seconds processor time on a Symbolics 3600 Lisp Machine to absorb all the definitions and prove all the lemmas and the main goals; user interaction is also required in converting the problem formulation and lemmas into conditional rewrite rules with the Noetherian property.

It is interesting to note that Zhang has also been able, using his implementation of conditional rewriting and cover set induction in RRL, to carry out a proof of one of the larger Boyer-Moore theorem prover benchmarks: the unique factorization theorem for integers.

3.2 Inductionless Induction

An alternative method [32] of proof of inductive properties can sometimes be used when functions are defined by equations. This method has come to be known as "inductionless induction" [26], [27] (because there is no direct appeal to an inductive rule of inference) or "proof by consistency" [17] (because it associates the truth of an equation with the question of consistency of that equation with the axiomatization

```
;;                          Equations for numbers.
[ 0 : num]
[ suc : num -> num]
[ + : num, num -> num]
  x + 0 := x
  x + suc(y) := suc(x + y)
[< : num, num -> bool]
  0 < suc(x) := true
  x < 0 := false
  suc(x) < suc(y) := x < y
;;                          'cond' stands for the three-place IF function.
[cond: bool, univ, univ -> univ]
  cond(true, x, y) == x
  cond(false, x, y) == y
  cond(x, y, y) == y
;;                          Equations for lists of numbers.
[nl : list]
[cons : univ, list -> list]
[append : list, list -> list]
  append(nl, y) := y
  append(cons(x, y), z) := cons(x, append(y, z))
[split_below : num, list -> list]
  split_below(xa, nl) := nl
  split_below(xa, cons(ya, x)) :=
cond((ya < xa), cons(ya, split_below(xa, x)), split_below(xa, x))
[split_above: num, list -> list]
  split_above(xa, nl) := nl
  split_above(xa, cons(ya, x)) :=
cond((ya < xa), split_above(xa, x), cons(ya, split_above(xa, x)))
[all_gte: num, list -> bool]
  all_gte(xa, nl) := true
  all_gte(xa, cons(ya, x)) := cond((xa < ya), false, all_gte(xa, x))
[all_lte: num, list -> bool]
  all_lte(xa, nl) := true
  all_lte(xa, cons(ya, x)) := cond((ya < xa), false, all_lte(xa, x))
[occur: num, list -> num]
  occur(xa, nl) := 0
  occur(xa, cons(ya, x)) := cond((xa = ya), suc(occur(xa, x)), occur(xa, x))
[qsort: list -> list]
  qsort(nl) := nl
  qsort(cons(xa, x)) := append(qsort(split_below(xa, x)),
                               cons(xa, qsort(split_above(xa, x))))
[sortedp: list -> bool]
  sortedp(nl) := true
  sortedp(cons(xa, nl)) := true
  sortedp(cons(xa, cons(ya, x))) :=
     cond((ya < xa), false, sortedp(cons(ya, x)))
;;                          At this point the proofs begin
;; (the series of sixteen intermediate lemmas required is not shown)
  prove sortedp(qsort(x))
  prove occur(xa, qsort(x)) == occur(xa, x)
```

Figure 1: Input to RRL for Quick Sort Problem

of the functions). Improvements of the method, relaxing some of the requirements made in [32] on the form of the axioms, have been found by a number of authors; [17] includes a survey. More recent work includes [18], [16], [21], and [41].

A particularly interesting recent formulation is Toyama's [41], in terms of determining equivalence of two rewriting sets of equations and viewing inductive definitions as "reachability" conditions.

4 Interactive Proof Management

Since long, complex proofs are often required when reasoning about computations, it is useful to have a model of the proof process which helps to visualize and manage the potentially large number of theorems and subgoals that can arise. In this section I describe the particular proof management model supported in the Affirm-85 verification system [40], [23], [33] and in the following section I continue with a discussion of how this or similar models support an approach to synthesis of computations from proofs.

4.1 Proof Management in Affirm-85

Affirm-85 permits the user both to state specifications and perform theorem proving in a first order logical language extended with abstract data types. Data types are defined axiomatically, and may be created or edited by the user as necessary. Typically, one states and proves general properties and then calls upon these facts as lemmas when demonstrating more complex theorems (such as protocol invariants, or program or hardware design verification conditions).

The style of proof in Affirm-85 is highly interactive, with the system depending on the user for strategy. Theorems, or really, conjectures, are set up as top-level goals, and Affirm-85 provides various means of reducing these goals to simpler subgoals. The subgoaling process terminates when subgoals are reached that Affirm-85 can "finish off" just by simplification to the symbol **true** (using rewrite rule techniques, including some of the simpler Knuth-Bendix techniques discussed in previous sections). The simplification process generally involves not only simplification rules for standard logical and arithmetic operations, but also rules derived from data type axioms input by the user. One of the most common activities in carrying out a proof is to state new goals and use them as lemmas in proving the original goals. These lemmas must themselves be proved by the same subgoaling process: Affirm-85 indicates in status information any uses of unproved lemmas, and also disallows any attempt at circular reasoning (using the original theorem in the proof of the lemma, or any less direct circularities).

Affirm-85 maintains a transcript of all user input and its own output, and one record of a proof is the transcript of user-system interaction in finding it. However, the main record is a tree structure representation of the goals and subgoals generated by the user-system interaction. These tree structures (really a forest) are maintained internally and can be displayed at any time (indentation of text lines is used to show the structure). There are also numerous facilities for moving a "current step pointer" around in the trees, affording the user a great deal of flexibility in deciding what to work on next. Affirm-85 also makes it easy to recover from wrong turns made in exploring the space of proof attempts, since previous steps are "undoable" as part of a general facility for recording, redoing, or undoing user commands (inherited mainly from the underlying Interlisp system).

It is also possible to change the axioms in data type specifications, even while a proof is in progress. This allows experimentation with different axiomatizations to see if they permit simpler proofs, but is a potential source of unsoundness. To aid the user in keeping track of the need to reprove those goals that might be affected by changing axioms, the theorem prover maintains a record of what data type specifications are currently known, called a "set of support," as part of each deduction step in a proof tree. These sets of support are then used to calculate whether all steps of the proof of a goal were done with the same set of support or with ones that were merely extensions of others, in which case the proof is said to be "certified." This proof certification facility [33] is particularly important in maintaining and reusing theorems and proofs that Affirm-85 saves on, and restores from, files.

4.2   Proof Forests

When a goal is broken down into one or more subgoals, a logical dependency is established. This dependency relation on the set of goals corresponds to a directed acyclic graph, which is used to represent the state of a proof attempt. Nodes (which correspond to logical formulas) are annotated to indicate the Affirm-85 command which transformed them. If a node has several children, each arc may be labeled to indicate it by case. For example, an induction on a sequence data type might result in the following tree fragment:

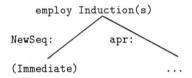

NewSeq and apr ("append an element to the right") are constructors for the Sequence data type and are used as labels of the basis and induction step cases of the induction argument. The leaf label (Immediate) indicates that the proof of the NewSeq case follows directly by reduction, using the axioms, to the symbol true.

If we suppose that no two cases of a proof are identical, then the proof state of a theorem may be visualized as a tree, rooted in the theorem. This is the model presented to the user. As the proof progresses, structure is added at the leaves (except for backtracking that results from undoing). When all leaves are the constant true, the proof is complete. (If two subgoals in the forest happen to be identical, they have a common proof.)

In presenting a proof, it is useful to regard any auxiliary facts as atomic; while their status is relevant, details of their proofs are not. Affirm-85 therefore denotes certain nodes as **theorems**: formulas of interest (thus we have both "proven theorems" and "unproven theorems"; sometimes, unproven theorems are in fact found to be false and get discarded). These include the user's original goals and verification conditions, as well as all subgoals entered by the user; the latter are generally of simple form and are called lemmas. Any logical dependencies upon theorems are recorded separately from the goal-subgoal dependency graph. Each theorem therefore is the root of a (possibly degenerate) proof tree in this graph, which is called the proof forest. When summarizing a proof tree, Affirm-85 notes any lemmas used (along with their status) but leaves their proofs separate. An example of an Affirm proof structure is given in the following section.

5 Proof-Based Computation Synthesis

An automated theorem prover that produces a proof structure—one such as Affirm-85 proof forests which at least reflect explicit induction arguments, case-splitting, and instantiations—can be used as the basis for a paradigm for producing formally verified programs or hardware circuit designs that might be called *proof-based computation synthesis*:

- specify the problem via a theorem

- construct a proof of the theorem

- extract an algorithm or circuit design from the proof

In this paradigm, the central activity of the software or hardware developer is stating theorems (or really, conjectures) and finding proofs (or discovering that the conjectures are not provable, in which case they must be restated). In the case of specifications of algorithms or procedures, the appropriate conjectures are those stating the existence of the outputs to be computed; these could be called *existence-conjectures*. In the proof-based approach, existence-conjectures, stated in a rigorous logical language, are the form that specifications take, and finding proofs is the first step toward producing computations that meet the specifications. If a proof that is found is recorded in a proof structure such as that described in the previous discussion of Affirm-85, then there is enough information in the proof that algorithm can be extracted from it in a relatively straightforward way.

Hoare, Dijkstra, Gries, and others have long advocated an approach to programming in which as part of the process of developing an algorithm one thinks about how to prove that the algorithm does what it is supposed to, rather than proof being an after the fact activity. In the proof-based approach one goes a step farther, and makes conjecturing and proving the first and central activities; obtaining algorithms comes later.

5.1 Examples of Algorithm Synthesis

To take a small programming example, the task of sorting a list of numbers might be specified by saying that the output list must be in increasing order and must contain the same elements as the input list. This would be put in the form of a conjecture (stated in a logical language) that, given any input list, another list exists with elements in increasing order and containing the same elements as the given list. One then attempts to find a constructive proof of this conjecture, for the next step is to *extract* the construction used in the proof as the basis of an algorithm or procedure for sorting. Note that one does not start with an algorithm and attempt to prove it correct (first having to figure out how to specify exactly what it is supposed to do); rather, one proceeds from existence-conjecture to constructive proof to algorithm.

Of course, a conjecture that is a theorem can have many proofs, and different constructive proofs of an existence-conjecture will yield different algorithms. In the sorting example, perhaps the simplest proof of the existence-conjecture is: let L be a list of all permutations of the elements of the given list, and let P be any member of L in which all of the elements are in increasing order (the fact that such a member exists must be proved as a lemma); P then is the list that is sought. Of course, the algorithm one would extract from this would be ridiculously inefficient. Instead one wants a constructive proof in which the construction that can be extracted is an efficient algorithm.

As an example of such a "good" constructive proof, one might have a proof of the following nature for sorting: if the given list has only one element, the answer is the same list; otherwise, split the given list into two equal parts (or with one part larger by one). By induction on the size of lists, for each of the two parts one can find a list that is in increasing order and a permutation of the original elements. One can them apply a lemma about *merging* that says: Given two lists, each of which is in increasing order, there exists a list in increasing order whose elements are a permutation of all the elements of the two given lists. Assuming this lemma, which is actually another existence-conjecture, is proved with the obvious efficient construction, one obtains an overall construction corresponding to *merge-sort*, one of the most efficient sorting algorithms.

Applying a lemma in a constructive proof, in order to show existence of some quantity needed as an ingredient to the main existence demonstration, corresponds to calling another procedure. Similarly, the induction done in this proof yields a recursive control structure in the algorithm. This is true in general, although in some cases the recursion could be transformed into iteration (e.g., tail-recursions). Case splitting in the proof yields conditional control structures in the algorithm. Proof statements like "Let P be ..." correspond to assignment statements or parameter passing.

5.2 Variants of the Proof-Based Approach

A proof-based paradigm for program development has been advocated in a form very similar to that discussed here (but with somewhat different terminology) by Manna and Waldinger [30], and in a substantially different form within a constructive type theory framework investigated earlier by Martin-Löf [31]. Work at Cornell University on Nuprl by Constable *et al* [1], [5], at INRIA by Coquand and Huet[13], and at Odyssey Research by Platek *et al* has also followed the constructive type theory framework.

Proof-based computation synthesis is, however, not necessarily tied to constructive type theory. Constructive type theory has been one of the main frameworks in which arguments have been made for making mathematics more constructive, but these arguments have little to do with the goal of producing software, because in *any* framework for producing *algorithms* one is, at least implicitly, proving existence-conjectures constructively. Such proofs can be done in other notations besides constructive type theory; for example, proofs such as have been discussed above for sorting specifications can easily be formalized in first-order logic. A demonstration of this is summarized below, making use of the previously discussed proof management facilities of Affirm-85, starting with existence-conjectures about sorting.

Manna and Waldinger's work has concentrated mainly on a proof-based approach in which proofs are found by a fully automatic theorem prover; thus their goal is a form of automatic programming. This is a highly desirable long term goal, but I believe an approach in which an interactive prover is used is likely to be more useful over at least the next decade.

To summarize this part of the discussion, I am advocating developing a existence-conjecture / constructive proof / construction-extraction paradigm in which the proofs are carried out interactively. The real issue is not whether or not the proofs are constructive (they naturally will be) but whether enough proof assistance can be provided to make this a natural and productive way to specify and implement programs or hardware circuit designs.

5.3   An Exercise with Affirm-85

We start with by giving the following specifications relating to sorting can be given in Affirm-85 (as part of a larger specification of a `SequenceOfInteger` type):

```
axioms
    count(NewSeq) == true,
    count(s apr i, j)
        == if i=j then count(s, j) + 1 else count(s,j),
    ordered(NewSeq) == true,
    ordered(s apr i)
        == ((s = NewSeq) or ((Last(s) <= i) and ordered(s)));
define
    s perm s1
        == all i (count(s, i) = count(s1,i)),
    s is_sorted_as s1
        == (ordered(s1) and s perm s1);
```

The existence-conjecture specifying sorting is then:

```
            all s (some s1 (s is_sorted_as s1));
```

The proof we construct is one from which the *insertion-sort* algorithm can be extracted. Though this is an order n^2 algorithm, it can be useful in some contexts where only short sequences need to be sorted, or when applied to a sequence that is already mostly in order (insertion-sort becomes almost linear in that case).

The proof is by an induction on the apr–structure of s. In the induction step, the key step of the proof (and the one which gives a proof structure from which insertion-sort is naturally extractable) is to state and prove another existence-conjecture:

```
    all s,i (some s1 (ordered(s) implies (s apr i) is_sorted_as s1));
```

i.e., if s is already in order and we have one more element i, we can obtain an s1 such that s1 is ordered and is a permutation of the elements in s plus i.

This second existence-conjecture is also proved by induction on the apr–structure of s, but this time there are several cases to consider, corresponding to whether s is NewSeq or not, and if not, how i compares with the last element of s. The case-splitting in the proof yields the different branches of conditional statements in the extracted algorithm. Part of the proof structure created interactively with Affirm-85 for this theorem is shown in Figure 2.

Altogether, the exercise of proving the sorting and one-element-insertion conjectures with Affirm-85 was quite involved. Fifteen lemmas other were required, for various properties of permutations and orderings (see Figure 3) that turned out to be needed to support the two main goals. The transcript of the proof session runs over sixty pages, including about ten pages to print the final proof structures.

Figure 4 shows which of these lemmas are referenced in the proofs of other lemmas. Note that some, like count_non_negative and perm_preserved, are used several times.

Several observations can be made about the results of this sorting proof exercise. First, regarding the size of the proofs and the amount of user interaction required,

```
theorem sort1_exists, sortgoal1(s, i);
sort1_exists uses perm_1!, perm_not_NewSequenceOfInteger!, perm_maximum!,
maximum_of_ordered!, perm_preserved!, perm_swap!, and perm_transitive!.

proof tree:
44:!    sort1_exists
            employ Induction(s)
45:     NewSequenceOfInteger:
            26 invoke sortgoal1
46:         28 put s1 = NewSequenceOfInteger apr i
47:         29 invoke is_sorted_as
48:         30 invoke perm
48:         (proven!)
49:     apr: {sort1_exists}
            27 suppose ii' <= i
50:         yes:
                32 invoke sortgoal1
51:             34 put s1 = (ss' apr ii') apr i
52:             35 invoke is_sorted_as
53:             36 invoke perm
53:             (proven!)
54:         no: {sort1_exists, apr:}
                33 invoke IH
55:             38 invoke sortgoal1
56:             39 put i'=i
59:             40 put s1' = s1(i) apr ii'
60:             41 suppose as' = NewSequenceOfInteger
61:             yes:
                    42 invoke is_sorted_as
62:                 44 replace
63:                 45 apply perm_1
64:                 47 put (i'=i) and (s = s1(i))
65:                 48 replace
66:                 49 invoke perm
66:                 50 cases
66:                 51 normint
->                  (proven!)
67:             no: {sort1_exists, apr:, no:}
                    43 invoke is_sorted_as
68:                 52 apply perm_not_NewSequenceOfInteger
69:                 54 put    s=ss'
                            and (i'=i) and (s1' = s1(i))
78:                 55 split
79:                 first:
                        56 apply perm_maximum
80:                     59 put  s = ss' apr i
                                and s1' = s1(i)
80:                     60 cases
81:                     61 apply maximum_of_ordered
82:                     63 put s=ss'
89:                     64 apply maximum_of_ordered
90:                     65 put s = s1(i)
91:                     66 replace
->                      (proven!)
    . . .
```

Figure 2: Part of Affirm-85 Proof Structure for Sort_1 Existence Conjecture

```
sort_exists:
    all s [some s1 (s is_sorted_as s1)];

sort1_exists:
    all s [some s1 (ordered(s) implies (s apr i) is_sorted_as s1)];

perm_not_new:
    all s, i [(s apr i) perm s1 implies (s1 /= NewSeq)];

count_nonnegative:
    all s, i [count(s, i) >= 0];

maximum_of_ordered:
    all s [ordered(s) and (s /= NewSeq) implies (maximum(s) = last(s))];

perm_swap:
    all s, i, j [((s apr i) apr j) perm ((s apr j) apr i)];

perm_1:
    all s, i [(NewSeq apr i) perm s implies (s = NewSeq apr i)];

count_all_zero:
    all s [all i [count(s, i) = 0] implies (s = NewSeq)];

perm_maximum:
    all s,s1 [(s /= NewSeq) and (s1 /=NewSeq) and s perm s1
                implies (maximum(s) = maximum(s1))];

maximum_in:
    all s [(s /= NewSeq) implies (maximum(s) in s)];

lemma1_of_perm_maximum:
    all s,i [(i in s) implies (i <= maximum(s))];

perm_in_eqv:
    all s,s1,i [s perm s1 implies (i in s eqv i in s1)];

perm_in:
    all s,s1,i [s perm s1 and i in s implies i in s1];

perm_commutes:
    all s,s1 [s perm s1 eqv s1 perm s];

in_count:
    all s,i [i in s eqv count(s, i) >= 1];
```

Figure 3: Lemmas in Affirm-85 Proof of Insertion-Sort Existence Conjectures

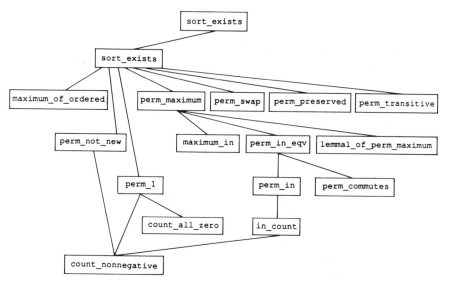

Figure 4: Usage Tree of Insertion-Sort Lemmas

certainly this represents a great deal of work to obtain one sorting algorithm (and not one of the best algorithms at that), but examination of the lemmas shows that:

- Many of the lemmas would be reusable in the proofs of other sorting algorithms or even in a much wider class of other algorithms involving permutations or ordering relations.

- The algorithms to be extracted from the proof structure (the insertion sort algorithm from the **sort_exists** lemma and the one-element-insertion algorithm from the **sort1_exists** lemma) "reside near the top" of the proof trees; the rest of the proof structure below the existence lemmas is merely supporting material.

On the second point, as Bates and Constable observed [1], by proceeding to develop the proof structure in a top down manner (as was actually done in this exercise), one could extract the algorithms even if the complete proof were not carried out. While completing the proof gives the maximum assurance of the correctness of the algorithms, just the fact that this approach requires stating the specifications as theorems, and the structure of the algorithms is obtained by a formal process, gives more assurance than one has with conventional algorithm construction.

Even when proofs are to be carried out in full, one may want to interrupt the proof finding session to extract algorithms from the incomplete proof structure and try testing them on a few examples; these tests might reveal errors in the formulation of (unproved) lemmas that might otherwise be caught only much later, deep into attempts to prove them.

References

[1] Bates, J. and Constable, R., "Proofs as Programs," *ACM Trans. Prog. Lang. Sys.*, v. 7, n. 1, 1985, 113-136.

[2] Boyer, R. and Moore, J, *A Computational Logic*, Academic Press, New York, 1980.

[3] Bundy, A., *The Computer Modelling of Mathematical Reasoning*, Academic Press, London, 1983.

[4] Burstall, R., "Proving properties of Programs by Structural Induction," *Computer Jour.*, 12(1), 41-48, 1969.

[5] Constable, R., Allen, S., Bromley, H., *et al*, *Implementing Mathematics with the Nuprl Proof Development System*, Prentice Hall, 1986.

[6] Dershowitz, N., "Orderings for term-rewriting systems," *Theoretical Computer Science* 17, 1982, 279-301.

[7] Dershowitz, N., "Termination of Rewriting," *J. Symbolic Comp.*, 1987, 3, 69-116.

[8] Dershowitz, N. and Manna, Z., "Proving Termination with Multiset Orderings," *Comm. ACM* 22, 1979, 465-475.

[9] Dershowitz, N., Okada, M., and Sivakumar, G., "Canonical Conditional Rewrite Systems," in *Proceedings of the 9th Int. Conf. on Automated Deduction*, Argonne, Ill., 1988, LNCS 310, Springer Verlag, NY, 538-549.

[10] Evans, T., "The Word Problem for Abstract Algebras," *J. of the London Math. Soc.* 26, 1951, 64-71.

[11] Fortenbacher, A., "An Algebraic Approach to Unification under Associativity and Commutativity," in *Rewriting Techniques and Applications*, Jean-Pierre Jouannaud, ed., LNCS 202, Springer Verlag, Berlin, 1985, 381-397.

[12] Hsiang, J., and Dershowitz, N., "Rewrite methods for clausal and non-clausal theorem proving," in *Proc. 10th EATCS Int. Colloq. on Automata, Languages, and Programming*, Barcelona, Spain, 1983.

[13] Coquand, T. and Huet, G., "Constructions: A Higher Order Proof System for Mechanizing Mathematics," *EUROCAL 85* Linz, Austria, April 1985.

[14] Huet, G., and Lankford, D.S., "On the Uniform Halting Problem for Term Rewriting Systems," Rapport Laboria 283, INRIA, Paris, 1978.

[15] Huet, G. and Oppen, D., "Equations and Rewrite Rules: a Survey," in *Formal Languages: Perspectives and Open Problems* (R. Book, ed.), Academic Press, New York, 1980.

[16] Jouannaud, J.-P., and Kounalis, E., "Proofs by Induction in Equational Theories with Constructors," *Proc. IEEE Symp. on Logic in Comp. Sci.*, Cambridge, Mass., June 1986.

[17] Kapur, D. and Musser, D.R., "Proof by Consistency," *Artificial Intelligence* 31 (1987) 125-157.

[18] Kapur, D. and Musser, D.R., "Inductive Reasoning with Incomplete Specifications," *Proc. Logic in Computer Science Conf.*, June, 1986.

[19] Kapur, D., Musser, D. and Narendran, P., "Only Prime Superpositions Need be Considered in the Knuth-Bendix Completion Procedure," to appear in *J. Symbolic Computation.*

[20] Kapur, D. and Narendran, P., "An Equational Approach to Theorem Proving in First-Order Predicate Calculus," in *9th Int. Joint Conf. on Artificial Intelligence*, Los Angeles, California, 1985.

[21] Kapur, D., Narendran, P., and Zhang, H., "Proof by Induction Using Test Sets," in *Proceedings of the 8th Int. Conf. on Automated Deduction*, Oxford, U.K., LNCS 230, Springer-Verlag, Berlin, 1986.

[22] Kapur, D., Sivakumar, G., and Zhang, H., "RRL: a Rewrite Rule Laboratory," in *Proceedings of the 8th Int. Conf. on Automated Deduction*, Oxford, U.K., LNCS 230, Springer-Verlag, Berlin, 1986.

[23] Kemmerer, R., ed., *Verification Assessment Study Final Report, Volume III, The Affirm System*, National Computer Security Center, Fort George G. Meade, Maryland, 1986.

[24] Knuth, D.E., and Bendix, P.B., "Simple Word Problems in Universal Algebras," in *Computational Problems in Abstract Algebra* (J. Leech, ed.), Pergamon Press, Oxford, 1970, pp. 263-297.

[25] Lankford, D.S., "Canonical Inference," Tech. Rep. ATP-25, Univ. of Texas, Austin, 1975.

[26] Lankford, D.S., "Some Remarks on Inductionless Induction," Tech. Rep. MTP-11, Louisiana Tech Univ., Ruston, 1980.

[27] Lankford, D.S., "A Simple Explanation of Inductionless Induction," Tech. Rep. MTP-14, Louisiana Tech Univ., Ruston, 1981.

[28] Lankford, D.S., and Ballantyne, A.M., "Decision Procedures for Simple Equational Theories with Commutative-Associative Axioms: Complete Sets of Commutative-Associative Reductions," Tech. Rep. ATP-39, Univ. of Texas, Austin, 1979.

[29] Lescanne, P., "Computer Experiments with the REVE Term Rewriting System Generator," *Proc. 10th Principles of Programming Languages Conf.*, Austin, Texas, 1983.

[30] Manna, Z., and Waldinger, R., "A Deductive Approach to Program Synthesis," *ACM Trans. Prog. Lang. Sys.*, 2, 1, 1980, 90-121.

[31] Martin-Löf, P., "An Intuistionistic Theory of Types: Predicative Part, in *Logic Colloquium '73*, Rose, H. and Shepherdson, eds., North-Holland, Amsterdam, 1973, 73-118.

[32] Musser, D.R., "On Proving Inductive Properties of Abstract Data Types," *Proc. of the 7th ACM Symposium on Principles of Programming Languages*, Las Vegas, Nevada, January 1980.

[33] Musser, D., "Aids to Hierarchical Specification and Reusing Theorems in Affirm-85," *Software Engineering News*, August 1985.

[34] Musser, D.R., Narendran, P., and Premelani, W.J., "BIDS: A Method of Specifying and Verifying Bidirectional Hardware Devices," in *VLSI Specification, Verification, and Synthesis*, G. Birtwistle and P. Subrahmanyam, eds. (Proc. of a workshop held in Calgary, Alberta, Canada, January 1987), Kluwer, 1987.

[35] Narendran, P., and Stillman, J., "Hardware Verification in the Interactive VHDL Workstation," in *VLSI Specification, Verification, and Synthesis*, G. Birtwistle and P. Subrahmanyam, eds. (Proc. of a workshop held in Calgary, Alberta, Canada, January 1987), Kluwer, 1987.

[36] Narendran, P., and Stillman, J., "Formal Verification of the Sobel Chip," in this volume.

[37] Newman, M.H.A., "On Theories with a Combinatorial Definition of "Equivalence", *Annals of Math.* 43/2, 1942, 223-243.

[38] Plaisted, D.A., "A simple non-termination test for the Knuth-Bendix method," in *Proceedings of the 8th Int. Conf. on Automated Deduction*, Oxford, U.K., 1986, LNCS 230, Springer Verlag, NY, 79-88.

[39] Stickel, M.E., "A Complete Unification Algorithm for Associative-Commutative Functions," Advance papers, *4th Int. Joint Conf. on Artificial Intelligence,* Tbilisi, USSR, 1985, 71-76.

[40] Thompson, D.H., and Erickson, R.W. (eds.), *AFFIRM Reference Manual*, USC Information Sciences Institute, Marina Del Rey, California, 1981.

[41] Toyama, Y. "How to Prove Equivalence of Term Rewriting Systems without Induction," in *Proc. of 8th Int. Conf. on Automated Deduction*, Oxford, U.K., LNCS 230, Springer Verlag, NY, 1986.

[42] Zhang, H. and Kapur, D., "First-Order Theorem Proving Using Conditional Rewrite Rules," in *Proceedings of the 9th Int. Conf. on Automated Deduction*, Argonne, IL, LNCS 310, Springer Verlag, NY, 1988, 1-20.

[43] Zhang, H., Kapur, D., and Krishnamoorthy, M., "A Mechanizable Induction Principle for Equational Specifications," in *Proceedings of the 9th Int. Conf. on Automated Deduction*, Argonne, IL, LNCS 310, Springer Verlag, NY, 1988, 162-181.

[44] Zhang, H., and Remy, J.L., "Contextual Rewriting," in *Rewriting Techniques and Applications*, Jean-Pierre Jouannaud, ed., LNCS 202, Springer Verlag, Berlin, 1985, 46-62.

What Do Computer Architects Design Anyway?

A. L. Davis

Hewlett-Packard Laboratories

Palo Alto, CA 94304

September 22, 1988

ABSTRACT

This paper is an attempt to provide a tutorial answer to the question: *What do computer architects design anyway?* The target readers are people concerned with automatic verification of computer hardware. The goal is to help this audience understand both the nature of the circuits which are being designed today and some of the factors that influence the design process. At the very least, automatic verification techniques must be mature enough to be useful in proving correctness of some sufficiently large component of contemporary designs before designers can be expected to make much use of the technology. Additionally, history has shown that mere functional correctness is an insufficient reason to adopt a new method. New methods must be able to build upon present practices and techniques. If a new method requires abandoning the bulk of present practice, it is seldom accepted. The paper presents a skeptical view of automatic verification systems. The domain, in practice, is both too imprecise and too complex to generate much confidence in the short-term future existence of useful verification systems. The hardware design domain covers a broad scope from theory to the thousands of people who design and build countless boards, chips, and systems every year. Until verification systems are capable of supporting a very broad piece of the overall spectrum, they will remain merely an interesting puzzle for the research community. On the other hand, when verifiers exist that are useful in broad practice, they will be a tremendous asset to the designer's tool kit. The paper attempts to impart some aspects of both design and specification which should be part of the focus of the activity in automatic verification. Finally the paper presents a small catalog of real design fragments. It is hoped that these fragments will help verification efforts by being perverse cases which will serve as interesting small scale proof targets for new verification systems.

1   Verification, Specification and Design

A common definition of automatic verification is *the automated proof that a design meets its specification*. Clearly it is easier to define something than to make an automatic verification system. This definition raises three serious questions:

- What constitutes an adequate specification, and how is it represented?

- What constitutes a design, and how is it represented?

- Is it possible to prove that the latter representation satisfies the former?

This paper does not attempt to speculate on the answer to the final question, but does provide some commentary in the areas posed by the first two questions. The field of digital design is diverse and contains a large variety of design styles and techniques. Some designs are simple logically but require complex analog interface circuitry. Other designs have a functional behavior which would be impossible to determine from circuit descriptions since their function is predominately dependent upon information stored in a ROM or RAM. The choice to use some form of small ROM or RAM is typically made by the designer and is not fundamentally imposed by the intended behavior of the circuit. It is similarly difficult to decide what level of precision is required to describe a design accurately. For instance a VLSI chip design description could include:

- A full geometric description of the designed circuit.

- A specification of how that geometry is aliased during the mask making process.

- A complete characterization of the process recipe.

- A description of how the process recipe is aliased by individual pieces of equipment on the line.

- A detailed fab trace of what actually happened during a specific run.

- A chemical characterization of all the raw materials used in the run.

- A description of the package and its electrical properties.

- A description of the intended operational domain, i.e. source supply levels, input signal characteristics, etc.

In theory all of this information could be considered to be **the design.** In an ideal world it is possible we would in fact care to know all of this detail. In today's real world we don't know what to do with all of this information except to ignore most of it. The hard task is to know what is important for the purpose at hand and what can be ignored. For example, it is clear that transistor size is not an aspect of VLSI geometry that can be ignored but it may be the case that only the W/L ratio is really important since it is gain that we are really interested in, or it may be that the entire size is important to control other electrical aspects such as parasitic capacitance. If the design description is too detailed then it loses on several counts:

- The description itself takes too long to prepare. This is a moot point if the preparation is automatically generated from work that the designer has to do anyway, but there is little hope that designers will gracefully accept a large amount of what they would feel to be redundant effort.

- Analysis of too detailed a description could be potentially rendered impossible due to the added complexity.

- The more complex anything becomes the greater the chance for error.

Hence it is important to not over describe the design, but in order to do that the designer would have to significantly modify what is normally specified in present day design. In some sense all of the above aspects may be part of what the designer wants. If it is not then the designer must describe that it is not important. There is an inherent *Catch-22* situation here and unless verification efforts come up with some useful compromise to the problem it will be difficult to get designers to seriously consider using their verification systems.

The behavioral specification of a digital design presents a similar problem. It is clearly not enough to simply view a legal behavior as one which takes some sequence of inputs and produces some other sequence of outputs. There is inherently no functional relation between inputs and outputs that does not also include the state of the design. The range of possible states may be combinatorially huge but the problem gets worse when other important aspects of the behavior of the system are time, electrical properties, etc. Clearly the specification problem is equally open ended. Somehow a practical specification method must be found which is adequate to describe precisely what is constrained and conversely what is unconstrained.

The author's personal view is that formal specification techniques have had tremendous impact on the state of computer science literature and absolutely no effect on digital design techniques - past or present!! The reason is that the formal specification of a real design is baroque and takes longer than direct implementation. The problem is exacerbated somewhat by the sorry state of an art where historically specifications have not been very precise. Most designers start the design process with a set of goals and the general attitude that some goals are more important than others. The designers then commence to do their best to come as close as they can to actually achieving these goals. This historical scenario is too imprecise to be the basis of an accurate formal method such as automatic verification, yet verification needs to be able to benefit from practice – another *Catch-22*. The benefits of automatic verification are dubious at present but they do have a real promise.

The main point is that in order for verification systems to be viable tools, they must be used by real designers and they must not inherently limit the designer since often the design task is hard enough as it is. A common tune is to develop a system first for the easy case, learn from it, and then enhance the system to meet real needs. This is fine, but often the focus gets shifted to supporting esoteric aspects of toy problems, and the system never really evolves to meet actual needs. Digital designers do not need verification systems for easy designs. Experience suffices in this case and there is no need to make an easy job take longer. We need verification systems for hard designs, especially those which require that we push techniques in new areas, and consequently can not rely on our experience. It is important that the basic approach to verification be extensible to include the hard *realistic* systems that designers are struggling to produce today. These hard designs inherently involve complex specification and design descriptions. Somehow this complexity must be organized and mechanisms must be developed to remove this complexity from the designer or system specification's direct view.

Clearly the view presented here is that the hard part is not just the development of proof techniques. It will be at least as hard to properly describe the correct aspects of a design and to specify the correct behavior. The remainder of the paper attempts to describe aspects of both design descriptions and specifications that should be included in any system which has a hope of being useful in practice. A small catalog of small circuits is presented as an attempt to help verification researchers by providing some small but difficult test cases.

Note: Since this is primarily a philosophy paper which hopefully has some tutorial value, there are no bibliographic citations. Clearly, the opinions presented here have been influenced by many of the author's colleagues.

2 Aspects of Design Specification

A design specification must include at least the following parameters:

1. **Function** - at the very least the design must receive a sequence of logical inputs and produce the right set of logical outputs. Ideally it would be nice to specify things either by enumeration or as a transfer function – unfortunately with digital circuits containing feedback this is seldom possible. Enumeration is too time consuming and too clumsy to be practical, and the level of history sensitivity caused by storage (feedback) removes the possibility of characterizing the functional behavior of the circuit by a simple transfer function.

2. **I/O Characterization** - There is a significant difference between logical and physical I/O. Physical interfaces are often constrained severely by their environment. Three sub-domains of characterization exist:

   - Logical - inputs are grouped in logical subsets, with defined ranges of values that may be placed on these sets, and a correspondence between individual values and valid electrical levels.

- Temporal - sets of inputs may be constrained about when values must be stable and when they may be changed.

- Electrical - a variety of electrical requirements are necessary if the interface is to interact properly with its environment, e.g. binary or multi-valued signal levels, minimum and maximum voltages for a particular level, domain constraints of absolute level swings, current levels, rise and fall times, etc.

3. **Cost** - the system cost may be constrained in a variety of ways. One useful model is to say that system cost is represented by a polynomial where the terms are things like dollars, total pin count, total board area, circuit count, manpower, calendar units from now to delivery, weight, watts, etc. The coefficients of each term vary with each design as does the maximum permissable total value.

4. **Speed** - most systems must meet certain performance criterion. In certain applications, performance and functional requirements are of equal importance.

5. **Power** - does the total system stay within its power budget in all phases of its operation, and does it only use the specified supply - e.g. a NASA contract might specify the usual 28 volt space shuttle supply at less than 5 amps for the vision and motion detection system combined.

6. **Volume** - physical size is an important limit for some designs.

7. **Packaging** - the nature of the design may be constrained to be packaged in a certain way, e.g. a 64-pin package, a triple VME card, etc.

8. **Reliability** - some systems are designed with reliability specifications, while there is some statistical theory that can be used to combinatorially predict reliability - the complexity is staggering and practically useful simplifications need to be developed.

All of these specification aspects present their own class of worst case nightmare. It is certainly true that in the worst case very few designers are able to meet the specification. Hence it would not be so surprising to find out that verification systems were incapable of proving whether a design met the specification or not. However the point remains that it is very difficult to prove correctness of these complex aspects of system specification. When faced with new designs some of these aspects are a part of the goal set. In practice, sometimes it is acceptable to miss these goals by some amount. It is often the case with research architectures to find out that your best effort falls way short of the desired goal. How is a verification system going to deal with all this imprecision or all of the inherent complexity? Proving functional correctness is not such a big problem and simulators do it reasonably well now (not without reasonable pain unfortunately). It is all of the other aspects and their cumulative complexity that is difficult. Neither simulator or verification efforts have made much of a dent yet.

3 Design Components

This purpose of this section is to present a set of design aspects that need to be described before realistic verification of the design description can be done. The design space is huge and the following lists are an attempt to cover the aspects that control the first and second order effects of behavior. Some of them are inherently difficult to describe and to date no single description method has been able to cover all of the points. Common practice is to specify the set with an almost equally large set of descriptive techniques. It will be a challenge to formal description efforts to provide adequate descriptive coverage in a single semantically consistent representation.

A reasonable first order partition is whether the design is of a board or a chip. For simplicity, an assumption is made here that collections of boards are essentially the same problem as single board design. Clearly issues such as transmission lines, cooling, and power distribution take on new meanings but since these problems exist at the single board design level as well, the assumption is not so far out of line.

The following aspects are all a part of the design of VLSI circuitry:

- Analog Circuits - analog circuitry exists in every chip, and as circuit sizes shrink the role played by analog circuit is increasing. Part of this is the fact that our normal path of shrinking linear dimensions to gain a quadratic improvement in area (cost) and a linear speed increase is slowing down. It is easy to scoff at analog techniques as being overly optimized hacks that would gladly be dispensed with in order to gain the ability to use high quality verification tools. This is not true - since analog techniques typically are a big win in terms of both speed and area, they will be used. Even more importantly there are some functional needs that can only be met by resorting to analog techniques - one of them is the mutual exclusion (ME) element shown in the catalog. Analog circuits are here to stay and the importance of their role is increasing rapidly. In order to understand and model analog circuits, a simple binary algebra and some functions are insufficient tools. Some analog circuits are inherently part of every chip, i.e. pads, boot-strapping circuits, sense amplifiers, etc. These traditional analog components have always been there and no realistic verifier can ignore their existence.

- Digital Circuits - a potential decomposition into sequential and combinatorial forms can be made.

  1. Combinational designs range from highly optimized transistor level circuits such as the 4 transistor XOR circuit shown in the catalog to regular geometric structures such as a PLA. The use of these general structures is usually an indication that the designer has decided that this particular piece of function is neither speed or space critical, or that flexibility is more important than anything else. It is possible to view a PLA structure as a particular combinatorial function. On the other hand it is also possible to view the PLA as a piece of area that can compute

any logical function of 3 variables and 2 outputs. Folding of course mitigates this claim but the point is still that a PLA can be potentially an area of flexibility.

2. Sequential designs may be full custom or use a similarly general technique of storing the state table in either a writable or hard-wired memory area. In the case of a RAM based state table, the physical implementation of the chip will not be sufficient to characterize the design since the contents of these small control memories must be known. In this case the situation might be even worse since the RAM image may be changed during operation rather than being simply loaded as part of some initialization sequence. In this case the function of the state machine would change. This technique is not uncommon and currently, the analysis of such circuits has to be done separately. Perhaps hybrid verification systems which can do piece-wise proofs are also necessary.

- Temporal Discipline - a variety of temporal design styles exist but for now the following two categories should suffice:

1. Synchronous - traditionally systems have been synchronous and this discipline simplifies a lot of electrical complexity due to some global assumptions about when signals should be stable with respect to a critical clock edge. These assumptions must be enforced. Failure to do so will result in flip-flop metastability and subsequent malfunction in other parts of the circuit. Another assumption is that each clocked device sees the proper set of stable inputs on each clock as opposed to the previous or the next set. As chips are built where it takes many clock ticks for a signal to traverse the physical dimensions of the chip, the issue of clock skew becomes more serious and the design difficulty increases proportionately. Clock distribution on large chips is a tremendous problem.

2. Asynchronous - due to the difficulty of enforcing the synchronous timing rules on large VLSI devices, designers are forced to employ a wide variety of asynchronous techniques. Present day high-speed DRAM devices are all either hot-clocked or self-timed. The variety of asynchronous methods is large but in general each individual transition on control wires is important. The result is the need to design circuits which change their outputs cleanly without generating runt pulses whenever their inputs obey the same rules. Traditional models of boolean time-stepped activity have failed as the basis for self-timed circuit simulators and will undoubtedly fail as well for verification systems.

- Aggregate System Parameters - system speed, power consumption, heat dissipation, electrical I/O characteristics, etc. must all be represented as part of system description.

Board level designs generally include all of the aspects of IC circuit design with some different focus in a few areas. Analog circuita play a reduced role in providing logic functionality but may play a major role in signal conditioning, or transmission line balancing. As system speeds increase the transmission line problem using PC board technology gets much worse and noise tolerance and suppression become larger factors of board level design. There are some areas however that are additional headaches in board level designs.

- Programmed Functionality - ROM's and PAL's are used commonly in board level designs. The function of the device is the combination of the program and the physical circuit.

- In cases where components are used from different processes, signal conditioning may be required. These conditioning circuits are primarily analog in nature as well.

- Due to the large physical extent of modern PC boards, and the rather large current levels, power distribution is potentially a big problem. In ECL designs it is a major piece of the overall design effort. Often ECL designs fail to meet their behavior specifications purely due to poor power distribution. They may also fail to operate correctly because of board level noise problems. Neither of these aspects are normally considered as part of the logical design of the board but they are intrinsically part of the total design.

As is the case with specification, a design is a complex collection of diverse considerations. Real designs incorporate many of these aspects routinely and most good designers have developed a lot of expertise in a subset of these areas. Verification systems will need to work from a similarly large subset before they will be useful.

4   Test Case Catalog

This section contains a set of small circuit designs which should give automatic verification systems considerable problems. Many of them give simulation systems trouble as well. The set has been restricted to contain circuits which are representative of commonly used techniques and which are small enough to be understood quickly. The previous philosophy sections have presented a view of complexity that will be a problem in the long term, but this section should present immediate verification problems that are by nature both simple and fundamental.

The first class of test cases all share the same basic property which makes them difficult to prove - namely their correct operation depends on strong transistors overpowering weak ones.

Figure 1 shows a standard, commonly used, 6 transistor RAM cell. Its behavior is to use the dual-rail bit-lines in a bidirectional manner. If the bit lines are driven they must be driven stronger than the cross coupled inverters drive the RAM cell on a write. On a read, the bit lines are undriven and they are read by a sense amplifier. Driving the word select line high addresses the cell. Clearly details of the driving circuitry for the bit lines would need to be part of the design description.

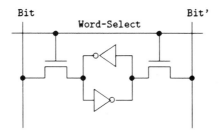

Figure 1: Standard 6 Transistor Static RAM Cell

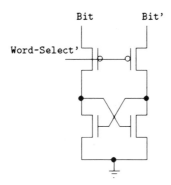

Figure 2: A 4 Transistor Static RAM Cell

If the 6 transistor cell can be verified then try the 4 transistor RAM cell shown in Figure 2. Except for a complemented word-select line, the behavioral specification is the same. Both are static RAM cells. In order for the 4 transistor cell to function as a static memory cell, it is necessary for the voltage on the word-select line to be held such that the two P-types will be biased in their sub-threshold region. Hence the static-mode voltage level of the select-line must be part of the design specification.

Figure 3 shows a 3-input multiplexor circuit which illustrates several interesting properties:

- The circuitry that generates the select signals Sa, Sb, and Sc will assert at most one of them at any given time. It will be important for any verification circuitry to detect this and understand that the overpowering competition is NOT between any of the 3 input N-types or between the N-types and the P-type BUT between the P-type and exactly 1 N-type.

- When OUT is 0 then N1 is 1. The 0 turns on the P-type which holds N1 high even when all of the N-type are off. When one of the N-types turns on and passes an input 0, then that N-type will overpower the P-type enough for the inverter to switch the

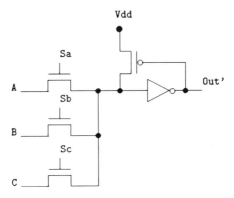

Figure 3: Three-Input Multiplexor

output to a 1 which turns off the P-type and the N-type wins completely. Clearly the individual transistor sizes and their drive capability must be understood to get this one right.

- Finally it is interesting that when one of the input transistors turns on to pass a 1, the high voltage level is degraded through the N-type by a threshold. This degradation does not last for long at N1 since after the inverter switches its output to 0 the P-type turns on hard and solidly yanks N1 to Vdd.

Figure 4 shows both gate- and transistor-level designs for a Muller C-element. The functional specification is that the output changes to a 1 when both inputs are a 1 and to a 0 when both inputs are 0 and maintain its previous state otherwise. This circuit is as common to asynchronous designs as flip-flops are to synchronous circuits. The gate-level circuit shows that the dumb version would require 18 transistors. An 8 transistor version is possible if overpowering is used. When A=B either the N-types will be turned on or the P-types. They will clamp the input to the strong (S) inverter, overpower the weak (W) inverter, and the output will change to the proper level. When A and B are different neither a pull-down or a pull-up path will exist and the previous output will be maintained by the latch formed by the W and S inverters. The use of *trickle* or weak inverters is a common technique found in many designs.

Figure 5 shows the common Schmitt Trigger element which is used for cleaning up edges, buffering inputs, converting between minor differences in signal levels, providing delays in self-timed circuits, etc. The overpowering is done at nodes N1 and N2. With Vin at 0, both Q1 and Q2 are solidly on and the output is at Vdd. Q3 is off and there is no conflict at N1. Q4 and Q5 are off and Q6 is on leaving N2 high but degraded by a threshold voltage. The trigger point comes when Q4 first turns on. At this point Q5 is on the edge of saturation, and the feedback voltage at N2 is dependent on the ratio between Q6 and Q5. Due to the degraded source voltage Q6 is not strong and when the trigger voltage is

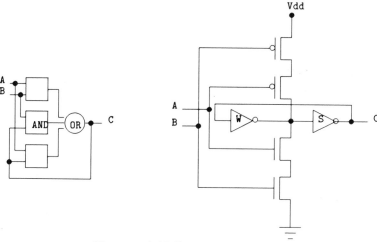

Figure 4: A Muller C-Element

exceeded N2 snaps abruptly low. In reality this phenomenon is better explained by the gain around the Q4/Q6 loop being greater than 1 but the result is the same. The converse is true as Vin is lowered. The difference is that the high-going transition of the output is controlled by the P-types. The difference in threshold voltages of the P and N types gives the expected hysteresis property. Glasser's book describes a few more esoteric characteristics of the design:

- Due to an inherent gain-bandwidth trade-off a Schmitt trigger is typically slower than an inverter of the same power. This hypothesis would be interesting to prove automatically.

- The Schmitt trigger is also a power hog since several wide devices are turned on simultaneously for a relatively long time. In systems where meeting a power budget was an important part of the specification, this would be an important thing to notice.

Figure 6 shows a standard CMOS 4 transistor XOR gate. Unlike the previous designs there is no analog trickery here. It is similar in nature to many clever and highly optimized designs that have been reduced to pass transistor logic rather than simple logic gates. The problem that many simulation systems and potentially verification systems have is that the output N1 is potentially driven by any of the 4 transistors. The operation is simple (no tricks at all) and won't be belabored here.

Figure 7 shows a circuit for a fast complementary set/reset latch (CSRL) designed originally by Carver Mead. The operation is that of a straight forward latch. However there are a few extra transistors that enhance the design significantly. Q4 is a power down transistor that causes Vss to float during loading at the same time Q3 is turned on to short the dual rail data lines. Q3 becomes a leg in a resistive ladder which when the load goes back

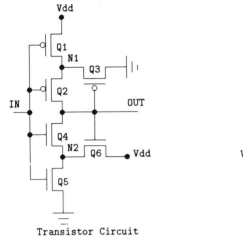

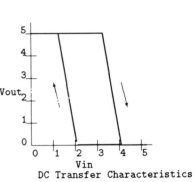

Transistor Circuit

DC Transfer Characteristics

Figure 5: A Schmitt Trigger

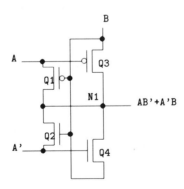

Figure 6: A 4 Transistor XOR Circuit

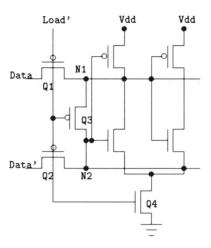

Figure 7: A Fast Complementary Set/Reset Latch

high leaves a few millivolts of charge differential on nodes N1 and N2. With the positive gain of the cross-coupled inverters, the differential is sufficient to swing the latch to the appropriate full logic level.

Figure 8 is a standard mutual exclusion circuit. The specification is that A2 and and A1 must be asserted mutually exclusive of each other. A1 must be asserted after R1 is asserted and deasserted after R1 is deasserted. R1 cannot be asserted until A1 is deasserted. A similar sequence constraint holds for R2 and A2. The circuit is also an example where an analog circuit is required, i.e. there is no way to design a digital circuit that exhibits this behavior. The design works by placing an analog metastability detector on the output of a latch. If R1 and R2 are asserted with sufficient separation then the values at N1 and N2 are complementary and either Q1 or Q2 will be turned on to assert the appropriate acknowledge. If sufficient temporal separation does not exist between the assertions of R1 and R2 then the latch cannot make the decision and will go metastable. When the latch is metastable, nodes N1 and N2 will oscillate about the switching threshold *in phase* and with a separation of less than a threshold voltage. If this is the case then Vgs (the gate to source voltage) of Q1 and Q2 is also less than a threshold and they will be turned off. In this state no current flows through Q3 and Q4 and the outputs are pulled high in the deasserted state. When the latch leaves the metastable region, then N1 and N2 will be separated by more than a threshold and the gain of the latch will pull to a clean assertion of one of the acknowledge lines.

The final two cases take a slightly different tack. Neither actually meet their specification but will most likely appear to bad simulator or verification systems as if they do work. The reasons for failure vary but this class of designs is equally important. Most designers would much rather have a verification system fail to prove things which were known to work than to indicate correctness of an incorrect circuit.

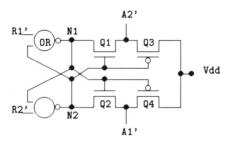

Figure 8: A Mutual Exclusion Circuit

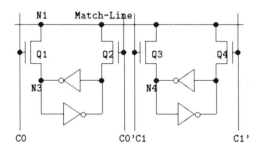

Figure 9: A Faulty 2-bit Comparator

Figure 9 shows a pair of partial compare cells. The circuitry for writing into the latches has been omitted for clarity. The values on the dual rail compare inputs are matched against the contents of the latch. Prior to driving the C lines the Match line is precharged high. N3 and N4 are the stored latch values. Let N3 and N4 be 0, and C0 be 0, and C1 be 1. The match line will remain high if the values are equal. Q1, Q2 and Q4 are off and Q3 is on. Q3 starts to pull the match line down until N1's potential gets low enough to look like a source for Q2 which then pulls it back up to a weak high. The bidirectional nature of the MOS transistor makes this cell invalid under the intended composition, even though a single cell does perform properly.

Figure 10 should cause problems for systems that do not understand path delays and runt free asynchronous transition specifications. The specification for this circuit is that it compute $BC'+A'C$ in a way that cannot generate runt pulses or false transitions when only a single input level is changed. Subsequent changes on the inputs are only permitted when outputs are stable. The wrong case clearly fails if C is changed and the speeds of paths N1 and N2 differ widely. For instance let C change from 0 to 1. This change turns off the top AND gate and turns on the bottom AND gate. If N2 is slow then the output will temporarily indicate a false 0. The correct version that is speed independent and cleanly obeys the previously stated specification is shown on the right.

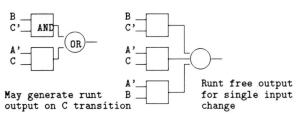

Figure 10: Combinatorial Glitches

5 Conclusions

This paper has presented some philosophy which will hopefully influence the hardware verification community to broaden its scope, to consider at least in principle the entire design
process. While it is important to make progress and by nature this may require a limited
scope, it is equally important to keep the entire problem in focus. Since philosophy in any
discipline is cheap, the meat of the paper is in the test-case catalog. Hopefully the catalog
will serve the function of an initial *strawman* that will inspire others to add their own fa-
vorite small but perverse circuits. Every designer has a few circuits that are part of their
standard library that work properly in every respect, yet are considered to be incorrect by
almost every tool in the tool-suite. The catalog effort was a group effort by the author and
two of his colleagues at HP labs. Thanks to Bill Coates and Bic Schediwy for their excellent
ideas and review.

Index